BRIEF
CALCULUS
& ITS APPLICATIONS
TWELFTH EDITION

Larry J. Goldstein
Goldstein Educational Technologies

David C. Lay
University of Maryland

David I. Schneider
University of Maryland

Nakhlé H. Asmar
University of Missouri

Prentice Hall
is an imprint of

Upper Saddle River, NJ 07458

Library of Congress Cataloging-in-Publication Data

Brief calculus & its aplications / Larry J. Goldstein ... [et al.]. -- 12th ed.
    p. cm.
  Includes index.
  ISBN-13: 978-0-321-56856-4
  ISBN-10: 0-321-56856-7
  1.   Calculus–Textbooks.   I. Goldstein, Larry Joel. II. Title: Brief calculus and its
applications.
  QA303.2.G65 2010
  515--dc22

                                                                    2009000253

Editor in Chief: *Deirdre Lynch*
Acquisitions Editor: *Chuck Synovec*
Executive Project Manager: *Christine O'Brien*
Editorial Assistant: *Joanne Wendelken*
Executive Marketing Manager: *Jeff Weidenaar*
Marketing Assistant: *Jon Connelly*
Production Management: *Bob Walters, Prepress Management, Inc.*
Senior Managing Editor: *Linda Mihatov Behrens*
Associate Managing Editor: *Bayani Mendoza de Leon*
Operations Specialist: *Ilene Kahn*
Senior Operations Supervisor: *Diane Peirano*
Supplements Production: *Katherine Roz*
Art Director: *Maureen Eide*
Cover Director: *Heather Scott*
Interior/Cover Designer: *Suzanne Behnke*
Cover Art: *Origami by Ilene Kahn*
Art Editor: *Thomas Benfatti*
Senior Technical Art Specialist: *Joe Vetere*
Art Studio: *ICC Macmillan, Inc.*
Compositor: *Dennis Kletzing*
Media Producer: *Audra Walsh*
Software Development: *Eric Gregg*, MathXL; *Marty Wright*, TestGen
Rights and Permissions Advisor: *Shannon Barbe*

Prentice Hall
is an imprint of

PEARSON

Copyright © 2010, 2007, 2004, 2001, 1999, 1996, 1993, 1990, 1987, 1984, 1980, 1977
by Pearson Education, Inc.
Pearson Prentice Hall
Pearson Education, Inc.
Upper Saddle River, NJ 07458

Printed in the United States of America
10 9 8 7 6 5 4 3 2

ISBN 10:   0-321-56856-7
ISBN 13: 978-0-321-56856-4

Pearson Education Ltd., *London*
Pearson Education Singapore, Pte. Ltd.
Pearson Education Canada, Inc.
Pearson Education–Japan
Pearson Education Australia PTY, Limited
Pearson Education North Asia, Lt., *Hong Kong*
Pearson Educación de Mexico, S.A. de C.V.
Pearson Educacion Malaysia, Pte. Ltd.
Pearson Education Upper Saddle River, New Jersey

CONTENTS

---

*Sections preceded by a * are optional in the sense that they are not prerequisites for later material.

8 The Trigonometric Functions 393

9 Techniques of Integration 419

PREFACE

We have been very pleased with the enthusiastic response to the previous editions of *Calculus & Its Applications* and its Brief version by teachers and students alike. The present work incorporates many of the suggestions they have put forward.

Although there are many changes, we have preserved the approach and the flavor. Our goals remain the same: to begin the calculus as soon as possible; to present calculus in an intuitive yet intellectually satisfying way; and to illustrate the many applications of calculus to business, life sciences, and social sciences.

The distinctive order of topics has proven over the years to be successful—easier for students to learn, and more interesting because students see significant applications early. For instance, the derivative is explained geometrically before the analytic material on limits is presented. To allow you to reach the applications in Chapter 2 quickly, we present only the differentiation rules and the curve sketching needed for those applications. Advanced topics come later when they are needed. Other aspects of this student-oriented approach follow below.

The Series

This text is part of a series consisting of three texts: *Finite Mathematics & Its Applications*, *Calculus & Its Applications*, and *Brief Calculus & Its Applications*. All three titles are available for purchase in three formats: printed text, eBook and as an eBook within the MyMathLab online course.

Topics Included

This edition contains more material than can be covered in most two-semester courses. Optional sections are starred in the table of contents. In addition, the level of theoretical material may be adjusted to the needs of the students. For instance, Section 1.4 may be omitted entirely, if the instructor does not wish to present the notion of limit beyond the required material that is contained in Section 1.3.

Review of Prerequisites

In Chapter 0, we review those concepts that the reader needs to study calculus. Some important topics, such as the laws of exponents, are reviewed again when they are used in a later chapter. Section 0.6 prepares students for applied problems that appear throughout the text. A reader familiar with the content of Chapter 0 should begin with Chapter 1 and use Chapter 0 as a reference, whenever needed.

Trusted Features

In this edition, we attempt to maintain our popular student-oriented approach throughout and, in particular, by the following features:

Relevant and Varied Applications

We provide realistic applications that illustrate the uses of calculus in other disciplines. See the Index of Applications at the back of the text. Wherever possible, we have attempted to use applications to motivate the mathematics.

Plentiful Examples

The text includes many more worked examples than is customary. Furthermore, we have included computational details to enhance readability. "Warning!" notes provide tips on common pitfalls and mistakes and appear at relevant times throughout the text.

Exercises to Meet All Student Needs

The exercises comprise about one-quarter of the text—the most important part of the text in our opinion. The exercises at the ends of the sections are usually arranged in the order in which the text proceeds, so that the homework assignments may easily be made after only part of a section is discussed. Interesting applications and more challenging problems tend to be located near the ends of the exercise sets. Supplementary exercises at the end of each chapter expand the other exercise sets and include problems that require skills from earlier chapters.

Practice Problems

The Practice Problems have proven to be a popular and useful feature. Practice Problems are carefully selected questions located at the end of each section, just before the exercise set. Complete solutions are given following the exercise set. The Practice Problems often focus on points that are potentially confusing or are likely to be overlooked. We recommend the reader work them and study their solutions before moving on to the exercises. In effect, the Practice Problems constitute a built-in workbook.

New To This Edition

Among the many changes in this edition, the following are the most significant:

1. New and Revised Exercises and Examples

162 new and updated exercises feature a wide range of levels from those designed to aid students with a limited math background to exercises that challenge the more capable students.

2. Revised presentation of crucial topics

a. The Derivative

Some content of Section 1.3 has been moved to Section 1.4 and, as a result, Section 1.3 presents the derivative in an intuitive way, before the expanded presentation of limits that appears in Section 1.4. The new presentation offers the instructor the option of spending less time on limits (and therefore more time on applications) by not emphasizing Section 1.4.

b. Differential equations

We rewrote Section 10.3 on first-order linear differential equations in order to lead directly into the step-by-step method for solving this type of equation. This new presentation offers a faster track to the applications of equations presented in Section 10.4.

3. Enhancing pedagogy with graphing calculators

As in previous editions, the use of graphing calculators is not required for this text; however, graphing calculators are very useful tools that can be used to simplify computations, draw graphs, and sometimes enhance understanding of the fundamental topics of calculus. Helpful information about the use of calculators appears at the end of most sections in subsections titled Incorporating Technology. The examples in this text use the family of TI-83/84 calculators, which include TI-83, TI-84, TI-83+, TI-84+, TI-83+ Silver Edition, and TI-84+ Silver Edition. Additional graphing calculator support material can be downloaded from pearsoned.com/Goldstein.

Program Supplements

All supplements support both *Calculus & Its Applications* and *Brief Calculus & Its Applications*.

For Students

Student's Solutions Manual (ISBN: 0-321-59901-2)

Fully worked solutions to selected exercises and study tips.

DVD Lecture Videos (ISBN: 0-321-57735-3)

A comprehensive set of topical videos, in which examples for each topic are worked out by an instructor. The videos provide excellent support for students who require additional assistance, for distance learning and self-paced programs, or for students who missed class.

For Instructors

Instructor's Edition (ISBN: 0-321-60012-6)

This version of the text includes answers to all exercises presented in the book.

Instructor Resource Center

All instructor resources can be downloaded from www.pearsonhighered.com/irc. This is a password-protected site that requires instructors to set up an account or, alternatively, instructor resources can be ordered from your Pearson Higher Education sales representative. Instructors may use their instructor's log-in from MyMathLab to access the Instructor Resource Center.

Instructor's Solutions Manual (ISBN: 0-321-59900-4)

Fully worked solutions to every textbook exercise.

TestGen®

TestGen (www.pearsonhighered.com/testgen) enables instructors to build, edit, print, and administer tests using a computerized bank of questions developed to cover all of the text objectives. TestGen is algorithmically based, allowing instructors to create multiple but equivalent versions of the same question or test with the click of a button. Instructors can also modify test bank questions or add new ones. Tests can be printed or administered online. The software and testbank are available for download from Pearson Education's online catalog.

PowerPoint Lecture Slides

Fully editable and printable slides that follow the textbook. Use during lecture or post to a Web site in an online course. For download from the Instructor Resource Center and available within MyMathLab.

MyMathLab® Online Course (access code required)

MyMathLab is a series of text-specific, easily customizable online courses for Pearson Education's textbooks in mathematics and statistics. Powered by CourseCompass™ (our online teaching and learning environment) and MathXL® (our online homework, tutorial, and assessment system), MyMathLab gives you the tools you need to deliver all or a portion of your course online, whether students are in a lab setting or working from home. MyMathLab provides a rich and flexible set of course materials, featuring free-response exercises that are algorithmically generated for unlimited practice and mastery. Students can also use online tools, such as video lectures, and a multimedia textbook, to independently improve their understanding and performance. Instructors can use MyMathLab's homework and test managers to select and assign online exercises correlated to the textbook, and they can also create and assign their own online exercises and import TestGen tests for added flexibility. MyMathLab's online gradebook—designed specifically for mathematics and statistics—automatically tracks students' homework and test results and gives the instructor control over how to calculate final grades. Instructors can also add offline (paper-and-pencil) grades to the gradebook. MyMathLab also includes access to the **Pearson Tutor Center** (www.pearsontutorservices.com). The Tutor Center is staffed by qualified mathematics instructors who provide textbook-specific tutoring for students via toll-free phone, fax, email, and interactive Web sessions. MyMathLab is available to qualified adopters. For more information, visit our website at www.mymathlab.com or contact your sales representative.

MathXL® Online Course (access code required)

MathXL® is a powerful online homework, tutorial, and assessment system that accompanies Pearson Education's textbooks in mathematics or statistics. With MathXL, instructors can create, edit, and assign online homework and tests using algorithmically generated exercises correlated to the textbook's objectives. They can also create and assign their own online exercises and import TestGen tests for added flexibility. All student work is tracked in MathXL's online gradebook. Students can take chapter tests in MathXL and receive personalized study plans based on their test results. The study plan diagnoses weaknesses and links students directly to tutorial exercises for the objectives they need to study and retest. Students can also access supplemental video clips directly from selected exercises. MathXL is available to qualified adopters. For more information, visit our website at www.mathxl.com, or contact your Pearson sales representative.

Visual Calculus

Built from the ground up by David Schneider (University of Maryland), the Visual Calculus software enhances the visualization and therefore the understanding of the *ideas* behind the calculus. With this software students can: analyze a function and its derivatives, examine solutions of differential equations, graph functions, calculate definite integrals, evaluate finite and infinite sums, graph Taylor approximations, plot cosine and sine functions and much more. This software adds further value for you for constructing graphs for exams. Visual Calculus is available for download from the Website www.pearsoned.com/goldstein.

Acknowledgments

The following is a list of reviewers from this and previous editions. We apologize for any omissions. While writing this book, we have received assistance from many persons, and our heartfelt thanks goes out to them all. Especially, we would like to thank our reviewers, who took the time and energy to share their ideas, preferences, and often their enthusiasm with us.

James V. Balch, *Middle Tennessee State University*

Jack R. Barone, *Baruch College, CUNY*

Michael J. Berman, *James Madison University*

Dennis Bertholf, *Oklahoma State University*

Fred Brauer, *University of Wisconsin*

Dennis Brewer, *University of Arkansas*

James Brewer, *Florida Atlantic University*

Todd Brost, *South Dakota State University*

Robert Brown, *University of California, Los Angeles*

Robert Brown, *University of Kansas*

Alan Candiotti, *Drew University*

Der-Chen Chang, *Georgetown University*

Charles Clever, *South Dakota State University*

W. E. Conway, *University of Arizona*

Biswa Datta, *Northern Illinois University*

Karabi Datta, *Northern Illinois University*

Dennis DeTurck, *University of Pennsylvania*

Brenda Diesslin, *Iowa State University*

Bruce Edward, *University of Florida*

Janice Epstein, *Texas A&M University*

Albert G. Fadell, *SUNY Buffalo*

Betty Fein, *Oregon State University*

Howard Frisinger, *Colorado State University*

Larry Gerstein, *University of California, Santa Barbara*

Shirley A. Goldman, *University of California, Davis*

Jack E. Graves, *Syracuse University*

Harvey Greenwald, *California State Polytechnic University*

David Harbater, *University of Pennsylvania*

James L. Heitsch, *University of Illinois, Chicago*

Donald Hight, *Kansas State College of Pittsburg*

Charles Himmelberg, *University of Kansas*

W. R. Hintzman, *San Diego State University*

James E. Honeycutt, *North Carolina University*

E. John Hornsby, Jr., *University of New Orleans*

James A. Huckaba, *University of Missouri*

Samuel Jasper, *Ohio University*

Shujuan Ji, *Columbia University*

James Kaplan, *Boston University*

Judy B. Kidd, *James Madison University*

W. T. Kyner, *University of New Mexico*

T. Y. Lam, *University of California, Berkeley*

Lawrence J. Lardy, *Syracuse University*

Melvin D. Lax, *California State University, Long Beach*

Russell Lee, *Allan Hancock College*

Joyce Longman, *Villanova University*

Roy Lowman, *University of Illinois, Chicago*

Gabriel Lugo, *University of North Carolina, Wilmington*

Gordon Lukesh, *University of Texas, Austin*

John H. Mathews, *California State University, Fullerton*

William McCord, *University of Missouri*

Ann McGaw, *University of Texas, Austin*

Albert J. Milani, *University of Wisconsin-Milwaukee*

Robert A. Miller, *City University New York*

Carl David Minda, *University of Cincinnati*

Donald E. Myers, *University of Arizona, Tempe*

Dana Nimic, *Southeast Community College-Lincoln*

David W. Penico, *Virginia Commonwealth University*

Georgia B. Pyrros, *University of Delaware*

H. Suey Quan, *Golden West College*

Jens Rademacher, *University of British Columbia*

Heath K. Riggs, *University of Vermont*

Ronald Rose, *American River College*

Arthur Rosenthal, *Salem State College*

Murray Schechter, *Lehigh University*

Claude Schochet, *Wayne State University*

Arthur J. Schwartz, *University of Michigan*

Robert Seeley, *University of Massachusetts, Boston*

Arlene Sherburne, *Montgomery College, Rockville*

James Sochacki, *James Madison University*

Edward Spanier, *University of California, Berkeley*

H. Keith Stumpff, *Central Missouri State University*

Bruce Swenson, *Foothill College*

Geraldine Taiani, *Pace University*

Joan M. Thomas, *University of Oregon*

Frankl Warner, *University of Pennsylvania*

Dennis White, *University of Minnesota*

Dennis A. Widup, *University of Wisconsin-Parkside*

W. R. Wilson, *Central Piedmont Community College*

Carla Wofsky, *University of New Mexico*

Wallace A. Wood, *Bryant University*

The authors would like to thank the many people at Pearson Education who have contributed to the success of our books over the years. We appreciate the tremendous efforts of the production, art, manufacturing, and marketing departments. Special thanks go to Bob Walters, who managed production of this book. Many thanks to Jason Aubrey for his contributions to the technology sections. We also thank the accuracy checkers: Damon Demas, Blaise DeSesa, Doug Ewert, Bev Fusfield, Debra McGivney, Theresa Schille, Lauri Semarne, and Tom Wegleitner. The expert skills of our typesetter, Dennis Kletzing, have once again eased the burden of preparing this new edition.

Thanks also to Gracia Nabhane, M.D., for her help with the applications in biology and medicine.

The authors would like to thank our editor, Chuck Synovec, who helped us plan this edition.

Larry J. Goldstein
larrygoldstein@comcast.net

David C. Lay
lay@math.umd.edu

David I. Schneider
dis@math.umd.edu

Nakhlé H. Asmar
nakhle@math.missouri.edu

INTRODUCTION

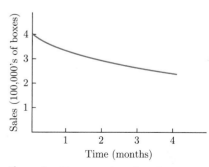

Often it is possible to give a succinct and revealing description of a situation by drawing a graph. For example, Fig. 1 describes the amount of money in a bank account drawing 5% interest, compounded daily. The graph shows that, as time passes, the amount of money in the account grows. Figure 2 depicts the weekly sales of a breakfast cereal at various times after advertising has ceased. The graph shows that the longer the time since the last advertisement, the fewer the sales.

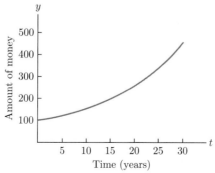

Figure 1. Growth of money in a savings account.

Figure 2. Decrease in sales of breakfast cereal.

Figure 3. Growth of a bacteria culture.

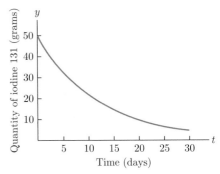

Figure 4. Decay of radioactive iodine.

Figure 3 shows the size of a bacteria culture at various times. The culture grows larger as time passes. But there is a maximum size that the culture cannot exceed. This maximum size reflects the restrictions imposed by food supply, space, and similar factors. The graph in Fig. 4 describes the decay of the radioactive isotope iodine 131. As time passes, less and less of the original radioactive iodine remains.

Each graph in Figs. 1 to 4 describes a change that is taking place. The amount of money in the bank is changing, as are the sales of cereal, the size of the bacteria culture, and the amount of iodine. Calculus provides mathematical tools to study each change in a quantitative way.

FUNCTIONS

Each graph in Figs. 1 to 4 of the Introduction depicts a relationship between two quantities. For example, Fig. 4 illustrates the relationship between the quantity of iodine (measured in grams) and time (measured in days). The basic quantitative tool for describing such relationships is a *function*. In this preliminary chapter, we develop the concept of a function and review important algebraic operations on functions used later in the text.

0.1 Functions and Their Graphs

Real Numbers

Most applications of mathematics use real numbers. For purposes of such applications (and the discussions in this text), it suffices to think of a real number as a decimal. A *rational* number is one that may be written as a finite or infinite repeating decimal or as a fraction, such as

$$-\frac{5}{2} = -2.5, \qquad 1, \qquad \frac{13}{3} = 4.333\ldots \qquad \text{(rational numbers)}.$$

An *irrational* number has an infinite decimal representation whose digits form no repeating pattern, such as

$$-\sqrt{2} = -1.414214\ldots, \qquad \pi = 3.14159\ldots \qquad \text{(irrational numbers)}.$$

The real numbers are described geometrically by a *number line*, as in Fig. 1. Each number corresponds to one point on the line, and each point determines one real number.

Figure 1. The real number line.

We use four types of inequalities to compare real numbers.

$$x < y \qquad x \text{ is less than } y$$
$$x \le y \qquad x \text{ is less than or equal to } y$$
$$x > y \qquad x \text{ is greater than } y$$
$$x \ge y \qquad x \text{ is greater than or equal to } y$$

The double inequality $a < b < c$ is shorthand for the pair of inequalities $a < b$ and $b < c$. Similar meanings are assigned to other double inequalities, such as $a \le b < c$. Three numbers in a double inequality, such as $1 < 3 < 4$ or $4 > 3 > 1$, should have the same relative positions on the number line as in the inequality (when read left to right or right to left). Thus $3 < 4 > 1$ is never written because the numbers are "out of order."

Geometrically, the inequality $x \le b$ means that either x equals b or x lies to the left of b on the number line. The set of real numbers x that satisfies the double inequality $a \le x \le b$ corresponds to the line segment between a and b, including the endpoints. This set is sometimes denoted by $[a, b]$ and is called the *closed interval* from a to b. If a and b are removed from the set, the set is written as (a, b) and is called the *open interval* from a to b. The notation for various line segments is listed in Table 1.

The symbols ∞ ("infinity") and $-\infty$ ("minus infinity") do not represent actual real numbers. Rather, they indicate that the corresponding line segment extends infinitely far to the right or left. An inequality that describes such an infinite interval may be written in two ways. For instance, $a \le x$ is equivalent to $x \ge a$.

TABLE 1 Intervals on the Number Line

Inequality	Geometric Description	Interval Notation
$a \le x \le b$		$[a, b]$
$a < x < b$		(a, b)
$a \le x < b$		$[a, b)$
$a < x \le b$		$(a, b]$
$a \le x$		$[a, \infty)$
$a < x$		(a, ∞)
$x \le b$		$(-\infty, b]$
$x < b$		$(-\infty, b)$

EXAMPLE 1   Describe each of the following intervals both graphically and in terms of inequalities.

(a) $(-1, 2)$ (b) $[-2, \pi]$ (c) $(2, \infty)$ (d) $(-\infty, \sqrt{2}\,]$

Solution   The line segments corresponding to the intervals are shown in Figs. 2(a)–(d). Note that an interval endpoint that is included (e.g., both endpoints of $[a, b]$) is drawn as a solid circle, whereas an endpoint not included (e.g., the endpoint a in $(a, b]$) is drawn as an unfilled circle.   ■

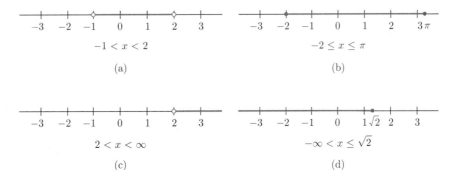

Figure 2. Line segments.

EXAMPLE 2   The variable x describes the profit that a company anticipates earning in the current fiscal year. The business plan calls for a profit of at least 5 million dollars. Describe this aspect of the business plan in the language of intervals.

Solution   The phrase "at least" means "greater than or equal to." The business plan requires that $x \geq 5$ (where the units are millions of dollars). This is equivalent to saying that x lies in the infinite interval $[5, \infty)$.   ■

Functions   A *function* of a variable x is a *rule* f that assigns to each value of x a unique number $f(x)$, called *the value of the function at* x. [We read "$f(x)$" as "f of x."] The variable x is called the *independent variable*. The set of values that the independent variable is allowed to assume is called the *domain* of the function. The domain of a function may be explicitly specified as part of the definition of a function or it may be understood from context. (See the following discussion.) The *range* of a function is the set of values that the function assumes.

The functions we shall meet in this book will usually be defined by algebraic formulas. For example, the domain of the function

$$f(x) = 3x - 1$$

consists of all real numbers x. This function is the rule that takes a number, multiplies it by 3, and then subtracts 1. If we specify a value of x, say $x = 2$, then we find the value of the function at 2 by substituting 2 for x in the formula:

$$f(2) = 3(2) - 1 = 5.$$

EXAMPLE 3   Let f be the function with domain all real numbers x and defined by the formula

$$f(x) = 3x^3 - 4x^2 - 3x + 7.$$

Find $f(2)$ and $f(-2)$.

Solution To find $f(2)$ we substitute 2 for every occurrence of x in the formula for $f(x)$:

$$f(2) = 3(2)^3 - 4(2)^2 - 3(2) + 7$$
$$= 3(8) - 4(4) - 3(2) + 7$$
$$= 24 - 16 - 6 + 7$$
$$= 9.$$

To find $f(-2)$ we substitute (-2) for each occurrence of x in the formula for $f(x)$. The parentheses ensure that the -2 is substituted correctly. For instance, x^2 must be replaced by $(-2)^2$, not -2^2.

$$f(-2) = 3(-2)^3 - 4(-2)^2 - 3(-2) + 7$$
$$= 3(-8) - 4(4) - 3(-2) + 7$$
$$= -24 - 16 + 6 + 7$$
$$= -27$$ ■

EXAMPLE 4

Temperature scales If x represents the temperature of an object in degrees Celsius, then the temperature in degrees Fahrenheit is a function of x, given by $f(x) = \frac{9}{5}x + 32$.

(a) Water freezes at $0°$C (C = Celsius) and boils at $100°$C. What are the corresponding temperatures in degrees Fahrenheit (F = Fahrenheit)?

(b) Aluminum melts at $660°$C. What is its melting point in degrees Fahrenheit?

Solution (a) $f(0) = \frac{9}{5}(0) + 32 = 32$. Water freezes at $32°$F.

$f(100) = \frac{9}{5}(100) + 32 = 180 + 32 = 212$. Water boils at $212°$F.

(b) $f(660) = \frac{9}{5}(660) + 32 = 1188 + 32 = 1220$. Aluminum melts at $1220°$F.   ■

EXAMPLE 5

A voting model Let x be the proportion of the total popular vote that a Democratic candidate for president receives in a U.S. national election (so x is a number between 0 and 1). Political scientists have observed that a good estimate of the proportion of seats in the House of Representatives going to Democratic candidates is given by the function

$$f(x) = \frac{x^3}{x^3 + (1-x)^3}, \qquad 0 \le x \le 1,$$

whose domain is the interval $[0, 1]$. This formula is called the *cube law*. Compute $f(.6)$ and interpret the result.

Solution We must substitute .6 for every occurrence of x in $f(x)$:

$$f(.6) = \frac{(.6)^3}{(.6)^3 + (1 - .6)^3} = \frac{(.6)^3}{(.6)^3 + (.4)^3}$$

$$= \frac{.216}{.216 + .064} = \frac{.216}{.280} \approx .77.$$

This calculation shows that the cube law function predicts that if .6 (or 60%) of the total popular vote is for the Democratic candidate for president, then approximately .77 (or 77%) of the seats in the House of Representatives will be won by Democratic candidates; that is, about 335 of the 435 seats will be won by Democrats. Note that for $x = .5$ we have $f(.5) = .5$. Is this what you would expect?   ■

In the preceding examples, the functions had domains consisting of all real numbers or an interval. For some functions, the domain may consist of several intervals, with a different formula defining the function on each interval. Here is an illustration of this phenomenon.

EXAMPLE 6

A leading brokerage firm charges a 6% commission on gold purchases in amounts from \$50 to \$300. For purchases exceeding \$300, the firm charges 2% of the amount purchased plus \$12.00. Let x denote the amount of gold purchased (in dollars) and let $f(x)$ be the commission charge as a function of x.

(a) Describe $f(x)$.

(b) Find $f(100)$ and $f(500)$.

Solution **(a)** The formula for $f(x)$ depends on whether $50 \leq x \leq 300$ or $300 < x$. When $50 \leq x \leq 300$, the charge is $.06x$ dollars. When $300 < x$, the charge is $.02x + 12$. The domain consists of the values of x in one of the two intervals $[50, 300]$ and $(300, \infty)$. In each of these intervals, the function is defined by a separate formula:

$$f(x) = \begin{cases} .06x & \text{for } 50 \leq x \leq 300 \\ .02x + 12 & \text{for } 300 < x. \end{cases}$$

Note that an alternative description of the domain is the interval $[50, \infty)$. That is, the value of x may be any real number greater than or equal to 50.

(b) Since $x = 100$ satisfies $50 \leq x \leq 300$, we use the first formula for $f(x)$: $f(100) = .06(100) = 6$. Since $x = 500$ satisfies $300 < x$, we use the second formula for $f(x)$: $f(500) = .02(500) + 12 = 22$. ∎

In calculus, it is often necessary to substitute an algebraic expression for x and simplify the result, as illustrated in the following example.

EXAMPLE 7

If $f(x) = (4 - x)/(x^2 + 3)$, what is $f(a)$? $f(a + 1)$?

Solution Here a represents some number. To find $f(a)$, we substitute a for x wherever x appears in the formula defining $f(x)$:

$$f(a) = \frac{4 - a}{a^2 + 3}.$$

To evaluate $f(a + 1)$, substitute $a + 1$ for each occurrence of x in the formula for $f(x)$:

$$f(a + 1) = \frac{4 - (a + 1)}{(a + 1)^2 + 3}.$$

The expression for $f(a + 1)$ may be simplified, using the fact that $(a + 1)^2 = (a + 1)(a + 1) = a^2 + 2a + 1$:

$$f(a + 1) = \frac{4 - (a + 1)}{(a + 1)^2 + 3} = \frac{4 - a - 1}{a^2 + 2a + 1 + 3} = \frac{3 - a}{a^2 + 2a + 4}. \quad \blacksquare$$

More about the Domain of a Function   When defining a function, it is necessary to specify the domain of the function, which is the set of acceptable values of the variable. In the preceding examples, we explicitly specified the domains of the functions considered. However, throughout the remainder of the text, we will usually mention functions without specifying domains. In such circumstances, we will understand the intended domain to consist of all numbers for which the defining formula(s) makes sense. For example, consider the function

$$f(x) = x^2 - x + 1.$$

The expression on the right may be evaluated for any value of x. So, in the absence of any explicit restrictions on x, the domain is understood to consist of all numbers. As a second example, consider the function

$$f(x) = \frac{1}{x}.$$

Here x may be any number except zero. (Division by zero is not permissible.) So the domain intended is the set of nonzero numbers. Similarly, when we write

$$f(x) = \sqrt{x},$$

we understand the domain of $f(x)$ to be the set of all nonnegative numbers, since the square root of a number x is defined if and only if $x \geq 0$.

Graphs of Functions Often it is helpful to describe a function f geometrically, using a rectangular xy-coordinate system. Given any x in the domain of f, we can plot the point $(x, f(x))$. This is the point in the xy-plane whose y-coordinate is the value of the function at x. The set of *all* such points $(x, f(x))$ usually forms a curve in the xy-plane and is called the *graph of the function $f(x)$*.

It is possible to approximate the graph of $f(x)$ by plotting the points $(x, f(x))$ for a representative set of values of x and joining them by a smooth curve. (See Fig. 3.) The more closely spaced the values of x, the closer the approximation.

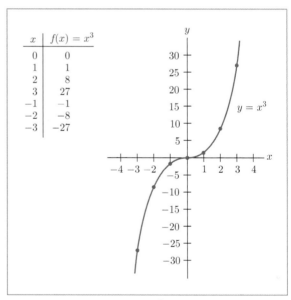

x	$f(x) = x^3$
0	0
1	1
2	8
3	27
−1	−1
−2	−8
−3	−27

Figure 4. Graph of $f(x) = x^3$.

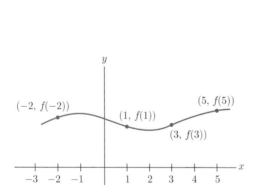

Figure 3

EXAMPLE 8

Sketch the graph of the function $f(x) = x^3$.

Solution The domain consists of all numbers x. We choose some representative values of x and tabulate the corresponding values of $f(x)$. We then plot the points $(x, f(x))$ and draw a smooth curve through the points. (See Fig. 4.) ■

EXAMPLE 9   Sketch the graph of the function $f(x) = 1/x$.

Solution The domain of the function consists of all numbers except zero. The table in Fig. 5 lists some representative values of x and the corresponding values of $f(x)$. A function often has interesting behavior for x near a number not in the domain. So when we chose representative values of x from the domain, we included some values close to zero. The points $(x, f(x))$ are plotted and the graph sketched in Fig. 5. ∎

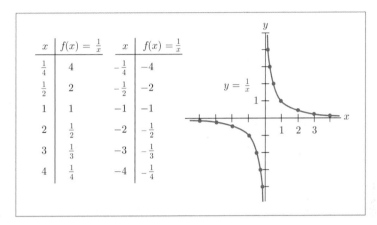

x	$f(x) = \frac{1}{x}$	x	$f(x) = \frac{1}{x}$
$\frac{1}{4}$	4	$-\frac{1}{4}$	-4
$\frac{1}{2}$	2	$-\frac{1}{2}$	-2
1	1	-1	-1
2	$\frac{1}{2}$	-2	$-\frac{1}{2}$
3	$\frac{1}{3}$	-3	$-\frac{1}{3}$
4	$\frac{1}{4}$	-4	$-\frac{1}{4}$

Figure 5. Graph of $f(x) = \frac{1}{x}$.

Now that graphing calculators and computer graphing programs are widely available, we seldom need to sketch graphs by hand-plotting large numbers of points on graph paper. However, to use such a calculator or program effectively, we must know in advance which part of a curve to display. Critical features of a graph may be missed or misinterpreted if, for instance, the scale on the x- or y-axis is inappropriate.

An important use of calculus is to identify key features of a function that should appear in its graph. In many cases, only a few points need be plotted, and the general shape of the graph is easy to sketch by hand. For more complicated functions, a graphing program is helpful. Even then, calculus provides a way of checking that the graph on the computer screen has the correct shape. Algebraic calculations are usually part of the analysis. The appropriate algebraic skills are reviewed in this chapter.

Note also that analytic solutions to problems typically provide more precise information than graphing calculators and can provide insight into the behavior of the functions involved in the solution.

The connection between a function and its graph is explored in this section and in Section 0.6.

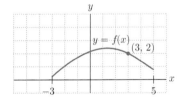

Figure 6

EXAMPLE 10   Suppose that f is the function whose graph is given in Fig. 6. Notice that the point $(x, y) = (3, 2)$ is on the graph of f.

(a) What is the value of the function when $x = 3$?

(b) Find $f(-2)$.

(c) What is the domain of f?

Solution **(a)** Since $(3, 2)$ is on the graph of f, the y-coordinate 2 must be the value of f at the x-coordinate 3. That is, $f(3) = 2$.

(b) To find $f(-2)$, we look at the y-coordinate of the point on the graph where $x = -2$. From Fig. 6 we see that $(-2, 1)$ is on the graph of f. Thus $f(-2) = 1$.

(c) The points on the graph of $f(x)$ all have x-coordinates between -3 and 5 inclusive; and for each value of x between -3 and 5 there is a point $(x, f(x))$ on the graph. So the domain consists of those x for which $-3 \leq x \leq 5$.   ■

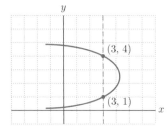

Figure 7. A curve that is **not** the graph of a function.

To every x in the domain, a function assigns one and only one value of y, that is, the function value $f(x)$. This implies, among other things, that not every curve is the graph of a function. To see this, refer first to the curve in Fig. 6, which *is* the graph of a function. It has the following important property: For each x between -3 and 5 inclusive, there is a *unique* y such that (x, y) is on the curve. The variable y is called the *dependent variable*, since its value depends on the value of the independent variable x. Refer to the curve in Fig. 7. It cannot be the graph of a function because a function f must assign to each x in its domain a *unique* value $f(x)$. However, for the curve of Fig. 7 there corresponds to $x = 3$ (for example) more than one y-value: $y = 1$ and $y = 4$.

The essential difference between the curves in Figs. 6 and 7 leads us to the following test.

The Vertical Line Test   A curve in the xy-plane is the graph of a function if and only if each vertical line cuts or touches the curve at no more than one point.

EXAMPLE 11

Which of the curves in Fig. 8 are graphs of functions?

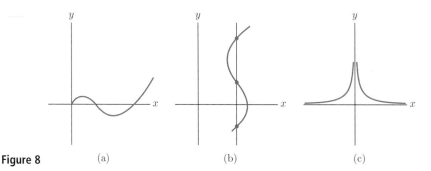

Figure 8   (a)   (b)   (c)

Solution   The curve in (a) is the graph of a function. It appears that vertical lines to the left of the y-axis do not touch the curve at all. This simply means that the function represented in (a) is defined only for $x \geq 0$. The curve in (b) is *not* the graph of a function because some vertical lines cut the curve in three places. The curve in (c) is the graph of a function whose domain is all nonzero x. [There is no point on the curve in (c) whose x-coordinate is 0.]   ■

There is another notation for functions that we will find useful. Suppose that $f(x)$ is a function. When $f(x)$ is graphed on an xy-coordinate system, the values of $f(x)$ give the y-coordinates of points of the graph. For this reason, the function is often abbreviated by the letter y, and we find it convenient to speak of "the function $y = f(x)$." For example, the function $y = 2x^2 + 1$ refers to the function $f(x)$ for which $f(x) = 2x^2 + 1$. The graph of a function $f(x)$ is often called *the graph of the equation* $y = f(x)$.

INCORPORATING TECHNOLOGY

Graphing Functions Using Technology   The use of graphing calculators is not required for the study of this text; however, graphing calculators are very useful tools that can be used to simplify computations, draw graphs, and sometimes enhance understanding of the fundamental topics of calculus. Helpful information about the use of calculators will appear at the ends of some sections in subsections titled Incorporating Technology. The examples in this text use TI-83/84 calculators,[*] and the keystrokes will be identical on every available version of these calculators. Other models and brands of graphing calculators should function similarly.

EXAMPLE 12

Graphing functions   Consider the function $f(x) = x^3 - 2$.

(a) Use your calculator to graph f.

(b) Change the parameters for the viewing window to get a different view of the graph.

Solution (a) **Step 1**   Press $\boxed{Y=}$. We will define our function in the calculator as Y_1. Move the cursor up if necessary so that it is placed directly after the expression "$\backslash Y_1 =$". Press $\boxed{\text{CLEAR}}$ to ensure that no formulas are entered for Y_1.

Step 2   Enter $X\char`^3 - 2$. The variable X may be entered using the $\boxed{\text{X,T,}\Theta,n}$ key. [See Fig. 9(a).]

(a)   (b)

Figure 9

Step 3   Press $\boxed{\text{GRAPH}}$. [See Fig. 9(b).]

(b) **Step 1**   Press $\boxed{\text{WINDOW}}$.

Step 2   Change the parameters to their desired values.

Figure 10

One of the most important tasks in using a graphing calculator is to determine the viewing window that shows the features of interest. For this example, we will simply set the view to $[-3, 3]$ by $[-29, 29]$. We will also set the value of $Yscl$ to 5. The parameter $Yscl$ and its sibling $Xscl$ set the distance between tick marks on their respective axes. To do this, set the window parameters on your calculator to match those in Fig. 10. The value of $Xres$ sets the screen resolution, and we will leave it at its defauly setting.

Figure 11

Step 3   Press $\boxed{\text{GRAPH}}$ to see the results. (See Fig. 11.)

---

[*]The designation TI-83/84 refers to the TI-83, TI-84, TI-83+, TI-84+, TI-83+ Silver Edition and TI-84+ Silver Edition calculators.

Practice Problems 0.1

1. Is the point $(3, 12)$ on the graph of the function $g(x) = x^2 + 5x - 10$?

2. Sketch the graph of the function $h(t) = t^2 - 2$.

EXERCISES 0.1*

Draw the following intervals on the number line.

1. $[-1, 4]$ **2.** $(4, 3\pi)$ **3.** $[-2, \sqrt{2})$

4. $[1, \frac{3}{2}]$ **5.** $(-\infty, 3)$ **6.** $(4, \infty)$

Use intervals to describe the real numbers satisfying the inequalities in Exercises 7–12.

7. $2 \leq x < 3$ **8.** $-1 < x < \frac{3}{2}$

9. $x < 0$ and $x \geq -1$ **10.** $x \geq -1$ and $x < 8$

11. $x < 3$ **12.** $x \geq \sqrt{2}$

13. If $f(x) = x^2 - 3x$, find $f(0)$, $f(5)$, $f(3)$, and $f(-7)$.

14. If $f(x) = 9 - 6x + x^2$, find $f(0)$, $f(2)$, $f(3)$, and $f(-13)$.

15. If $f(x) = x^3 + x^2 - x - 1$, find $f(1)$, $f(-1)$, $f(\frac{1}{2})$, and $f(a)$.

16. If $g(t) = t^3 - 3t^2 + t$, find $g(2)$, $g(-\frac{1}{2})$, $g(\frac{2}{3})$, and $g(a)$.

17. If $h(s) = s/(1 + s)$, find $h(\frac{1}{2})$, $h(-\frac{3}{2})$, and $h(a + 1)$.

18. If $f(x) = x^2/(x^2 - 1)$, find $f(\frac{1}{2})$, $f(-\frac{1}{2})$, and $f(a + 1)$.

19. If $f(x) = x^2 - 2x$, find $f(a + 1)$ and $f(a + 2)$.

20. If $f(x) = x^2 + 4x + 3$, find $f(a - 1)$ and $f(a - 2)$.

21. An office supply firm finds that the number of fax machines sold in year x is given approximately by the function $f(x) = 50 + 4x + \frac{1}{2}x^2$, where $x = 0$ corresponds to 1990.

    (a) What does $f(0)$ represent?

    (b) Find the number of fax machines sold in 1992.

22. Response of a Muscle When a solution of acetylcholine is introduced into the heart muscle of a frog, it diminishes the force with which the muscle contracts. The data from experiments of the biologist A. J. Clark are closely approximated by a function of the form

$$R(x) = \frac{100x}{b + x}, \qquad x \geq 0,$$

where x is the concentration of acetylcholine (in appropriate units), b is a positive constant that depends on the particular frog, and $R(x)$ is the response of the muscle to the acetylcholine, expressed as a percentage of the maximum possible effect of the drug.

    (a) Suppose that $b = 20$. Find the response of the muscle when $x = 60$.

    (b) Determine the value of b if $R(50) = 60$, that is, if a concentration of $x = 50$ units produces a 60% response.

In Exercises 23–26, describe the domain of the function.

23. $f(x) = \dfrac{8x}{(x - 1)(x - 2)}$ **24.** $f(t) = \dfrac{1}{\sqrt{t}}$

25. $g(x) = \dfrac{1}{\sqrt{3 - x}}$ **26.** $g(x) = \dfrac{4}{x(x + 2)}$

In Exercises 27–32, decide which curves are graphs of functions.

27.

28.

29.

30.

31.

32.

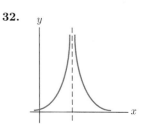

Exercises 33–42 relate to the function whose graph is sketched in Fig. 12.

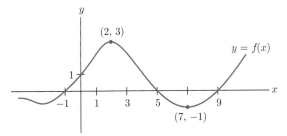

Figure 12

33. Find $f(0)$.

34. Find $f(7)$.

35. Find $f(2)$.

36. Find $f(-1)$.

37. Is $f(4)$ positive or negative?

38. Is $f(6)$ positive or negative?

39. Is $f(-\frac{1}{2})$ positive or negative?

40. Is $f(1)$ greater than $f(6)$?

41. For what values of x is $f(x) = 0$?

42. For what values of x is $f(x) \geq 0$?

Exercises 43–46 relate to Fig. 13. When a drug is injected into a person's muscle tissue, the concentration y of the drug in the blood is a function of the time elapsed since the injection. The graph of a typical time–concentration function f is given in Fig. 13, where $t = 0$ corresponds to the time of the injection.

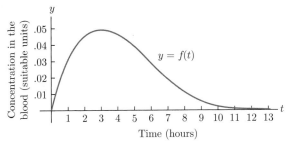

Figure 13.   Drug time–concentration curve.

43. What is the concentration of the drug when $t = 1$?

44. What is the value of the time–concentration function f when $t = 6$?

45. Find $f(5)$.

46. At what time does $f(t)$ attain its largest value?

47. Is the point $(3, 12)$ on the graph of the function $f(x) = (x - \frac{1}{2})(x + 2)$?

48. Is the point $(-2, 12)$ on the graph of the function $f(x) = x(5 + x)(4 - x)$?

49. Is the point $(\frac{1}{2}, \frac{2}{5})$ on the graph of the function $g(x) = (3x - 1)/(x^2 + 1)$?

50. Is the point $(\frac{2}{3}, \frac{5}{3})$ on the graph of the function $g(x) = (x^2 + 4)/(x + 2)$?

51. Find the y-coordinate of the point $(a+1,\ )$ if this point lies on the graph of the function $f(x) = x^3$.

52. Find the y-coordinate of the point $(2+h,\ )$ if this point lies on the graph of the function $f(x) = (5/x) - x$.

In Exercises 53–56, compute $f(1)$, $f(2)$, and $f(3)$.

53. $f(x) = \begin{cases} \sqrt{x} & \text{for } 0 \leq x < 2 \\ 1 + x & \text{for } 2 \leq x \leq 5 \end{cases}$

54. $f(x) = \begin{cases} 1/x & \text{for } 1 \leq x \leq 2 \\ x^2 & \text{for } 2 < x \end{cases}$

55. $f(x) = \begin{cases} \pi x^2 & \text{for } x < 2 \\ 1 + x & \text{for } 2 \leq x \leq 2.5 \\ 4x & \text{for } 2.5 < x \end{cases}$

56. $f(x) = \begin{cases} 3/(4 - x) & \text{for } x < 2 \\ 2x & \text{for } 2 \leq x < 3 \\ \sqrt{x^2 - 5} & \text{for } 3 \leq x \end{cases}$

57. If the brokerage firm in Example 6 decides to keep the commission charges unchanged for purchases up to and including \$600, but to charge only 1.5% plus \$15 for gold purchases exceeding \$600, express the brokerage commission as a function of the amount x of gold purchased.

58. Figure 14(a) shows the number 2 on the x-axis and the graph of a function. Let h represent a positive number and label a possible location for the number $2+h$. Plot the point on the graph whose first coordinate is $2 + h$ and label the point with its coordinates.

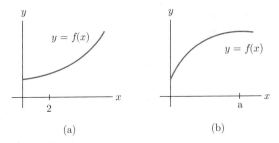

(a)   (b)

Figure 14

59. Figure 14(b) shows the number a on the x-axis and the graph of a function. Let h represent a negative number and label a possible location for the number $a+h$. Plot the point on the graph whose first coordinate is $a + h$ and label the point with its coordinates.

Technology Exercises

60. What is wrong with entering the function $f(x) = \dfrac{1}{x + 1}$ into a graphing utility as $\mathbf{Y_1} = \mathbf{1/X} + \mathbf{1}$?

61. What is wrong with entering the function $f(x) = x^{3/4}$ into a graphing utility as $\mathbf{Y_1} = \mathbf{X^\wedge 3/4}$?

In Exercises 62–65, graph the function with the specified viewing window setting.

62. $f(x) = x^3 - 33x^2 + 120x + 1500$; $[-8, 30]$ by $[-2000, 2000]$.

63. $f(x) = -x^2 + 2x + 2$; $[-2, 4]$ by $[-8, 5]$

64. $f(x) = \sqrt{x+1}$; $[0, 10]$ by $[-1, 4]$

65. $f(x) = \dfrac{1}{x^2 + 1}$; $[-4, 4]$ by $[-.5, 1.5]$

Solutions to Practice Problems 0.1

1. If $(3, 12)$ is on the graph of $g(x) = x^2 + 5x - 10$, we must have $g(3) = 12$. This is not the case, however, because

$$g(3) = 3^2 + 5(3) - 10$$
$$= 9 + 15 - 10 = 14.$$

Thus $(3, 12)$ is *not* on the graph of $g(x)$.

2. Choose some representative values for t, say $t = 0$, ± 1, ± 2, ± 3. For each value of t, calculate $h(t)$ and plot the point $(t, h(t))$. See Fig. 15.

t	$h(t) = t^2 - 2$
0	-2
1	-1
2	2
3	7
-1	-1
-2	2
-3	7

Figure 15.   Graph of $h(t) = t^2 - 2$.

0.2 Some Important Functions

In this section we introduce some of the functions that will play a prominent role in our discussion of calculus.

Linear Functions   As we shall see in Chapter 1, a knowledge of the algebraic and geometric properties of straight lines is essential for the study of calculus. Every straight line is the graph of a linear equation of the form

$$cx + dy = e,$$

where c, d, and e are given constants, with c and d not both zero. If $d \neq 0$, then we may solve the equation for y to obtain an equation of the form

$$y = mx + b, \tag{1}$$

for appropriate numbers m and b. If $d = 0$, then we may solve the equation for x to obtain an equation of the form

$$x = a, \tag{2}$$

for an appropriate number a. So every straight line is the graph of an equation of type (1) or (2). The graph of an equation of the form (1) is a nonvertical line [Fig. 1(a)], whereas the graph of (2) is a vertical line [Fig. 1(b)].

The straight line of Fig. 1(a) is the graph of the function $f(x) = mx + b$. Such a function, which is defined for all x, is called a *linear function*. Note that the straight line of Fig. 1(b) is not the graph of a function, since the vertical line test is violated.

An important special case of a linear function occurs if the value of m is zero; that is, $f(x) = b$ for some number b. In this case, $f(x)$ is called a *constant function*, since it assigns the same number b to every value of x. Its graph is the horizontal line whose equation is $y = b$. (See Fig. 2.)

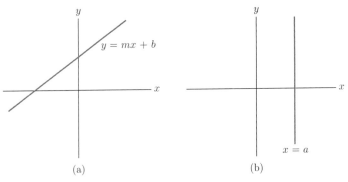

(a)

(b)

Figure 1

Figure 2. Graph of the constant function $f(x) = b$.

Linear functions often arise in real-life situations, as the first two examples show.

EXAMPLE 1   When the U.S. Environmental Protection Agency found a certain company dumping sulfuric acid into the Mississippi River, it fined the company $125,000, plus $1000 per day until the company complied with federal water pollution regulations. Express the total fine as a function of the number x of days that the company continued to violate the federal regulations.

Solution   The variable fine for x days of pollution, at $1000 per day, is $1000x$ dollars. The total fine is therefore given by the function

$$f(x) = 125{,}000 + 1000x.$$ ■

Since the graph of a linear function is a line, we may sketch it by locating any two points on the graph and drawing the line through them. For example, to sketch the graph of the function $f(x) = -\frac{1}{2}x + 3$, we may select two convenient values of x, say 0 and 4, and compute $f(0) = -\frac{1}{2}(0) + 3 = 3$ and $f(4) = -\frac{1}{2}(4) + 3 = 1$. The line through the points $(0, 3)$ and $(4, 1)$ is the graph of the function. (See Fig. 3.)

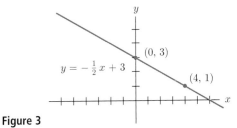

Figure 3

EXAMPLE 2   **Cost**   A simple cost function for a business consists of two parts: the *fixed costs*, such as rent, insurance, and business loans, which must be paid no matter how many items of a product are produced, and the *variable costs*, which depend on the number of items produced.

Suppose that a computer software company produces and sells a new spreadsheet program at a cost of $25 per copy, and the company has fixed costs of $10,000 per month. Express the total monthly cost as a function of the number of copies sold, x, and compute the cost when $x = 500$.

Solution The monthly variable cost is $25x$ dollars. Thus

$$[\text{total cost}] = [\text{fixed costs}] + [\text{variable costs}]$$
$$C(x) = 10{,}000 + 25x.$$

When sales are at 500 copies per month, the cost is

$$C(500) = 10{,}000 + 25(500) = 22{,}500 \text{ (dollars)}.$$

(See Fig. 4.) ■

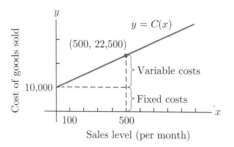

Figure 4. A linear cost function.

The point at which the graph of a linear function intersects the y-axis is called the y-*intercept* of the graph. The point at which the graph intersects the x-axis is called the x-*intercept*. The next example shows how to determine the intercepts of a linear function.

EXAMPLE 3   Determine the intercepts of the graph of the linear function $f(x) = 2x + 5$.

Solution Since the y-intercept is on the y-axis, its x-coordinate is 0. The point on the line with x-coordinate zero has y-coordinate

$$f(0) = 2(0) + 5 = 5.$$

So the y-intercept is $(0, 5)$. Since the x-intercept is on the x-axis, its y-coordinate is 0. Since $f(x)$ gives the y-coordinate, we must have

$$2x + 5 = 0$$
$$2x = -5$$
$$x = -\tfrac{5}{2}.$$

So $\left(-\tfrac{5}{2}, 0\right)$ is the x-intercept. (See Fig. 5.) ■

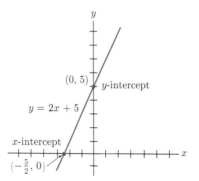

Figure 5. Graph of $f(x) = 2x + 5$.

The function in the next example is described by two expressions. Functions described by more than one expression are said to be *piecewise defined*.

EXAMPLE 4

Sketch the graph of the following function.

$$f(x) = \begin{cases} \frac{5}{2}x - \frac{1}{2} & \text{for } -1 \leq x \leq 1 \\ \frac{1}{2}x - 2 & \text{for } x > 1 \end{cases}$$

Solution This function is defined for $x \geq -1$. But it is defined by means of two distinct linear functions. We graph the two linear functions $\frac{5}{2}x - \frac{1}{2}$ and $\frac{1}{2}x - 2$. Then the graph of $f(x)$ consists of that part of the graph of $\frac{5}{2}x - \frac{1}{2}$ for which $-1 \leq x \leq 1$ plus that part of the graph of $\frac{1}{2}x - 2$ for which $x > 1$. (See Fig. 6.) ∎

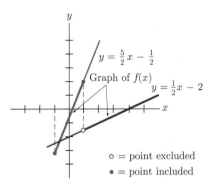

Figure 6. Graph of a function specified by two expressions.

o = point excluded
● = point included

Quadratic Functions Economists utilize average cost curves that relate the average unit cost of manufacturing a commodity to the number of units to be produced. (See Fig. 7.) Ecologists use curves that relate the net primary production of nutrients in a plant to the surface area of the foliage. (See Fig. 8.) Each curve is bowl-shaped, opening either up or down. The simplest functions whose graphs resemble these curves are the quadratic functions.

Figure 7. Average cost curve.

Figure 8. Production of nutrients.

A *quadratic function* is a function of the form

$$f(x) = ax^2 + bx + c,$$

where a, b, and c are constants and $a \neq 0$. The domain of such a function consists of all numbers. The graph of a quadratic function is called a *parabola*. Two typical parabolas are drawn in Figs. 9 and 10. We shall develop techniques for sketching graphs of quadratic functions after we have some calculus at our disposal.

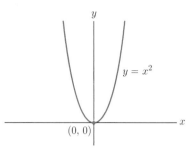

Figure 9.   Graph of $f(x) = x^2$.

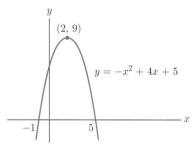

Figure 10.   Graph of $f(x) = -x^2 + 4x + 5$.

Polynomial and Rational Functions   A *polynomial function* $f(x)$ is of the form

$$f(x) = a_n x^n + a_{n-1} x^{n-1} + \cdots + a_0,$$

where n is a nonnegative integer and $a_0, a_1, \ldots, a_n$ are given numbers. Some examples of polynomial functions are

$$f(x) = 5x^3 - 3x^2 - 2x + 4$$
$$g(x) = x^4 - x + 1.$$

Of course, linear and quadratic functions are special cases of polynomial functions. The domain of a polynomial function consists of all numbers.

A function expressed as the quotient of two polynomials is called a *rational function*. Some examples are

$$h(x) = \frac{x^2 + 1}{x}$$

$$k(x) = \frac{x + 3}{x^2 - 4}.$$

The domain of a rational function excludes all values of x for which the denominator is zero. For example, the domain of $h(x)$ excludes $x = 0$, whereas the domain of $k(x)$ excludes $x = 2$ and $x = -2$. As we shall see, both polynomial and rational functions arise in applications of calculus.

Rational functions are used in environmental studies as *cost–benefit* models. The cost of removing a pollutant from the atmosphere is estimated as a function of the percentage of the pollutant removed. The higher the percentage removed, the greater the "benefit" to the people who breathe that air. The issues here are complex, of course, and the definition of "cost" is debatable. The cost to remove a small percentage of pollutant may be fairly low. But the removal cost of the final 5% of the pollutant, for example, may be terribly expensive.

EXAMPLE 5

A cost–benefit model   Suppose that a cost–benefit function is given by

$$f(x) = \frac{50x}{105 - x}, \qquad 0 \le x \le 100,$$

where x is the percentage of some pollutant to be removed and $f(x)$ is the associated cost (in millions of dollars). (See Fig. 11.) Find the costs to remove 70%, 95%, and 100% of the pollutant.

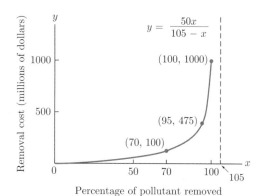

Figure 11. A cost–benefit model.

Solution The cost to remove 70% is

$$f(70) = \frac{50(70)}{105 - 70} = 100 \text{ (million dollars)}.$$

Similar calculations show that

$$f(95) = 475 \quad \text{and} \quad f(100) = 1000.$$

Observe that the cost to remove the last 5% of the pollutant is $f(100) - f(95) = 1000 - 475 = 525$ million dollars. This is more than five times the cost to remove the first 70% of the pollutant! ∎

Power Functions Functions of the form $f(x) = x^r$ are called *power functions*. The meaning of x^r is obvious when r is a positive integer. However, the power function $f(x) = x^r$ may be defined for any number r. We delay until Section 0.5 a discussion of power functions, where we will review the meaning of x^r in the case when r is a rational number.

The Absolute Value Function The absolute value of a number x is denoted by $|x|$ and is defined by

$$|x| = \begin{cases} x & \text{if } x \text{ is positive or zero} \\ -x & \text{if } x \text{ is negative.} \end{cases}$$

For example, $|5| = 5$, $|0| = 0$, and $|-3| = -(-3) = 3$.

The function defined for all numbers x by

$$f(x) = |x|$$

is called the *absolute value function*. Its graph coincides with the graph of the equation $y = x$ for $x \geq 0$ and with the graph of the equation $y = -x$ for $x < 0$. (See Fig. 12.)

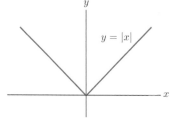

Figure 12. Graph of the absolute value function.

EXAMPLE 6

Evaluating functions Consider the quadratic function $f(x) = -x^2 + 4x + 5$. Use your graphing calculator to compute the value of $f(-5)$.

Solution *Step 1* Press $\boxed{\text{Y=}}$, enter the expression $-X^2 + 4X + 5$ for Y_1, and return to the home screen. (Recall that the variable X may be entered with the $\boxed{\text{X,T,}\Theta,n}$ key; to enter the expression $-X^2$, use the following key sequence: $\boxed{(-)}$ $\boxed{\text{X,T,}\Theta,n}$ $\boxed{x^2}$.)

INCORPORATING TECHNOLOGY

Step 2 From the home screen, press $\boxed{\text{VARS}}$ to access the variables menu, and then press $\boxed{\triangleright}$ to access the Y-VARS submenu. Next press $\boxed{1}$. You will be presented with a list of the y-variables Y_1, Y_2, and so on. Select Y_1.

Step 3   Now press $\boxed{(}$, enter -5, press $\boxed{)}$, and finally press $\boxed{\text{ENTER}}$.

The output (see Fig. 13) shows that $f(-5) = -40$. As an alternative, you can first assign the value -5 to X and then ask for the value of Y_1 (see Fig. 14). To assign the value -5 to X, use $\boxed{(-)}\ \boxed{5}\ \boxed{\text{STO} \triangleright}\ \boxed{X,T,\Theta,n}$.   ∎

| Figure 13 | Figure 14 |

Practice Problems 0.2

1. A photocopy service has a fixed cost of $2000 per month (for rent, depreciation of equipment, etc.) and variable costs of $0.04 for each page it reproduces for customers. Express its total cost as a (linear) function of the number of pages copied per month.

2. Determine the intercepts of the graph of

$$f(x) = -\tfrac{3}{8}x + 6.$$

EXERCISES 0.2

Graph the following functions.

1. $f(x) = 2x - 1$ **2.** $f(t) = 3$

3. $f(x) = 3x + 1$ **4.** $f(x) = -\tfrac{1}{2}x - 4$

5. $f(x) = -2x + 3$ **6.** $f(u) = \tfrac{1}{4}$

Determine the intercepts of the graphs of the following functions.

7. $f(x) = 9x + 3$ **8.** $f(x) = -\tfrac{1}{2}x - 1$

9. $f(v) = 5$ **10.** $f(x) = 14$

11. $f(x) = -\tfrac{1}{4}x + 3$ **12.** $f(w) = 6w - 4$

13. Enzyme Kinetics In biochemistry, such as in the study of enzyme kinetics, we encounter a linear function of the form $f(x) = (K/V)x + 1/V$, where K and V are constants.

   (a) If $f(x) = .2x + 50$, find K and V so that $f(x)$ may be written in the form $f(x) = (K/V)x + 1/V$.

   (b) Find the x-intercept and y-intercept of the line $y = (K/V)x + 1/V$ (in terms of K and V).

14. The constants K and V in Exercise 13 are often determined from experimental data. Suppose that a line is drawn through data points and has x-intercept $(-500, 0)$ and y-intercept $(0, 60)$. Determine K and V so that the line is the graph of the function $f(x) = (K/V)x + 1/V$. [*Hint:* Use Exercise 13(b).]

15. Cost of Car Rentals In some cities you can rent a car for $18 per day and $.20 per mile.

   (a) Find the cost of renting the car for one day and driving 200 miles.

   (b) If the car is to be rented for one day, express the total rental expense as a function of the number x of miles driven. (Assume that for each fraction of a mile driven the same fraction of $0.20 is charged.)

16. Right to Drill A gas company will pay a property owner $5000 for the right to drill on his land for natural gas and $.10 for each thousand cubic feet of gas extracted from the land. Express the amount of money the landowner will receive as a function of the amount of gas extracted from the land.

17. Medical Expense In 1998, a patient paid $300 per day for a semiprivate hospital room and $1500 for an appendectomy operation. Express the total amount paid for an appendectomy as a function of the number of days of hospital confinement.

18. Velocity of a Baseball When a baseball thrown at 85 miles per hour is hit by a bat swung at x miles per hour, the ball travels $6x - 40$ feet.[*] (This formula assumes that $50 \le x \le 90$ and that the bat is 35 inches long, weighs 32 ounces, and strikes a waist-high pitch so that the plane of the swing lies at $35°$ from the hor-

---

[*]Robert K. Adair, *The Physics of Baseball* (New York: Harper & Row, 1990).

izontal.) How fast must the bat be swung for the ball to travel 350 feet?

19. Cost–Benefit Let $f(x)$ be the cost–benefit function from Example 5. If 70% of the pollutant has been removed, what is the added cost to remove another 5%? How does this compare with the cost to remove the final 5% of the pollutant? (See Example 5.)

20. Suppose that the cost (in millions of dollars) to remove x percent of a certain pollutant is given by the cost–benefit function

$$f(x) = \frac{20x}{102 - x} \qquad \text{for } 0 \le x \le 100.$$

(a) Find the cost to remove 85% of the pollutant.

(b) Find the cost to remove the final 5% of the pollutant.

Each quadratic function in Exercises 21–26 has the form $y = ax^2 + bx + c$. Identify a, b, and c.

21. $y = 3x^2 - 4x$

22. $y = \dfrac{x^2 - 6x + 2}{3}$

23. $y = 3x - 2x^2 + 1$

24. $y = 3 - 2x + 4x^2$

25. $y = 1 - x^2$

26. $y = \frac{1}{2}x^2 + \sqrt{3}\,x - \pi$

Sketch the graphs of the following functions.

27. $f(x) = \begin{cases} 3x & \text{for } 0 \le x \le 1 \\ \frac{9}{2} - \frac{3}{2}x & \text{for } x > 1 \end{cases}$

28. $f(x) = \begin{cases} 1 + x & \text{for } x \le 3 \\ 4 & \text{for } x > 3 \end{cases}$

29. $f(x) = \begin{cases} 3 & \text{for } x < 2 \\ 2x + 1 & \text{for } x \ge 2 \end{cases}$

30. $f(x) = \begin{cases} \frac{1}{2}x & \text{for } 0 \le x < 4 \\ 2x - 3 & \text{for } 4 \le x \le 5 \end{cases}$

31. $f(x) = \begin{cases} 4 - x & \text{for } 0 \le x < 2 \\ 2x - 2 & \text{for } 2 \le x < 3 \\ x + 1 & \text{for } x \ge 3 \end{cases}$

32. $f(x) = \begin{cases} 4x & \text{for } 0 \le x < 1 \\ 8 - 4x & \text{for } 1 \le x < 2 \\ 2x - 4 & \text{for } x \ge 2 \end{cases}$

Evaluate each of the functions in Exercises 33–38 at the given value of x.

33. $f(x) = x^{100}$, $x = -1$

34. $f(x) = x^5$, $x = \frac{1}{2}$

35. $f(x) = |x|$, $x = 10^{-2}$

36. $f(x) = |x|$, $x = \pi$

37. $f(x) = |x|$, $x = -2.5$

38. $f(x) = |x|$, $x = -\frac{2}{3}$

Technology Exercises

In Exercises 39–42, use your graphing calculator to find the value of the given function at the indicated values of x.

39. $f(x) = 3x^3 + 8$; $x = -11$, $x = 10$

40. $f(x) = x^4 + 2x^3 + x - 5$; $x = -\frac{1}{2}$, $x = 3$

41. $f(x) = \frac{1}{2}x^2 + \sqrt{3}\,x - \pi$; $x = -2$, $x = 20$

42. $f(x) = \dfrac{2x - 1}{x^3 + 3x^2 + 4x + 1}$; $x = 2$, $x = 6$

Solutions to Practice Problems 0.2

1. If x represents the number of pages copied per month, then the variable cost is $.04x$ dollars. Now [total cost] = [fixed cost] + [variable cost]. If we define

$$f(x) = 2000 + .04x,$$

then $f(x)$ gives the total cost per month.

2. To find the y-intercept, evaluate $f(x)$ at $x = 0$.

$$f(0) = -\tfrac{3}{8}(0) + 6 = 0 + 6 = 6$$

To find the x-intercept, set $f(x) = 0$ and solve for x.

$$-\tfrac{3}{8}x + 6 = 0$$
$$\tfrac{3}{8}x = 6$$
$$x = \tfrac{8}{3} \cdot 6 = 16$$

Therefore, the y-intercept is $(0, 6)$ and the x-intercept is $(16, 0)$.

0.3 The Algebra of Functions

Many functions encountered later in the text can be viewed as combinations of other functions. For example, let $P(x)$ represent the profit a company makes on the sale of x units of some commodity. If $R(x)$ denotes the revenue received from the sale of x units, and if $C(x)$ is the cost of producing x units, then

$$P(x) = R(x) - C(x)$$
$$[\text{profit}] = [\text{revenue}] - [\text{cost}].$$

Writing the profit function in this way makes it possible to predict the behavior of $P(x)$ from properties of $R(x)$ and $C(x)$. For instance, we can determine when the profit $P(x)$ is positive by observing whether $R(x)$ is greater than $C(x)$. (See Fig. 1.)

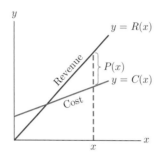

Figure 1. Profit equals revenue minus cost.

The next four examples review the algebraic techniques needed to combine functions by addition, subtraction, multiplication, and division.

EXAMPLE 1

Let $f(x) = 3x + 4$ and $g(x) = 2x - 6$. Find

$$f(x) + g(x), \quad f(x) - g(x), \quad \frac{f(x)}{g(x)}, \quad \text{and} \quad f(x)g(x).$$

Solution For $f(x) + g(x)$ and $f(x) - g(x)$, we add or subtract corresponding terms:

$$f(x) + g(x) = (3x + 4) + (2x - 6) = 3x + 4 + 2x - 6 = 5x - 2$$
$$f(x) - g(x) = (3x + 4) - (2x - 6) = 3x + 4 - 2x + 6 = x + 10.$$

To compute $\dfrac{f(x)}{g(x)}$ and $f(x)g(x)$, we first substitute the formulas for $f(x)$ and $g(x)$:

$$\frac{f(x)}{g(x)} = \frac{3x + 4}{2x - 6}$$

$$f(x)g(x) = (3x + 4)(2x - 6).$$

The expression for $\dfrac{f(x)}{g(x)}$ is already in simplest form. To simplify the expression for $f(x)g(x)$, we carry out the multiplication indicated in $(3x + 4)(2x - 6)$. We must be careful to multiply each term of $3x + 4$ by each term of $2x - 6$. A common order for multiplying such expressions is (1) the First terms, (2) the Outer terms, (3) the Inner terms, and (4) the Last terms. (This procedure may be remembered by the word FOIL.)

$$f(x)g(x) = \overbrace{(3x + 4)}^{1}\overbrace{(2x - 6)}^{4} = 6x^2 - 18x + 8x - 24$$
$$= 6x^2 - 10x - 24.\quad\blacksquare$$

EXAMPLE 2

Let

$$g(x) = \frac{2}{x} \quad \text{and} \quad h(x) = \frac{3}{x - 1}.$$

Express $g(x) + h(x)$ as a rational function.

Solution First we have

$$g(x) + h(x) = \frac{2}{x} + \frac{3}{x-1}, \qquad x \neq 0, 1.$$

The restriction $x \neq 0, 1$ comes from the fact that $g(x)$ is defined only for $x \neq 0$ and $h(x)$ is defined only for $x \neq 1$. (A rational function is not defined for values of the variable for which the denominator is 0.) To add two fractions, their denominators must be the same. A common denominator for $\frac{2}{x}$ and $\frac{3}{x-1}$ is $x(x-1)$. If we multiply $\frac{2}{x}$ by $\frac{x-1}{x-1}$, we obtain an equivalent expression whose denominator is $x(x-1)$. Similarly, if we multiply $\frac{3}{x-1}$ by $\frac{x}{x}$, we obtain an equivalent expression whose denominator is $x(x-1)$. Thus

$$\frac{2}{x} + \frac{3}{x-1} = \frac{2}{x} \cdot \frac{x-1}{x-1} + \frac{3}{x-1} \cdot \frac{x}{x}$$

$$= \frac{2(x-1)}{x(x-1)} + \frac{3x}{x(x-1)}$$

$$= \frac{2(x-1) + 3x}{x(x-1)}$$

$$= \frac{5x-2}{x(x-1)}.$$

So

$$g(x) + h(x) = \frac{5x-2}{x(x-1)}. \qquad \blacksquare$$

EXAMPLE 3   Find $f(t)g(t)$, where

$$f(t) = \frac{t}{t-1} \quad \text{and} \quad g(t) = \frac{t+2}{t+1}.$$

Solution To multiply rational functions, multiply numerator by numerator and denominator by denominator:

$$f(t)g(t) = \frac{t}{t-1} \cdot \frac{t+2}{t+1} = \frac{t(t+2)}{(t-1)(t+1)}.$$

An alternative way of expressing $f(t)g(t)$ is obtained by carrying out the indicated multiplications:

$$f(t)g(t) = \frac{t^2 + 2t}{t^2 + t - t - 1} = \frac{t^2 + 2t}{t^2 - 1}.$$

The choice of which expression to use for $f(t)g(t)$ depends on the particular application. $\blacksquare$

EXAMPLE 4   Find

$$\frac{f(x)}{g(x)}, \quad \text{where } f(x) = \frac{x}{x-3} \text{ and } g(x) = \frac{x+1}{x-5}.$$

Solution The function $f(x)$ is defined only for $x \neq 3$, and $g(x)$ is defined only for $x \neq 5$. The quotient $f(x)/g(x)$ is therefore not defined for $x = 3, 5$. Moreover, the quotient is not defined for values of x for which $g(x)$ is equal to 0; that is, $x = -1$. Thus, the quotient is defined for $x \neq -1, 3, 5$. To divide $f(x)$ by $g(x)$, we multiply $f(x)$ by the reciprocal of $g(x)$:

$$\frac{f(x)}{g(x)} = \frac{x}{x-3} \cdot \frac{x-5}{x+1} = \frac{x(x-5)}{(x-3)(x+1)} = \frac{x^2 - 5x}{x^2 - 2x - 3}, \quad x \neq -1, 3, 5. \quad \blacksquare$$

Composition of Functions Another important way of combining two functions $f(x)$ and $g(x)$ is to substitute the function $g(x)$ for every occurrence of the variable x in $f(x)$. The resulting function is called the *composition* (or *composite*) of $f(x)$ and $g(x)$ and is denoted by $f(g(x))$.

EXAMPLE 5

Let $f(x) = x^2 + 3x + 1$ and $g(x) = x - 5$. What is $f(g(x))$?

Solution We substitute $g(x)$ in place of each x in $f(x)$:

$$
\begin{aligned}
f(g(x)) &= [g(x)]^2 + 3g(x) + 1 \\
&= (x-5)^2 + 3(x-5) + 1 \\
&= (x^2 - 10x + 25) + (3x - 15) + 1 \\
&= x^2 - 7x + 11.
\end{aligned}
$$

$\blacksquare$

Later in the text we shall need to study expressions of the form $f(x+h)$, where $f(x)$ is a given function and h represents some number. The meaning of $f(x+h)$ is that $x+h$ is to be substituted for each occurrence of x in the formula for $f(x)$. In fact, $f(x+h)$ is just a special case of $f(g(x))$, where $g(x) = x + h$.

EXAMPLE 6

If $f(x) = x^3$, find $f(x+h) - f(x)$.

Solution

$$
\begin{aligned}
f(x+h) &= (x+h)^3 = x^3 + 3x^2h + 3xh^2 + h^3 \\
f(x+h) - f(x) &= (x^3 + 3x^2h + 3xh^2 + h^3) - x^3 \\
&= 3x^2h + 3xh^2 + h^3.
\end{aligned}
$$

$\blacksquare$

EXAMPLE 7

In a certain lake, the bass feed primarily on minnows, and the minnows feed on plankton. If the size of the bass population is a function $f(n)$ of the number n of minnows in the lake, and the number of minnows is a function $g(x)$ of the amount x of plankton in the lake, express the size of the bass population as a function of the amount of plankton, if $f(n) = 50 + \sqrt{n/150}$ and $g(x) = 4x + 3$.

Solution We have $n = g(x)$. Substituting $g(x)$ for n in $f(n)$, we find that the size of the bass population is given by

$$f(g(x)) = 50 + \sqrt{\frac{g(x)}{150}} = 50 + \sqrt{\frac{4x+3}{150}}.$$

$\blacksquare$

EXAMPLE 8

INCORPORATING TECHNOLOGY

Algebraic combinations of functions Consider the functions

$$f(x) = x^2 \quad \text{and} \quad g(x) = x + 3.$$

Use your graphing calculator to graph the function $f(g(x))$.

Solution *Step 1*   Press $\boxed{\text{Y=}}$ and set $Y_1 = X^2$ and $Y_2 = X + 3$.

Step 2   Set $Y_3 = Y_1(Y_2)$ by accessing the Y-VARS submenu of $\boxed{\text{VARS}}$.

Step 3   To graph our composite function $Y_3 = Y_1(Y_2)$ without also graphing Y_1 and Y_2, we must first deselect the functions Y_1 and Y_2. To deselect Y_1, place your cursor on top of the equal sign after \Y_1 and press $\boxed{\text{ENTER}}$. Similarly, deselect Y_2; your screen should look like Fig. 2(a). Finally, press $\boxed{\text{GRAPH}}$. [See Fig. 2(b).] ■

(a)

(b)

Figure 2

Practice Problems 0.3

1. Let $f(x) = x^5$, $g(x) = x^3 - 4x^2 + x - 8$.
   (a) Find $f(g(x))$.   **(b)** Find $g(f(x))$.

2. Let $f(x) = x^2$. Calculate $\dfrac{f(1+h) - f(1)}{h}$ and simplify.

EXERCISES 0.3

Let $f(x) = x^2 + 1$, $g(x) = 9x$, and $h(x) = 5 - 2x^2$. Calculate the following functions.

1. $f(x) + g(x)$
2. $f(x) - h(x)$
3. $f(x)g(x)$
4. $g(x)h(x)$
5. $\dfrac{f(t)}{g(t)}$
6. $\dfrac{g(t)}{h(t)}$

In Exercises 7–12, express $f(x) + g(x)$ as a rational function. Carry out all multiplications.

7. $f(x) = \dfrac{2}{x-3}$, $g(x) = \dfrac{1}{x+2}$

8. $f(x) = \dfrac{3}{x-6}$, $g(x) = \dfrac{-2}{x-2}$

9. $f(x) = \dfrac{x}{x-8}$, $g(x) = \dfrac{-x}{x-4}$

10. $f(x) = \dfrac{-x}{x+3}$, $g(x) = \dfrac{x}{x+5}$

11. $f(x) = \dfrac{x+5}{x-10}$, $g(x) = \dfrac{x}{x+10}$

12. $f(x) = \dfrac{x+6}{x-6}$, $g(x) = \dfrac{x-6}{x+6}$

Let $f(x) = \dfrac{x}{x-2}$, $g(x) = \dfrac{5-x}{5+x}$, and $h(x) = \dfrac{x+1}{3x-1}$. Express the following as rational functions.

13. $f(x) - g(x)$
14. $f(t) - h(t)$
15. $f(x)g(x)$
16. $g(x)h(x)$

17. $\dfrac{f(x)}{g(x)}$
18. $\dfrac{h(s)}{f(s)}$
19. $f(x+1)g(x+1)$
20. $f(x+2) + g(x+2)$
21. $\dfrac{g(x+5)}{f(x+5)}$
22. $f\left(\dfrac{1}{t}\right)$
23. $g\left(\dfrac{1}{u}\right)$
24. $h\left(\dfrac{1}{x^2}\right)$

Let $f(x) = x^6$, $g(x) = \dfrac{x}{1-x}$, and $h(x) = x^3 - 5x^2 + 1$. Calculate the following functions.

25. $f(g(x))$
26. $h(f(t))$
27. $h(g(x))$
28. $g(f(x))$
29. $g(h(t))$
30. $f(h(x))$

31. If $f(x) = x^2$, find $f(x+h) - f(x)$ and simplify.

32. If $f(x) = 1/x$, find $f(x+h) - f(x)$ and simplify.

33. If $g(t) = 4t - t^2$, find $\dfrac{g(t+h) - g(t)}{h}$ and simplify.

34. If $g(t) = t^3 + 5$, find $\dfrac{g(t+h) - g(t)}{h}$ and simplify.

35. Cost   After t hours of operation, an assembly line has assembled $A(t) = 20t - \frac{1}{2}t^2$ power lawn mowers, $0 \le t \le 10$. Suppose that the factory's cost of manufacturing x units is $C(x)$ dollars, where $C(x) = 3000 + 80x$.
   (a) Express the factory's cost as a (composite) function of the number of hours of operation of the assembly line.
   (b) What is the cost of the first 2 hours of operation?

36. Cost During the first $\frac{1}{2}$ hour, the employees of a machine shop prepare the work area for the day's work. After that, they turn out 10 precision machine parts per hour, so the output after t hours is $f(t)$ machine parts, where $f(t) = 10\left(t - \frac{1}{2}\right) = 10t - 5$, $\frac{1}{2} \le t \le 8$. The total cost of producing x machine parts is $C(x)$ dollars, where $C(x) = .1x^2 + 25x + 200$.

(a) Express the total cost as a (composite) function of t.

(b) What is the cost of the first 4 hours of operation?

37. Conversion Scales Table 1 shows a conversion table for men's hat sizes for three countries. The function $g(x) = 8x + 1$ converts from British sizes to French sizes, and the function $f(x) = \frac{1}{8}x$ converts from French sizes to U.S. sizes. Determine the function $h(x) = f(g(x))$ and give its interpretation.

Technology Exercises

38. Let $f(x) = x^2$. Graph the functions $f(x+1)$, $f(x-1)$, $f(x+2)$, and $f(x-2)$. Make a guess about the relationship between the graph of a general function $f(x)$ and the graph of $f(g(x))$, where $g(x) = x + a$ for some constant a. Test your guess on the functions $f(x) = x^3$

and $f(x) = \sqrt{x}$.

39. Let $f(x) = x^2$. Graph the functions $f(x)+1$, $f(x)-1$, $f(x) + 2$, and $f(x) - 2$. Make a guess about the relationship between the graph of a general function $f(x)$ and the graph of $f(x) + c$ for some constant c. Test your guess on the functions $f(x) = x^3$ and $f(x) = \sqrt{x}$.

40. Based on the results of Exercises 38 and 39, sketch the graph of $f(x) = (x - 1)^2 + 2$ without using a graphing calculator. Check your result with a graphing calculator.

41. Based on the results of Exercises 38 and 39, sketch the graph of $f(x) = (x + 2)^2 - 1$ without using a graphing calculator. Check your result with a graphing calculator.

42. Let $f(x) = x^2 + 3x + 1$ and let $g(x) = x^2 - 3x - 1$. Graph the two functions $f(g(x))$ and $g(f(x))$ together in the window $[-4, 4]$ by $[-10, 10]$ and determine if they are the same function.

43. Let $f(x) = \dfrac{x}{x - 1}$ and graph the function $f(f(x))$ in the window $[-15, 15]$ by $[-10, 10]$. Trace to examine the coordinates of several points on the graph and then determine the formula for $f(f(x))$.

TABLE 1   Conversion Table for Men's Hat Sizes

Britain	$6\frac{1}{2}$	$6\frac{5}{8}$	$6\frac{3}{4}$	$6\frac{7}{8}$	7	$7\frac{1}{8}$	$7\frac{1}{4}$	$7\frac{3}{8}$
France	53	54	55	56	57	58	59	60
United States	$6\frac{5}{8}$	$6\frac{3}{4}$	$6\frac{7}{8}$	7	$7\frac{1}{8}$	$7\frac{1}{4}$	$7\frac{3}{8}$	$7\frac{1}{2}$

Solutions to Practice Problems 0.3

1. (a) $f(g(x)) = [g(x)]^5 = (x^3 - 4x^2 + x - 8)^5$

(b) $g(f(x)) = [f(x)]^3 - 4[f(x)]^2 + f(x) - 8$

$= (x^5)^3 - 4(x^5)^2 + x^5 - 8$

$= x^{15} - 4x^{10} + x^5 - 8.$

2. $\dfrac{f(1+h) - f(1)}{h} = \dfrac{(1+h)^2 - 1^2}{h}$

$= \dfrac{1 + 2h + h^2 - 1}{h}$

$= \dfrac{2h + h^2}{h} = 2 + h.$

0.4 Zeros of Functions---The Quadratic Formula and Factoring

A *zero* of a function $f(x)$ is a value of x for which $f(x) = 0$. For instance, the function $f(x)$ whose graph is shown in Fig. 1 has $x = -3$, $x = 3$, and $x = 7$ as zeros. Throughout this book we shall need to determine zeros of functions or, what amounts to the same thing, to solve the equation $f(x) = 0$.

In Section 0.2 we found zeros of linear functions. In this section our emphasis is on zeros of quadratic functions.

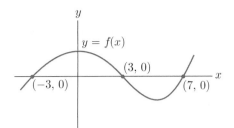

Figure 1.   Zeros of a function.

The Quadratic Formula   Consider the quadratic function $f(x) = ax^2 + bx + c$, $a \neq 0$. The zeros of this function are precisely the solutions of the quadratic equation

$$ax^2 + bx + c = 0.$$

One way of solving such an equation is by using the *quadratic formula*.

> The solutions of the equation $ax^2 + bx + c = 0$ are
>
> $$x = \frac{-b \pm \sqrt{b^2 - 4ac}}{2a}.$$

The $\pm$ sign tells us to form two expressions, one with $+$ and one with $-$. The quadratic formula implies that a quadratic equation has at most two roots. It will have none if the expression $b^2 - 4ac$ is negative and one if $b^2 - 4ac$ equals 0. The quadratic formula is derived at the end of this section.

EXAMPLE 1

Solve the quadratic equation $3x^2 - 6x + 2 = 0$.

Solution   Here $a = 3$, $b = -6$, and $c = 2$. Substituting these values into the quadratic formula, we find that

$$\sqrt{b^2 - 4ac} = \sqrt{(-6)^2 - 4(3)(2)} = \sqrt{36 - 24} = \sqrt{12} = \sqrt{4 \cdot 3} = 2\sqrt{3}$$

and

$$\begin{aligned} x &= \frac{-b \pm \sqrt{b^2 - 4ac}}{2a} \\ &= \frac{-(-6) \pm 2\sqrt{3}}{2(3)} \\ &= \frac{6 \pm 2\sqrt{3}}{6} \\ &= 1 \pm \frac{\sqrt{3}}{3}. \end{aligned}$$

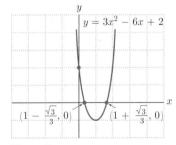

Figure 2

The solutions of the equation are $1 + \sqrt{3}/3$ and $1 - \sqrt{3}/3$. (See Fig. 2.)   ■

EXAMPLE 2

Find the zeros of the following quadratic functions.

(a) $f(x) = 4x^2 - 4x + 1$ **(b)** $f(x) = \frac{1}{2}x^2 - 3x + 5$

Solution   **(a)** We must solve $4x^2 - 4x + 1 = 0$. Here $a = 4$, $b = -4$, and $c = 1$, so

$$\sqrt{b^2 - 4ac} = \sqrt{(-4)^2 - 4(4)(1)} = \sqrt{0} = 0.$$

Thus there is only one zero:

$$x = \frac{-(-4) \pm 0}{2(4)} = \frac{4}{8} = \frac{1}{2}.$$

The graph of $f(x)$ is sketched in Fig. 3.

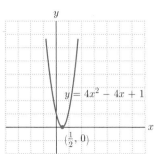

Figure 3   **Figure 4**

(b) We must solve $\frac{1}{2}x^2 - 3x + 5 = 0$. Here $a = \frac{1}{2}$, $b = -3$, and $c = 5$, so

$$\sqrt{b^2 - 4ac} = \sqrt{(-3)^2 - 4(\tfrac{1}{2})(5)} = \sqrt{9 - 10} = \sqrt{-1}.$$

The square root of a negative number is undefined, so we conclude that $f(x)$ has no zeros. The reason for this is clear from Fig. 4. The graph of $f(x)$ lies entirely above the x-axis and has no x-intercepts.   ■

The common problem of finding where two curves intersect amounts to finding the zeros of a function.

EXAMPLE 3   Find the points of intersection of the graphs of the functions $y = x^2 + 1$ and $y = 4x$. (See Fig. 5.)

Solution   If a point (x, y) is on both graphs, then its coordinates must satisfy both equations. That is, x and y must satisfy $y = x^2 + 1$ and $y = 4x$. Equating the two expressions for y, we have

$$x^2 + 1 = 4x.$$

To use the quadratic formula, we rewrite the equation in the form

$$x^2 - 4x + 1 = 0.$$

From the quadratic formula,

$$x = \frac{4 \pm \sqrt{16 - 4}}{2} = \frac{4 \pm \sqrt{12}}{2} = \frac{4 \pm 2\sqrt{3}}{2} = 2 \pm \sqrt{3}.$$

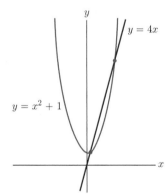

Figure 5. Points of intersection of two graphs.

Thus the x-coordinates of the points of intersection are $2 + \sqrt{3}$ and $2 - \sqrt{3}$. To find the y-coordinates, we substitute these values of x into either equation, $y = x^2 + 1$ or $y = 4x$. The second equation is simpler. We obtain $y = 4(2 + \sqrt{3}) = 8 + 4\sqrt{3}$ and $y = 4(2 - \sqrt{3}) = 8 - 4\sqrt{3}$. Thus the points of intersection are $(2 + \sqrt{3}, 8 + 4\sqrt{3})$ and $(2 - \sqrt{3}, 8 - 4\sqrt{3})$.   ■

EXAMPLE 4   **Profit and break-even points**   A cable television company estimates that with x thousand subscribers its monthly revenue and cost (in thousands of dollars) are

$$R(x) = 32x - .21x^2$$

$$C(x) = 195 + 12x.$$

Determine the company's *break-even points*; that is, find the number of subscribers at which the revenue equals the cost. (See Fig. 6.)

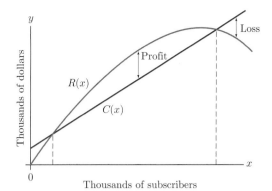

Figure 6. Break-even points.

Solution Let $P(x)$ be the profit function.

$$P(x) = R(x) - C(x)$$
$$= (32x - .21x^2) - (195 + 12x)$$
$$= -.21x^2 + 20x - 195.$$

The break-even points occur where the profit is zero. Thus we must solve

$$-.21x^2 + 20x - 195 = 0.$$

From the quadratic formula,

$$x = \frac{-20 \pm \sqrt{20^2 - 4(-.21)(-195)}}{2(-.21)} = \frac{-20 \pm \sqrt{236.2}}{-.42}$$
$$\approx 47.62 \pm 36.59 = 11.03 \text{ and } 84.21.$$

The break-even points occur where the company has 11,030 or 84,210 subscribers. Between those two levels, the company will be profitable. ■

Factoring If $f(x)$ is a polynomial, we can often write $f(x)$ as a product of linear factors (i.e., factors of the form $ax + b$). If this can be done, then the zeros of $f(x)$ can be determined by setting each linear factor equal to zero and solving for x. (The reason is that the product of numbers can be zero only when one of the numbers is zero.)

EXAMPLE 5   Factor the following quadratic polynomials.

(a) $x^2 + 7x + 12$ (b) $x^2 - 13x + 12$
(c) $x^2 - 4x - 12$ (d) $x^2 + 4x - 12$

Solution Note first that, for any numbers c and d,

$$(x + c)(x + d) = x^2 + (c + d)x + cd.$$

In the quadratic on the right, the constant term is the product cd, whereas the coefficient of x is the sum $c + d$.

(a) Think of all integers c and d such that $cd = 12$. Then choose the pair that satisfies $c + d = 7$; that is, take $c = 3$, $d = 4$. Thus

$$x^2 + 7x + 12 = (x + 3)(x + 4).$$

(b) We want $cd = 12$. Since 12 is positive, c and d must be both positive or both negative. We must also have $c + d = -13$. These facts lead us to

$$x^2 - 13x + 12 = (x - 12)(x - 1).$$

(c) We want $cd = -12$. Since -12 is negative, c and d must have opposite signs. Also, they must sum to give -4. We find that

$$x^2 - 4x - 12 = (x - 6)(x + 2).$$

(d) This is almost the same as part (c).

$$x^2 + 4x - 12 = (x + 6)(x - 2).$$ ■

EXAMPLE 6

Factor the following polynomials.

(a) $x^2 - 6x + 9$ **(b)** $x^2 - 25$
(c) $3x^2 - 21x + 30$ **(d)** $20 + 8x - x^2$

Solution **(a)** We look for $cd = 9$ and $c + d = -6$. The solution is $c = d = -3$, and

$$x^2 - 6x + 9 = (x - 3)(x - 3) = (x - 3)^2.$$

In general,

$$x^2 - 2cx + c^2 = (x - c)(x - c) = (x - c)^2.$$

(b) We use the identity

$$x^2 - c^2 = (x + c)(x - c).$$

Hence

$$x^2 - 25 = (x + 5)(x - 5).$$

(c) We first factor out a common factor of 3 and then use the method of Example 5.

$$3x^2 - 21x + 30 = 3(x^2 - 7x + 10)$$
$$= 3(x - 5)(x - 2).$$

(d) We first factor out -1 to make the coefficient of x^2 equal to $+1$.

$$20 + 8x - x^2 = (-1)(x^2 - 8x - 20)$$
$$= (-1)(x - 10)(x + 2).$$ ■

EXAMPLE 7

Factor the following polynomials.

(a) $x^2 - 8x$ **(b)** $x^3 + 3x^2 - 18x$ **(c)** $x^3 - 10x$

Solution In each case we first factor out a common factor of x.

(a) $x^2 - 8x = x(x - 8)$.
(b) $x^3 + 3x^2 - 18x = x(x^2 + 3x - 18) = x(x + 6)(x - 3)$.
(c) $x^3 - 10x = x(x^2 - 10)$. To factor $x^2 - 10$, we use the identity $x^2 - c^2 = (x + c)(x - c)$, where $c^2 = 10$ and $c = \sqrt{10}$. Thus

$$x^3 - 10x = x(x^2 - 10) = x(x + \sqrt{10})(x - \sqrt{10}).$$ ■

EXAMPLE 8

Solve the following equations.

(a) $x^2 - 2x - 15 = 0$ **(b)** $x^2 - 20 = x$ **(c)** $\dfrac{x^2 + 10x + 25}{x + 1} = 0$

Solution **(a)** The equation $x^2 - 2x - 15 = 0$ may be written in the form

$$(x - 5)(x + 3) = 0.$$

The product of two numbers is zero if one or the other of the numbers (or both) is zero. Hence

$$x - 5 = 0 \quad \text{or} \quad x + 3 = 0.$$

That is,

$$x = 5 \quad \text{or} \quad x = -3.$$

(b) First we must rewrite the equation $x^2 - 20 = x$ in the form $ax^2 + bx + c = 0$; that is,

$$x^2 - x - 20 = 0$$
$$(x - 5)(x + 4) = 0.$$

We conclude that

$$x - 5 = 0 \quad \text{or} \quad x + 4 = 0;$$

that is,

$$x = 5 \quad \text{or} \quad x = -4.$$

(c) A rational function will be zero only if the numerator is zero. Thus

$$x^2 + 10x + 25 = 0$$
$$(x + 5)^2 = 0$$
$$x + 5 = 0.$$

That is,

$$x = -5.$$

Since the denominator is not 0 at $x = -5$, we conclude that $x = -5$ is the solution. ∎

Derivation of the Quadratic Formula

$$ax^2 + bx + c = 0 \qquad (a \neq 0)$$

$$ax^2 + bx = -c$$

$$4a^2x^2 + 4abx = -4ac \qquad \text{(both sides multiplied by } 4a)$$

$$4a^2x^2 + 4abx + b^2 = b^2 - 4ac \qquad (b^2 \text{ added to both sides to complete the square)}$$

Now note that $4a^2x^2 + 4abx + b^2 = (2ax + b)^2$. To check this, simply multiply out the right side. Therefore,

$$(2ax + b)^2 = b^2 - 4ac$$

$$2ax + b = \pm\sqrt{b^2 - 4ac}$$

$$2ax = -b \pm \sqrt{b^2 - 4ac}$$

$$x = \frac{-b \pm \sqrt{b^2 - 4ac}}{2a}.$$

EXAMPLE 9

Computing the zeros of a function   Find the values of x at which $x^3 - 2x + 1 = 0$.

INCORPORATING TECHNOLOGY

Solution   ***Step 1***   Press $\boxed{Y=}$ and enter the formula for the polynomial for Y_1. Press $\boxed{\text{GRAPH}}$ to graph the function. The zeros of this polynomial appear to occur in the interval $[-2, 2]$, so press $\boxed{\text{WINDOW}}$ and set $Xmin = -2$ and $Xmax = 2$.

Step 2   Now press $\boxed{\text{2nd}}$ [CALC] and then $\boxed{2}$ to begin the process of finding the zeros. When prompted, enter -2 for the left bound, -1 for the right bound, and $-.5$ for the guess. We can enter any value between our left and right bounds for the guess. The calculator then computes for us that $x^3 - 2x + 1 = 0$ when $x = -1.618034$. (See Fig. 7.)

We can now repeat this process with different bounds to find the other value of x at which $x^3 - 2x + 1 = 0$. Finally, notice that the menu behind $\boxed{\text{2nd}}$ [CALC] includes a number of useful routines. Of particular interest for this section is option **5: intersect**, which will compute for us the intersection points of two functions, Y_1 and Y_2. ■

Figure 7

One problem in graphing a function is to find a domain (Xmin to Xmax) that contains all the zeros. For a polynomial function, there is an easy solution to this problem. Write the polynomial in the form $c(x^n + a_{n-1}x^{n-1} + \cdots + a_0)$ and let M be the number that is one more than the largest magnitude of the coefficients $a_{n-1}, \ldots, a_0$. Then the interval $[-M, M]$ will contain all the zeros of the polynomial. For instance, all the zeros of the polynomial $x^3 - 10x^2 + 9x + 8$ lie in the interval $[-11, 11]$. After you view the polynomial on the domain $[-M, M]$, you can usually find a smaller domain that also contains all the zeros.

Practice Problems 0.4

1. Solve the equation $x - \dfrac{14}{x} = 5$.

2. Use the quadratic formula to solve $7x^2 - 35x + 35 = 0$.

EXERCISES 0.4

Use the quadratic formula to find the zeros of the functions in Exercises 1–6.

1. $f(x) = 2x^2 - 7x + 6$ **2.** $f(x) = 3x^2 + 2x - 1$

3. $f(t) = 4t^2 - 12t + 9$ **4.** $f(x) = \frac{1}{4}x^2 + x + 1$

5. $f(x) = -2x^2 + 3x - 4$ **6.** $f(a) = 11a^2 - 7a + 1$

Use the quadratic formula to solve the equations in Exercises 7–12.

7. $5x^2 - 4x - 1 = 0$ **8.** $x^2 - 4x + 5 = 0$

9. $15x^2 - 135x + 300 = 0$ **10.** $z^2 - \sqrt{2}\,z - \frac{5}{4} = 0$

11. $\frac{3}{2}x^2 - 6x + 5 = 0$ **12.** $9x^2 - 12x + 4 = 0$

Factor the polynomials in Exercises 13–24.

13. $x^2 + 8x + 15$ **14.** $x^2 - 10x + 16$

15. $x^2 - 16$ **16.** $x^2 - 1$

17. $3x^2 + 12x + 12$ **18.** $2x^2 - 12x + 18$

19. $30 - 4x - 2x^2$ **20.** $15 + 12x - 3x^2$

21. $3x - x^2$ **22.** $4x^2 - 1$

23. $6x - 2x^3$ **24.** $16x + 6x^2 - x^3$

Find the points of intersection of the pairs of curves in Exercises 25–32.

25. $y = 2x^2 - 5x - 6$, $y = 3x + 4$

26. $y = x^2 - 10x + 9$, $y = x - 9$

27. $y = x^2 - 4x + 4$, $y = 12 + 2x - x^2$

28. $y = 3x^2 + 9$, $y = 2x^2 - 5x + 3$

29. $y = x^3 - 3x^2 + x$, $y = x^2 - 3x$

30. $y = \frac{1}{2}x^3 - 2x^2$, $y = 2x$

31. $y = \frac{1}{2}x^3 + x^2 + 5$, $y = 3x^2 - \frac{1}{2}x + 5$

32. $y = 30x^3 - 3x^2$, $y = 16x^3 + 25x^2$

Solve the equations in Exercises 33–38.

33. $\dfrac{21}{x} - x = 4$ **34.** $x + \dfrac{2}{x - 6} = 3$

35. $x + \dfrac{14}{x + 4} = 5$ **36.** $1 = \dfrac{5}{x} + \dfrac{6}{x^2}$

37. $\dfrac{x^2 + 14x + 49}{x^2 + 1} = 0$ **38.** $\dfrac{x^2 - 8x + 16}{1 + \sqrt{x}} = 0$

39. Break-Even Points Suppose that the cable television company's cost function in Example 4 changes to $C(x) = 275 + 12x$. Determine the new break-even points.

40. Velocity When a car is moving at x miles per hour and the driver decides to slam on the brakes, the car will travel $x + \frac{1}{20}x^2$ feet.* If a car travels 175 feet after the driver decides to stop, how fast was the car moving?

Technology Exercises

In Exercises 41–44, find the zeros of the function. (Use the specified viewing window.)

41. $f(x) = x^2 - x - 2$; $[-4, 5]$ by $[-4, 10]$

42. $f(x) = x^3 - 3x + 2$; $[-3, 3]$ by $[-10, 10]$

43. $f(x) = \sqrt{x + 2} - x + 2$; $[-2, 7]$ by $[-2, 4]$

44. $f(x) = \dfrac{x}{x + 2} - x^2 + 1$; $[-1.5, 2]$ by $[-2, 3]$

In Exercises 45–48, find the points of intersection of the graphs of the functions. (Use the specified viewing window.)

45. $f(x) = 2x - 1$; $g(x) = x^2 - 2$; $[-4, 4]$ by $[-6, 10]$

46. $f(x) = -x - 2$; $g(x) = -4x^2 + x + 1$; $[-2, 2]$ by $[-5, 2]$

47. $f(x) = 3x^4 - 14x^3 + 24x - 3$; $g(x) = 2x - 30$; $[-3, 5]$ by $[-80, 30]$

48. $f(x) = \dfrac{1}{x}$; $g(x) = \sqrt{x^2 - 1}$; $[0, 4]$ by $[-1, 3]$

In Exercises 49–52, find a good window setting for the graph of the function. The graph should show all the zeros of the polynomial.

49. $f(x) = x^3 - 22x^2 + 17x + 19$

50. $f(x) = x^4 - 200x^3 - 100x^2$

51. $f(x) = 3x^3 + 52x^2 - 12x - 12$

52. $f(x) = 2x^5 - 24x^4 - 24x + 2$

Solutions to Practice Problems 0.4

1. Multiply both sides of the equation by x. Then $x^2 - 14 = 5x$. Now take the term $5x$ to the left side of the equation and solve by factoring.

$$x^2 - 5x - 14 = 0$$
$$(x - 7)(x + 2) = 0$$
$$x = 7 \quad \text{or} \quad x = -2$$

2. In this case, each coefficient is a multiple of 7. To simplify the arithmetic, we divide both sides of the equation by 7 before using the quadratic formula.

$$x^2 - 5x + 5 = 0$$
$$\sqrt{b^2 - 4ac} = \sqrt{(-5)^2 - 4(1)(5)} = \sqrt{5}$$
$$x = \frac{-b \pm \sqrt{b^2 - 4ac}}{2a} = \frac{5 \pm \sqrt{5}}{2 \cdot 1} = \frac{5}{2} \pm \frac{1}{2}\sqrt{5}.$$

0.5 Exponents and Power Functions

In this section we review the operations with exponents that occur frequently throughout the text. We begin with the definition of b^r for various types of numbers b and r.

For any nonzero number b and any positive integer n, we have by definition that

$$b^n = \underbrace{b \cdot b \cdot \cdots \cdot b}_{n \text{ times}}$$

$$b^{-n} = \frac{1}{b^n},$$

and

$$b^0 = 1.$$

For example, $2^4 = 2 \cdot 2 \cdot 2 \cdot 2 = 16$, $2^{-4} = \dfrac{1}{2^4} = \dfrac{1}{16}$, and $2^0 = 1$.

---

*The general formula is $f(x) = ax + bx^2$, where the constant a depends on the driver's reaction time and the constant b depends on the weight of the car and the type of tires. This mathematical model was analyzed in D. Burghes, I. Huntley, and J. McDonald, *Applying Mathematics: A Course in Mathematical Modelling* (New York: Halstead Press, 1982), pp. 57–60.

Next we consider numbers of the form $b^{1/n}$, where n is a positive integer. For instance,

$2^{1/2}$ is the positive number whose square is 2: $2^{1/2} = \sqrt{2}$

$2^{1/3}$ is the positive number whose cube is 2: $2^{1/3} = \sqrt[3]{2}$

$2^{1/4}$ is the positive number whose fourth power is 2: $2^{1/4} = \sqrt[4]{2}$

and so on. In general, when b is zero or positive, $b^{1/n}$ is zero or the positive number whose nth power is b.

If n is even, there is no number whose nth power is b if b is negative. Therefore, $b^{1/n}$ is not defined if n is even and $b < 0$. When n is odd, we may permit b to be negative as well as positive. For example, $(-8)^{1/3}$ is the number whose cube is -8; that is,

$$(-8)^{1/3} = -2.$$

Thus, when b is negative and n is odd, we again define $b^{1/n}$ to be the number whose nth power is b.

Finally, let us consider numbers of the form $b^{m/n}$ and $b^{-m/n}$, where m and n are positive integers. We may assume that the fraction m/n is in lowest terms (so that m and n have no common factor). Then we define

$$b^{m/n} = (b^{1/n})^m$$

whenever $b^{1/n}$ is defined, and

$$b^{-m/n} = \frac{1}{b^{m/n}}$$

whenever $b^{m/n}$ is defined and is not zero. For example,

$$8^{5/3} = (8^{1/3})^5 = (2)^5 = 32,$$

$$8^{-5/3} = \frac{1}{8^{5/3}} = \frac{1}{32},$$

$$(-8)^{5/3} = \left[(-8)^{1/3}\right]^5 = (-2)^5 = -32.$$

Exponents may be manipulated algebraically according to the following rules:

Laws of Exponents

1. $b^r b^s = b^{r+s}$

2. $b^{-r} = \dfrac{1}{b^r}$

3. $\dfrac{b^r}{b^s} = b^r \cdot b^{-s} = b^{r-s}$

4. $(b^r)^s = b^{rs}$

5. $(ab)^r = a^r b^r$

6. $\left(\dfrac{a}{b}\right)^r = \dfrac{a^r}{b^r}$

EXAMPLE 1

Use the laws of exponents to calculate the following quantities.

(a) $2^{1/2}\, 50^{1/2}$ **(b)** $(2^{1/2}\, 2^{1/3})^6$ **(c)** $\dfrac{5^{3/2}}{\sqrt{5}}$

Solution **(a)** $2^{1/2}\, 50^{1/2} = (2 \cdot 50)^{1/2}$ (Law 5)

$$= \sqrt{100}$$

$$= 10.$$

(b) $\left(2^{1/2}\, 2^{1/3}\right)^6 = \left(2^{(1/2)+(1/3)}\right)^6$ (Law 1)

$\qquad\qquad = \left(2^{5/6}\right)^6$

$\qquad\qquad = 2^{(5/6)6}$ (Law 4)

$\qquad\qquad = 2^5$

$\qquad\qquad = 32.$

(c) $\dfrac{5^{3/2}}{\sqrt{5}} = \dfrac{5^{3/2}}{5^{1/2}}$

$\qquad\quad = 5^{(3/2)-(1/2)}$ (Law 3)

$\qquad\quad = 5^1$

$\qquad\quad = 5.$ ∎

Now that we know the definition of x^r for r rational, we can examine some calculations and applications involving power functions.

EXAMPLE 2   Simplify the following expressions.

(a) $\dfrac{1}{x^{-4}}$ **(b)** $\dfrac{x^2}{x^5}$ **(c)** $\sqrt{x}\,(x^{3/2} + 3\sqrt{x})$

Solution **(a)** $\dfrac{1}{x^{-4}} = x^{-(-4)}$ (Law 2 with $r = -4$)

$\qquad\qquad = x^4.$

(b) $\dfrac{x^2}{x^5} = x^{2-5}$ (Law 3)

$\qquad\quad = x^{-3}.$

It is also correct to write this answer as $\dfrac{1}{x^3}$.

(c) $\sqrt{x}\,(x^{3/2} + 3\sqrt{x}) = x^{1/2}\left(x^{3/2} + 3x^{1/2}\right)$

$\qquad\qquad\qquad\qquad = x^{1/2}x^{3/2} + 3x^{1/2}x^{1/2}$

$\qquad\qquad\qquad\qquad = x^{(1/2)+(3/2)} + 3x^{(1/2)+(1/2)}$ (Law 1)

$\qquad\qquad\qquad\qquad = x^2 + 3x.$ ∎

A *power function* is a function of the form

$$f(x) = x^r$$

for some number r.

EXAMPLE 3   Let $f(x)$ and $g(x)$ be the power functions

$$f(x) = x^{-1} \quad \text{and} \quad g(x) = x^{1/2}.$$

Determine the following functions.

(a) $\dfrac{f(x)}{g(x)}$ **(b)** $f(x)g(x)$ **(c)** $\dfrac{g(x)}{f(x)}$

Solution **(a)** $\dfrac{f(x)}{g(x)} = \dfrac{x^{-1}}{x^{1/2}} = x^{-1-(1/2)} = x^{-3/2} = \dfrac{1}{x^{3/2}}$

(b) $f(x)g(x) = x^{-1}x^{1/2} = x^{-1+(1/2)} = x^{-1/2} = \dfrac{1}{x^{1/2}} = \dfrac{1}{\sqrt{x}}$

(c) $\dfrac{g(x)}{f(x)} = \dfrac{x^{1/2}}{x^{-1}} = x^{(1/2)-(-1)} = x^{3/2}$ ■

Compound Interest   The subject of compound interest provides a significant application of exponents. We introduce this topic at this point with a view toward using it as a source of applied problems throughout the book.

When money is deposited in a savings account, interest is paid at stated intervals. If this interest is added to the account and thereafter earns interest itself, then the interest is called *compound interest*. The original amount deposited is called the *principal amount*. The principal amount plus the compound interest is called the *compound amount*. The interval between interest payments is referred to as the *interest period*. In formulas for compound interest, the interest rate is expressed as a decimal rather than a percent. Thus 6% is written as .06.

If $1000 is deposited at 6% annual interest, compounded annually, the compound amount at the end of the first year will be

$$A_1 = \underset{\text{principal}}{1000} + \underset{\text{interest}}{1000(.06)} = 1000(1 + .06).$$

At the end of the second year the compound amount will be

$$A_2 = \underset{\substack{\text{compound} \\ \text{amount}}}{A_1} + \underset{\text{interest}}{A_1(.06)} = A_1(1 + .06)$$

$$= [1000(1 + .06)](1 + .06) = 1000(1 + .06)^2.$$

At the end of 3 years

$$A_3 = A_2 + A_2(.06) = A_2(1 + .06)$$
$$= \left[1000(1 + .06)^2\right](1 + .06) = 1000(1 + .06)^3.$$

After n years the compound amount will be

$$A = 1000(1 + .06)^n.$$

In this example the interest period was 1 year. The important point to note, however, is that at the end of each interest period the amount on deposit grew by a factor of $(1 + .06)$. In general, if the interest rate is i instead of .06, the compound amount will grow by a factor of $(1 + i)$ at the end of each interest period.

If a principal amount P is invested at a compound interest rate i per interest period, for a total of n interest periods, the compound amount A at the end of the nth period will be

$$A = P(1 + i)^n. \tag{1}$$

EXAMPLE 4   If $5000 is invested at 8% per year, with interest compounded annually, what is the compound amount after 3 years?

Solution   Substituting $P = 5000$, $i = .08$, and $n = 3$ into formula (1), we have

$$A = 5000(1 + .08)^3 = 5000(1.08)^3$$
$$= 5000(1.259712) = 6298.56 \text{ dollars.} \qquad ■$$

It is common practice to state the interest rate as a percentage per year ("per annum"), even though each interest period is often shorter than 1 year. If the annual rate is r and if interest is paid and compounded m times per year, the interest rate i for each period is given by

$$[\text{rate per period}] = i = \frac{r}{m} = \frac{[\text{annual interest rate}]}{[\text{periods per year}]}.$$

Many banks pay interest quarterly. If the stated annual rate is 5%, then $i = .05/4 = .0125$.

If interest is compounded for t years, with m interest periods each year, there will be a total of mt interest periods. If in formula (1) we replace n by mt and replace i by r/m, we obtain the following formula for the compound amount:

$$A = P\left(1 + \frac{r}{m}\right)^{mt},$$

where P = principal amount   (2)
r = interest rate per annum
m = number of interest periods per year
t = number of years.

EXAMPLE 5   Suppose that $1000 is deposited in a savings account that pays 6% per annum, compounded quarterly. If no additional deposits or withdrawals are made, how much will be in the account at the end of 1 year?

Solution   We use formula (2) with $P = 1000$, $r = .06$, $m = 4$, and $t = 1$.

$$A = 1000\left(1 + \frac{.06}{4}\right)^4 = 1000(1.015)^4$$

$$\approx 1000(1.06136355) \approx 1061.36 \text{ dollars} \qquad \blacksquare$$

Note that the $1000 in Example 5 earned a total of $61.36 in (compound) interest. This is 6.136% of $1000. Savings institutions sometimes advertise this rate as the *effective* annual interest rate. That is, the savings institutions mean that, *if* they paid interest only once a year, they would have to pay a rate of 6.136% to produce the same earnings as their 6% rate compounded quarterly. The stated rate of 6% is often called the *nominal rate*.

The effective annual rate can be increased by compounding the interest more often. Some savings institutions compound interest monthly or even daily.

EXAMPLE 6   If the interest in Example 5 were compounded monthly, how much would be in the account at the end of 1 year? What about the case when 6% annual interest is compounded daily?

Solution   For monthly compounding, $m = 12$. From formula (2) we have

$$A = 1000\left(1 + \frac{.06}{12}\right)^{12} = 1000(1.005)^{12} \approx 1061.68 \text{ dollars}.$$

The effective rate in this case is 6.168%.

A "bank year" usually consists of 360 days (to simplify calculations). So, for daily compounding, we take $m = 360$. Then

$$A = 1000 \left(1 + \frac{.06}{360} \right)^{360} \approx 1000(1.00016667)^{360}$$

$$\approx 1000(1.06183251) \approx 1061.83 \text{ dollars.}$$

With daily compounding, the effective rate is 6.183%. ■

EXAMPLE 7   A corporation issues a bond costing $200 and paying interest compounded monthly. The interest is accumulated until the bond reaches maturity. (A security of this sort is called a *zero coupon bond.*) If after 5 years the bond is worth $500, what is the annual interest rate?

Solution   Let r denote the annual interest rate. The value A of the bond after 5 years $= 60$ months is given by the compound interest formula:

$$A = 200 \left(1 + \frac{r}{12} \right)^{60}.$$

We must find r that satisfies

$$500 = 200 \left(1 + \frac{r}{12} \right)^{60}$$

$$2.5 = \left(1 + \frac{r}{12} \right)^{60}.$$

Raise both sides to the power $\frac{1}{60}$ and apply the laws of exponents to obtain

$$(2.5)^{1/60} = \left[\left(1 + \frac{r}{12} \right)^{60} \right]^{\frac{1}{60}} = \left(1 + \frac{r}{12} \right)^{60 \cdot \frac{1}{60}} = 1 + \frac{r}{12}$$

$$r = 12 \cdot \left((2.5)^{1/60} - 1 \right).$$

Using a calculator, we see that $r \approx .18466$. That is, the annual interest rate is 18.466%. (A bond paying a rate of interest this high is generally called a *junk bond.*) ■

INCORPORATING TECHNOLOGY

By default, your TI-83/84-type calculator will display the result of a calculation in 10 digits. For example, if asked to calculate 1/3, your calculator will return .3333333333, and 7/3 returns 2.3333333333. In each case, 10 digits are given in the answer.

However, if the answer cannot display 10 digits (or if the absolute value is less than .001), your calculator will express the answer in scientific notation. Scientific notation expresses numbers in two parts. The significant digits display with one digit to the left of the decimal. The appropriate power of 10 displays to the right of **E**, as in **2.5E-4**. This stands for 2.5×10^{-4} or 0.00025. Similarly, **1E12** stands for 1×10^{12} or 1,000,000,000,000. (*Note:* Multiplying a number by 10^{-4} moves the decimal point four places to the left and multiplying by 10^{12} moves the decimal point 12 places to the right.) ■

Practice Problems 0.5

1. Compute the following.

    (a) -5^2 **(b)** $16^{.75}$

2. Simplify the following.

    (a) $(4x^3)^2$ **(b)** $\dfrac{\sqrt[3]{x}}{x^3}$ **(c)** $\dfrac{2 \cdot (x+5)^6}{x^2 + 10x + 25}$

EXERCISES 0.5

In Exercises 1–28, compute the numbers.

1. 3^3 **2.** $(-2)^3$ **3.** 1^{100}

4. 0^{25} **5.** $(.1)^4$ **6.** $(100)^4$

7. -4^2 **8.** $(.01)^3$ **9.** $(16)^{1/2}$

10. $(27)^{1/3}$ **11.** $(.000001)^{1/3}$ **12.** $\left(\dfrac{1}{125}\right)^{1/3}$

13. 6^{-1} **14.** $\left(\dfrac{1}{2}\right)^{-1}$ **15.** $(.01)^{-1}$

16. $(-5)^{-1}$ **17.** $8^{4/3}$ **18.** $16^{3/4}$

19. $(25)^{3/2}$ **20.** $(27)^{2/3}$ **21.** $(1.8)^0$

22. $9^{1.5}$ **23.** $16^{.5}$ **24.** $(81)^{.75}$

25. $4^{-1/2}$ **26.** $\left(\dfrac{1}{8}\right)^{-2/3}$ **27.** $(.01)^{-1.5}$

28. $1^{-1.2}$

In Exercises 29–40, use the laws of exponents to compute the numbers.

29. $5^{1/3} \cdot 200^{1/3}$ **30.** $(3^{1/3} \cdot 3^{1/6})^6$ **31.** $6^{1/3} \cdot 6^{2/3}$

32. $(9^{4/5})^{5/8}$ **33.** $\dfrac{10^4}{5^4}$ **34.** $\dfrac{3^{5/2}}{3^{1/2}}$

35. $(2^{1/3} \cdot 3^{2/3})^3$ **36.** $20^{.5} \cdot 5^{.5}$ **37.** $\left(\dfrac{8}{27}\right)^{2/3}$

38. $(125 \cdot 27)^{1/3}$ **39.** $\dfrac{7^{4/3}}{7^{1/3}}$ **40.** $(6^{1/2})^0$

In Exercises 41–70, use the laws of exponents to simplify the algebraic expressions. Your answer should not involve parentheses or negative exponents.

41. $(xy)^6$ **42.** $(x^{1/3})^6$ **43.** $\dfrac{x^4 \cdot y^5}{xy^2}$

44. $\dfrac{1}{x^{-3}}$ **45.** $x^{-1/2}$ **46.** $(x^3 \cdot y^6)^{1/3}$

47. $\left(\dfrac{x^4}{y^2}\right)^3$ **48.** $\left(\dfrac{x}{y}\right)^{-2}$ **49.** $(x^3y^5)^4$

50. $\sqrt{1+x}\,(1+x)^{3/2}$ **51.** $x^5 \cdot \left(\dfrac{y^2}{x}\right)^3$

52. $x^{-3} \cdot x^7$ **53.** $(2x)^4$ **54.** $\dfrac{-3x}{15x^4}$

55. $\dfrac{-x^3y}{-xy}$ **56.** $\dfrac{x^3}{y^{-2}}$ **57.** $\dfrac{x^{-4}}{x^3}$

58. $(-3x)^3$ **59.** $\sqrt[3]{x} \cdot \sqrt[3]{x^2}$ **60.** $(9x)^{-1/2}$

61. $\left(\dfrac{3x^2}{2y}\right)^3$ **62.** $\dfrac{x^2}{x^5y}$ **63.** $\dfrac{2x}{\sqrt{x}}$

64. $\dfrac{1}{yx^{-5}}$ **65.** $(16x^8)^{-3/4}$ **66.** $(-8y^9)^{2/3}$

67. $\sqrt{x}\left(\dfrac{1}{4x}\right)^{5/2}$ **68.** $\dfrac{(25xy)^{3/2}}{x^2y}$

69. $\dfrac{(-27x^5)^{2/3}}{\sqrt[3]{x}}$ **70.** $(-32y^{-5})^{3/5}$

The expressions in Exercises 71–74 may be factored as shown. Find the missing factors.

71. $\sqrt{x} - \dfrac{1}{\sqrt{x}} = \dfrac{1}{\sqrt{x}}(\quad)$

72. $2x^{2/3} - x^{-1/3} = x^{-1/3}(\quad)$

73. $x^{-1/4} + 6x^{1/4} = x^{-1/4}(\quad)$

74. $\sqrt{\dfrac{x}{y}} - \sqrt{\dfrac{y}{x}} = \sqrt{xy}\,(\quad)$

75. Explain why $\sqrt{a} \cdot \sqrt{b} = \sqrt{ab}$.

76. Explain why $\sqrt{a}/\sqrt{b} = \sqrt{a/b}$.

In Exercises 77–84, evaluate $f(4)$.

77. $f(x) = x^2$ **78.** $f(x) = x^3$

79. $f(x) = x^{-1}$ **80.** $f(x) = x^{1/2}$

81. $f(x) = x^{3/2}$ **82.** $f(x) = x^{-1/2}$

83. $f(x) = x^{-5/2}$ **84.** $f(x) = x^0$

Calculate the compound amount from the given data in Exercises 85–92.

85. principal = \$500, compounded annually, 6 years, annual rate = 6%

86. principal = \$700, compounded annually, 8 years, annual rate = 8%

87. principal = \$50,000, compounded quarterly, 10 years, annual rate = 9.5%

88. principal = \$20,000, compounded quarterly, 3 years, annual rate = 12%

89. principal = \$100, compounded monthly, 10 years, annual rate = 5%

90. principal = \$500, compounded monthly, 1 year, annual rate = 4.5%

91. principal = $1500, compounded daily, 1 year, annual rate = 6%

92. principal = $1500, compounded daily, 3 years, annual rate = 6%

93. Annual Compound Assume that a couple invests $1000 upon the birth of their daughter. Assume that the investment earns 6.8% compounded annually. What will the investment be worth on the daughter's 18th birthday?

94. Annual Compound with Deposits Assume that a couple invests $4000 each year for 4 years in an investment that earns 8% compounded annually. What will the value of the investment be 8 years after the first amount is invested?

95. Quarterly Compound Assume that a $500 investment earns interest compounded quarterly. Express the value of the investment after 1 year as a polynomial in the annual rate of interest r.

96. Semiannual Compound Assume that a $1000 investment earns interest compounded semiannually. Express the value of the investment after 2 years as a polynomial in the annual rate of interest r.

97. Velocity When a car's brakes are slammed on at a speed of x miles per hour, the stopping distance is $\frac{1}{20}x^2$ feet. Show that when the speed is doubled the stopping distance increases fourfold.

Technology Exercises

In Exercises 98–101, convert the numbers from graphing calculator form to standard form (that is, without E).

98. 5E–5

99. 8.103E–4

100. 1.35E13

101. 8.23E–6

Solutions to Practice Problems 0.5

1. (a) $-5^2 = -25$. [Note that -5^2 is the same as $-(5^2)$. This number is different from $(-5)^2$, which equals 25. Whenever there are no parentheses, apply the exponent first and then apply the other operations.]

(b) Since $.75 = \frac{3}{4}$, $16^{.75} = 16^{3/4} = (\sqrt[4]{16})^3 = 2^3 = 8$.

2. (a) Apply Law 5 with $a = 4$ and $b = x^3$. Then use Law 4.

$$(4x^3)^2 = 4^2 \cdot (x^3)^2 = 16 \cdot x^6$$

[A common error is to forget to square the 4. If that had been our intent, we would have asked for $4\left(x^3\right)^2$.]

(b) $\dfrac{\sqrt[3]{x}}{x^3} = \dfrac{x^{1/3}}{x^3} = x^{(1/3)-3} = x^{-8/3}$

[The answer can also be given as $1/x^{8/3}$.] When simplifying expressions involving radicals, it is usually a good idea to first convert the radicals to exponents.

(c) $\dfrac{2(x+5)^6}{x^2 + 10x + 25} = \dfrac{2 \cdot (x+5)^6}{(x+5)^2} = 2(x+5)^{6-2}$

$$= 2(x+5)^4$$

[Here the third law of exponents was applied to $(x + 5)$. The laws of exponents apply to any algebraic expression.]

0.6 Functions and Graphs in Applications

The key step in solving many applied problems in this text is to construct appropriate functions or equations. Once this is done, the remaining mathematical steps are usually straightforward. This section focuses on representative applied problems and reviews skills needed to set up and analyze functions, equations, and their graphs.

Geometric Problems Many examples and exercises in the text involve dimensions, areas, or volumes of objects similar to those in Fig. 1. When a problem involves a plane figure, such as a rectangle or circle, we must distinguish between the *perimeter* and the *area* of the figure. The perimeter of a figure, or "distance around" the figure, is a *length* or *sum of lengths*. Typical units, if specified, are inches, feet, centimeters, meters, and so on. Area involves the *product of two lengths*, and the units are *square* inches, *square* feet, *square* centimeters, and so on.

EXAMPLE 1

Cost   Suppose that the longer side of the rectangle in Fig. 1 has twice the length of the shorter side, and let x denote the length of the shorter side.

(a) Express the perimeter of the rectangle as a function of x.

(b) Express the area of the rectangle as a function of x.

(c) If the rectangle represents a kitchen countertop to be constructed of a durable material costing \$25 per square foot, write a function $C(x)$ that expresses the cost of the material as a function of x, where lengths are in feet.

Figure 1.   Geometric figures.   Rectangle   Rectangular box   Cylinder

Solution   (a) The rectangle is shown in Fig. 2. The length of the longer side is $2x$. If the perimeter is denoted by P, then P is the sum of the lengths of the four sides of the rectangle: $x + 2x + x + 2x$. That is, $P = 6x$.

(b) The area A of the rectangle is the product of the lengths of two adjacent sides. That is, $A = x \cdot 2x = 2x^2$.

(c) Here the area is measured in square feet. The basic principle for this part is

$$\begin{bmatrix} \text{cost of} \\ \text{materials} \end{bmatrix} = \begin{bmatrix} \text{cost per} \\ \text{square foot} \end{bmatrix} \cdot \begin{bmatrix} \text{number of} \\ \text{square feet} \end{bmatrix}$$

$$C(x) \quad = \quad 25 \quad \cdot \quad 2x^2$$

$$= 50x^2 \text{ (dollars)}. \qquad \blacksquare$$

Figure 2

When a problem involves a three-dimensional object, such as a box or cylinder, we must distinguish between the *surface area* of the object and the *volume* of the object. Surface area is an area, of course, so it is measured in *square* units. Typically, the surface area is a *sum of areas* (each area is a product of two lengths). The volume of an object is often a *product of three lengths* and is measured in *cubic* units.

EXAMPLE 2

Surface area   A rectangular box has a square copper base, wooden sides, and a wooden top. The copper costs \$21 per square foot and the wood costs \$2 per square foot.

(a) Write an expression giving the surface area (that is, the sum of the areas of the bottom, the top, and the four sides of the box) in terms of the dimensions of the box. Also, write an expression giving the volume of the box.

(b) Write an expression giving the total cost of the materials used to make the box in terms of the dimensions.

Solution   (a) The first step is to assign letters to the dimensions of the box. Denote the length of one (and therefore every) side of the square base by x, and denote the height of the box by h. (See Fig. 3.)

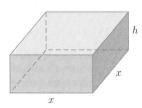

Figure 3. Closed box.

The top and bottom each have area x^2, and each of the four sides has area xh. Therefore, the surface area is $2x^2 + 4xh$. The volume of the box is the product of the length, width, and height. Because the base is square, the volume is x^2h.

(b) When the various surfaces of the box have different costs per square foot, the cost of each is computed separately:

$$[\text{cost of bottom}] = [\text{cost per sq. ft.}] \cdot [\text{area of bottom}] = 21x^2;$$
$$[\text{cost of top}] = [\text{cost per sq. ft.}] \cdot [\text{area of top}] = 2x^2;$$
$$[\text{cost of one side}] = [\text{cost per sq. ft.}] \cdot [\text{area of one side}] = 2xh.$$

The total cost is

$$C = [\text{cost of bottom}] + [\text{cost of top}] + 4 \cdot [\text{cost of one side}]$$
$$= 21x^2 + 2x^2 + 4 \cdot 2xh = 23x^2 + 8xh.$$ ■

Business Problems Many applications in this text involve cost, revenue, and profit functions.

EXAMPLE 3

Cost A toy manufacturer has fixed costs of $3000 (such as rent, insurance, and business loans) that must be paid no matter how many toys are produced. In addition, there are variable costs of $2 per toy. At a production level of x toys, the variable costs are $2 \cdot x$ (dollars) and the total cost is

$$C(x) = 3000 + 2x \quad \text{(dollars)}.$$

(a) Find the cost of producing 2000 toys.

(b) What additional cost is incurred if the production level is raised from 2000 toys to 2200 toys?

(c) To answer the question "How many toys may be produced at a cost of $5000?", should you compute $C(5000)$ or should you solve the equation $C(x) = 5000$?

Solution **(a)** $C(2000) = 3000 + 2(2000) = 7000$ (dollars).

(b) The total cost when $x = 2200$ is $C(2200) = 3000 + 2(2200) = 7400$ (dollars). So the *increase* in cost when production is raised from 2000 to 2200 toys is

$$C(2200) - C(2000) = 7400 - 7000 = 400 \quad \text{(dollars)}.$$

(c) This is an important type of question. The phrase "how many toys" implies that the quantity x is unknown. Therefore, the answer is found by solving $C(x) = 5000$ for x:

$$3000 + 2x = 5000$$
$$2x = 2000$$
$$x = 1000 \quad \text{(toys)}.$$

Another way to analyze this problem is to look at the types of units involved. The input x of the cost function is the *quantity* of toys, and the output of the cost function is the *cost*, measured in dollars. Since the question involves 5000 *dollars*, it is the *output* that is specified. The input x is unknown. ■

EXAMPLE 4

Cost, profit, and revenue The toys in Example 3 sell for $10 apiece. When x toys are sold, the revenue (amount of money received) $R(x)$ is $10x$ dollars. Given the same cost function, $C(x) = 3000 + 2x$, the profit (or loss) $P(x)$ generated by the x toys will be

$$P(x) = R(x) - C(x)$$
$$= 10x - (3000 + 2x) = 8x - 3000.$$

(a) To determine the revenue generated by 8000 toys, should you compute $R(8000)$ or should you solve the equation $R(x) = 8000$?

(b) If the revenue from the production and sale of some toys is $7000, what is the corresponding profit?

Solution **(a)** The revenue is unknown, but the input to the revenue function is known. So compute $R(8000)$ to find the revenue.

(b) The profit is unknown, so we want to compute the value of $P(x)$. Unfortunately, we don't know the value of x. However, the fact that the revenue is $7000 enables us to solve for x. Thus the solution has two steps:

(i)   Solve $R(x) = 7000$ to find x.

$$10x = 7000$$
$$x = 700 \quad \text{(toys)}$$

(ii) Compute $P(x)$ when $x = 700$.

$$P(x) = 8(700) - 3000$$
$$= 2600 \quad \text{(dollars)} \qquad \blacksquare$$

Functions and Graphs   When a function arises in an applied problem, the graph of the function provides useful information. Every statement or task involving a function corresponds to a feature or task involving its graph. This "graphical" point of view will broaden your understanding of functions and strengthen your ability to work with them.

Modern graphing calculators and calculus computer software provide excellent tools for thinking geometrically about functions. Most popular graphing calculators and programs provide a *cursor* or *cross hairs* that may be moved to any point on the screen, with the x- and y-coordinates of the cursor displayed somewhere on the screen. The next example shows how geometric calculations with the graph of a function correspond to the more familiar numerical computations. This example is worth reading even if a computer (or calculator) is unavailable.

EXAMPLE 5   To plan for future growth, a company analyzes production costs for one of its products and estimates that the cost (in dollars) of operating at a production level of x units per hour is given by the function

$$C(x) = 150 + 59x - 1.8x^2 + .02x^3.$$

Suppose that the graph of this function is available, either displayed on the screen of a graphing utility or perhaps printed on graph paper in a company report. (See Fig. 4.)

(a) The point $(16, 715.12)$ is on the graph. What does that say about the cost function $C(x)$?

(b) The equation $C(x) = 900$ may be solved graphically by finding a certain point on the graph and reading its x- and y-coordinates. Describe how to locate the point. How do the coordinates of the point provide the solution to the equation $C(x) = 900$?

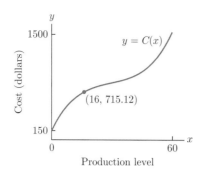

Figure 4. Graph of a cost
function.

(c) The task "Find $C(45)$" may be completed graphically by finding a point on
the graph. Describe how to locate the point. How do the coordinates of the
point provide the value of $C(45)$?

Solution (a) The fact that $(16, 715.12)$ is on the graph of $C(x)$ means that $C(16) = 715.12$.
That is, if the production level is 16 units per hour, the cost is $715.12.

(b) To solve $C(x) = 900$ graphically, locate 900 on the y-axis and move to the
right until you reach the point $(?, 900)$ on the graph of $C(x)$. (See Fig. 5.) The
x-coordinate of the point is the solution of $C(x) = 900$. Estimate x graphically.
(With a graphing utility, find the coordinates of the point of intersection of the
graph with the horizontal line $y = 900$; on graph paper, use a ruler to find x
on the x-axis. To two decimal places, $x = 39.04$.)

(c) To find $C(45)$ graphically, locate 45 on the x-axis and move up until you reach
the point $(45, ?)$ on the graph of $C(x)$. (See Fig. 6.) The y-coordinate of the
point is the value of $C(45)$. [In fact, $C(45) = 982.50$.]   ■

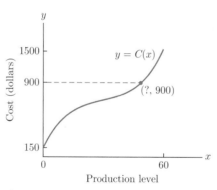

Figure 5

Figure 6

The two final examples illustrate how to extract information about a function
by examining its graph.

EXAMPLE 6

Figure 7 is the graph of the function $R(x)$, the revenue obtained from selling x
bicycles.

(a) What is the revenue from the sale of 1000 bicycles?

(b) How many bicycles must be sold to achieve a revenue of $102,000?

(c) What is the revenue from the sale of 1100 bicycles?

(d) What additional revenue is derived from the sale of 100 more bicycles if the
current sales level is 1000 bicycles?

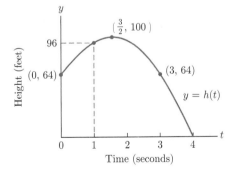

Figure 7. Graph of a revenue function.

Solution (a) Since $(1000, 150{,}000)$ is on the graph of $R(x)$, the revenue from the sale of 1000 bicycles is $150,000.

(b) The horizontal line at $y = 102{,}000$ intersects the graph at the point with x-coordinate 600. Therefore, the revenue from the sale of 600 bicycles is $102,000.

(c) The vertical line at $x = 1100$ intersects the graph at the point with y-coordinate 159,500. Therefore, $R(1100) = 159{,}500$ and the revenue is $159,500.

(d) When the value of x increases from 1000 to 1100, the revenue increases from 150,000 to 159,500. Therefore, the additional revenue is $9500. ∎

EXAMPLE 7

Height of a ball A ball is thrown straight up into the air from the top of a 64-foot tower. The function $h(t)$, the height of the ball (in feet) after t seconds, has the graph shown in Fig. 8. (*Note*: This graph is not a picture of the physical path of the ball; the ball is thrown vertically into the air.)

(a) What is the height of the ball after 1 second?

(b) After how many seconds does the ball reach its greatest height, and what is this height?

(c) After how many seconds does the ball hit the ground?

(d) When is the height 64 feet?

Figure 8. Graph of a height function.

Solution (a) Since the point $(1, 96)$ is on the graph of $h(t)$, $h(1) = 96$. Therefore, the height of the ball after 1 second is 96 feet.

(b) The highest point on the graph of the function has coordinates $\left(\frac{3}{2}, 100\right)$. Therefore, after $\frac{3}{2}$ seconds the ball achieves its greatest height, 100 feet.

(c) The ball hits the ground when the height is 0. This occurs after 4 seconds.

(d) The height of 64 feet occurs twice, at times $t = 0$ and $t = 3$ seconds. ∎

Table 1 summarizes most of the concepts in Examples 3 to 7. Although stated here for a profit function, the concepts will arise later for many other types of functions as well. Each statement about the profit is translated into a statement about $f(x)$ and a statement about the graph of $f(x)$. The graph in Fig. 9 illustrates each statement.

TABLE 1   Translating an Applied Problem

Assume that $f(x)$ is the profit in dollars at production level x.

Applied Problem	Function	Graph
When production is at 2 units, the profit is $7.	$f(2) = 7$.	The point $(2, 7)$ is on the graph.
Determine the number of units that generate a profit of $12.	Solve $f(x) = 12$ for x.	Find the x-coordinate(s) of the point(s) on the graph whose y-coordinate is 12.
Determine the profit when the production level is 4 units.	Evaluate $f(4)$.	Find the y-coordinate of the point on the graph whose x-coordinate is 4.
Find the production level that maximizes the profit.	Find x such that $f(x)$ is as large as possible.	Find the x-coordinate of the highest point, M, on the graph.
Determine the maximum profit.	Find the maximum value of $f(x)$.	Find the y-coordinate of the highest point on the graph.
Determine the change in profit when the production level is changed from 6 to 7 units.	Find $f(7) - f(6)$.	Determine the difference in the y-coordinates of the points with x-coordinates 7 and 6.
The profit decreases when the production level is changed from 6 to 7 units.	The function value decreases when x changes from 6 to 7.	The point on the graph with x-coordinate 6 is higher than the point with x-coordinate 7.

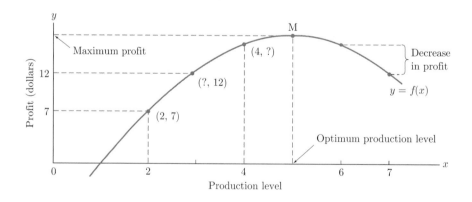

Figure 9. Graph of a profit function.

EXAMPLE 8

Approximating maximum or minimum values   For a certain commodity, it is found that when x thousands of units are sold, the revenue (in thousands of dollars) is given by

$$R(x) = -x^2 + 6.233x \quad \text{for } 0 \le x \le 6.233.$$

Approximate the number of units that must be sold to maximize revenue and find the approximate maximum revenue.

Solution We can approximate the maximum value of this function using at least two methods. We begin by entering the function into the calculator. Press $\boxed{Y=}$ and enter the formula for $R(x)$ for Y_1. Press $\boxed{\text{WINDOW}}$ and set $Xmin = 0$ and $Xmax = 6.233$.

The first method uses the $\boxed{\text{TRACE}}$ command. After entering the function into the calculator, press $\boxed{\text{TRACE}}$. We use the $\boxed{\triangleleft}$ and $\boxed{\triangleright}$ to trace the cursor along the graph. As we do so, the values of X and Y are updated; and by observing the value of X, we conclude that maximum revenue of approximately \$9713 is achieved when approximately 3117 units are sold. (See Fig. 10.)

The second method uses the built in **fMax** routine. We begin by pressing $\boxed{\text{2nd}}$ [CALC] and then $\boxed{4}$ to start the **fMax** routine. Enter 2 for the left bound, 4 for the right bound, and 3 for the guess. The result is shown in Fig. 11 and also indicates that maximum revenue of approximately \$9713 is achieved when approximately 3117 units are sold.

Similarly, we can use $\boxed{\text{TRACE}}$ or **fMin** to find the minimum value of a function. ∎

Figure 10

Figure 11

Practice Problems 0.6

Consider the cylinder shown in Fig. 12.

1. Assign letters to the dimensions of the cylinder.

2. The girth of the cylinder is the circumference of the colored circle in the figure. Express the girth in terms of the dimensions of the cylinder.

3. What is the area of the bottom (or top) of the cylinder?

4. What is the surface area of the side of the cylinder? [*Hint*: Imagine cutting the side of the cylinder and unrolling the cylinder to form a rectangle.]

Figure 12

EXERCISES 0.6

In Exercises 1–6, assign letters to the dimensions of the geometric object.

1.

Rectangle with
height $= 3 \cdot$ width

2.

Norman window: Rectangle
topped with a semicircle

3.

Rectangular box with
square base

4.

Rectangular box with
height $= \frac{1}{2} \cdot$ length

5.

Cylinder

6.

Cylinder with
height = diameter

Exercises 7–14 refer to the letters assigned to the figures in Exercises 1–6.

7. **Perimeter, Area** Consider the rectangle in Exercise 1. Write an expression for the perimeter. If the area is 25 square feet, write this fact as an equation.

8. **Perimeter, Area** Consider the rectangle in Exercise 1. Write an expression for the area. Write an equation expressing the fact that the perimeter is 30 centimeters.

9. **Area, Circumference** Consider a circle of radius r. Write an expression for the area. Write an equation expressing the fact that the circumference is 15 centimeters.

10. **Perimeter, Area** Consider the Norman window of Exercise 2. Write an expression for the perimeter. Write an equation expressing the fact that the area is 2.5 square meters.

11. **Volume, Surface Area** Consider the rectangular box in Exercise 3, and suppose that it has no top. Write an expression for the volume. Write an equation expressing the fact that the surface area is 65 square inches.

12. **Surface Area, Volume** Consider the closed rectangular box in Exercise 4. Write an expression for the surface area. Write an equation expressing the fact that the volume is 10 cubic feet.

13. **Volume, Surface Area, and Cost** Consider the cylinder of Exercise 5. Write an equation expressing the fact that the volume is 100 cubic inches. Suppose that the material to construct the left end costs $5 per square inch, the material to construct the right end costs $6 per square inch, and the material to construct the side costs $7 per square inch. Write an expression for the total cost of material for the cylinder.

14. **Surface Area, Volume** Consider the cylinder of Exercise 6. Write an equation expressing the fact that the surface area is 30π square inches. Write an expression for the volume.

15. **Fencing a Rectangular Corral** Consider a rectangular corral with a partition down the middle, as shown in Fig. 13. Assign letters to the outside dimensions of the corral. Write an equation expressing the fact that 5000 feet of fencing is needed to construct the corral (including the partition). Write an expression for the total area of the corral.

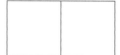

Figure 13

16. **Fencing a Rectangular Corral** Consider a rectangular corral with two partitions, as in Fig. 14. Assign letters to the outside dimensions of the corral. Write an equation expressing the fact that the corral has a total area of 2500 square feet. Write an expression for the amount of fencing needed to construct the corral (including both partitions).

Figure 14

17. **Cost of Fencing** Consider the corral of Exercise 16. If the fencing for the boundary of the corral costs $10 per foot and the fencing for the inner partitions costs $8 per foot, write an expression for the total cost of the fencing.

18. **Cost of an Open Box** Consider the rectangular box of Exercise 3. Assume that the box has no top, the material needed to construct the base costs $5 per square foot, and the material needed to construct the sides costs $4 per square foot. Write an equation expressing the fact that the total cost of materials is $150. (Use the dimensions assigned in Exercise 3.)

19. If the rectangle in Exercise 1 has a perimeter of 40 cm, find the area of the rectangle.

20. If the cylinder in Exercise 6 has a volume of 54π cubic inches, find the surface area of the cylinder.

21. **Cost** A specialty shop prints custom slogans and designs on T-shirts. The shop's total cost at a daily sales level of x T-shirts is $C(x) = 73 + 4x$ dollars.

   (a) At what sales level will the cost be $225?

   (b) If the sales level is at 40 T-shirts, how much will the cost rise if the sales level changes to 50 T-shirts?

22. **Cost, Revenue, and Profit** A college student earns income by typing term papers on a computer, which she leases (along with a printer). The student charges $4 per page for her work, and she estimates that her monthly cost when typing x pages is $C(x) = .10x + 75$ dollars.

   (a) What is the student's profit if she types 100 pages in 1 month?

   (b) Determine the change in profit when the typing business rises from 100 to 101 pages per month.

23. **Profit** A frozen yogurt stand makes a profit of $P(x) = .40x - 80$ dollars when selling x scoops of yogurt per day.

   (a) Find the break-even sales level, that is, the level at which $P(x) = 0$.

   (b) What sales level generates a daily profit of $30?

   (c) How many more scoops of yogurt will have to be sold to raise the daily profit from $30 to $40?

24. **Profit** A cellular telephone company estimates that, if it has x thousand subscribers, its monthly profit is $P(x)$ thousand dollars, where $P(x) = 12x - 200$.

   (a) How many subscribers are needed for a monthly profit of 160 thousand dollars?

   (b) How many new subscribers would be needed to raise the monthly profit from 160 to 166 thousand dollars?

25. **Cost, Revenue, and Profit** An average sale at a small florist shop is $21, so the shop's weekly revenue function is $R(x) = 21x$, where x is the number of sales in 1 week. The corresponding weekly cost is $C(x) = 9x + 800$ dollars.

(a) What is the florist shop's weekly profit function?

(b) How much profit is made when sales are at 120 per week?

(c) If the profit is $1000 for 1 week, what is the revenue for the week?

26. A catering company estimates that, if it has x customers in a typical week, its expenses will be approximately $C(x) = 550x + 6500$ dollars, and its revenue will be approximately $R(x) = 1200x$ dollars.

(a) How much profit will the company earn in 1 week when it has 12 customers?

(b) How much profit is the company making each week if the weekly costs are running at a level of $14,750?

Exercises 27–32 refer to the function $f(r)$, which gives the cost (in cents) of constructing a 100-cubic-inch cylinder of radius r inches. The graph of $f(r)$ is shown in Fig. 15.

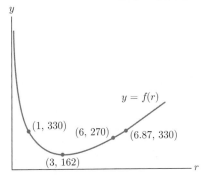

Figure 15. Cost of a cylinder.

27. What is the cost of constructing a cylinder of radius 6 inches?

28. For what value(s) of r is the cost 330 cents?

29. Interpret the fact that the point $(3, 162)$ is on the graph of the function.

30. Interpret the fact that the point $(3, 162)$ is the lowest point on the graph of the function. What does this say in terms of cost versus radius?

31. What is the additional cost of increasing the radius from 3 inches to 6 inches?

32. How much is saved by increasing the radius from 1 inch to 3 inches?

Exercises 33–36 refer to the cost and revenue functions in Fig. 16. The cost of producing x units of goods is $C(x)$ dollars and the revenue from selling x units of goods is $R(x)$ dollars.

33. What are the revenue and cost from the production and sale of 30 units of goods?

34. At what level of production is the revenue $1400?

35. At what level of production is the cost $1400?

36. What is the profit from the manufacture and sale of 30 units of goods?

Exercises 37–40 refer to the cost function in Fig. 17.

37. The point $(1000, 4000)$ is on the graph of the function. Restate this fact in terms of the function $C(x)$.

38. Translate the task "solve $C(x) = 3500$ for x" into a task involving the graph of the function.

39. Translate the task "find $C(400)$" into a task involving the graph.

40. If 500 units of goods are produced, what is the cost of producing 100 more units of goods?

Exercises 41–44 refer to the profit function in Fig. 18.

41. The point $(2500, 52,500)$ is the highest point on the graph of the function. What does this say in terms of profit versus quantity?

42. The point $(1500, 42,500)$ is on the graph of the function. Restate this fact in terms of the function $P(x)$.

43. Translate the task "solve $P(x) = 30,000$" into a task involving the graph of the function.

44. Translate the task "find $P(2000)$" into a task involving the graph.

A ball is thrown straight up into the air. The function $h(t)$ gives the height of the ball (in feet) after t seconds. In Exercises 45–50, translate the task into both a statement in-

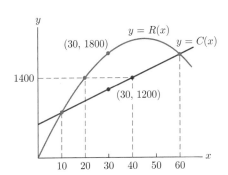

Figure 16. Cost and revenue functions.

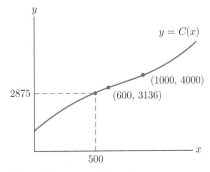

Figure 17. A cost function.

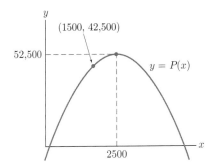

Figure 18. A profit function.

volving the function and a statement involving the graph of the function.

45. Find the height of the ball after 3 seconds.

46. Find the time at which the ball attains its greatest height.

47. Find the greatest height attained by the ball.

48. Determine when the ball will hit the ground.

49. Determine when the height of the ball is 100 feet.

50. Find the height of the ball when it is first released.

Technology Exercises

51. A ball thrown straight up into the air has height $-16x^2 + 80x$ feet after x seconds.

   (a) Graph the function in the window

$$[0, 6] \text{ by } [-30, 120].$$

   (b) What is the height of the ball after 3 seconds?

   (c) At what times will the height be 64 feet?

   (d) At what time will the ball hit the ground?

   (e) When will the ball reach its greatest height? What is that height?

52. The daily cost (in dollars) of producing x units of a certain product is given by the function $C(x) = 225 + 36.5x - .9x^2 + .01x^3$.

   (a) Graph $C(x)$ in the window $[0, 70]$ by $[-400, 2000]$.

   (b) What is the cost of producing 50 units of goods?

   (c) Consider the situation as in part (b). What is the additional cost of producing one more unit of goods?

   (d) At what production level will the daily cost be $510?

53. A store estimates that the total revenue (in dollars) from the sale of x bicycles per year is given by the function $R(x) = 250x - .2x^2$.

   (a) Graph $R(x)$ in the window

$$[200, 500] \text{ by } [42000, 75000].$$

   (b) What sales level produces a revenue of $63,000?

   (c) What revenue is received from the sale of 400 bicycles?

   (d) Consider the situation of part (c). If the sales level were to decrease by 50 bicycles, by how much would revenue fall?

   (e) The store believes that if it spends $5000 in advertising it can raise the total sales from 400 to 450 bicycles next year. Should it spend the $5000? Explain your conclusion.

Solutions to Practice Problems 0.6

1. Let r be the radius of the circular base and let h be the height of the cylinder.

2. Girth $= 2\pi r$ (the circumference of the circle).

3. Area of the bottom $= \pi r^2$.

4. The cylinder is a rolled-up rectangle of height h and base $2\pi r$ (the circumference of the circle). The area is $2\pi rh$. (See Fig. 19.)

$2\pi r$

Figure 19.   Unrolled side of a cylinder.

REVIEW OF FUNDAMENTAL CONCEPTS

1. Explain the relationships and differences among real numbers, rational numbers, and irrational numbers.

2. What are the four types of inequalities and what do they each mean?

3. What is the difference between an open interval and a closed interval from a to b?

4. What is a function?

5. What is meant by "the value of a function at x"?

6. What is meant by the domain and range of a function?

7. What is the graph of a function and how is it related to vertical lines?

8. What is a linear function? Constant function? Give examples.

9. What are the x- and y-intercepts of a function and how are they found?

10. What is a quadratic function? What shape does its graph have?

11. Define and give an example of each of the following types of functions.

(a) quadratic function   (b) polynomial function
(c) rational function   (d) power function

12. What is meant by the absolute value of a number?

13. What five operations on functions are discussed in this chapter? Give an example of each.

14. What is a zero of a function?

15. Give two methods for finding the zeros of a quadratic function.

16. State the six laws of exponents.

17. In the formula $A = P(1+i)^n$, what do A, P, i, and n represent?

18. Explain how to solve $f(x) = b$ geometrically from the graph of $y = f(x)$.

19. Explain how to find $f(a)$ geometrically from the graph of $y = f(x)$.

SUPPLEMENTARY EXERCISES

1. Let $f(x) = x^3 + \dfrac{1}{x}$. Evaluate $f(1)$, $f(3)$, $f(-1)$, $f(-\frac{1}{2})$, and $f(\sqrt{2})$.

2. Let $f(x) = 2x + 3x^2$. Evaluate $f(0)$, $f(-\frac{1}{4})$, and $f(1/\sqrt{2})$.

3. Let $f(x) = x^2 - 2$. Evaluate $f(a-2)$.

4. Let $f(x) = [1/(x+1)] - x^2$. Evaluate $f(a+1)$.

Determine the domains of the following functions.

5. $f(x) = \dfrac{1}{x(x+3)}$

6. $f(x) = \sqrt{x-1}$

7. $f(x) = \sqrt{x^2 + 1}$

8. $f(x) = \dfrac{1}{\sqrt{3x}}$

9. Is the point $(\frac{1}{2}, -\frac{3}{5})$ on the graph of the function $h(x) = (x^2 - 1)/(x^2 + 1)$?

10. Is the point $(1, -2)$ on the graph of the function $k(x) = x^2 + (2/x)$?

Factor the polynomials in Exercises 11–14.

11. $5x^3 + 15x^2 - 20x$

12. $3x^2 - 3x - 60$

13. $18 + 3x - x^2$

14. $x^5 - x^4 - 2x^3$

15. Find the zeros of the quadratic function $y = 5x^2 - 3x - 2$.

16. Find the zeros of the quadratic function $y = -2x^2 - x + 2$.

17. Find the points of intersection of the curves $y = 5x^2 - 3x - 2$ and $y = 2x - 1$.

18. Find the points of intersection of the curves $y = -x^2 + x + 1$ and $y = x - 5$.

Let $f(x) = x^2 - 2x$, $g(x) = 3x - 1$, and $h(x) = \sqrt{x}$. Find the following functions.

19. $f(x) + g(x)$

20. $f(x) - g(x)$

21. $f(x)h(x)$

22. $f(x)g(x)$

23. $f(x)/h(x)$

24. $g(x)h(x)$

Let $f(x) = x/(x^2 - 1)$, $g(x) = (1 - x)/(1 + x)$, and $h(x) = 2/(3x + 1)$. Express the following as rational functions.

25. $f(x) - g(x)$

26. $f(x) - g(x+1)$

27. $g(x) - h(x)$

28. $f(x) + h(x)$

29. $g(x) - h(x-3)$

30. $f(x) + g(x)$

Let $f(x) = x^2 - 2x + 4$, $g(x) = 1/x^2$, and $h(x) = 1/(\sqrt{x}-1)$. Determine the following functions.

31. $f(g(x))$

32. $g(f(x))$

33. $g(h(x))$

34. $h(g(x))$

35. $f(h(x))$

36. $h(f(x))$

37. Simplify $(81)^{3/4}$, $8^{5/3}$, and $(.25)^{-1}$.

38. Simplify $(100)^{3/2}$ and $(.001)^{1/3}$.

39. The population of a city is estimated to be $750 + 25t + .1t^2$ thousand people t years from the present. Ecologists estimate that the average level of carbon monoxide in the air above the city will be $1 + .4x$ ppm (parts per million) when the population is x thousand people. Express the carbon monoxide level as a function of the time t.

40. The revenue $R(x)$ (in thousands of dollars) that a company receives from the sale of x thousand units is given by $R(x) = 5x - x^2$. The sales level x is in turn a function $f(d)$ of the number d of dollars spent on advertising, where

$$f(d) = 6\left(1 - \frac{200}{d + 200}\right).$$

Express the revenue as a function of the amount spent on advertising.

In Exercises 41–44, use the laws of exponents to simplify the algebraic expressions.

41. $(\sqrt{x+1})^4$

42. $\dfrac{xy^3}{x^{-5}y^6}$

43. $\dfrac{x^{3/2}}{\sqrt{x}}$

44. $\sqrt[3]{x}\,(8x^{2/3})$

Figure 1. Yeast cells in a culture.

1.1 The Slope of a Straight Line

As we shall see later, the study of straight lines is crucial for the study of the steepness of curves. So this section is devoted to a discussion of the geometric and algebraic properties of straight lines. Since a vertical line is not the graph of a function (it fails the vertical line test), we focus our discussion on nonvertical lines.

> **Equations of Nonvertical Lines**   A nonvertical line L has an equation of the form
> $$y = mx + b. \tag{1}$$
> The number m is called the *slope* of L and the point $(0, b)$ is called the *y-intercept*. The equation (1) is called the *slope–intercept equation* of L.

If we set $x = 0$, we see that $y = b$, so $(0, b)$ is on the line L. Thus the y-intercept tells us where the line L crosses the y-axis. The slope measures the steepness of the line. In Fig. 1 we give three examples of lines with slope $m = 2$. In Fig. 2 we give three examples of lines with slope $m = -2$.

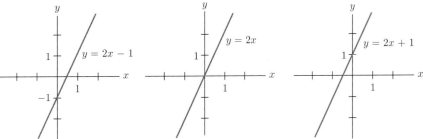

Figure 1. Three lines of slope 2.

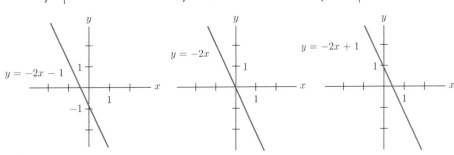

Figure 2. Three lines of slope −2.

To conceptualize the meaning of slope, think of walking along a line from left to right. On lines of positive slope we will be walking uphill; the greater the slope, the steeper the ascent. On lines of negative slope we will be walking downhill; the more negative the slope, the steeper the descent. Walking on lines of zero slope corresponds to walking on level ground. In Fig. 3 we have graphed lines with $m = 3, 1, \frac{1}{3}, 0, -\frac{1}{3}, -1, -3$, all having $b = 0$. The reader can readily verify our conceptualization of slope for these lines.

The slope and y-intercept of a straight line often have physical interpretations, as the following three examples illustrate.

Figure 3

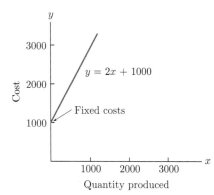

Figure 4. A cost function.

EXAMPLE 1

A manufacturer finds that the total cost of producing x units of a commodity is $2x + 1000$ dollars. What is the economic significance of the y-intercept and the slope of the line $y = 2x + 1000$? (See Fig. 4.)

Solution The y-intercept is $(0, 1000)$. In other words, when $x = 0$ (no units produced), the cost is still $y = 1000$ dollars. The number 1000 represents the *fixed costs* of the manufacturer—those overhead costs, such as rent and insurance, that must be paid no matter how many items are produced.

The slope of the line is 2. This number represents the cost of producing each additional unit. To see this, we can calculate some typical costs.

Quantity Produced	Total Cost
$x = 1500$	$y = 2(1500) + 1000 = 4000$
$x = 1501$	$y = 2(1501) + 1000 = 4002$
$x = 1502$	$y = 2(1502) + 1000 = 4004$

Each time x is increased by 1, the value of y increases by 2. The number 2 is called the *marginal cost*. It is equal to the additional cost that is incurred when the production level is increased by 1 unit, from x to $x + 1$. ■

EXAMPLE 2

An apartment complex has a storage tank to hold its heating oil. The tank was filled on January 1, but no more deliveries of oil will be made until some time in March. Let t denote the number of days after January 1, and let y denote the number of gallons of oil in the tank. Current records of the apartment complex show that y and t are related approximately by the equation

$$y = 30{,}000 - 400t. \tag{2}$$

What interpretation can be given to the y-intercept and slope of this line? (See Fig. 5.)

Solution   The y-intercept is $(0, 30,000)$. This value of y corresponds to $t = 0$, so there were 30,000 gallons of oil in the tank on January 1. Let us examine how fast the oil is removed from the tank.

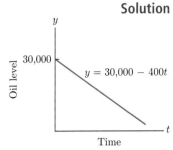

30,000

$y = 30,000 - 400t$

Oil level

Time

t

y

Figure 5.   Amount of heating oil in a tank.

Days After January 1	Gallons of Oil in the Tank
$t = 0$	$y = 30,000 - 400(0) = 30,000$
$t = 1$	$y = 30,000 - 400(1) = 29,600$
$t = 2$	$y = 30,000 - 400(2) = 29,200$
$t = 3$	$y = 30,000 - 400(3) = 28,800$
$\vdots$	$\vdots$

The oil level in the tank drops by 400 gallons each day; that is, the oil is being used at the rate of 400 gallons per day. The slope of the line (2) is -400. Thus the slope gives the rate at which the level of oil in the tank is changing. The negative sign on the -400 indicates that the oil level is decreasing rather than increasing. ■

EXAMPLE 3   For tax purposes, businesses are allowed to regard equipment as decreasing in value (or depreciating) each year. The amount of depreciation may be taken as an income tax deduction.

If the value y of a piece of equipment x years after its purchase is given by

$$y = 500,000 - 50,000x,$$

interpret the y-intercept and the slope of the graph.

Solution   The y-intercept is $(0, 500,000)$ and corresponds to the value of y when $x = 0$. That is, the y-intercept gives the original value, \$500,000, of the equipment. The slope indicates the rate at which the equipment is changing in value. Thus the value of the equipment is decreasing at the rate of 50,000 dollars per year. ■

Properties of the Slope of a Line   Let us now examine several useful properties of the slope of a straight line. At the end of the section we shall explain why these properties are valid.

Slope Property 1   If we start at a point on a line of slope m and move 1 unit to the right, then we must move m units in the y-direction in order to return to the line.

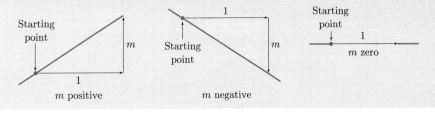

Starting point

m

1

m positive

Starting point

1

m

m negative

Starting point

1

m zero

Slope Property 2   We can compute the slope of a line by knowing two points on the line. If (x_1, y_1) and (x_2, y_2) are on the line, the slope of the line is

$$m = \frac{y_2 - y_1}{x_2 - x_1}.$$

Slope Property 3   The equation of a line can be obtained if we know the slope and one point on the line. If the slope is m and if (x_1, y_1) is on the line, the equation of the line is

$$y - y_1 = m(x - x_1). \tag{3}$$

This equation is called the *point–slope form* of the equation of the line.

Slope Property 4   Distinct lines of the same slope are parallel. Conversely, if two non-vertical lines are parallel, they have the same slope.

Slope Property 5   When two lines are perpendicular, excluding the case of the vertical and horizontal line, the product of their slopes is -1.

Calculations Involving Slope of a Line

EXAMPLE 4   Find the slope and the y-intercept of the line whose equation is $2x + 3y = 6$.

Solution   We put the equation into slope–intercept form by solving for y in terms of x.

$$3y = -2x + 6$$

$$y = -\frac{2}{3}x + 2$$

The slope is $-\frac{2}{3}$ and the y-intercept is $(0, 2)$.   ◼

EXAMPLE 5   Sketch the graph of the line

(a) passing through $(2, -1)$ with slope 3,

(b) passing through $(2, 3)$ with slope $-\frac{1}{2}$.

Solution   We use Slope Property 1. (See Fig. 6.) In each case, we begin at the given point, move 1 unit to the right, and then move m units in the y-direction (upward for positive m, downward for negative m). The new point reached will also be on the line. Draw the straight line through these two points.   ◼

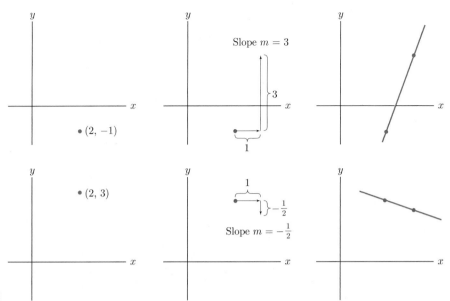

Figure 6

EXAMPLE 6   Find the slope of the line passing through the points $(6, -2)$ and $(9, 4)$.

Solution We apply Slope Property 2 with $(x_1, y_1) = (6, -2)$ and $(x_2, y_2) = (9, 4)$. Then

$$\frac{y_2 - y_1}{x_2 - x_1} = \frac{4 - (-2)}{9 - 6} = \frac{6}{3} = 2.$$

Thus the slope is 2. [We would have reached the same answer if we had let $(x_1, y_1) = (9, 4)$ and $(x_2, y_2) = (6, -2)$.] The slope is just the difference of the y-coordinates divided by the difference of the x-coordinates, with each difference formed in the same order. ∎

EXAMPLE 7   Find an equation of the line passing through $(-1, 2)$ with slope 3.

Solution We let $(x_1, y_1) = (-1, 2)$ and $m = 3$, and we use Slope Property 3. The equation of the line is

$$y - 2 = 3[x - (-1)]$$

or

$$y - 2 = 3(x + 1).$$

If desired, this equation can be put into the form $y = mx + b$:

$$y - 2 = 3(x + 1) = 3x + 3$$
$$y = 3x + 5.$$

∎

EXAMPLE 8   Find an equation of the line passing through the points $(1, -2)$ and $(2, -3)$.

Solution By Slope Property 2, the slope of the line is

$$\frac{-3 - (-2)}{2 - 1} = \frac{-3 + 2}{1} = -1.$$

Since $(1, -2)$ is on the line, we can use Property 3 to get the equation of the line:

$$y - (-2) = (-1)(x - 1)$$
$$y + 2 = -x + 1$$
$$y = -x - 1.$$

■

EXAMPLE 9   Find an equation of the line passing through $(5, 3)$ parallel to the line $2x + 5y = 7$.

Solution We first find the slope of the line $2x + 5y = 7$.

$$2x + 5y = 7$$
$$5y = 7 - 2x$$
$$y = -\frac{2}{5}x + \frac{7}{5}$$

The slope of this line is $-\frac{2}{5}$. By Slope Property 4, any line parallel to this line will also have slope $-\frac{2}{5}$. Using the given point $(5, 3)$ and Slope Property 3, we get the desired equation:

$$y - 3 = -\frac{2}{5}(x - 5).$$

This equation can also be written as

$$y = -\frac{2}{5}x + 5.$$

■

The Slope as a Rate of Change

Let us revisit the notion of rate of change that we discussed in Example 2 and see how it ties in with the notion of slope. Given a linear function $y = L(x) = mx + b$, as x varies from x_1 to x_2, y varies from $y_1 = L(x_1)$ to $y_2 = L(x_2)$. The *rate of change* of $y = L(x)$ over the interval from x_1 to x_2 is the change in y divided by the change in x:

$$\left[\begin{array}{c} \text{rate of change of } y = L(x) \\ \text{over the interval from } x_1 \text{ to } x_2 \end{array} \right] = \frac{[\text{Change in } y]}{[\text{Change in } x]}$$

$$= \frac{y_2 - y_1}{x_2 - x_1} = m,$$

by Slope Property 2. Thus the rate of change of a linear function over any interval is constant and equal to the slope m. As we did in Example 2, we will refer to the slope as *the* rate of change of the function, without mentioning an underlying interval. The fact that the rate of change of a linear function is constant characterizes linear functions (see Exercise 63) and has many important applications.

EXAMPLE 10   **Impact of mad cow disease on Canadian beef exports**   The discovery of one case of bovine spongiform encephalopathy, or mad cow disease, in May 2003 in Canada led to an immediate ban on all Canadian beef exports. At the beginning of September 2003, the ban was eased and exports of Canadian beef rose at a steady rate of 42.5 million dollars per month.* Express the value of the monthly exports of Canadian beef as a function of time for the period starting September 1, 2003. What was the value of the monthly exports at the end of December 2003, when presumably the exports regained their normal level?

Source: International Trade Division Statistics, Canada.

Solution Let y denote the value of monthly exports in millions of dollars and x the number of months that have elapsed since September 1, 2003. Since the rate of change of y is constant, we conclude that y is a linear function of x whose slope is equal to its rate of change, $m = 42.5$. On September 1 (when $x = 0$), the value of monthly exports was 0 dollars, since the ban had just ended. So the point $(0, 0)$ is on the graph of y. To write y in terms of x, we use the point–slope form of the equation of a line and get

$$y - 0 = 42.5(x - 0)$$
$$y = 42.5x.$$

The end of December 2003 corresponds to $x = 4$, at which time the exports had reached the value $y = (42.5) \cdot (4) = 170$ million dollars (Fig. 7). ■

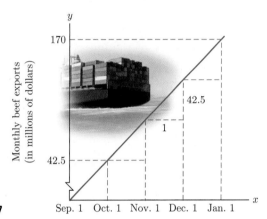

Figure 7

EXAMPLE 11

Temperature in a city As a cold front moved through a midwestern city, the temperature dropped at the rate of $3°\text{F}$ per hour between noon and 8 P.M. Express the temperature as a function of time and find the temperature at noon, given that at 1 P.M. the temperature was $47°\text{F}$.

Solution Let t denote time measured in hours since noon and $T(t)$ the temperature of the town at time t. The fact that the temperature was dropping at a (constant) rate of $3°\text{F}$ per hour tells us that the temperature is a linear function whose graph is falling. Thus its slope is $m = -3$, and so $T(t) = -3t + b$. To determine b, we use the fact that at 1 P.M. the temperature was $47°\text{F}$ or $T(1) = 47$:

$$(-3)(1) + b = 47$$
$$b = 50.$$

So $T(t) = -3t + 50$. The graph of the temperature is shown in Fig. 8. To determine the temperature at noon, we set $t = 0$ and get $T(0) = 50°\text{F}$. ■

Note the importance of the sign of the slope in Examples 10 and 11. An increasing linear function has a positive rate of change and thus a positive slope. A decreasing linear function has a negative rate of change and thus a negative slope.

Verification of the Properties of Slope It is convenient to verify the properties of slope in the order 2, 3, 1. Verifications of Properties 4 and 5 are outlined in Exercises 61 and 62.

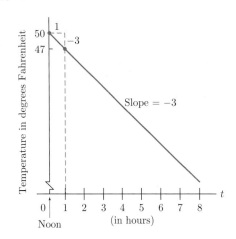

Figure 8. Temperature in a city.

Verification of Property 2 Suppose that the equation of the line is $y = mx + b$. Then, since (x_2, y_2) is on the line, we have $y_2 = mx_2 + b$. Similarly, since (x_1, y_1) is on the line, we have $y_1 = mx_1 + b$. Subtracting, we see that

$$y_2 - y_1 = mx_2 - mx_1$$
$$y_2 - y_1 = m(x_2 - x_1),$$

so

$$\frac{y_2 - y_1}{x_2 - x_1} = m.$$

This is the formula for the slope m stated in Property 2.

Verification of Property 3 The equation $y - y_1 = m(x - x_1)$ may be put into the form

$$y = mx + \underbrace{(y_1 - mx_1)}_{b}. \tag{4}$$

This is the equation of a line with slope m. Furthermore, the point (x_1, y_1) is on this line because equation (4) remains true when we plug in x_1 for x and y_1 for y. Thus the equation $y - y_1 = m(x - x_1)$ corresponds to the line of slope m passing through (x_1, y_1).

Verification of Property 1 Let $P = (x_1, y_1)$ be on a line l, and let y_2 be the y-coordinate of the point R on l obtained by moving P 1 unit to the right and then moving vertically to return to the line. (See Fig. 9.) By Property 2, the slope m of this line l satisfies

$$m = \frac{y_2 - y_1}{(x_1 + 1) - x_1} = \frac{y_2 - y_1}{1} = y_2 - y_1.$$

Thus the difference of the y-coordinates of R and Q is m. This is Property 1.

Figure 9

INCORPORATING TECHNOLOGY

At the end of each section in Chapter 0, we reviewed in detail many very useful techniques for graphing functions using a TI-83/84 type of calculator. All those techniques can be used to analyze the graphs of nonlinear functions. We encourage you to review those sections before attempting to solve the Technology Exercises for this section. ■

Practice Problems 1.1

Find the slopes of the following lines.

1. The line whose equation is $x = 3y - 7$

2. The line going through the points $(2, 5)$ and $(2, 8)$

EXERCISES 1.1*

Find the slopes and y-intercepts of the following lines.

1. $y = 3 - 7x$

2. $y = \dfrac{3x + 1}{5}$

3. $x = 2y - 3$

4. $y = 6$

5. $y = \dfrac{x}{7} - 5$

6. $4x + 9y = -1$

Find the equations of the following lines.

7. Slope is -1; $(7, 1)$ on line.

8. Slope is 2; $(1, -2)$ on line.

9. Slope is $\frac{1}{2}$; $(2, 1)$ on line.

10. Slope is $\frac{7}{3}$; $\left(\frac{1}{4}, -\frac{2}{5}\right)$ on line.

11. $\left(\frac{5}{7}, 5\right)$ and $\left(-\frac{5}{7}, -4\right)$ on line.

12. $\left(\frac{1}{2}, 1\right)$ and $(1, 4)$ on line.

13. $(0, 0)$ and $(1, 0)$ on line.

14. $\left(-\frac{1}{2}, -\frac{1}{7}\right)$ and $\left(\frac{2}{3}, 1\right)$ on line.

15. Horizontal through $(2, 9)$.

16. x-intercept is 1; y-intercept is -3.

17. x-intercept is $-\pi$; y-intercept is 1.

18. Slope is 2; x-intercept is -3.

19. Slope is -2; x-intercept is -2.

20. Horizontal through $(\sqrt{7}, 2)$.

21. Parallel to $y = x$; $(2, 0)$ on line.

22. Parallel to $x + 2y = 0$; $(1, 2)$ on line.

23. Parallel to $y = 3x + 7$; x-intercept is 2.

24. Parallel to $y - x = 13$; y-intercept is 0.

25. Perpendicular to $y + x = 0$; $(2, 0)$ on line.

26. Perpendicular to $y = -5x + 1$; $(1, 5)$ on line.

In Exercises 27–30, we specify a line by giving the slope and one point on the line. Start at the given point and use Slope Property 1 to sketch the graph of the line.

27. $m = 1$, $(1, 0)$ on line.   **28.** $m = \frac{1}{2}$, $(-1, 1)$ on line.

29. $m = -\frac{1}{3}$, $(1, -1)$ on line.

30. $m = 0$, $(0, 2)$ on line.

31. Each of lines (A), (B), (C), and (D) in Fig. 10 is the graph of one of the equations (a), (b), (c), and (d). Match each equation with its graph.

(a) $x + y = 1$

(b) $x - y = 1$

(c) $x + y = -1$

(d) $x - y = -1$

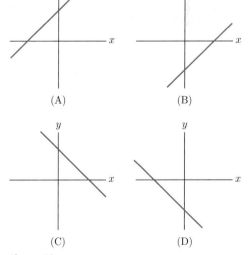

Figure 10

32. The line through the points $(-1, 2)$ and $(3, b)$ is parallel to $x + 2y = 0$. Find b.

---

*A complete solution for every third odd-numbered exercise is given in the *Study Guide* for this text.

In Exercises 33–36, refer to a line of slope m. If you begin at a point on the line and move h units in the x-direction, how many units must you move in the y-direction to return to the line?

33. $m = \frac{1}{3}$, $h = 3$ **34.** $m = 2$, $h = \frac{1}{2}$

35. $m = -3$, $h = .25$ **36.** $m = \frac{2}{3}$, $h = \frac{1}{2}$

In each of Exercises 37–38, we specify a line by giving the slope and one point on the line. We give the first coordinate of some points on the line. Without deriving the equation of the line, find the second coordinate of each point.

37. Slope is 2, $(1, 3)$ on line; $(2, \ )$; $(3, \ )$; $(0, \ )$.

38. Slope is -3, $(2, 2)$ on line; $(3, \ )$; $(4, \ )$; $(1, \ )$.

39. If $f(x)$ is a linear function, $f(1) = 0$, and $f(2) = 1$, what is $f(3)$?

40. Is the line through the points $(3, 4)$ and $(-1, 2)$ parallel to the line $2x + 3y = 0$? Justify your answer.

For each pair of lines in the following figures, determine the one with the greater slope.

41.

42.

Find the equation and sketch the graph of the following lines.

43. With slope -2 and y-intercept $(0, -1)$.

44. With slope $\frac{1}{3}$ and y-intercept $(0, 1)$.

In Exercises 45 and 46, two lines intersect the graph of a function $y = f(x)$ as shown in the figure. Find a and $f(a)$.

45.

46.

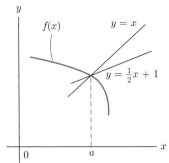

47. **(Marginal Cost)** Let $C(x) = 12x + 1100$ denote the total cost (in dollars) of manufacturing x units of a certain commodity per day.

    (a) What is the total cost if the production is set at 10 units per day?

    (b) What is the marginal cost?

    (c) Use (b) to determine the additional cost of raising the daily production level from 10 to 11 units.

48. Refer to Exercise 47. Use the formula for $C(x)$ to show directly that $C(x + 1) - C(x) = 12$. Interpret your result as it pertains to the marginal cost.

49. **Price of Gasoline** The price of 1 gallon of unleaded gasoline at the pump reached $4.89 on January 1, 2009, and continued to rise at the rate of 6 cents per month for the next 9 months. Express the price of 1 gallon of unleaded gasoline as a function of time for the period starting January 1, 2009. What was the price of 15 gallons of gasoline on April 1, 2009? On September 1, 2009?

50. **Marginal Cost** Let $C(x)$ denote the cost (in thousands of dollars) of manufacturing x thousand computer chips. Suppose that $C(x)$ is a linear function. If the marginal cost is 100, what is the additional cost of increasing the current production level by 4 thousand chips?

51. An online bookstore charges $5 plus 3% of the purchase price of books for shipping and handling. Find a function $C(x)$ that expresses the shipping and handling charge for a book order that costs x dollars.

52. **Quit Ratio** In industry, the relationship between wages and the quit ratio of employees is defined to be the percentage of employees that quit within 1 year of employment. The quit ratio of a large restaurant chain that paid its employees the minimum hourly wage ($6.55 per hour) was .2 or 20 employees per 100. When the company raised the hourly wage to $8, the quit ratio dropped to .18, or 18 employees per 100.

    (a) Assuming a linear relationship between the quite ratio $Q(x)$ and the hourly wage x, find an expression for $Q(x)$.

    (b) What should the hourly wage be for the quit ratio to drop to 10 employees per 100?

53. Price Affects Sales When the owner of a gas station sets the price of 1 gallon of unleaded gasoline at $4.10, he can sell approximately 1500 gallons per day. When he sets the price at $4.25 per gallon, he can sell approximately 1250 gallons per day. Let $G(x)$ denote the number of gallons of unleaded gasoline sold per day when the price is x dollars. Assume that $G(x)$ is a linear function of x. Approximately how many gallons will be sold per day if the price is set at $4.35 per gallon?

54. Refer to the previous exercise. Where should the owner set the price if he wants to sell 2200 gallons per day?

55. Marginal Cost Analysis A company manufactures and sells fishing rods. The company has a fixed cost of $1500 per day and a total cost of $2200 per day when the production is set at 100 rods per day. Assume that the total cost $C(x)$ is linearly related to the daily production level x.

    (a) Express the total cost as a function of the daily production level.

    (b) What is the marginal cost at production level $x = 100$?

    (c) What is the additional cost of raising the daily production level from 100 to 101 rods? Answer this question in two different ways: (1) by using the marginal cost and (2) by computing $C(101) - C(100)$.

56. A salesperson's weekly pay depends on the volume of sales. If she sells x units of goods, her pay is $y = 5x + 60$ dollars. Give an interpretation of the slope and the y-intercept of this straight line.

57. The demand equation for a monopolist is $y = -.02x + 7$, where x is the number of units produced and y is the price. That is, to sell x units of goods, the price must be $y = -.02x + 7$ dollars. Interpret the slope and y-intercept of this line.

58. Converting Fahrenheit to Celsius Temperatures of $32°$F and $212°$F correspond to temperatures of $0°$C and $100°$C. The linear equation $y = mx + b$ converts Fahrenheit temperatures to Celsius temperatures. Find m and b. What is the Celsius equivalent of $98.6°$F?

59. Intravenous Injection A drug is administered to a patient through an IV (intravenous) injection at the rate of 6 milliliters (mL) per minute. Assuming that the patient's body already contained 1.5 mL of this drug at the beginning of the infusion, find an expression for the amount of the drug in the body x minutes from the start of the infusion.

60. Refer to Exercise 59. If the patient's body eliminates the drug at the rate of 2 mL per hour, find an expression for the amount of the drug in the body x minutes from the start of the infusion.

61. Prove Slope Property 4 of straight lines. [*Hint*: If $y = mx + b$ and $y = m'x + b'$ are two lines, they have a point in common if and only if the equation $mx + b = m'x + b'$ has a solution x.]

62. Prove Slope Property 5 of straight lines. [*Hint*: Without loss of generality, assume that both lines pass through the origin. Use Slope Property 1 and the Pythagorean theorem. See Fig. 11.]

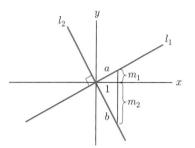

Figure 11

63. If, for some constant m,

$$\frac{f(x_2) - f(x_1)}{x_2 - x_1} = m$$

for all $x_1 \neq x_2$, show that $f(x) = mx + b$, where b is some constant. [*Hint*: Fix x_1 and take $x = x_2$; then solve for $f(x)$.]

64. (a) Draw the graph of any function $f(x)$ that passes through the point $(3, 2)$.

    (b) Choose a point to the right of $x = 3$ on the x-axis and label it $3 + h$.

    (c) Draw the straight line through the points $(3, f(3))$ and $(3 + h, f(3 + h))$.

    (d) What is the slope of this straight line (in terms of h)?

Technology Exercises

65. Let y denote the percentage of the world population that is urban x years after 1980. According to recently published data, y has been a linear function of x since 1980. The percent of the world population that is urban was 39.5 in 1980 and 45.2 in 1995.

    (a) Determine y as a function of x.

    (b) Graph this function in the window $[0, 40]$ by $[0, 100]$.

    (c) Interpret the slope in applied terms.

    (d) Determine graphically the percent of the world population that was urban in 1990.

    (e) Determine graphically the year in which 50% of the world population will be urban.

    (f) By what amount does the percent of the world population that is urban increase every 5 years?

66. Let y denote the average amount claimed for itemized deductions on a tax return reporting x dollars of income. According to Internal Revenue Service data, y is a linear function of x. Moreover, in a recent year income tax returns reporting $20,000 of income averaged $729 in itemized deductions, while returns reporting $50,000 averaged $1380.

(a) Determine y as a function of x.

(b) Graph this function in the window $[0, 75000]$ by $[0, 2000]$.

(c) Give an interpretation of the slope in applied terms.

(d) Determine graphically the average amount of itemized deductions on a return reporting $75,000.

(e) Determine graphically the income level at which the average itemized deductions are $1600.

(f) If the income level increases by $15,000, by how much do the average itemized deductions increase?

Solutions to Practice Problems 1.1

1. We solve for y in terms of x.

$$y = \frac{1}{3}x + \frac{7}{3}$$

The slope of the line is the coefficient of x, that is, $\frac{1}{3}$.

2. The line passing through these two points is a vertical line; therefore, its slope is undefined.

1.2 The Slope of a Curve at a Point

To extend the concept of slope from straight lines to more general curves, we must first discuss the notion of the tangent line to a curve at a point.

We have a clear idea of what is meant by the tangent line to a circle at a point P. It is the straight line that touches the circle at just the one point P. Let us focus on the region near P, designated by the dashed rectangle shown in Fig. 1. The enlarged portion of the circle looks almost straight, and the straight line that it resembles is the tangent line. Further enlargements would make the circle near P look even straighter and have an even closer resemblance to the tangent line. In this sense, the tangent line to the circle at the point P is the straight line through P that best approximates the circle near P. In particular, the tangent line at P reflects the steepness of the circle at P. Thus it seems reasonable to define the *slope* of the circle at P to be the slope of the tangent line at P.

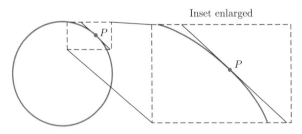

Figure 1. Enlarged portion of a circle.

Similar reasoning leads us to a suitable definition of slope for an arbitrary curve at a point P. Consider the three curves drawn in Fig. 2. We have drawn an enlarged version of the dashed box around each point P. Notice that the portion of each curve lying in the boxed region looks almost straight. If we further magnify the curve near P, it would appear even straighter. Indeed, if we apply higher and higher magnification, the portion of the curve near P would approach a certain straight line more and more exactly. (See Fig. 3.) This straight line is called the *tangent line to the curve at P*. This line best approximates the curve near P and leads us to the following definition.

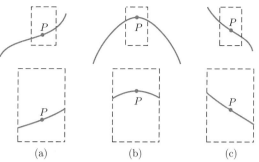

Figure 2. Enlarged portions of curves.

Figure 3. Tangent lines to curves.

> **DEFINITION Slope of a Curve** The *slope of a curve at a point* P is defined to be the slope of the tangent line to the curve at P.

Let us consider some examples and applications involving slopes of curves and tangent lines. Later, in the next section, we shall describe a process for constructing the tangent line and computing its slope.

EXAMPLE 1

Slope of a graph The graph of $f(x) = x^2$ and the tangent line at the point $P = (1, 1)$ are shown in Fig. 4. Find the slope of the graph at P.

Solution The slope of the graph at a point is by definition the slope of the tangent line at that point. Figure 4 shows that the tangent line at P rises 2 units for each unit change in x. Thus the slope of the tangent line at $P = (1, 1)$ is

$$\big[\text{slope of tangent line at } P\big] = \frac{[\text{change in } y]}{[\text{change in } x]} = \frac{2}{1} = 2$$

(see Fig. 4), and hence the slope of the graph at P is 2. ∎

Slope of a Graph as a Rate of Change

We learned in the previous section that the slope of a linear function measures its rate of increase or decrease. We have just defined the slope of a curve at a point P to be the slope of the tangent line to the curve at P. Now, the portion of the curve near P can be, at least within an approximation, replaced by the tangent line at P. So, thinking of the curve as being approximated by the tangent line near P, we obtain the following important interpretation of the slope of a curve.

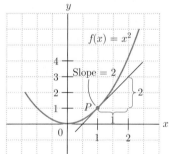

Figure 4. Tangent line and slope at the point P.

> **Slope of a Curve as a Rate of Change** The slope of a curve at a point P—that is, the slope of the tangent line at P—measures the rate of change (the rate of increase or decrease) of the curve as it passes through P.

EXAMPLE 2

In Example 2 of Section 1.1, the apartment complex used approximately 400 gallons of oil per day. Suppose that we keep a continuous record of the oil level in the storage tank. The graph for a typical 2-day period appears in Fig. 5. What is the physical significance of the slope of the graph at the point P?

Solution The curve near P is closely approximated by its tangent line. So think of the curve as being replaced by its tangent line near P. Then the slope at P is just the rate of decrease of the oil level at 7 A.M. on March 5. ∎

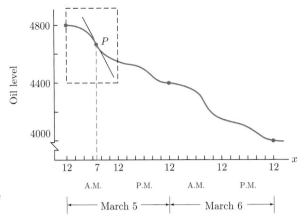

Figure 5. Oil level in a storage tank.

Notice that during the entire day of March 5, the graph in Fig. 5 seems to be the steepest at 7 A.M. That is, the oil level is falling the fastest at that time. This corresponds to the fact that most people awake around 7 A.M., turn up their thermostats, take showers, and so on. You can estimate the rate of consumption of oil at 7 A.M. by estimating the slope of the tangent line at P in Fig. 5.

EXAMPLE 3   **Slope of a graph**   In Fig. 6, we show an enlarged version of a portion of the graph designated by a dashed rectangle in Fig. 5. Estimate the slope of the graph at the point P and give an interpretation of your result.

Solution   The slope of the graph at P is by definition the slope of the tangent line at P. From Fig. 6 we see that the tangent line at P falls approximately 250 units for a 4-unit change in x. So the slope of the graph at P is

$$\left[\text{slope of tangent line at } P\right] = \frac{[\text{change in } y]}{[\text{change in } x]} = \frac{-250}{4} = -62.5.$$

Since the slope at P is the rate of increase or decrease of the oil level, we conclude that the oil level at 7 A.M. is decreasing at the rate of 62.5 gallons per hour.   ■

Figure 6. Oil level in a storage tank around 7 A.M.

Examples 2 and 3 provide typical illustrations of the manner in which slopes can be interpreted as rates of change. We shall return to this important idea in later sections.

Slope Formulas

We know that the slope of a straight line is constant and does not depend on where we are on the line. This is clearly not the case with other curves. In general, the slope of a curve at a point will depend on the point. We saw in Example 1 that the slope of the tangent line to the graph of $y = x^2$ at the point $(1, 1)$ is 2. In fact, we can show that the slope of the tangent line to the graph of $y = x^2$ at the point $(3, 9)$ is 6; and the slope at $\left(-\frac{5}{2}, \frac{25}{4}\right)$ is -5. The various tangent lines are shown in Fig. 7. As we will demonstrate in the next section, in calculus we can usually compute the slopes by using formulas. For the parabola $y = x^2$, we notice that the slope at each point is two times the x-coordinate of the point. This is a general fact for this graph, which is expressed by the following *slope formula*:

$$\left[\text{ slope of the graph of } y = x^2 \text{ at the point } (x, y) \right] = 2x$$

(see Fig. 8). This simple formula will be derived in the next section. For now we will use it to get further insight into the topics of slopes of curves and tangent lines.

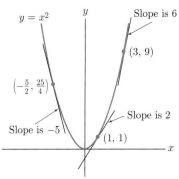

Figure 7. Graph of $y = x^2$.

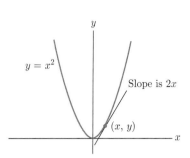

Figure 8. Slope of tangent line to $y = x^2$.

EXAMPLE 4

(a) What is the slope of the graph of $y = x^2$ at the point $\left(\frac{3}{4}, \frac{9}{16}\right)$?

(b) Write the equation of the tangent line to the graph of $y = x^2$ at the point $\left(\frac{3}{4}, \frac{9}{16}\right)$.

Solution

(a) The x-coordinate of $\left(\frac{3}{4}, \frac{9}{16}\right)$ is $\frac{3}{4}$, so the slope of $y = x^2$ at this point is $2\left(\frac{3}{4}\right) = \frac{3}{2}$.

(b) We shall write the equation of the tangent line in the point–slope form. The point is $\left(\frac{3}{4}, \frac{9}{16}\right)$ and the slope is $\frac{3}{2}$ by part (a). Hence the equation is

$$y - \frac{9}{16} = \frac{3}{2}\left(x - \frac{3}{4}\right). \qquad \blacksquare$$

INCORPORATING TECHNOLOGY

Approximating the Slope of a Graph at a Point The slope of a graph at a point can be approximated by zooming in on the graph. In Fig. 9 we have the graph the of the curve $y = 3^x$, with window setting ZDecimal, obtained with ⌐ZOOM⌐ ⌐4⌐. We now zoom in on the point $(0, 1)$ by hitting ⌐TRACE⌐ to find the point $(0, 1)$ and then ⌐ZOOM⌐ ⌐2⌐ ⌐ENTER⌐ to zoom in. If we zoom in close enough (Fig. 10), the graph starts to look like a straight line. We approximate the slope of this line by measuring the ratio of rise over run in the window in Fig. 10(b), and find

$$\frac{3^{.1} - 3^{-.1}}{.1 - (-.1)} = \frac{1.116 - 0.896}{.2} = \frac{.22}{.2} = 1.1.$$

Figure 9

Hence the slope of the curve at the point $(0, 1)$ is approximately 1.1. (We will learn how to calculate this slope in Chapter 4. To find a point on the graph, enter ⌐TRACE⌐ and then enter the x-value.) $\blacksquare$

(a)

(b)

Figure 10

Practice Problems 1.2

1. Refer to Fig. 11.

   (a) What is the slope of the curve at $(3, 4)$?

   (b) What is the equation of the tangent line at the point where $x = 3$?

2. What is the equation of the tangent line to the graph of $y = \frac{1}{2}x + 1$ at the point $(4, 3)$?

Figure 11

EXERCISES 1.2

Estimate the slope of each of the following curves at the designated point P.

1.

2.

3.

4.

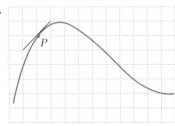

Exercises 5–10 refer to the points in Fig. 12. Assign one of the following descriptors to each point: large positive slope, small positive slope, zero slope, small negative slope, large negative slope.

5. A **6.** B **7.** C **8.** D **9.** E **10.** F

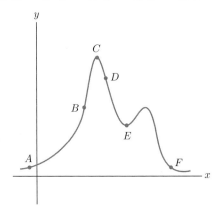

Figure 12

11. **Price of Crude Oil** Figure 13 shows the price of one barrel of crude oil on the New York stock exchange from June 1, 2007, to April 1, 2008. Determine the price increase from June 1 to December 1 in 2007. Also determine whether the price was rising, falling, or holding steady on these days.

12. Refer to Fig. 13. Do you agree with the statement that the price of one barrel of crude oil rose on July 1, 2007, and then fell on August 1, 2007, at approximately the same rate? Justify your answer.

13. Refer to Fig. 14, which shows an enlarged version of one portion of the curve in Fig. 13. Estimate the price of one barrel of crude oil on April 3, 2008, and the rate at which the price was rising on that day. (Your answer for the rate should be in dollars per day.)

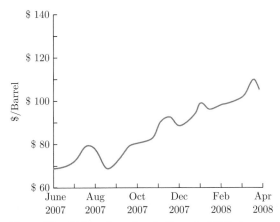

Figure 13. Price of crude oil, June 1, 2007, to April 1, 2008.

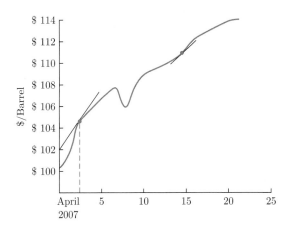

Figure 14. Price of crude oil.

14. Refer to Fig. 14. Estimate the price of one barrel of crude oil on April 14, 2008, and the rate at which it was rising on that day.

15. U.S. Federal Debt The federal debt is the total amount of money owed by the U.S. federal government. Figure 15 shows the graph of the federal debt in trillions of dollars for the years 1940 to 2004 (with projections for the year 2006).* Trace the graph onto another sheet of paper and draw the tangent line at the point corresponding to the year 1990. Use the tangent line to approximate the annual rate of increase of the federal debt in 1990.

16. Repeat Exercise 15 using the year 2002 instead of 1990.

17. Debt per Capita The debt per capita measures the average amount of money owed by one person in the United States. Figure 16 shows the graph of the debt per capita in thousands of dollars for the years 1950 to

2004. Use the graph to answer the following questions.

(a) Approximate the debt per capita in 1950, 1990, 2000, and 2004.

(b) Trace the graph onto another sheet of paper and draw the tangent line at the point corresponding to the year 2000. Use the tangent line to approximate the annual rate of increase of the debt per capita in 2000.

18. Refer to Fig. 16 to decide whether the following statements about the debt per capita are correct or not. Justify your answers.

(a) The debt per capita rose at a faster rate in 1980 than in 2000.

(b) The debt per capita was almost constant from 1950 to the mid-1970s and then rose at an almost constant rate from the mid-1970s to the mid-1980s.

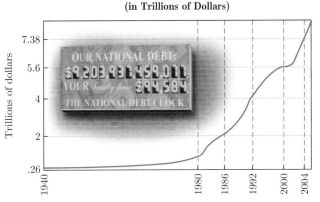

Figure 15. U.S. federal debt.

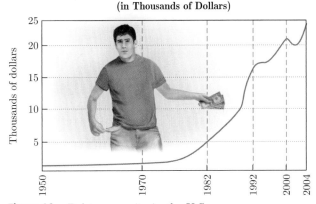

Figure 16. Debt per capita in the U.S.

---

*Exercises 15 and 17 are based on data from the U.S. Treasury Department.

In Exercises 19–21, find the slope of the tangent line to the graph of $y = x^2$ at the point indicated and then write the corresponding equation of the tangent line.

19. $(-.5, .16)$ **20.** $(-2, 4)$ **21.** $\left(\frac{1}{3}, \frac{1}{9}\right)$

22. Find the slope of the tangent line to the graph of $y = x^2$ at the point where $x = -\frac{1}{4}$.

23. Write the equation of the tangent line to the graph of $y = x^2$ at the point where $x = 2.5$.

24. The line $y = 5x$ intersects the parabola $y = x^2$ at two points, A and B.

   (a) Find the coordinates of the points A and B.

   (b) Find the equations of the tangent lines to the parabola at the points A and B.

25. Find the point on the graph of $y = x^2$ where the curve has slope $\frac{7}{2}$.

26. Find the point on the graph of $y = x^2$ where the curve has slope -6.

27. Find the point on the graph of $y = x^2$ where the tangent line is parallel to the line $2x + 3y = 4$.

28. Find the point on the graph of $y = x^2$ where the tangent line is parallel to the line $3x - 2y = 2$.

In the next section we shall see that the tangent line to the graph of $y = x^3$ at the point (x, y) has slope $3x^2$. See Fig. 17. Using this result, find the slope of the curve at the points in Exercises 29–31.

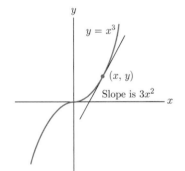

Figure 17. Slope of tangent line to $y = x^3$.

29. $(2, 8)$ **30.** $\left(\frac{3}{2}, \frac{27}{8}\right)$ **31.** $\left(-\frac{1}{2}, -\frac{1}{8}\right)$

32. Write the equation of the line tangent to the graph of $y = x^3$ at the point where $x = -1$.

In Exercises 33–34, you are shown the tangent line to the graph of $f(x) = x^2$ at the point $(a, f(a))$. Find a, $f(a)$, and the slope of the parabola at $(a, f(a))$.

33.

34.

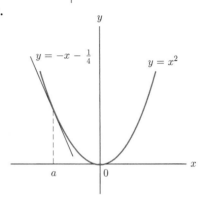

35. Find the point(s) on the graph in Fig. 17 where the slope is equal to $\frac{3}{2}$.

36. Find the points on the graph in Fig. 17 where the tangent line is parallel to $y = 2x$.

37. Let l be the line through the points P and Q in Fig. 18.

   (a) If $P = (2, 4)$ and $Q = (5, 13)$, find the slope of the line l and the length of the line segment d.

   (b) As the point Q moves toward P, does the slope of the line l increase or decrease?

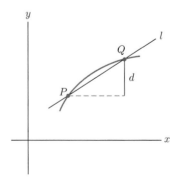

Figure 18. The slope of a secant line.

38. In Fig. 19, h represents a positive number, and $3 + h$ is the number h units to the right of 3. Draw line segments on the graph having the following lengths.

   (a) $f(3)$ **(b)** $f(3 + h)$

   (c) $f(3 + h) - f(3)$ **(d)** h

(e) Draw a line of slope $\dfrac{f(3+h) - f(3)}{h}$.

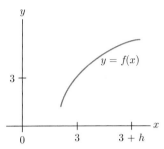

Figure 19. Geometric representation of values.

In Exercises 39–42, you are given a function and a point on the graph of the function. Zoom in on the graph at the given point until it starts to look like a straight line. Estimate the slope of the graph at the point indicated; then use this information to estimate the value of the derivative.

39. $f(x) = 2x^2 - 3x + 2$, $(0, 2)$, $f'(0)$

40. $f(x) = \dfrac{x-1}{x+1}$, $(1, 0)$, $f'(1)$

41. $f(x) = \sqrt{x+3}$, $(1, 2)$, $f'(1)$

42. $f(x) = \sqrt[3]{x+6}$, $(2, 2)$, $f'(2)$

Solutions to Practice Problems 1.2

1. (a) The slope of the curve at the point $(3, 4)$ is, by definition, the slope of the tangent line at $(3, 4)$. Note that the point $(4, 6)$ is also on the line. Therefore, the slope is

$$\frac{6-4}{4-3} = \frac{2}{1} = 2.$$

(b) Use the point–slope formula. The equation of the line passing through the point $(3, 4)$ and having slope 2 is

$$y - 4 = 2(x - 3)$$

or

$$y = 2x - 2.$$

2. The tangent line at $(4, 3)$ is, by definition, the line that best approximates the curve at $(4, 3)$. Since the "curve" in this case is itself a line, the curve and its tangent line at $(4, 3)$ (and at every other point) must be the same. Therefore, the equation is $y = \frac{1}{2}x + 1$.

1.3 The Derivative

Suppose that a curve is the graph of a function $f(x)$ and that we have a tangent line at every point on the graph. We saw in the previous section that it is possible to find a formula that gives the slope of the curve $y = f(x)$ at any point. This slope formula is called the *derivative* of $f(x)$ and is written $f'(x)$. For each value of x, $f'(x)$ gives the slope of the curve $y = f(x)$ at the point with first coordinate x.* (See Fig. 1.)

In the previous section we considered some interesting examples of slope formulas. Here we add a few more examples and state them in the terminology of derivatives. At the end of the section, we use the fact that the derivative is a slope formula for the tangent line to describe a geometric construction of the tangent line and hence of the derivative. This construction will lead us naturally into the topic of limits, which will be treated in detail in the following section. The process of computing $f'(x)$ for a given function $f(x)$ is called **differentiation**.

Examples of Derivatives: The Power Rule

The case of a linear function $f(x) = mx + b$ is particularly simple. The graph of $y = mx + b$ is a straight line L of slope m. Recall that, by definition, the tangent

---

*As we shall see, some curves do not have tangent lines at every point. At values of x corresponding to such points, the derivative $f'(x)$ is not defined. For the sake of the current discussion, which is designed to develop an intuitive feeling for the derivative, let us assume that the graph of $f(x)$ has a tangent line for each x in the domain of f.

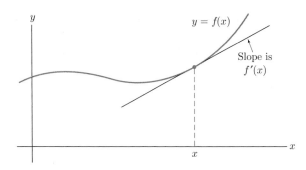

Figure 2. Definition of $f'(x)$.

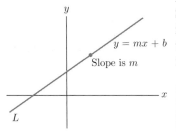

Figure 1. Derivative of a linear function.

Figure 3. Derivative of a constant function.

line to a curve at a point is the line that best approximates the curve at that point. Since the curve L is itself a line, the tangent line to L at any point is the line L itself. So the slope of the graph is m at every point. (See Fig. 2.) In other words, the value of the derivative $f'(x)$ is always equal to m. We summarize this fact as follows:

If $f(x) = mx + b$, then we have

$$f'(x) = m. \tag{1}$$

Set $m = 0$ in $f(x) = mx + b$. Then the function becomes $f(x) = b$, which has the value b for each value of x. The graph is a horizontal line of slope 0, so $f'(x) = 0$ for all x. (See Fig. 3.) Thus we have the following fact:

The derivative of a constant function $f(x) = b$ is zero; that is,

$$f'(x) = 0. \tag{2}$$

Next, consider the function $f(x) = x^2$. As we stated in Section 1.2 (and will prove at the end of this section), the slope of the graph of $y = x^2$ at the point (x, y) is equal to $2x$. That is, the value of the derivative $f'(x)$ is $2x$:

If $f(x) = x^2$, then its derivative is the function $2x$. That is,

$$f'(x) = 2x. \tag{3}$$

In Exercises 29–32 of Section 1.2, we made use of the fact that the slope of the graph of $y = x^3$ at the point (x, y) is $3x^2$. This can be restated in terms of derivatives as follows:

If $f(x) = x^3$, then the derivative is $3x^2$. That is,

$$f'(x) = 3x^2. \tag{4}$$

One reason that calculus is so useful is because it provides general techniques that can be easily used to determine derivatives. One such general rule, which contains formulas (3) and (4) as special cases, is the power rule.

Power Rule   Let r be any number and let $f(x) = x^r$. Then $f'(x) = rx^{r-1}$.

Indeed, if $r = 2$, then $f(x) = x^2$ and $f'(x) = 2x^{2-1} = 2x$, which is formula (3). If $r = 3$, then $f(x) = x^3$ and $f'(x) = 3x^{3-1} = 3x^2$, which is (4). We shall prove the power rule in Chapter 4. Until then, we shall use it to calculate derivatives.

EXAMPLE 1   Let $f(x) = \sqrt{x}$. What is $f'(x)$?

Solution   Recall that $\sqrt{x} = x^{1/2}$. We may apply the power rule with $r = \frac{1}{2}$.

$$f(x) = x^{1/2}$$
$$f'(x) = \tfrac{1}{2}x^{1/2-1} = \tfrac{1}{2}x^{-1/2}$$
$$= \frac{1}{2} \cdot \frac{1}{x^{1/2}} = \frac{1}{2\sqrt{x}}$$

Another important special case of the power rule occurs for $r = -1$, corresponding to $f(x) = x^{-1}$. In this case, $f'(x) = (-1)x^{-1-1} = -x^{-2}$. However, since $x^{-1} = 1/x$ and $x^{-2} = 1/x^2$, the power rule for $r = -1$ may also be written as follows:*

> If $f(x) = \dfrac{1}{x}$, then $f'(x) = -\dfrac{1}{x^2}$ $(x \neq 0)$.   (5)

Geometric Meaning of the Derivative

We should, at this stage at least, keep the geometric meaning of the derivative clearly in mind. Figure 4 shows the graphs of x^2 and x^3 together with the interpretations of formulas (3) and (4) in terms of slope.

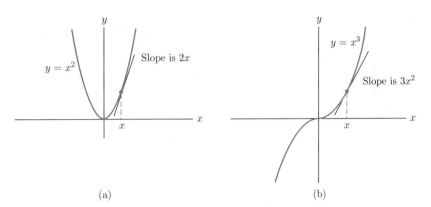

Figure 4. Derivatives of x^2 and x^3.

(a)   (b)

EXAMPLE 2   Find the slope of the curve $y = 1/x$ at $\left(2, \frac{1}{2}\right)$.

Solution   Set $f(x) = 1/x$. The point $\left(2, \frac{1}{2}\right)$ corresponds to $x = 2$, so to find the slope at this point we compute $f'(2)$. From (5) we find that

$$f'(2) = -\frac{1}{2^2} = -\frac{1}{4}.$$

Thus the slope of $y = 1/x$ at the point $\left(2, \frac{1}{2}\right)$ is $-\frac{1}{4}$. (See Fig. 5.)

---

*The formula gives $f'(x)$ for $x \neq 0$. The derivative of $f(x)$ is not defined at $x = 0$ since $f(x)$ itself is not defined there.

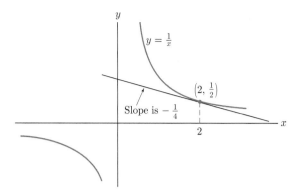

Figure 5. Derivative of $\frac{1}{x}$.

WARNING Do not confuse $f'(2)$, the value of the derivative at 2, with $f(2)$, the value of the y-coordinate at the point on the graph at which $x = 2$. In Example 2 we have $f'(2) = -\frac{1}{4}$, whereas $f(2) = \frac{1}{2}$. The number $f'(2)$ gives the *slope* of the graph at $x = 2$; the number $f(2)$ gives the *height* of the graph at $x = 2$. ■

Suppose that the graph of $y = f(x)$ has a tangent line at the point $(a, f(a))$. Obviously, one point on the tangent line is $(a, f(a))$ and, since the slope of the tangent line is $f'(a)$, we obtain the point–slope form of the equation of the tangent line at $(a, f(a))$:

Equation of the Tangent Line

$$y - f(a) = f'(a)(x - a) \tag{6}$$

As an application, let us derive the equation of the tangent line to the graph of $f(x) = 1/x$ at $(2, \frac{1}{2})$ (Fig. 5). Take $a = 2$, $f(a) = \frac{1}{2}$, and $f'(2) = -\frac{1}{4}$; then (6) yields the equation of the tangent line

$$y - \frac{1}{2} = -\frac{1}{4}(x - 2).$$

You do not need to memorize formula (6), but you should be able to derive it in a given situation; in particular, you should be able to distinguish all its different parts. Here is one more illustration.

EXAMPLE 3 Find the equation of the tangent line to the graph of $f(x) = \sqrt{x}$ at the point $(1, 1)$.

Solution From Example 1 we have $f'(x) = \frac{1}{2\sqrt{x}}$. Thus the slope of the tangent line at the point $(1, 1)$ is $f'(1) = \frac{1}{2\sqrt{1}} = \frac{1}{2}$. Since the tangent line passes through the point $(1, 1)$, its equation in point–slope form is

$$y - 1 = \frac{1}{2}(x - 1).$$

Note how this follows from (6) with $a = 1$, $f(a) = 1$, and $f'(a) = f'(1) = \frac{1}{2}$. Figure 6 shows the graph of $y = \sqrt{x}$ and the tangent line at the point $(1, 1)$. (To plot the tangent line, start at the point $(1, 1)$; move over 1 unit to the right and then up $\frac{1}{2}$ unit.) ◢

EXAMPLE 4 **Tangent line** The line $y = -\frac{1}{3}x + b$ is tangent to the graph of $y = \frac{1}{x}$ at the point $P = \left(a, \frac{1}{a}\right)$ $(a > 0)$. Find P and determine b (Fig. 7).

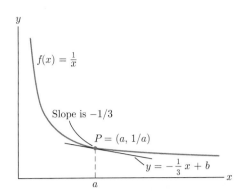

Figure 6. Tangent to $f(x) = \sqrt{x}$ at the point $(1,1)$. Slope of tangent line is $f'(1) = \frac{1}{2\sqrt{1}} = \frac{1}{2}$.

Figure 7

Solution Let $f(x) = \frac{1}{x}$; then $f'(x) = -\frac{1}{x^2}$. We know that the slope of the tangent line at any point $P = \left(a, \frac{1}{a}\right)$ is equal to the value of the derivative $f'(x)$ at $x = a$ or $-\frac{1}{a^2}$. Since the slope of the tangent line $y = -\frac{1}{3}x + b$ is $-\frac{1}{3}$, we conclude that

$$-\frac{1}{a^2} = -\frac{1}{3};$$
$$a^2 = 3$$
$$a = \pm\sqrt{3}.$$

We take $a = \sqrt{3}$, since by assumption $a > 0$. Hence $P = \left(\sqrt{3}, \frac{1}{\sqrt{3}}\right)$. To determine b, we use the fact that P is also on the tangent line, and so its coordinates must satisfy the equation $y = -\frac{1}{3}x + b$. So

$$\frac{1}{\sqrt{3}} = -\frac{\sqrt{3}}{3} + b;$$
$$1 = -1 + b\sqrt{3} \quad \text{(multiply both sides by } \sqrt{3}\text{)}$$
$$\frac{2}{\sqrt{3}} = b.$$

Hence the equation of the tangent line is $y = -\frac{1}{3}x + \frac{2}{\sqrt{3}}$. ∎

Notation The operation of forming a derivative $f'(x)$ from a function $f(x)$ is also indicated by the symbol $\dfrac{d}{dx}$ (read "the derivative with respect to x"). Thus

$$\frac{d}{dx}f(x) = f'(x).$$

For example,

$$\frac{d}{dx}(x^6) = 6x^5, \qquad \frac{d}{dx}(x^{5/3}) = \frac{5}{3}x^{2/3}, \qquad \frac{d}{dx}\left(\frac{1}{x}\right) = -\frac{1}{x^2}.$$

When working with an equation of the form $y = f(x)$, we often write $\dfrac{dy}{dx}$ as a symbol for the derivative $f'(x)$. For example, if $y = x^6$, we may write

$$\frac{dy}{dx} = 6x^5.$$

The Secant-Line Calculation of the Derivative

So far, we have said nothing about how to derive differentiation formulas such as (3), (4), or (5). Let us remedy that omission now. The derivative gives the slope of the tangent line, so we must describe a procedure for calculating that slope.

The fundamental idea for calculating the slope of the tangent line at a point P is to approximate the tangent line very closely by *secant lines*. A secant line at P is a straight line passing through P and a nearby point Q on the curve. (See Fig. 8.) By taking Q very close to P, we can make the slope of the secant line approximate the slope of the tangent line to any desired degree of accuracy. Let us see what this amounts to in terms of calculations.

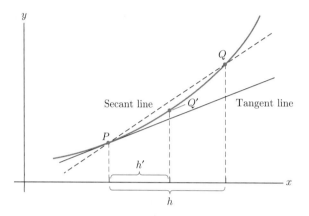

Figure 8. A secant line approximation to a tangent line.

Suppose that the point P is $(x, f(x))$. Also, Q is h horizontal units away from P. Then Q has x-coordinate $x + h$ and y-coordinate $f(x + h)$. The slope of the secant line through the points $P = (x, f(x))$ and $Q = (x + h, f(x + h))$ is simply

$$[\text{slope of secant line}] = \frac{f(x + h) - f(x)}{(x + h) - x} = \frac{f(x + h) - f(x)}{h}.$$

(See Fig. 9.)

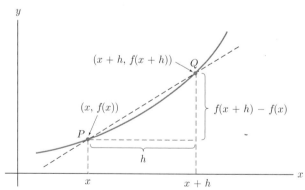

Figure 9. Computing the slope of a secant line.

To move Q close to P along the curve, we let h approach zero. Then the secant line approaches the tangent line, and so

$$[\text{slope of secant line}] \quad \text{approaches} \quad [\text{slope of tangent line}];$$

that is,

$$\frac{f(x + h) - f(x)}{h} \quad \text{approaches} \quad f'(x).$$

Since we can make the secant line as close to the tangent line as we wish by taking h sufficiently small, the quantity $[f(x + h) - f(x)]/h$ can be made to approximate $f'(x)$ to any desired degree of accuracy. Thus we arrive at the following method to compute the derivative $f'(x)$.

To calculate $f'(x)$:

1. Calculate the quotient $\dfrac{f(x + h) - f(x)}{h}$ for $h \neq 0$.

2. Let h approach zero.

3. The quantity $\dfrac{f(x + h) - f(x)}{h}$ will approach $f'(x)$.

The three-step method just outlined will be discussed in greater detail using the concept of limits in the following section. However, it is instructive to apply this method at this point and show how it can be used to compute a derivative.

EXAMPLE 5   **Verification of the power rule for $r = 2$**   Apply the three-step method to confirm formula (3); that is, show that the derivative of $f(x) = x^2$ is $f'(x) = 2x$.

Solution   Here $f(x) = x^2$, so the slope of the secant line is

$$\frac{f(x + h) - f(x)}{h} = \frac{(x + h)^2 - x^2}{h} \quad (h \neq 0).$$

However, by multiplying out, we have $(x + h)^2 = x^2 + 2xh + h^2$, so

$$\frac{f(x + h) - f(x)}{h} = \frac{x^2 + 2xh + h^2 - x^2}{h} = \frac{(2x + h)\cancel{h}}{\cancel{h}}$$
$$= 2x + h.$$

(*Note:* Since $h \neq 0$, we were able to divide both numerator and denominator by h.)

As h approaches 0 (that is, as the secant line approaches the tangent line), the quantity $2x + h$ approaches $2x$. Thus we have

$$f'(x) = 2x$$

which is formula (3). ◾

In the next section, we formulate the three-step method that we just described using limits and derive the *limit definition* of the derivative.

INCORPORATING TECHNOLOGY

Numerical Derivatives and Tangent Lines   TI-83/84 calculators have a numerical derivative routine, **nDeriv**, accessed by pressing [MATH] [8]. For example, to compute the derivative of $f(x) = \sqrt{x}$ at $x = 3$, we proceed as indicated in Fig. 10. We can also graph the tangent line to a function at a point. For example, to graph the tangent line $f(x) = \sqrt{x}$ at $x = 3$, we proceed as follows: First, enter $Y_1 = \sqrt{X}$ and press [GRAPH]. Now, from the graph window, press [2nd] [DRAW] [5] to select **Tangent**. Enter [1] to set $x = 3$ and press [ENTER]. The result is shown in Fig. 11. The equation of the tangent line is also shown in Fig. 11. Note how the slope is equal to the numerical derivative that we obtained in Fig. 10. ◼

Figure 10

Figure 11

Practice Problems 1.3

1. Consider the curve $y = f(x)$ in Fig. 12.
   (a) Find $f(5)$.   (b) Find $f'(5)$.

2. Let $f(x) = 1/x^4$.
   (a) Find its derivative.   (b) Find $f'(2)$.

Figure 12

EXERCISES 1.3

Use formulas (1) and (2) and the power rule to find the derivatives of the following functions.

1. $f(x) = 3x + 7$ 2. $f(x) = -2x$

3. $f(x) = \dfrac{3x}{4} - 2$ 4. $f(x) = \dfrac{2x - 6}{7}$

5. $f(x) = x^7$ 6. $f(x) = x^{-2}$

7. $f(x) = x^{2/3}$ 8. $f(x) = x^{-1/2}$

9. $f(x) = \dfrac{1}{\sqrt{x^5}}$ 10. $f(x) = \dfrac{1}{x^3}$

11. $f(x) = \sqrt[3]{x}$ 12. $f(x) = \dfrac{1}{\sqrt[5]{x}}$

13. $f(x) = \dfrac{1}{x^{-2}}$ 14. $f(x) = \sqrt[7]{x^2}$

15. $f(x) = 4^2$ 16. $f(x) = \pi$

In Exercises 17–24, find the derivative of $f(x)$ at the designated value of x.

17. $f(x) = x^3$ at $x = \frac{1}{2}$ 18. $f(x) = x^5$ at $x = \frac{3}{2}$

19. $f(x) = \dfrac{1}{x}$ at $x = \frac{2}{3}$ 20. $f(x) = \frac{1}{3}$ at $x = 2$

21. $f(x) = x + 11$ at $x = 0$ 22. $f(x) = x^{1/3}$ at $x = 8$

23. $f(x) = \sqrt{x}$ at $x = \frac{1}{16}$ 24. $f(x) = \dfrac{1}{\sqrt[5]{x^2}}$ at $x = 32$

25. Find the slope of the curve $y = x^4$ at $x = 2$.

26. Find the slope of the curve $y = x^5$ at $x = \frac{1}{3}$.

27. If $f(x) = x^3$, compute $f(-5)$ and $f'(-5)$.

28. If $f(x) = 2x + 6$, compute $f(0)$ and $f'(0)$.

29. If $f(x) = x^{1/3}$, compute $f(8)$ and $f'(8)$.

30. If $f(x) = 1/x^2$, compute $f(1)$ and $f'(1)$.

31. If $f(x) = 1/x^5$, compute $f(-2)$ and $f'(-2)$.

32. If $f(x) = x^{3/2}$, compute $f(16)$ and $f'(16)$.

In Exercises 33–40, find the equation of the tangent line to the graph of $y = f(x)$ at the given x. Do not apply formula (6), but proceed as we did in Example 3.

33. $f(x) = x^3$, $x = -2$ 34. $f(x) = x^2$, $x = -\frac{1}{2}$

35. $f(x) = 3x + 1$, $x = 4$ 36. $f(x) = 5$, $x = -2$

37. $f(x) = \sqrt{x}$, $x = \frac{1}{9}$ 38. $f(x) = \dfrac{1}{x}$, $x = .01$

39. $f(x) = \dfrac{1}{\sqrt{x}}$, $x = 1$ 40. $f(x) = \dfrac{1}{x^3}$, $x = 3$

41. The point–slope form of the equation of the tangent line to the graph of $y = x^4$ at the point $(1, 1)$ is $y - 1 = 4(x - 1)$. Explain how this equation follows from formula (6).

42. The tangent to the graph of $y = \frac{1}{x}$ at the point $P = \left(a, \frac{1}{a}\right)$, where $a > 0$, is perpendicular to the line $y = 4x + 1$. Find P.

43. The line $y = 2x + b$ is tangent to the graph of $y = \sqrt{x}$ at the point $P = (a, \sqrt{a})$. Find P and determine b.

44. The line $y = ax + b$ is tangent to the graph of $y = x^3$ at the point $P = (-3, -27)$. Find a and b.

45. (a) Find the point on the curve $y = \sqrt{x}$ where the tangent line is parallel to the line $y = \frac{x}{8}$.

(b) On the same axes, plot the curve $y = \sqrt{x}$, the line $y = \frac{x}{8}$, and the tangent line to $y = \sqrt{x}$ that is parallel to $y = \frac{x}{8}$.

46. There are two points on the graph of $y = x^3$ where the tangent lines are parallel to $y = x$. Find these points.

47. Is there any point on the graph of $y = x^3$ where the tangent line is perpendicular to $y = x$? Justify your answer.

48. The graph of $y = f(x)$ goes through the point $(2, 3)$ and the equation of the tangent line at that point is $y = -2x + 7$. Find $f(2)$ and $f'(2)$.

In Exercises 49–56, find the indicated derivative.

49. $\dfrac{d}{dx}\left(x^8\right)$

50. $\dfrac{d}{dx}\left(x^{-3}\right)$

51. $\dfrac{d}{dx}\left(x^{3/4}\right)$

52. $\dfrac{d}{dx}\left(x^{-1/3}\right)$

53. $\dfrac{dy}{dx}$ if $y = 1$

54. $\dfrac{dy}{dx}$ if $y = x^{-4}$

55. $\dfrac{dy}{dx}$ if $y = x^{1/5}$

56. $\dfrac{dy}{dx}$ if $y = \dfrac{x-1}{3}$

57. Consider the curve $y = f(x)$ in Fig. 13. Find $f(6)$ and $f'(6)$.

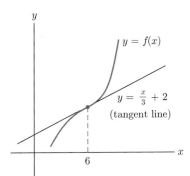

Figure 13

58. Consider the curve $y = f(x)$ in Fig. 14. Find $f(1)$ and $f'(1)$.

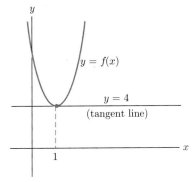

Figure 14

59. In Fig. 15, the straight line $y = \frac{1}{4}x + b$ is tangent to the graph of $f(x) = \sqrt{x}$. Find the values of a and b.

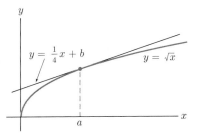

Figure 15

60. In Fig. 16, the straight line is tangent to the graph of $f(x) = 1/x$. Find the value of a.

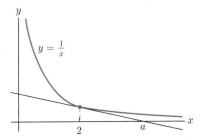

Figure 16

61. Consider the curve $y = f(x)$ in Fig. 17. Find a and $f(a)$. Estimate $f'(a)$.

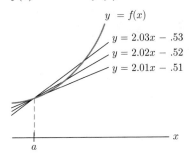

Figure 17

62. Consider the curve $y = f(x)$ in Fig. 18. Estimate $f'(1)$.

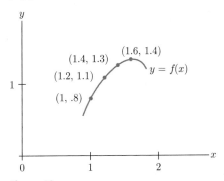

Figure 18

63. In Fig. 19, find the equation of the tangent line to $f(x)$ at the point A.

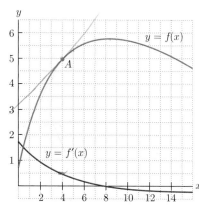

$f(x)$ $x = 4$

$y = 5$

$(4, 5)$

Figure 19

64. In Fig. 20, find the equation of the tangent line to $f(x)$ at the point P.

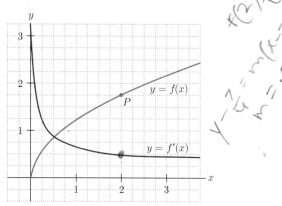

Figure 20

In Exercises 65–70, compute the difference quotient

$$\frac{f(x+h) - f(x)}{h}.$$

Simplify your answer as much as possible.

65. $f(x) = 2x^2$ **66.** $f(x) = x^2 - 7$

67. $f(x) = -x^2 + 2x$ **68.** $f(x) = -2x^2 + x + 3$

69. $f(x) = x^3$ [*Hint:* $(a+b)^3 = a^3 + 3a^2b + 3ab^2 + b^3$.]

70. $f(x) = 2x^3 + x^2$

In Exercises 71–76, apply the three-step method to compute the derivative of the given function.

71. $f(x) = -x^2$ **72.** $f(x) = 3x^2 - 2$

73. $f(x) = 7x^2 + x - 1$ **74.** $f(x) = x + 3$

75. $f(x) = x^3$ **76.** $f(x) = 2x^3 - x$

77. (a) Draw two graphs (of your choice) that represent a function $y = f(x)$ and its vertical shift $y = f(x) + 3$.

(b) Pick two points on these graphs with the same first coordinates: $(x, f(x))$ and $(x, f(x) + 3)$. Draw the tangent lines to the curves at these points and describe what you observe about the tangent lines.

(c) Based on your observation in part (b), explain why

$$\frac{d}{dx} f(x) = \frac{d}{dx}(f(x) + 3).$$

78. Use the approach of Exercise 77 to show that

$$\frac{d}{dx} f(x) = \frac{d}{dx}(f(x) + c),$$

for any constant c. [*Hint:* Compare the tangent lines to the graphs of $f(x)$ and $f(x) + c$.]

Technology Exercises

In Exercises 79–84, use a derivative routine to obtain the value of the derivative. Give the value to 5 decimal places.

79. $f'(0)$, where $f(x) = 2^x$

80. $f'(1)$, where $f(x) = \dfrac{1}{1+x^2}$

81. $f'(1)$, where $f(x) = \sqrt{1+x^2}$

82. $f'(3)$, where $f(x) = \sqrt{25 - x^2}$

83. $f'(2)$, where $f(x) = \dfrac{x}{1+x}$

84. $f'(0)$, where $f(x) = 10^{1+x}$

In Exercises 85 and 86, let Y_1 be the specified function and use a derivative routine to set Y_2 as its derivative. For instance, you might use $\mathbf{Y_2 = nDeriv(Y_1, X, X)}$ or $\mathbf{y2 = der1(y1, x, x)}$. Then graph Y_2 in the specified window and use TRACE to obtain the value of the derivative of Y_1 at $x = 2$.

85. $f(x) = 3x^2 - 5$, $[0, 4]$ by $[-5, 40]$
86. $f(x) = \sqrt{2x}$, $[0, 4]$ by $[-.5, 3]$

In Exercises 87–90, use paper and pencil to find the equation of the tangent line to the graph of the function at the designated point. Then graph both the function and the line to confirm it is indeed the sought-after tangent line.

87. $f(x) = \sqrt{x}$, $(9, 3)$ **88.** $f(x) = \dfrac{1}{x}$, $(.5, 2)$

89. $f(x) = \dfrac{1}{x^2}$, $(.5, 4)$ **90.** $f(x) = x^3$, $(1, 1)$

In Exercises 91–94, $g(x)$ is the tangent line to the graph of $f(x)$ at $x = a$. Graph $f(x)$ and $g(x)$ and determine the value of a.

91. $f(x) = \dfrac{5}{x}$, $g(x) = 5 - 1.25x$

92. $f(x) = \sqrt{8x}$, $g(x) = x + 2$

93. $f(x) = x^3 - 12x^2 + 46x - 50$, $g(x) = 14 - 2x$

94. $f(x) = (x - 3)^3 + 4$, $g(x) = 4$

Solutions to Practice Problems 1.3

1. **(a)** The number $f(5)$ is the y-coordinate of the point P. Since the tangent line passes through P, the coordinates of P satisfy the equation $y = -x + 8$. Since its x-coordinate is 5, its y-coordinate is $-5 + 8 = 3$. Therefore, $f(5) = 3$.

(b) The number $f'(5)$ is the slope of the tangent line at P, which is readily seen to be -1.

2. **(a)** The function $1/x^4$ can be written as the power function x^{-4}. Here $r = -4$. Therefore,

$$f'(x) = (-4)x^{(-4)-1} = -4x^{-5} = \frac{-4}{x^5}.$$

(b) $f'(2) = -4/2^5 = -4/32 = -\frac{1}{8}$.

1.4 Limits and the Derivative

The notion of a limit is one of the fundamental ideas of calculus. Indeed, any "theoretical" development of calculus rests on an extensive use of the theory of limits. Even in this book, where we have adopted an intuitive viewpoint, limit arguments are used occasionally (although in an informal way). In the previous section, limits were used to define the derivative. They arose naturally from our geometric construction of the tangent line as a *limit* of secant lines. In this section we continue our brief introduction to limits and state some additional properties that are required in our development of the derivative and several other areas of calculus. We start with a definition.

> **DEFINITION Limit of a Function** Let $g(x)$ be a function and a a number. We say that the number L *is the limit of* $g(x)$ *as x approaches a*, provided that $g(x)$ can be made arbitrarily close to L for all x sufficiently close (but not equal) to a. In this case we write
> $$\lim_{x \to a} g(x) = L.$$

If, as x approaches a, the values $g(x)$ do *not* approach a specific number, we say that the limit of $g(x)$ as x approaches a *does not exist*. We saw several examples of limits in the previous section. Let us give some further basic examples before turning to the properties of limits.

EXAMPLE 1 Determine $\lim_{x \to 2}(3x - 5)$.

Solution We make a table of values of x approaching 2 and the corresponding values of $3x - 5$:

x	$3x - 5$	x	$3x - 5$
2.1	1.3	1.9	.7
2.01	1.03	1.99	.97
2.001	1.003	1.999	.997
2.0001	1.0003	1.9999	.9997

As x approaches 2, we see that $3x - 5$ approaches 1. In terms of our notation,

$$\lim_{x \to 2}(3x - 5) = 1. \qquad \blacksquare$$

EXAMPLE 2 For each of the functions in Fig. 1, determine if $\lim_{x \to 2} g(x)$ exists. (The circles drawn on the graphs are meant to represent breaks in the graph, indicating that the functions under consideration are not defined at $x = 2$.)

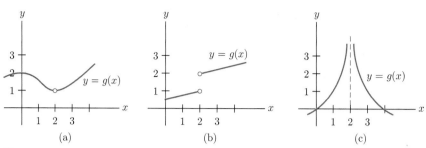

Figure 1. Functions with an undefined point.

Solution **(a)** $\lim_{x \to 2} g(x) = 1$. We can see that as x gets closer and closer to 2, the values of $g(x)$ get closer and closer to 1. This is true for values of x to both the right and the left of 2.

(b) $\lim_{x \to 2} g(x)$ does not exist. As x approaches 2 from the right, $g(x)$ approaches 2. However, as x approaches 2 from the left, $g(x)$ approaches 1. For a limit to exist, the function must approach the *same* value from each direction.

(c) $\lim_{x \to 2} g(x)$ does not exist. As x approaches 2, the values of $g(x)$ become larger and larger and do not approach a fixed number. ∎

The following limit theorems, which we cite without proof, allow us to reduce the computation of limits for combinations of functions to computations of limits involving the constituent functions.

Limit Theorems If $\lim_{x \to a} f(x)$ and $\lim_{x \to a} g(x)$ both exist, we have the following results.

(I) If k is a constant, then $\lim_{x \to a} k \cdot f(x) = k \cdot \lim_{x \to a} f(x)$.

(II) If r is a positive constant and $[f(x)]^r$ is defined for $x \neq a$, then $\lim_{x \to a} [f(x)]^r = \left[\lim_{x \to a} f(x) \right]^r$.

(III) $\lim_{x \to a} [f(x) + g(x)] = \lim_{x \to a} f(x) + \lim_{x \to a} g(x)$.

(IV) $\lim_{x \to a} [f(x) - g(x)] = \lim_{x \to a} f(x) - \lim_{x \to a} g(x)$.

(V) $\lim_{x \to a} [f(x) \cdot g(x)] = \left[\lim_{x \to a} f(x) \right] \cdot \left[\lim_{x \to a} g(x) \right]$.

(VI) If $\lim_{x \to a} g(x) \neq 0$, then $\lim_{x \to a} \dfrac{f(x)}{g(x)} = \dfrac{\lim_{x \to a} f(x)}{\lim_{x \to a} g(x)}$.

EXAMPLE 3 Use the limit theorems to compute the following limits.

(a) $\lim_{x \to 2} x^3$ **(b)** $\lim_{x \to 2} 5x^3$ **(c)** $\lim_{x \to 2} (5x^3 - 15)$

(d) $\lim_{x \to 2} \sqrt{5x^3 - 15}$ **(e)** $\lim_{x \to 2} \left(\sqrt{5x^3 - 15}/x^5 \right)$

Solution **(a)** Since $\lim_{x \to 2} x = 2$, we have by Limit Theorem II that

$$\lim_{x \to 2} x^3 = \left(\lim_{x \to 2} x \right)^3 = 2^3 = 8.$$

(b) $\displaystyle\lim_{x\to 2} 5x^3 = 5\lim_{x\to 2} x^3$ (Limit Theorem I with $k = 5$)

$\qquad\qquad = 5\cdot 8$ [by part (a)]

$\qquad\qquad = 40$

(c) $\displaystyle\lim_{x\to 2}(5x^3 - 15) = \lim_{x\to 2} 5x^3 - \lim_{x\to 2} 15$ (Limit Theorem IV)

Note that $\displaystyle\lim_{x\to 2} 15 = 15$. This is because the constant function $g(x) = 15$ always has the value 15, and so its limit as x approaches *any* number is 15. By part (b), $\displaystyle\lim_{x\to 2} 5x^3 = 40$. Thus

$$\lim_{x\to 2}(5x^3 - 15) = 40 - 15 = 25.$$

(d) $\displaystyle\lim_{x\to 2}\sqrt{5x^3 - 15} = \lim_{x\to 2}(5x^3 - 15)^{1/2}$

$$= \left[\lim_{x\to 2}(5x^3 - 15)\right]^{1/2} \quad \begin{bmatrix}\text{Limit Theorem II with}\\ r = \tfrac{1}{2},\ f(x) = 5x^3 - 15\end{bmatrix}$$

$$= 25^{1/2} \qquad\qquad \bigl[\text{by part (c)}\bigr]$$

$$= 5$$

(e) The limit of the denominator is $\displaystyle\lim_{x\to 2} x^5$, which is $2^5 = 32$, a nonzero number. So by Limit Theorem VI we have

$$\lim_{x\to 2}\frac{\sqrt{5x^3 - 15}}{x^5} = \frac{\displaystyle\lim_{x\to 2}\sqrt{5x^3 - 15}}{\displaystyle\lim_{x\to 2} x^5}$$

$$= \frac{5}{32} \quad \bigl[\text{by part (d)}\bigr]. \qquad\blacksquare$$

The following facts, which may be deduced by repeated applications of the various limit theorems, are extremely handy in evaluating limits.

Limit Theorems (Continued)

(VII) Limit of a Polynomial Function Let $p(x)$ be a polynomial function, a any number. Then

$$\lim_{x\to a} p(x) = p(a).$$

(VIII) Limit of a Rational Function Let $r(x) = p(x)/q(x)$ be a rational function, where $p(x)$ and $q(x)$ are polynomials. Let a be a number such that $q(a) \neq 0$. Then

$$\lim_{x\to a} r(x) = r(a).$$

In other words, to determine a limit of a polynomial or a rational function, simply evaluate the function at $x = a$, provided, of course, that the function is defined at $x = a$. For instance, we can rework the solution to Example 3(c) as follows:

$$\lim_{x\to 2}(5x^3 - 15) = 5(2)^3 - 15 = 25.$$

Many situations require algebraic simplifications before the limit theorems can be applied.

EXAMPLE 4

Compute the following limits.

(a) $\lim\limits_{x \to 3} \dfrac{x^2 - 9}{x - 3}$

(b) $\lim\limits_{x \to 0} \dfrac{\sqrt{x + 4} - 2}{x}$

Solution (a) The function $\dfrac{x^2 - 9}{x - 3}$ is not defined when $x = 3$, since

$$\frac{3^2 - 9}{3 - 3} = \frac{0}{0},$$

which is undefined. This causes no difficulty, since the limit as x approaches 3 depends only on the values of x *near* 3 and excludes considerations of the value at $x = 3$ itself. To evaluate the limit, note that $x^2 - 9 = (x - 3)(x + 3)$. So for $x \neq 3$,

$$\frac{x^2 - 9}{x - 3} = \frac{(x - 3)(x + 3)}{x - 3} = x + 3.$$

As x approaches 3, $x + 3$ approaches 6. Therefore,

$$\lim_{x \to 3} \frac{x^2 - 9}{x - 3} = 6.$$

(b) Since the denominator approaches zero when taking the limit, we may not apply Limit Theorem VI directly. However, if we first apply an algebraic "rationalization" trick, the limit may be evaluated. Multiply numerator and denominator by $\sqrt{x + 4} + 2$.

$$\frac{\sqrt{x + 4} - 2}{x} \cdot \frac{\sqrt{x + 4} + 2}{\sqrt{x + 4} + 2} = \frac{(x + 4) - 4}{x(\sqrt{x + 4} + 2)}$$

$$= \frac{x}{x(\sqrt{x + 4} + 2)}$$

$$= \frac{1}{\sqrt{x + 4} + 2}$$

Thus

$$\lim_{x \to 0} \frac{\sqrt{x + 4} - 2}{x} = \lim_{x \to 0} \frac{1}{\sqrt{x + 4} + 2}$$

$$= \frac{\lim\limits_{x \to 0} 1}{\lim\limits_{x \to 0} (\sqrt{x + 4} + 2)} \qquad \text{(Limit Theorem VI)}$$

$$= \frac{1}{4}. \qquad \blacksquare$$

Limit Definition of the Derivative

Our discussion of the derivative in the previous section was based on an intuitive geometric concept of the tangent line. This geometric approach led us to the three-step method for computing the derivative as a limit of a difference quotient. This process may be considered independently of its geometric interpretation and used to *define* $f'(x)$. In fact, we say that f is *differentiable* at x if

$$\frac{f(x + h) - f(x)}{h}$$

approaches some number as h approaches 0, and we denote this limiting number by $f'(x)$; in symbols, using the limit notation,

DEFINITION Derivative as a Limit

$$f'(x) = \lim_{h \to 0} \frac{f(x+h) - f(x)}{h} \tag{1}$$

If the *difference quotient*

$$\frac{f(x+h) - f(x)}{h}$$

does not approach any specific number as h approaches 0, we say that f is *non-differentiable* at x. Essentially all the functions in this text are differentiable at all points in their domain. A few exceptions are described in Section 1.5.

Using Limits to Calculate $f'(a)$

Step 1   Write the difference quotient $\dfrac{f(a+h) - f(a)}{h}$.

Step 2   Simplify the difference quotient.

Step 3   Find the limit as $h \to 0$. This limit is $f'(a)$.

EXAMPLE 5

Use limits to compute the derivative $f'(5)$ for the following functions.

(a) $f(x) = 15 - x^2$ (b) $f(x) = \dfrac{1}{2x - 3}$

Solution (a) $\dfrac{f(5+h) - f(5)}{h} = \dfrac{\left[15 - (5+h)^2\right] - \left(15 - 5^2\right)}{h}$ (step 1)

$$= \frac{15 - \left(25 + 10h + h^2\right) - (15 - 25)}{h} \quad \text{(step 2)}$$

$$= \frac{-10h - h^2}{h} = -10 - h$$

Therefore, $f'(5) = \lim\limits_{h \to 0}(-10 - h) = -10.$ (step 3)

(b) $\dfrac{f(5+h) - f(5)}{h} = \dfrac{\dfrac{1}{2(5+h) - 3} - \dfrac{1}{2(5) - 3}}{h}$ (step 1)

$$= \frac{\dfrac{1}{7 + 2h} - \dfrac{1}{7}}{h} = \frac{\dfrac{7 - (7 + 2h)}{(7 + 2h)7}}{h} \quad \text{(step 2)}$$

$$= \frac{-2h}{(7 + 2h)7 \cdot h} = \frac{-2}{(7 + 2h)7} = \frac{-2}{49 + 14h}$$

Therefore,

$$f'(5) = \lim_{h \to 0} \frac{-2}{49 + 14h} = -\frac{2}{49}. \quad \text{(step 3)}$$

■

REMARK ▸ When computing the limits in Example 5, we considered only values of h near zero (and not $h = 0$ itself). Therefore, we were freely able to divide both numerator and denominator by h. ■

It is important to distinguish between the derivative at a specific point (as in Example 5) and a derivative formula. A derivative at a specific point is a number, while a derivative formula is a function. The three-step method that we outlined can be used to find derivative formulas, as we now illustrate.

EXAMPLE 6   Find the derivative of $f(x) = x^2 + 2x + 2$.

Solution We have

$$\frac{f(x+h) - f(x)}{h} = \frac{(x+h)^2 + 2(x+h) + 2 - (x^2 + 2x + 2)}{h} \qquad \text{(step 1)}$$

$$= \frac{x^2 + 2xh + h^2 + 2x + 2h + 2 - x^2 - 2x - 2}{h} \qquad \text{(step 2)}$$

$$= \frac{2xh + h^2 + 2h}{h} = \frac{(2x + h + 2)h}{h}$$

$$= 2x + 2 + h.$$

Therefore,

$$f'(x) = \lim_{h \to 0} (2x + 2 + h) = 2x + 2. \qquad \text{(step 3)}$$

∎

Let us go back to the power rule in the previous section and verify it in two special cases: $r = -1$ and $r = \frac{1}{2}$.

EXAMPLE 7   Find the derivative of $f(x) = \frac{1}{x}$.

Solution We have

$$\frac{f(x+h) - f(x)}{h} = \frac{\frac{1}{x+h} - \frac{1}{x}}{h} \qquad \text{(step 1)}$$

$$= \frac{1}{h}\left[\frac{1}{x+h} - \frac{1}{x}\right] \qquad \text{(step 2)}$$

$$= \frac{1}{h}\left[\frac{x - (x+h)}{(x+h)x}\right] = \frac{1}{h}\left[\frac{-h}{(x+h)x}\right]$$

$$= \frac{-1}{(x+h)x}.$$

Step 3   Using Limit Theorem VIII, we find that, as h approaches 0, $\frac{-1}{(x+h)x}$ approaches $\frac{-1}{x^2}$. Hence

$$f'(x) = \frac{-1}{x^2}.$$

∎

In evaluating the next limit, we use the rationalization trick of Example 4(b).

EXAMPLE 8   Find the derivative of $f(x) = \sqrt{x}$.

Solution In simplifying the difference quotient, we use the algebraic identity $(a+b)(a-b) = a^2 - b^2$. So

$$(\sqrt{x+h} - \sqrt{x})(\sqrt{x+h} + \sqrt{x}) = (\sqrt{x+h})^2 - (\sqrt{x})^2 = x + h - x = h.$$

We have

$$\frac{f(x+h) - f(x)}{h} = \frac{\sqrt{x+h} - \sqrt{x}}{h} \qquad \text{(step 1)}$$

$$= \frac{\sqrt{x+h} - \sqrt{x}}{h} \cdot \frac{\sqrt{x+h} + \sqrt{x}}{\sqrt{x+h} + \sqrt{x}} \qquad \text{(step 2)}$$

$$= \frac{x+h-x}{h(\sqrt{x+h} + \sqrt{x})} = \frac{\not{h}}{\not{h}(\sqrt{x+h} + \sqrt{x})}$$

$$= \frac{1}{\sqrt{x+h} + \sqrt{x}}.$$

Therefore,

$$f'(x) = \lim_{h \to 0} \frac{1}{\sqrt{x+h} + \sqrt{x}} = \frac{1}{\sqrt{x} + \sqrt{x}} = \frac{1}{2\sqrt{x}}. \qquad \text{(step 3)}$$ ∎

Sometimes we can use our knowledge of differentiation to evaluate some difficult limits. Here is one illustration.

EXAMPLE 9   Match the limit

$$\lim_{x \to 0} \frac{(1+x)^{1/3} - 1}{x}$$

with a derivative and find its value.

Solution   The idea here is to identify the given limit as a derivative given by formula (1) for a specific choice of f and x. Toward this end, let us replace x by h in the given limit. This does not change the limit, since we are only renaming the variables. The limit becomes

$$\lim_{h \to 0} \frac{(1+h)^{1/3} - 1}{h}.$$

Now go back to (1). Take $f(x) = x^{1/3}$ and evaluate $f'(1)$ according to (1):

$$f'(1) = \lim_{h \to 0} \frac{(1+h)^{1/3} - 1}{h}.$$

On the right side, we have the desired limit; on the left side, $f'(1)$, can be computed using the power rule:

$$f'(x) = \frac{1}{3} x^{-2/3}$$

$$f'(1) = \frac{1}{3}(1)^{-2/3} = \frac{1}{3}.$$

Hence

$$\lim_{h \to 0} \frac{(1+h)^{1/3} - 1}{h} = f'(1) = \frac{1}{3}. \qquad ∎$$

Infinity and Limits   Consider the function $f(x)$ whose graph is sketched in Fig. 2. As x grows large, the value of $f(x)$ approaches 2. In this circumstance, we say that 2 is the *limit of $f(x)$ as x approaches infinity*. Infinity is denoted by the symbol ∞. The preceding limit statement is expressed in the following notation:

$$\lim_{x \to \infty} f(x) = 2.$$

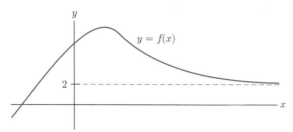

Figure 2.   Function with a limit as x approaches infinity.

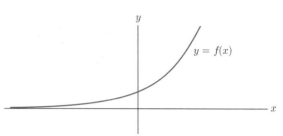

Figure 3.   Function with a limit as x approaches minus infinity.

In a similar vein, consider the function whose graph is sketched in Fig. 3. As x grows large in the negative direction, the value of $f(x)$ approaches 0. In this circumstance, we say that 0 is the *limit of $f(x)$ as x approaches minus infinity*. In symbols,

$$\lim_{x \to -\infty} f(x) = 0.$$

EXAMPLE 10   Calculate the following limits.

    (a) $\displaystyle\lim_{x \to \infty} \frac{1}{x^2 + 1}$ **(b)** $\displaystyle\lim_{x \to \infty} \frac{6x - 1}{2x + 1}$

Solution **(a)** As x increases without bound, so does $x^2 + 1$. Therefore, $1/(x^2 + 1)$ approaches zero as x approaches ∞.

    (b) Both $6x - 1$ and $2x + 1$ increase without bound as x does. To determine the limit of their quotient, we employ an algebraic trick. Divide both numerator and denominator by x to obtain

$$\lim_{x \to \infty} \frac{6x - 1}{2x + 1} = \lim_{x \to \infty} \frac{6 - \dfrac{1}{x}}{2 + \dfrac{1}{x}}.$$

As x increases without bound, $1/x$ approaches zero, so $6 - (1/x)$ approaches 6 and $2 + (1/x)$ approaches 2. Thus the desired limit is $6/2 = 3$. ∎

INCORPORATING TECHNOLOGY

Using Tables to Find Limits   Consider the function

$$y = \frac{x^2 - 9}{x - 3}$$

from Example 4(a). This function is undefined at $x = 3$, but if you examine its values near $x = 3$, as in Fig. 4, you will be convinced that

$$\lim_{x \to 3} \frac{x^2 - 9}{x - 3} = 6.$$

To generate the tables in Fig. 4, first press $\boxed{Y=}$ and assign the function $\frac{x^2 - 9}{x - 3}$ to Y_1. Press $\boxed{\text{2nd}}$ [TBLSET] and set **Indpnt** to **Ask**, leaving the other settings at their default values. Finally, press $\boxed{\text{2nd}}$ [TABLE] and enter the values shown for X. ∎

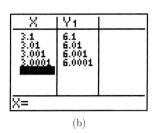

(a)   (b)

Figure 4

Practice Problems 1.4

Determine which of the following limits exist. Compute the limits that exist.

1. $\lim\limits_{x \to 6} \dfrac{x^2 - 4x - 12}{x - 6}$

2. $\lim\limits_{x \to 6} \dfrac{4x + 12}{x - 6}$

EXERCISES 1.4

For each of the following functions $g(x)$, determine whether or not $\lim\limits_{x \to 3} g(x)$ exists. If so, give the limit.

1.

2.

3.

4.

5.

6.

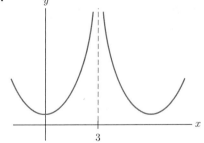

Determine which of the following limits exist. Compute the limits that exist.

7. $\lim\limits_{x \to 1} (1 - 6x)$

8. $\lim\limits_{x \to 2} \dfrac{x}{x - 2}$

9. $\lim\limits_{x \to 3} \sqrt{x^2 + 16}$

10. $\lim\limits_{x \to 4} \left(x^3 - 7\right)$

11. $\lim\limits_{x\to 5} \dfrac{x^2+1}{5-x}$

12. $\lim\limits_{x\to 6} \left(\sqrt{6x}+3x-\dfrac{1}{x}\right)\left(x^2-4\right)$

13. $\lim\limits_{x\to 7} \left(x+\sqrt{x-6}\right)\left(x^2-2x+1\right)$

14. $\lim\limits_{x\to 8} \dfrac{\sqrt{5x-4}-1}{3x^2+2}$

15. $\lim\limits_{x\to 9} \dfrac{\sqrt{x^2-5x-36}}{8-3x}$

16. $\lim\limits_{x\to 10} \left(2x^2-15x-50\right)^{20}$

17. $\lim\limits_{x\to 0} \dfrac{x^2+3x}{x}$

18. $\lim\limits_{x\to 1} \dfrac{x^2-1}{x-1}$

19. $\lim\limits_{x\to 2} \dfrac{-2x^2+4x}{x-2}$

20. $\lim\limits_{x\to 3} \dfrac{x^2-x-6}{x-3}$

21. $\lim\limits_{x\to 4} \dfrac{x^2-16}{4-x}$

22. $\lim\limits_{x\to 5} \dfrac{2x-10}{x^2-25}$

23. $\lim\limits_{x\to 6} \dfrac{x^2-6x}{x^2-5x-6}$

24. $\lim\limits_{x\to 7} \dfrac{x^3-2x^2+3x}{x^2}$

25. $\lim\limits_{x\to 8} \dfrac{x^2+64}{x-8}$

26. $\lim\limits_{x\to 9} \dfrac{1}{(x-9)^2}$

27. Compute the limits that exist, given that

$$\lim_{x\to 0} f(x) = -\frac{1}{2} \quad\text{and}\quad \lim_{x\to 0} g(x) = \frac{1}{2}.$$

(a) $\lim\limits_{x\to 0}\left(f(x)+g(x)\right)$ (b) $\lim\limits_{x\to 0}\left(f(x)-2g(x)\right)$

(c) $\lim\limits_{x\to 0} f(x)\cdot g(x)$ (d) $\lim\limits_{x\to 0} \dfrac{f(x)}{g(x)}$

28. Use the limit definition of the derivative to show that if $f(x) = mx + b$, then $f'(x) = m$.

Use limits to compute the following derivatives.

29. $f'(3)$, where $f(x) = x^2 + 1$

30. $f'(2)$, where $f(x) = x^3$

31. $f'(0)$, where $f(x) = x^3 + 3x + 1$

32. $f'(0)$, where $f(x) = x^2 + 2x + 2$

In Exercises 33–36, apply the three-step method to compute $f'(x)$ for the given function. Follow the steps that we used in Example 6. Make sure to simplify the difference quotient as much as possible before taking limits.

33. $f(x) = x^2 + 1$ **34.** $f(x) = -x^2 + 2$

35. $f(x) = x^3 - 1$ **36.** $f(x) = -3x^2 + 1$

In Exercises 37–48, use limits to compute $f'(x)$. [Hint: In Exercises 45–48, use the rationalization trick of Example 8.]

37. $f(x) = 3x + 1$ **38.** $f(x) = -x + 11$

39. $f(x) = x + \dfrac{1}{x}$ **40.** $f(x) = \dfrac{1}{x^2}$

41. $f(x) = \dfrac{x}{x+1}$ **42.** $f(x) = -1 + \dfrac{2}{x-2}$

43. $f(x) = \dfrac{1}{x^2+1}$ **44.** $f(x) = \dfrac{x}{x+2}$

45. $f(x) = \sqrt{x+2}$ **46.** $f(x) = \sqrt{x^2+1}$

47. $f(x) = \dfrac{1}{\sqrt{x}}$ **48.** $f(x) = x\sqrt{x}$

Each limit in Exercises 49–54 is a definition of $f'(a)$. Determine the function $f(x)$ and the value of a.

49. $\lim\limits_{h\to 0} \dfrac{(1+h)^2-1}{h}$

50. $\lim\limits_{h\to 0} \dfrac{(2+h)^3-8}{h}$

51. $\lim\limits_{h\to 0} \dfrac{\dfrac{1}{10+h}-.1}{h}$

52. $\lim\limits_{h\to 0} \dfrac{(64+h)^{1/3}-4}{h}$

53. $\lim\limits_{h\to 0} \dfrac{\sqrt{9+h}-3}{h}$

54. $\lim\limits_{h\to 0} \dfrac{(1+h)^{-1/2}-1}{h}$

In Exercises 55–60, match the given limit with a derivative and then find the limit by computing the derivative.

55. $\lim\limits_{h\to 0} \dfrac{(h+2)^2-4}{h}$

56. $\lim\limits_{h\to 0} \dfrac{(h-1)^3+1}{h}$

57. $\lim\limits_{h\to 0} \dfrac{\sqrt{h+2}-\sqrt{2}}{h}$

58. $\lim\limits_{h\to 0} \dfrac{\sqrt{h+4}-2}{h}$

59. $\lim\limits_{h\to 0} \dfrac{(8+h)^{1/3}-2}{h}$

60. $\lim\limits_{h\to 0} \dfrac{1}{h}\left[\dfrac{1}{h+1}-1\right]$

Compute the following limits.

61. $\lim\limits_{x\to\infty} \dfrac{1}{x^2}$ **62.** $\lim\limits_{x\to-\infty} \dfrac{1}{x^2}$

63. $\lim\limits_{x\to\infty} \dfrac{1}{x-8}$ **64.** $\lim\limits_{x\to\infty} \dfrac{5x+3}{3x-2}$

65. $\lim\limits_{x\to\infty} \dfrac{10x+100}{x^2-30}$ **66.** $\lim\limits_{x\to\infty} \dfrac{x^2+x}{x^2-1}$

In Exercises 67–72, refer to Figure 5 to find the given limit.

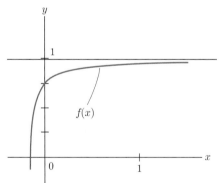

Figure 5

67. $\lim\limits_{x\to 0} f(x)$ **68.** $\lim\limits_{x\to\infty} f(x)$ **69.** $\lim\limits_{x\to 0} x f(x)$

70. $\lim\limits_{x\to\infty}\left(1+2f(x)\right)$ **71.** $\lim\limits_{x\to\infty}\left(1-f(x)\right)$ **72.** $\lim\limits_{x\to 0}\left[f(x)\right]^2$

Technology Exercises

Examine the graph of the function and evaluate the function at large values of x to guess the value of the limit.

73. $\lim\limits_{x\to\infty} \sqrt{25+x}-\sqrt{x}$ **74.** $\lim\limits_{x\to\infty} \dfrac{x^2}{2^x}$

75. $\lim\limits_{x\to\infty} \dfrac{x^2-2x+3}{2x^2+1}$ **76.** $\lim\limits_{x\to\infty} \dfrac{-8x^2+1}{x^2+1}$

Solutions to Practice Problems 1.4

1. The function under consideration is a rational function. Since the denominator has value 0 at $x = 6$, we cannot immediately determine the limit by just evaluating the function at $x = 6$. Also,

$$\lim_{x \to 6} (x - 6) = 0.$$

Since the function in the denominator has the limit zero, we cannot apply Limit Theorem VI. However, since the definition of limit considers only values of x different from 6, the quotient can be simplified by factoring and canceling.

$$\frac{x^2 - 4x - 12}{x - 6} = \frac{(x + 2)(x - 6)}{(x - 6)} = x + 2 \qquad \text{for } x \neq 6.$$

Now $\lim_{x \to 6} (x + 2) = 8$. Therefore,

$$\lim_{x \to 6} \frac{x^2 - 4x - 12}{x - 6} = 8.$$

2. No limit exists. It is easily seen that $\lim_{x \to 6} (4x + 12) = 36$ and $\lim_{x \to 6} (x - 6) = 0$. As x approaches 6, the denominator gets very small and the numerator approaches 36. For example, if $x = 6.00001$, the numerator is 36.00004 and the denominator is .00001. The quotient is 3,600,004. As x approaches 6 even more closely, the quotient gets arbitrarily large and cannot possibly approach a limit.

1.5 Differentiability and Continuity

In the preceding section we defined differentiability of $f(x)$ at $x = a$ in terms of a limit. If this limit does not exist, then we say that $f(x)$ is *nondifferentiable* at $x = a$. Geometrically, the nondifferentiability of $f(x)$ at $x = a$ can manifest itself in several different ways. First, the graph of $f(x)$ could have no tangent line at $x = a$. Second, the graph could have a vertical tangent line at $x = a$. (Recall that slope is not defined for vertical lines.) Some of the various geometric possibilities are illustrated in Fig. 1.

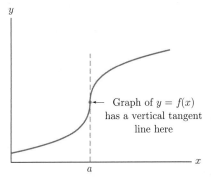

Figure 1. Functions that are nondifferentiable at $x = a$.

The following example illustrates how nondifferentiable functions can arise in practice.

EXAMPLE 1

A railroad company charges $10 per mile to haul a boxcar up to 200 miles and $8 per mile for each mile exceeding 200. In addition, the railroad charges a $1000 handling charge per boxcar. Graph the cost of sending a boxcar x miles.

Solution If x is at most 200 miles, then the cost $C(x)$ is given by $C(x) = 1000 + 10x$ dollars. The cost for 200 miles is $C(200) = 1000 + 2000 = 3000$ dollars. If x exceeds 200 miles, then the total cost will be

$$C(x) = \underbrace{3000}_{\substack{\text{cost of first} \\ \text{200 miles}}} + \underbrace{8(x-200)}_{\substack{\text{cost of miles in} \\ \text{excess of 200}}} = 1400 + 8x.$$

Figure 2. Cost of hauling a boxcar.

Thus

$$C(x) = \begin{cases} 1000 + 10x, & \text{for } 0 < x \le 200, \\ 1400 + 8x, & \text{for } x > 200. \end{cases}$$

The graph of $C(x)$ is sketched in Fig. 2. Note that $C(x)$ is not differentiable at $x = 200$. ∎

Closely related to the concept of differentiability is that of continuity. We say that a function $f(x)$ is *continuous* at $x = a$ provided that, roughly speaking, its graph has no breaks (or gaps) as it passes through the point $(a, f(a))$. That is, $f(x)$ is continuous at $x = a$ provided that we can draw the graph through $(a, f(a))$ without lifting our pencil from the paper. The functions whose graphs are drawn in Figs. 1 and 2 are continuous for all values of x. By contrast, however, the function whose graph is drawn in Fig. 3(a) is not continuous (we say it is *discontinuous*) at $x = 1$ and $x = 2$, since the graph has breaks there. Similarly, the function whose graph is drawn in Fig. 3(b) is discontinuous at $x = 2$.

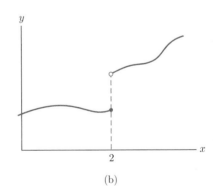

(a)   (b)

Figure 3. Functions with discontinuities.

Discontinuous functions can occur in applications, as the following example shows.

EXAMPLE 2

A manufacturing plant is capable of producing 15,000 units in one shift of 8 hours. For each shift worked, there is a fixed cost of $2000 (for light, heat, and the like). If the variable cost (the cost of labor and raw materials) is $2 per unit, graph the cost $C(x)$ of manufacturing x units.

Solution If $x \leq 15{,}000$, a single shift will suffice, so

$$C(x) = 2000 + 2x, \qquad 0 \leq x \leq 15{,}000.$$

If x is between 15,000 and 30,000, one extra shift will be required, and

$$C(x) = 4000 + 2x, \qquad 15{,}000 < x \leq 30{,}000.$$

If x is between 30,000 and 45,000, the plant will need to work three shifts, and

$$C(x) = 6000 + 2x, \qquad 30{,}000 < x \leq 45{,}000.$$

The graph of $C(x)$ for $0 \leq x \leq 45{,}000$ is drawn in Fig. 4. Note that the graph has breaks at two points. ■

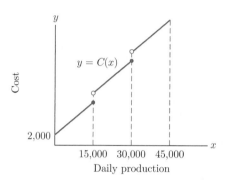

Figure 4. Cost function of a manufacturing plant.

The relationship between differentiability and continuity is this:

Theorem I If $f(x)$ is differentiable at $x = a$, then $f(x)$ is continuous at $x = a$.

Note, however, that the converse statement is definitely false: A function may be continuous at $x = a$ but still not be differentiable there. The functions whose graphs are drawn in Fig. 1 provide examples of this phenomenon. We prove Theorem I at the end of the section.

Just as with differentiability, the notion of continuity can be phrased in terms of limits. For $f(x)$ to be continuous at $x = a$, the values of $f(x)$ for all x near a must be close to $f(a)$ (otherwise, the graph would have a break at $x = a$). In fact, the closer x is to a, the closer $f(x)$ must be to $f(a)$ (again, to avoid a break in the graph). In terms of limits, we must therefore have

$$\lim_{x \to a} f(x) = f(a).$$

Conversely, an intuitive argument shows that, if the preceding limit relation holds, the graph of $y = f(x)$ has no break at $x = a$.

DEFINITION Continuity at a Point A function $f(x)$ is continuous at $x = a$ provided that the following limit relation holds:

$$\lim_{x \to a} f(x) = f(a). \tag{1}$$

For formula (1) to hold, three conditions must be fulfilled.

1. $f(x)$ must be defined at $x = a$.
2. $\lim_{x \to a} f(x)$ must exist.
3. The limit $\lim_{x \to a} f(x)$ must have the value $f(a)$.

A function will fail to be continuous at $x = a$ when any one of these conditions fails to hold. The various possibilities are illustrated in the next example.

EXAMPLE 3   Determine whether the functions whose graphs are drawn in Fig. 5 are continuous at $x = 3$. Use the limit definition.

Solution   **(a)** Here $\lim_{x \to 3} f(x) = 2$. However, $f(3) = 4$. So

$$\lim_{x \to 3} f(x) \neq f(3)$$

and $f(x)$ is not continuous at $x = 3$. (Geometrically, this is clear. The graph has a break at $x = 3$.)

(b) $\lim_{x \to 3} g(x)$ does not exist, so $g(x)$ is not continuous at $x = 3$.

(c) $\lim_{x \to 3} h(x)$ does not exist, so $h(x)$ is not continuous at $x = 3$.

(d) $f(x)$ is not defined at $x = 3$, so $f(x)$ is not continuous at $x = 3$.   ■

(a)

(b)

(c)

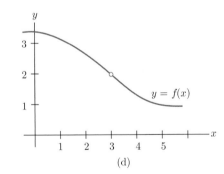

(d)

Figure 5

Using our result on the limit of a polynomial function (Section 1.4), we see that

$$p(x) = a_0 + a_1 x + \cdots + a_n x^n, \qquad a_0, \ldots, a_n \text{ constants,}$$

is continuous for all x. Similarly, a rational function

$$\frac{p(x)}{q(x)}, \qquad p(x), q(x) \text{ polynomials,}$$

is continuous at all x for which $q(x) \neq 0$.

Proof of Theorem I   Suppose that f is differentiable at $x = a$, and let us show that f is continuous at $x = a$. It is enough to show that

$$\lim_{x \to a} (f(x) - f(a)) = 0. \tag{2}$$

For $x \neq a$, write $x = a + h$, where $h \neq 0$, and note that x approaches a if and only if h approaches 0. Then

$$\lim_{x \to a} (f(x) - f(a)) = \lim_{h \to 0} \left[\overbrace{\frac{f(a+h) - f(a)}{h}}^{f(x)} \cdot h \right]$$

$$= \lim_{h \to 0} \left[\frac{f(a+h) - f(a)}{h} \right] \cdot \lim_{h \to 0} h,$$

by Limit Theorem V of the previous section. To evaluate the product of these limits, note that $\lim_{h \to 0} h = 0$ and, since f is differentiable at a,

$$\lim_{h \to 0} \frac{f(a+h) - f(a)}{h} = f'(a).$$

So the product of the limits is $f'(a) \cdot 0 = 0$ and hence formula (2) holds.

INCORPORATING TECHNOLOGY

Piecewise-Defined Functions   Let us graph the function

$$f(x) = \begin{cases} (5/2)x - 1/2 & \text{for } -1 \le x \le 1, \\ (1/2)x - 2 & \text{for } x > 1. \end{cases}$$

To do so, we will enter the function into our calculator as

$$Y_1 = (-1 \le X) * (X \le 1) * ((5/2)X - (1/2)) + (X > 1) * ((1/2)X - 2)$$

Figure 6

and then press $\boxed{\text{GRAPH}}$. The result is shown in Fig. 6 with window setting **ZDecimal** and graph mode set to **Dot**. The graph mode can be changed from the $\boxed{\text{MODE}}$ menu.

   To understand this expression, let us look closely at the first term. The inequality relations can be accessed and entered into the formula from the menu under $\boxed{\text{2nd}}$ [TEST]. The calculator gives the expression $-1 \le X$ the value 1, when the value of X is greater than or equal to -1, and it gives the expression the value 0 otherwise. Similarly, the calculator gives the expression $X \le 1$ the value 1, when the value of X is less than or equal to 1, and it gives the expression the value 0 otherwise. Therefore, $(-1 \le X) * (X \le 1)$ has value 1 when both inequalities are true and value 0 otherwise. And, in that case, the inequality $X > 1$ will have value 0. Hence, our formula for Y_1 takes on the value of

$$(5/2)X - (1/2)$$

when $-1 \le X \le 1$ and, following the same rules, our formula takes on the value of $(1/2)X - 2$ when $X > 1$.   ∎

Practice Problems 1.5

Let $f(x) = \begin{cases} \dfrac{x^2 - x - 6}{x - 3} & \text{for } x \neq 3 \\ 4 & \text{for } x = 3. \end{cases}$

1. Is $f(x)$ continuous at $x = 3$?

2. Is $f(x)$ differentiable at $x = 3$?

EXERCISES 1.5

Is the function whose graph is drawn in Fig. 7 continuous at the following values of x?

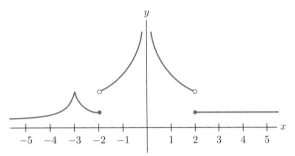

Figure 7

1. $x = 0$ **2.** $x = -3$ **3.** $x = 3$

4. $x = .001$ **5.** $x = -2$ **6.** $x = 2$

Is the function whose graph is drawn in Fig. 7 differentiable at the following values of x?

7. $x = 0$ **8.** $x = -3$ **9.** $x = 3$

10. $x = .001$ **11.** $x = -2$ **12.** $x = 2$

Determine whether each of the following functions is continuous and/or differentiable at $x = 1$.

13. $f(x) = x^2$ **14.** $f(x) = \dfrac{1}{x}$

15. $f(x) = \begin{cases} x + 2 & \text{for } -1 \le x \le 1 \\ 3x & \text{for } 1 < x \le 5 \end{cases}$

16. $f(x) = \begin{cases} x^3 & \text{for } 0 \le x < 1 \\ x & \text{for } 1 \le x \le 2 \end{cases}$

17. $f(x) = \begin{cases} 2x - 1 & \text{for } 0 \le x \le 1 \\ 1 & \text{for } 1 < x \end{cases}$

18. $f(x) = \begin{cases} x & \text{for } x \ne 1 \\ 2 & \text{for } x = 1 \end{cases}$

19. $f(x) = \begin{cases} \dfrac{1}{x - 1} & \text{for } x \ne 1 \\ 0 & \text{for } x = 1 \end{cases}$

20. $f(x) = \begin{cases} x - 1 & \text{for } 0 \le x < 1 \\ 1 & \text{for } x = 1 \\ 2x - 2 & \text{for } x > 1 \end{cases}$

The functions in Exercises 21–26 are defined for all x except for one value of x. If possible, define $f(x)$ at the exceptional point in a way that makes $f(x)$ continuous for all x.

21. $f(x) = \dfrac{x^2 - 7x + 10}{x - 5}, \ x \ne 5$

22. $f(x) = \dfrac{x^2 + x - 12}{x + 4}, \ x \ne -4$

23. $f(x) = \dfrac{x^3 - 5x^2 + 4}{x^2}, \ x \ne 0$

24. $f(x) = \dfrac{x^2 + 25}{x - 5}, \ x \ne 5$

25. $f(x) = \dfrac{(6 + x)^2 - 36}{x}, \ x \ne 0$

26. $f(x) = \dfrac{\sqrt{9 + x} - \sqrt{9}}{x}, \ x \ne 0$

27. The tax that you pay to the federal government is a percentage of your *taxable income*, which is what remains of your gross income after you subtract your allowed deductions. In a recent year there were five rates or brackets for a single taxpayer, as shown in Table 1.

TABLE 1	Single Taxpayer Rates	
Amount Over	**But Not Over**	**Tax Rate**
$0	$27,050	15%
$27,050	$65,550	27.5%
$65,550	$136,750	30.5%
$136,750	$297,350	35.5%
$297,350	. . .	39.1%

So, if you are single and your taxable income was less than $27,050, your tax is your taxable income times 15% (.15). The maximum amount of tax that you will pay on your income in this first bracket is 15% of $27,050, or (.15) × 27,050 = 4057.50 dollars. If your taxable income is more than $27,050 but less than $65,550, your tax is $4057.50 plus 27.5% of the amount in excess of $27,050. So, for example, if your taxable income is $50,000, your tax is $4057.5 + .275(50,000 − 27,050) = 4057.5 + .275 × 22,950 = 10,368.75 dollars. Let x denote your taxable income and $T(x)$ your tax.

(a) Find a formula for $T(x)$ if x is not over $136,750.

(b) Plot the graph of $T(x)$ for $0 \le x \le 136,750$.

(c) Find the maximum amount of tax that you will pay on the portion of your income in the second tax bracket. Express this amount as a difference between two values of $T(x)$.

28. Refer to Exercise 27.

(a) Find a formula for $T(x)$ for all taxable income x.

(b) Plot $T(x)$.

(c) Determine the maximum amount of tax that you will pay on the portion of your income in the fourth tax bracket.

29. The owner of a photocopy store charges 7 cents per copy for the first 100 copies and 4 cents per copy for each copy exceeding 100. In addition, there is a setup fee of $2.50 for each photocopying job.

(a) Determine $R(x)$, the revenue from selling x copies.

(b) If it costs the store owner 3 cents per copy, what is the profit from selling x copies? (Recall that profit is revenue minus cost.)

30. Do Exercise 29 if it costs 10 cents per copy for the first 50 copies and 5 cents per copy for each copy exceeding 50, and there is no setup fee.

31. The graph in Fig. 8 shows the total sales in thousands of dollars in a department store during a typical 24-hour period.

    (a) Estimate the rate of sales during the period between 8 A.M. and 10 A.M.

    (b) Which 2-hour interval in the day sees the highest rate of sales and what is this rate?

32. Refer to Exercise 31.

    (a) Which 2-hour time intervals have the same rate of sales and what is this rate?

    (b) What is the total amount of sales between midnight and 8 A.M.? Compare this amount to the total sales in the period between 8 A.M. and 10 A.M.

In Exercises 33 and 34, determine the value of a that makes the function $f(x)$ continuous at $x = 0$.

33. $f(x) = \begin{cases} 1 & \text{for } x \geq 0 \\ x + a & \text{for } x < 0 \end{cases}$

34. $f(x) = \begin{cases} 2(x - a) & \text{for } x \geq 0 \\ x^2 + 1 & \text{for } x < 0 \end{cases}$

Figure 8

Solutions to Practice Problems 1.5

1. The function $f(x)$ is defined at $x = 3$, that is, $f(3) = 4$. When computing $\lim_{x \to 3} f(x)$, we exclude consideration of $x = 3$; therefore, we can simplify the expression for $f(x)$ as follows:

$$f(x) = \frac{x^2 - x - 6}{x - 3} = \frac{(x - 3)(x + 2)}{x - 3} = x + 2$$

$$\lim_{x \to 3} f(x) = \lim_{x \to 3} (x + 2) = 5.$$

Since $\lim_{x \to 3} f(x) = 5 \neq 4 = f(3)$, $f(x)$ is not continuous at $x = 3$.

2. There is no need to compute any limits to answer this question. By Theorem I, since $f(x)$ is not continuous at $x = 3$, it cannot possibly be differentiable there.

1.6 Some Rules for Differentiation

Three additional rules of differentiation greatly extend the number of functions that we can differentiate.

1. Constant-multiple rule: $\frac{d}{dx}[k \cdot f(x)] = k \cdot \frac{d}{dx}[f(x)]$, k a constant.

2. Sum rule: $\frac{d}{dx}[f(x) + g(x)] = \frac{d}{dx}[f(x)] + \frac{d}{dx}[g(x)]$.

3. General power rule: $\frac{d}{dx}([g(x)]^r) = r \cdot [g(x)]^{r-1} \cdot \frac{d}{dx}[g(x)]$.

We shall discuss these rules, present some applications, and then prove the first two rules at the end of the section.

The Constant-Multiple Rule   Starting with a function $f(x)$, we can multiply it by a constant number k to obtain a new function $k \cdot f(x)$. For instance, if $f(x) = x^2 - 4x + 1$ and $k = 2$, then

$$2f(x) = 2(x^2 - 4x + 1) = 2x^2 - 8x + 2.$$

The constant-multiple rule says that the derivative of the new function $k \cdot f(x)$ is just k times the derivative of the original function.* In other words, when faced with the differentiation of a constant times a function, simply carry along the constant and differentiate the function.

EXAMPLE 1

Calculate.

(a) $\dfrac{d}{dx}\left(2x^5\right)$ **(b)** $\dfrac{d}{dx}\left(\dfrac{x^3}{4}\right)$ **(c)** $\dfrac{d}{dx}\left(-\dfrac{3}{x}\right)$ **(d)** $\dfrac{d}{dx}\left(5\sqrt{x}\right)$

Solution   **(a)** With $k = 2$ and $f(x) = x^5$, we have

$$\frac{d}{dx}\left(2 \cdot x^5\right) = 2 \cdot \frac{d}{dx}\left(x^5\right) = 2\left(5x^4\right) = 10x^4.$$

(b) Write $\dfrac{x^3}{4}$ in the form $\dfrac{1}{4} \cdot x^3$. Then

$$\frac{d}{dx}\left(\frac{x^3}{4}\right) = \frac{1}{4} \cdot \frac{d}{dx}\left(x^3\right) = \frac{1}{4}\left(3x^2\right) = \frac{3}{4}x^2.$$

(c) Write $-\dfrac{3}{x}$ in the form $(-3) \cdot \dfrac{1}{x}$. Then

$$\frac{d}{dx}\left(-\frac{3}{x}\right) = (-3) \cdot \frac{d}{dx}\left(\frac{1}{x}\right) = (-3) \cdot \frac{-1}{x^2} = \frac{3}{x^2}.$$

(d) $\dfrac{d}{dx}\left(5\sqrt{x}\right) = 5\dfrac{d}{dx}\left(\sqrt{x}\right) = 5\dfrac{d}{dx}\left(x^{1/2}\right) = \dfrac{5}{2}x^{-1/2}.$

This answer may also be written in the form $\dfrac{5}{2\sqrt{x}}$.   ■

The Sum Rule   To differentiate a sum of functions, differentiate each function individually and add together the derivatives.† Another way of saying this is "the derivative of a sum of functions is the sum of the derivatives."

EXAMPLE 2

Find each of the following.

(a) $\dfrac{d}{dx}\left(x^3 + 5x\right)$ **(b)** $\dfrac{d}{dx}\left(x^4 - \dfrac{3}{x^2}\right)$ **(c)** $\dfrac{d}{dx}\left(2x^7 - x^5 + 8\right)$

Solution   **(a)** Let $f(x) = x^3$ and $g(x) = 5x$. Then

$$\frac{d}{dx}\left(x^3 + 5x\right) = \frac{d}{dx}\left(x^3\right) + \frac{d}{dx}\left(5x\right) = 3x^2 + 5.$$

---

*More precisely, the constant-multiple rule asserts that, if $f(x)$ is differentiable at $x = a$, so is the function $k \cdot f(x)$, and the derivative of $k \cdot f(x)$ at $x = a$ may be computed using the given formula.
†More precisely, the sum rule asserts that, if both $f(x)$ and $g(x)$ are differentiable at $x = a$, so is $f(x) + g(x)$, and the derivative (at $x = a$) of the sum is then the sum of the derivatives (at $x = a$).

(b) The sum rule applies to differences as well as sums (see Exercise 46). Indeed, by the sum rule,

$$\frac{d}{dx}\left(x^4 - \frac{3}{x^2}\right) = \frac{d}{dx}(x^4) + \frac{d}{dx}\left(-\frac{3}{x^2}\right) \qquad \text{(sum rule)}$$

$$= \frac{d}{dx}(x^4) - 3\frac{d}{dx}(x^{-2}) \qquad \text{(constant-multiple rule)}$$

$$= 4x^3 - 3(-2x^{-3})$$

$$= 4x^3 + 6x^{-3}.$$

After some practice, we usually omit most or all of the intermediate steps and simply write

$$\frac{d}{dx}\left(x^4 - \frac{3}{x^2}\right) = 4x^3 + 6x^{-3}.$$

(c) We apply the sum rule repeatedly and use the fact that the derivative of a constant function is 0:

$$\frac{d}{dx}(2x^7 - x^5 + 8) = \frac{d}{dx}(2x^7) - \frac{d}{dx}(x^5) + \frac{d}{dx}(8)$$

$$= 2(7x^6) - 5x^4 + 0$$

$$= 14x^6 - 5x^4. \qquad \blacksquare$$

REMARK The differentiation of a function *plus* a constant is different from the differentiation of a constant *times* a function. Figure 1 shows the graphs of $f(x)$, $f(x) + 2$, and $2 \cdot f(x)$, where $f(x) = x^3 - \frac{3}{2}x^2$. For each x, the graphs of $f(x)$ and $f(x) + 2$ have the same slope. In contrast, for each x, the slope of the graph of $2 \cdot f(x)$ is twice the slope of the graph of $f(x)$. Upon differentiation, an added constant disappears, whereas a constant that multiplies a function is carried along. $\blacksquare$

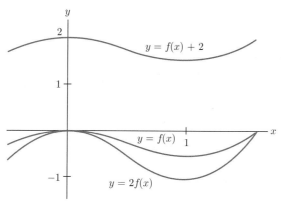

Figure 1. Two effects of a constant on the graph of $f(x)$.

The General Power Rule Frequently, we will encounter expressions of the form $[g(x)]^r$, for instance, $(x^3 + 5)^2$, where $g(x) = x^3 + 5$ and $r = 2$. The general power rule says that, to differentiate $[g(x)]^r$, we must first treat $g(x)$ as if it were simply an x, form $r[g(x)]^{r-1}$, and then multiply it by a "correction factor" $g'(x)$.* Thus

$$\frac{d}{dx}(x^3+5)^2 = 2(x^3+5)^1 \cdot \frac{d}{dx}(x^3+5)$$

$$= 2(x^3+5) \cdot (3x^2)$$

$$= 6x^2(x^3+5).$$

In this special case it is easy to verify that the general power rule gives the correct answer. We first expand $(x^3+5)^2$ and then differentiate.

$$(x^3+5)^2 = (x^3+5)(x^3+5) = x^6 + 10x^3 + 25$$

From the constant-multiple rule and the sum rule, we have

$$\frac{d}{dx}(x^3+5)^2 = \frac{d}{dx}(x^6 + 10x^3 + 25)$$

$$= 6x^5 + 30x^2 + 0$$

$$= 6x^2(x^3+5).$$

The two methods give the same answer.

Note that if we set $g(x) = x$ in the general power rule, we recover the power rule. So the general power rule contains the power rule as a special case.

EXAMPLE 3   Differentiate $\sqrt{1-x^2}$.

Solution

$$\frac{d}{dx}(\sqrt{1-x^2}) = \frac{d}{dx}[(1-x^2)^{1/2}] = \frac{1}{2}(1-x^2)^{-1/2} \cdot \frac{d}{dx}(1-x^2)$$

$$= \frac{1}{2}(1-x^2)^{-1/2} \cdot (-2x)$$

$$= \frac{-x}{(1-x^2)^{1/2}} = \frac{-x}{\sqrt{1-x^2}}$$

EXAMPLE 4   Differentiate

$$y = \frac{1}{x^3 + 4x}.$$

Solution

$$y = \frac{1}{x^3 + 4x} = (x^3 + 4x)^{-1}$$

$$\frac{dy}{dx} = (-1)(x^3 + 4x)^{-2} \cdot \frac{d}{dx}(x^3 + 4x)$$

$$= \frac{-1}{(x^3+4x)^2}(3x^2 + 4)$$

$$= -\frac{3x^2 + 4}{(x^3+4x)^2}$$

Some derivatives require more than one differentiation rule.

EXAMPLE 5   Differentiate $5\sqrt[3]{1+x^3}$.

---

*More precisely, the general power rule asserts that, if $g(x)$ is differentiable at $x = a$ and if $[g(x)]^{r-1}$ is defined at $x = a$, then $[g(x)]^r$ is also differentiable at $x = a$ and its derivative is given by the formula stated.

Solution $\dfrac{d}{dx}(5\sqrt[3]{1+x^3}) = 5\dfrac{d}{dx}\left[\sqrt[3]{1+x^3}\right]$ (constant-multiple rule)

$$= 5\dfrac{d}{dx}\left[(1+x^3)^{1/3}\right]$$

$$= 5\left(\dfrac{1}{3}\right)(1+x^3)^{\frac{1}{3}-1}\dfrac{d}{dx}\left[(1+x^3)\right]$$ (general power rule)

$$= \dfrac{5}{3}(1+x^3)^{-\frac{2}{3}}(3x^2) = 5x^2(1+x^3)^{-\frac{2}{3}}.$$ ■

Example 5 illustrates how the differentiation rules can be combined. For example, as a by-product of the constant multiple rule and the general power rule, we have

$$\dfrac{d}{dx}(k \cdot [g(x)]^r) = kr \cdot [g(x)]^{r-1}g'(x).$$

The Derivative as a Rate of Change

Suppose that $y = f(x)$ is a function and $P = (a, f(a))$ a point on its graph (Fig. 2). Recall two important properties of the tangent line to the graph at P.

- The tangent line at P has slope $f'(a)$.
- The tangent line at P is the line that best approximates the graph near P.

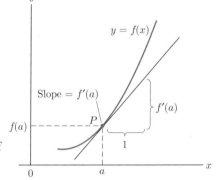

Figure 2. The slope of the graph of $y = f(x)$ at P is the slope of the tangent line at P. This slope measures the rate of change of f as its graph goes through P.

The second property allows us to think of the portion of the graph near P as being a straight line with slope $f'(a)$. Now we know from Section 1.1 that the rate of change of a linear function is equal to its slope. So if we think of the graph of f near P as being a line with slope $f'(a)$, it follows that the rate of change of the graph of f, as it goes through P, is equal to the slope $f'(a)$. This gives us the following important interpretation of the derivative:

$$f'(a) = \text{rate of change of } f(x) \text{ at } x = a.$$

Thus the derivative can be used to measure the rate of change of quantities. We will see many applications of this observation in this book.

EXAMPLE 6   **Declining sales**   At the end of the holiday season in January, the sales at a department store are expected to fall (Fig. 3). It is estimated that for the x day of

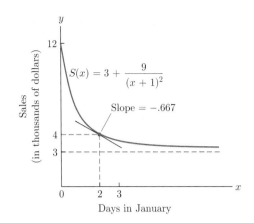

Figure 3. The sales decrease starting the first day of January.

January the sales will be

$$S(x) = 3 + \frac{9}{(x+1)^2} \text{ thousand dollars.}$$

Compute $S(2)$ and $S'(2)$ and interpret your results.

Solution We have

$$S(2) = 3 + \frac{9}{(2+1)^2} = 4 \text{ thousand dollars.}$$

To compute $S'(2)$, write $\frac{9}{(x+1)^2}$ as $9 \cdot (x+1)^{-2}$; then, using all three rules of differentiation, we find

$$S'(x) = \frac{d}{dx}[S(x)] = \overbrace{\frac{d}{dx}[3]}^{=0} + \frac{d}{dx}(9 \cdot (x+1)^{-2})$$

$$= 9 \cdot \frac{d}{dx}((x+1)^{-2})$$

$$= 9(-2)(x+1)^{-3}\frac{d}{dx}(x+1) = -18(x+1)^{-3}(1) = \frac{-18}{(x+1)^3};$$

$$S'(2) = \frac{-18}{(2+1)^3} = -\frac{2}{3} \approx -.667.$$

The equations $S(2) = 4$ and $S'(2) = -.667$ tell us that on January 2 the sales are \$4000 and are falling at the rate of .667 thousand dollars per day, or \$667 per day. ∎

EXAMPLE 7

Predicting sales Refer to Example 6. Use the values $S(2) = 4$ and $S'(2) = -.667$ to estimate the sales on January 3. Compare your result with the exact value of sales on January 3 by computing $S(3)$.

Solution On January 2 the sales are \$4000 and are falling at the rate of 667 dollars per day. Thus we expect the sales on January 3 to be \$667 less than the sales on January 2. Thus our estimate of the sales on January 3 is $4000 - 667 = \$3333$. To compare with the exact value of sales on January 3, we compute $S(3)$ from the formula in Example 6:

$$S(3) = 3 + \frac{9}{(3+1)^2} = 3 + \frac{9}{16} = \frac{57}{16} = 3.5625 \text{ thousand dollars,}$$

or \$3562.5. This is close to our estimate of \$3333. ∎

Proofs of the Let us verify both rules when x has the value a. Recall that if $f(x)$ is differentiable,
Constant-Multiple at $x = a$, then its derivative is the limit
and Sum Rules

$$\lim_{h \to 0} \frac{f(a+h) - f(a)}{h}.$$

Constant-Multiple Rule We assume that $f(x)$ is differentiable at $x = a$. We must prove that $k \cdot f(x)$ is differentiable at $x = a$ and that its derivative is $k \cdot f'(x)$. This amounts to showing that the limit

$$\lim_{h \to 0} \frac{k \cdot f(a+h) - k \cdot f(a)}{h}$$

exists and has the value $k \cdot f'(a)$. However,

$$\lim_{h \to 0} \frac{k \cdot f(a+h) - k \cdot f(a)}{h}$$

$$= \lim_{h \to 0} k \left[\frac{f(a+h) - f(a)}{h} \right]$$

$$= k \cdot \lim_{h \to 0} \frac{f(a+h) - f(a)}{h} \qquad \text{(by Limit Theorem I)}$$

$$= k \cdot \lim_{h \to 0} \frac{f(a+h) - f(a)}{h} = kf'(a)$$

$$\text{[since } f(x) \text{ is differentiable at } x = a\text{]},$$

which is what we desired to show.

Sum Rule We assume that both $f(x)$ and $g(x)$ are differentiable at $x = a$. We must prove that $f(x) + g(x)$ is differentiable at $x = a$ and that its derivative is $f'(a) + g'(a)$. That is, we must show that the limit

$$\lim_{h \to 0} \frac{[f(a+h) + g(a+h)] - [f(a) + g(a)]}{h}$$

exists and equals $f'(a) + g'(a)$. Using Limit Theorem III and the fact that $f(x)$ and $g(x)$ are differentiable at $x = a$, we have

$$\lim_{h \to 0} \frac{[f(a+h) + g(a+h)] - [f(a) + g(a)]}{h}$$

$$= \lim_{h \to 0} \left[\frac{f(a+h) - f(a)}{h} + \frac{g(a+h) - g(a)}{h} \right]$$

$$= \lim_{h \to 0} \frac{f(a+h) - f(a)}{h} + \lim_{h \to 0} \frac{g(a+h) - g(a)}{h}$$

$$= f'(a) + g'(a).$$

The general power rule will be proved as a special case of the chain rule presented in Chapter 3.

Practice Problems 1.6

1. Find the derivative $\dfrac{d}{dx}(x)$.

2. Differentiate the function $y = \dfrac{x + (x^5 + 1)^{10}}{3}$.

EXERCISES 1.6

Differentiate.

1. $y = x^3 + x^2$

2. $y = 3x^4$

3. $y = 2\sqrt{x}$

4. $y = \dfrac{1}{3x^3}$

5. $y = \dfrac{x}{2} - \dfrac{2}{x}$

6. $f(x) = 12 + \dfrac{1}{7^3}$

7. $f(x) = x^4 + x^3 + x$

8. $y = 4x^3 - 2x^2 + x + 1$

9. $y = (2x + 4)^3$

10. $y = (x^2 - 1)^3$

11. $y = (x^3 + x^2 + 1)^7$

12. $y = (x^2 + x)^{-2}$

13. $y = \dfrac{4}{x^2}$

14. $y = 4(x^2 - 6)^{-3}$

15. $y = 3\sqrt[3]{2x^2 + 1}$

16. $y = 2\sqrt{x + 1}$

17. $y = 2x + (x + 2)^3$

18. $y = (x - 1)^3 + (x + 2)^4$

19. $y = \dfrac{1}{5x^5}$

20. $y = (x^2 + 1)^2 + 3(x^2 - 1)^2$

21. $y = \dfrac{1}{x^3 + 1}$

22. $y = \dfrac{2}{x + 1}$

23. $y = x + \dfrac{1}{x + 1}$

24. $y = 2\sqrt[4]{x^2 + 1}$

25. $f(x) = 5\sqrt{3x^3 + x}$

26. $y = \dfrac{1}{x^3 + x + 1}$

27. $y = 3x + \pi^3$

28. $y = \sqrt{1 + x^2}$

29. $y = \sqrt{1 + x + x^2}$

30. $y = \dfrac{1}{2x + 5}$

31. $y = \dfrac{2}{1 - 5x}$

32. $y = \dfrac{7}{\sqrt{1 + x}}$

33. $y = \dfrac{45}{1 + x + \sqrt{x}}$

34. $y = (1 + x + x^2)^{11}$

35. $y = x + 1 + \sqrt{x + 1}$

36. $y = \pi^2 x$

37. $f(x) = \left(\dfrac{\sqrt{x}}{2} + 1 \right)^{3/2}$

38. $y = \left(x - \dfrac{1}{x} \right)^{-1}$

In Exercises 39 and 40, find the slope of the graph of $y = f(x)$ at the designated point.

39. $f(x) = 3x^2 - 2x + 1$, $(1, 2)$

40. $f(x) = x^{10} + 1 + \sqrt{1 - x}$, $(0, 2)$

41. Find the slope of the tangent line to the curve $y = x^3 + 3x - 8$ at $(2, 6)$.

42. Write the equation of the tangent line to the curve $y = x^3 + 3x - 8$ at $(2, 6)$.

43. Find the slope of the tangent line to the curve $y = (x^2 - 15)^6$ at $x = 4$. Then write the equation of this tangent line.

44. Find the equation of the tangent line to the curve $y = \dfrac{8}{x^2 + x + 2}$ at $x = 2$.

45. Differentiate the function $f(x) = (3x^2 + x - 2)^2$ in two ways.

    (a) Use the general power rule.

    (b) Multiply $3x^2 + x - 2$ by itself and then differentiate the resulting polynomial.

46. Using the sum rule and the constant-multiple rule, show that for any functions $f(x)$ and $g(x)$

$$\frac{d}{dx}[f(x) - g(x)] = \frac{d}{dx}f(x) - \frac{d}{dx}g(x).$$

47. Figure 4 contains the curves $y = f(x)$ and $y = g(x)$ and the tangent line to $y = f(x)$ at $x = 1$, with $g(x) = 3 \cdot f(x)$. Find $g(1)$ and $g'(1)$.

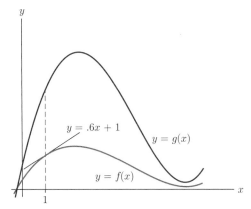

Figure 4. Graphs of $f(x)$ and $g(x) = 3f(x)$.

48. Figure 5 contains the curves $y = f(x)$, $y = g(x)$, and $y = h(x)$ and the tangent lines to $y = f(x)$ and $y = g(x)$ at $x = 1$, with $h(x) = f(x) + g(x)$. Find $h(1)$ and $h'(1)$.

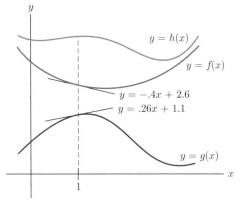

Figure 5. Graphs of $f(x)$, $g(x)$, and $h(x) = f(x) + g(x)$.

49. If $f(5) = 2$, $f'(5) = 3$, $g(5) = 4$, and $g'(5) = 1$, find $h(5)$ and $h'(5)$, where $h(x) = 3f(x) + 2g(x)$.

50. If $g(3) = 2$ and $g'(3) = 4$, find $f(3)$ and $f'(3)$, where $f(x) = 2 \cdot [g(x)]^3$.

51. If $g(1) = 4$ and $g'(1) = 3$, find $f(1)$ and $f'(1)$, where $f(x) = 5 \cdot \sqrt{g(x)}$.

52. If $h(x) = [f(x)]^2 + \sqrt{g(x)}$, determine $h(1)$ and $h'(1)$, given that $f(1) = 1$, $g(1) = 4$, $f'(1) = -1$, and $g'(1) = 4$.

53. The tangent line to the curve $y = \frac{1}{3}x^3 - 4x^2 + 18x + 22$ is parallel to the line $6x - 2y = 1$ at two points on the curve. Find the two points.

54. The tangent line to the curve $y = x^3 - 6x^2 - 34x - 9$ has slope 2 at two points on the curve. Find the two points.

55. In Fig. 6, the straight line is tangent to the graph of $f(x)$. Find $f(4)$ and $f'(4)$.

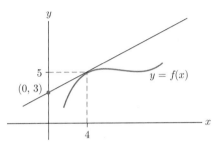

Figure 6

56. In Fig. 7, the straight line is tangent to the parabola. Find the value of b.

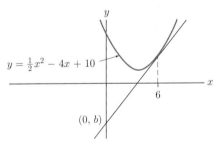

Figure 7

57. Let $S(x)$ represent the total sales (in thousands of dollars) for the month x in the year 2005 at a certain department store. Represent each following statement by an equation involving S or S'.

(a) The sales at the end of January reached $120,560 and were rising at the rate of $1500 per month.

(b) At the end of March, the sales for this month dropped to $80,000 and were falling by about $200 a day. (Use 1 month = 30 days.)

58. Comparing Rates of Change

(a) In Example 6, find the total sales for January 10 and determine the rate at which sales are falling on that day.

(b) Compare the rate of change of sales on January 2 (Example 6) to the rate on January 10. What can you infer about the rate of change of sales?

59. Predicting Sales Refer to Example 6.

(a) Compute $S(10)$ and $S'(10)$.

(b) Use the data in part (a) to estimate the total sales on January 11. Compare your estimate to the actual value given by $S(11)$.

60. Correcting a Prediction The financial analysts at the store in Example 6 corrected their projections and are now expecting the total sales for the x day of January to be

$$T(x) = \frac{24}{5} + \frac{36}{5(3x+1)^2} \text{ thousand dollars.}$$

(a) Let $S(x)$ be as in Example 6. Compute $T(1)$, $T'(1)$, $S(1)$, and $S'(1)$.

(b) Compare and interpret the data in part (a) as they pertain to the sales on January 1.

61. (a) Let $A(x)$ denote the number (in hundreds) of computers sold when x thousand dollars are spent on advertising. Represent the following statement by equations involving A or A': When 8 thousand dollars were spent on advertising, the number of computers sold was 1200 and it was rising at the rate of 50 computers for each 1000 dollars spent on advertising.

(b) Estimate the number of computers that will be sold if $9000 are spent on advertising.

62. A toy company introduces a new video game on the market. Let $S(x)$ denote the number of videos sold on the day x since the item was introduced. Let n be a positive integer. Interpret $S(n)$, $S'(n)$, and $S(n) + S'(n)$.

63. U.S. Federal Debt Based on data from the U.S. Treasury Department, the federal debt (in trillions of dollars) for the years 1995 to 2004 was given approximately by the formula

$$D(x) = 4.95 + .402x - .1067x^2 + .0124x^3 - .00024x^4,$$

where x is the number of years elapsed since the end of 1995. Estimate the federal debt at the end of 1999 $(x = 4)$ and the rate at which it was increasing at that time.

64. Refer to Exercise 63.

(a) Is it true that by the end of 2003 the federal debt was twice as large as it was at the end of 2001?

(b) Is it true that by the end of 2003 the federal debt was increasing at a rate that was more than twice the rate at the end of 2001?

Solutions to Practice Problems 1.6

1. The problem asks for the derivative of the function $y = x$, a straight line of slope 1. Therefore,

$$\frac{d}{dx}(x) = 1.$$

The result can also be obtained from the power rule with $r = 1$. If $f(x) = x^1$, then

$$\frac{d}{dx}(f(x)) = 1 \cdot x^{1-1} = x^0 = 1.$$

See Fig. 8.

Figure 8

2. All three rules are required to differentiate this function.

$$\frac{dy}{dx} = \frac{d}{dx}\frac{1}{3} \cdot [x + (x^5 + 1)^{10}]$$

$$= \frac{1}{3}\frac{d}{dx}[x + (x^5 + 1)^{10}] \qquad \begin{array}{l}\text{(constant-}\\\text{multiple}\\\text{rule)}\end{array}$$

$$= \frac{1}{3}\left[\frac{d}{dx}(x) + \frac{d}{dx}(x^5 + 1)^{10}\right] \qquad \text{(sum rule)}$$

$$= \frac{1}{3}[1 + 10(x^5 + 1)^9 \cdot (5x^4)] \qquad \begin{array}{l}\text{(general power}\\\text{rule)}\end{array}$$

$$= \frac{1}{3}[1 + 50x^4(x^5 + 1)^9]$$

1.7 More About Derivatives

In many applications it is convenient to use variables other than x and y. One might, for instance, study the function $f(t) = t^2$ instead of writing $f(x) = x^2$. In this case, the notation for the derivative involves t rather than x, but the concept of the derivative as a slope formula is unaffected. (See Fig. 1.) When the independent variable is t instead of x, we write $\dfrac{d}{dt}$ in place of $\dfrac{d}{dx}$. For instance,

$$\frac{d}{dt}(t^3) = 3t^2, \qquad \frac{d}{dt}(2t^2 + 3t) = 4t + 3.$$

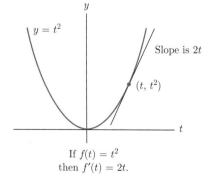

Figure 1. The same function but different variables.

If $f(x) = x^2$ then $f'(x) = 2x$.

If $f(t) = t^2$ then $f'(t) = 2t$.

Recall that if y is a function of x, say $y = f(x)$, we may write $\dfrac{dy}{dx}$ in place of $f'(x)$. We sometimes call $\dfrac{dy}{dx}$ "the derivative of y with respect to x." Similarly, if v is a function of t, the derivative of v with respect to t is written as $\dfrac{dv}{dt}$. For example, if $v = 4t^2$, then $\dfrac{dv}{dt} = 8t$.

Of course, other letters can be used to denote variables. The formulas

$$\frac{d}{dP}(P^3) = 3P^2, \qquad \frac{d}{ds}(s^3) = 3s^2, \qquad \frac{d}{dz}(z^3) = 3z^2$$

all express the same basic fact that the slope formula for the cubic curve $y = x^3$ is given by $3x^2$.

EXAMPLE 1

Compute.

(a) $\dfrac{ds}{dp}$ if $s = 3(p^2 + 5p + 1)^{10}$
(b) $\dfrac{d}{dt}(at^2 + St^{-1} + S^2)$

Solution (a) $\dfrac{d}{dp} 3(p^2 + 5p + 1)^{10} = 30(p^2 + 5p + 1)^9 \cdot \dfrac{d}{dp}(p^2 + 5p + 1)$

$$= 30(p^2 + 5p + 1)^9 (2p + 5)$$

(b) Although the expression $at^2 + St^{-1} + S^2$ contains several letters, the notation $\dfrac{d}{dt}$ indicates that, for purposes of calculating the derivative with respect to t, all letters except t are to be considered as constants. Hence

$$\frac{d}{dt}(at^2 + St^{-1} + S^2) = \frac{d}{dt}(at^2) + \frac{d}{dt}(St^{-1}) + \frac{d}{dt}(S^2)$$

$$= a \cdot \frac{d}{dt}(t^2) + S \cdot \frac{d}{dt}(t^{-1}) + 0$$

$$= 2at - St^{-2}.$$

$$\left[\text{The derivative } \frac{d}{dt}(S^2) \text{ is zero because } S^2 \text{ is a constant.} \right] \qquad \blacksquare$$

The Second Derivative   When we differentiate a function $f(x)$, we obtain a new function $f'(x)$ that is a formula for the slope of the curve $y = f(x)$. If we differentiate the function $f'(x)$, we obtain what is called the *second derivative* of $f(x)$, denoted by $f''(x)$. That is,

$$\frac{d}{dx} f'(x) = f''(x).$$

EXAMPLE 2

Find the second derivatives of the following functions.

(a) $f(x) = x^3 + (1/x)$ (b) $f(x) = 2x + 1$ (c) $f(t) = t^{1/2} + t^{-1/2}$

Solution (a) $f(x) = x^3 + (1/x) = x^3 + x^{-1}$

$$f'(x) = 3x^2 - x^{-2}$$

$$f''(x) = 6x + 2x^{-3}$$

(b) $f(x) = 2x + 1$

$f'(x) = 2$ (a constant function whose value is 2)

$f''(x) = 0$ (the derivative of a constant function is zero)

(c) $f(t) = t^{1/2} + t^{-1/2}$

$f'(t) = \frac{1}{2}t^{-1/2} - \frac{1}{2}t^{-3/2}$

$f''(t) = -\frac{1}{4}t^{-3/2} + \frac{3}{4}t^{-5/2}$ ◾

The first derivative of a function $f(x)$ gives the slope of the graph of $f(x)$ at any point. The second derivative of $f(x)$ gives important additional information about the shape of the curve near any point. We shall examine this subject carefully in the next chapter.

Other Notation for Derivatives Unfortunately, the process of differentiation does not have a standardized notation. Consequently, it is important to become familiar with alternative terminology.

If y is a function of x, say $y = f(x)$, then we may denote the first and second derivatives of this function in several ways.

Prime Notation	$\frac{d}{dx}$ Notation
$f'(x)$	$\frac{d}{dx}f(x)$
y'	$\frac{dy}{dx}$
$f''(x)$	$\frac{d^2}{dx^2}f(x)$
y''	$\frac{d^2y}{dx^2}$

The notation $\frac{d^2}{dx^2}$ is purely symbolic. It reminds us that the second derivative is obtained by differentiating $\frac{d}{dx}f(x)$; that is,

$$f'(x) = \frac{d}{dx}f(x)$$

$$f''(x) = \frac{d}{dx}\left[\frac{d}{dx}f(x)\right].$$

If we evaluate the derivative $f'(x)$ at a specific value of x, say $x = a$, we get a number $f'(a)$ that gives the slope of the curve $y = f(x)$ at the point $(a, f(a))$. Another way of writing $f'(a)$ is

$$\left.\frac{dy}{dx}\right|_{x=a}.$$

If we have a second derivative $f''(x)$, then its value when $x = a$ is written

$$f''(a) \quad \text{or} \quad \left.\frac{d^2y}{dx^2}\right|_{x=a}.$$

EXAMPLE 3   If $y = x^4 - 5x^3 + 7$, find

$$\left.\frac{d^2 y}{dx^2}\right|_{x=3}.$$

Solution

$$\frac{dy}{dx} = \frac{d}{dx}(x^4 - 5x^3 + 7) = 4x^3 - 15x^2$$

$$\frac{d^2 y}{dx^2} = \frac{d}{dx}(4x^3 - 15x^2) = 12x^2 - 30x$$

$$\left.\frac{d^2 y}{dx^2}\right|_{x=3} = 12(3)^2 - 30(3) = 108 - 90 = 18$$ ∎

EXAMPLE 4   If $s = t^3 - 2t^2 + 3t$, find

$$\left.\frac{ds}{dt}\right|_{t=-2} \quad \text{and} \quad \left.\frac{d^2 s}{dt^2}\right|_{t=-2}.$$

Solution

$$\frac{ds}{dt} = \frac{d}{dt}(t^3 - 2t^2 + 3t) = 3t^2 - 4t + 3$$

$$\left.\frac{ds}{dt}\right|_{t=-2} = 3(-2)^2 - 4(-2) + 3 = 12 + 8 + 3 = 23$$

To find the value of the second derivative at $t = -2$, we must first differentiate $\dfrac{ds}{dt}$.

$$\frac{d^2 s}{dt^2} = \frac{d}{dt}(3t^2 - 4t + 3) = 6t - 4$$

$$\left.\frac{d^2 s}{dt^2}\right|_{t=-2} = 6(-2) - 4 = -12 - 4 = -16$$ ∎

The Marginal Concept in Economics

In the previous section, we talked about the derivative as a rate of change. Let us revisit this topic in a setting that arises in economics. For the sake of this discussion, suppose that $C(x)$ is a cost function (the cost of producing x units of a commodity) measured in thousands of dollars.

A problem of interest to economists is to approximate the quantity $C(a + 1) - C(a)$, which is the additional cost that is incurred if the production level is increased by 1 unit from $x = a$ to $x = a + 1$. Note that $C(a + 1) - C(a)$ is also the cost of producing the $(a + 1)$ unit.

To be more specific, if you are given $C(2) = 4$ and $C'(2) = .5$, can you approximate the additional cost $C(3) - C(2)$?

To tackle this problem, we turn to the equation $C'(2) = .5$, which says that the slope of the graph of $C(x)$ is .5 when $x = 2$ (Fig. 2). That is, as the graph of $y = C(x)$ passes through the point $P = (2, 4)$, it is rising at the rate of .5 unit for each unit increase in x. If $C(x)$ were a linear function (equal to its tangent line at P), the additional cost would be exactly .5 thousand dollars, or 500 dollars (Fig. 2). In general, the tangent line approximates the graph, and all we can say is that the additional cost is *approximately* .5 thousand dollars. In symbols,

$$\text{additional cost} = C(3) - C(2) \approx .5 \text{ thousand dollars.}$$

Figure 2. Approximating the additional cost by using the derivative.

Recall that $C'(2) = .5$, and so

$$\text{additional cost} = C(3) - C(2) \approx C'(2).$$

The argument that we used at production level $x = 2$ works at any level $x = a$ and shows that

$$\text{Additional cost} = C(a+1) - C(a) \approx C'(a), \qquad (1)$$

where the symbol $\approx$ is used to indicate that, in general, we have an approximation and not an equality (Fig. 3). Economists refer to the derivative $C'(a)$ as the *marginal cost at production level a* or the *marginal cost of producing a units* of a commodity. In summary, we have the following important result.

Figure 3. Approximating the change in a cost function.

> **DEFINITION   Marginal Cost**   If $C(x)$ is a cost function, then the *marginal cost function* is $C'(x)$. The marginal cost of producing a units, $C'(a)$, is approximately equal to $C(a+1) - C(a)$, which is the additional cost that is incurred when the production level is increased by 1 unit from a to $a+1$.

Before we give an example, let us note that if $C(x)$ is measured in dollars, where x is the number of items, then $C'(x)$, being a rate of change, is measured in dollars per item.

EXAMPLE 5

Marginal cost   Suppose that the cost of producing x items is $C(x) = .005x^3 - .5x^2 + 28x + 300$ dollars, and daily production is 50 items.

(a) What is the extra cost of increasing daily production from 50 to 51 items?

(b) What is the marginal cost when $x = 50$?

Solution (a) The change in cost when daily production is raised from 50 to 51 items is $C(51) - C(50)$, which equals

$$[.005(51)^3 - .5(51)^2 + 28(51) + 300] - [.005(50)^3 - .5(50)^2 + 28(50) + 300]$$
$$= 1090.755 - 1075$$
$$= 15.755.$$

(b) The marginal cost at production level 50 is $C'(50)$.

$$C'(x) = .015x^2 - x + 28$$
$$C'(50) = 15.5$$

Notice that 15.5 is close to the actual cost in part (a) of increasing production by one item.   ■

Our discussion of cost and marginal cost applies as well to other economic quantities such as profit and revenue. In fact, in economics the derivatives are often described by the adjective "marginal." Here are two more definitions of marginal functions and their interpretations.

> **DEFINITION   Marginal Revenue and Marginal Profit**   If $R(x)$ is the revenue generated from the production of x units of a certain commodity and $P(x)$ is the corresponding profit, the *marginal revenue function* is $R'(x)$ and the *marginal profit function* is $P'(x)$.

The marginal revenue of producing a units, $R'(a)$, is an approximation of the additional revenue that results from increasing the production level by 1 unit, from a to $a + 1$:

$$R(a+1) - R(a) \approx R'(a). \tag{2}$$

Similarly, for the marginal profit, we have

$$P(a+1) - P(a) \approx P'(a). \tag{3}$$

The following example illustrates how marginal functions help in the decision-making process in economics.

EXAMPLE 6   **Predicting profits**   Let $R(x)$ denote the revenue (in thousands of dollars) generated from the production of x units of a certain commodity.

(a) Given that $R(4) = 7$ and $R'(4) = -.5$, estimate the additional revenue that results from increasing the production level by 1 unit from $x = 4$ to $x = 5$.

(b) Estimate the revenue generated from the production of 5 units.

(c) Is it profitable to raise the production to 5 units if the cost (in thousands of dollars) to produce x units is given by $C(x) = x + \frac{4}{x+1}$?

Solution   (a) Using formula (2) with $a = 4$, we find

$$R(5) - R(4) \approx R'(4) = -.5 \text{ thousand dollars.}$$

Thus, if the production level is raised to 5 units, the revenue will drop by about $500 dollars.

(b) The revenue that results from the production of 4 units is $R(4) = 7$, or $7000. By part (a), this revenue will drop by about $500 if we raise the production from 4 to 5 units. Thus, at production level $x = 5$, the revenue will be approximately $7000 - 500 = \$6500$.

(c) Let $P(x)$ denote the profit (in thousands of dollars) that results from the production of x units. We have $P(x) = R(x) - C(x)$. At production level $x = 5$, we have

$$C(5) = 5 + \frac{4}{5+1} = \frac{17}{3} \approx 5.667 \text{ thousand dollars or \$5667;}$$

also, from part (b), the revenue is $R(5) \approx \$6500$. So the profit at production level $x = 5$ is $P(5) \approx 6500 - 5667 = \$833$. So it is still profitable to raise the production level to $x = 5$, even though the cost will rise and the revenue will fall.   ∎

INCORPORATING TECHNOLOGY

Although functions can be specified (and differentiated) in graphing calculators with letters other than X, only functions in X can be graphed. Therefore, we will always use X as the variable.

Figure 4 contains the graphs of $f(x) = x^2$ and its first and second derivatives. These three graphs can be obtained with any of the function editor settings in Figs. 5(a) and 5(b).   ∎

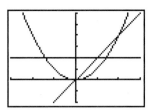

Figure 4.   Graph of x^2 and its first two derivatives.

(a) TI-83

(b) TI-83

Figure 5

Practice Problems 1.7

1. Let $f(t) = t + (1/t)$. Find $f''(2)$.

2. Differentiate $g(r) = 2\pi rh$.

EXERCISES 1.7

Find the first derivatives.

1. $f(t) = (t^2 + 1)^5$

2. $f(P) = P^3 + 3P^2 - 7P + 2$

3. $v(t) = 4t^2 + 11\sqrt{t} + 1$

4. $g(y) = y^2 - 2y + 4$

5. $y = T^5 - 4T^4 + 3T^2 - T - 1$

6. $x = 16t^2 + 45t + 10$

7. Find $\dfrac{d}{dP}(3P^2 - \frac{1}{2}P + 1)$.   **8.** Find $\dfrac{d}{ds}\sqrt{s^2 + 1}$.

9. Find $\dfrac{d}{dt}(a^2t^2 + b^2t + c^2)$.   **10.** Find $\dfrac{d}{dP}(T^2 + 3P)^3$.

Find the first and second derivatives.

11. $y = x + 1$ **12.** $y = (x + 12)^3$

13. $y = \sqrt{x}$ **14.** $y = 100$

15. $y = \sqrt{x + 1}$ **16.** $v = 2t^2 + 3t + 11$

17. $f(r) = \pi r^2$ **18.** $y = \pi^2 + 3x^2$

19. $f(P) = (3P + 1)^5$ **20.** $T = (1 + 2t)^2 + t^3$

Compute the following.

21. $\dfrac{d}{dx}(2x + 7)^2\Big|_{x=1}$ **22.** $\dfrac{d}{dt}\left(t^2 + \dfrac{1}{t+1}\right)\Big|_{t=0}$

23. $\dfrac{d}{dz}(z^2 + 2z + 1)^7\Big|_{z=-1}$ **24.** $\dfrac{d^2}{dx^2}(3x^4 + 4x^2)\Big|_{x=2}$

25. $\dfrac{d^2}{dx^2}(3x^3 - x^2 + 7x - 1)\Big|_{x=2}$

26. $\dfrac{d}{dx}\left(\dfrac{dy}{dx}\right)\Big|_{x=1}$, where $y = x^3 + 2x - 11$

27. $f'(1)$ and $f''(1)$, when $f(t) = \dfrac{1}{2 + t}$

28. $g'(0)$ and $g''(0)$, when $g(T) = (T + 2)^3$

29. $\dfrac{d}{dt}\left(\dfrac{dv}{dt}\right)\Big|_{t=32}$, where $v = 20t + 12$

30. $\dfrac{d}{dt}\left(\dfrac{dv}{dt}\right)$, where $v = 2t^2 + \dfrac{1}{t+1}$

31. A company finds that the revenue R generated by spending x dollars on advertising is given by $R = 1000 + 80x - .02x^2$, for $0 \le x \le 2000$. Find $\dfrac{dR}{dx}\Big|_{x=1500}$.

32. A supermarket finds that its average daily volume of business V (in thousands of dollars) and the number of hours t that the store is open for business each day are approximately related by the formula

$$V = 20\left(1 - \dfrac{100}{100 + t^2}\right), \qquad 0 \le t \le 24.$$

Find $\dfrac{dV}{dt}\Big|_{t=10}$.

33. The *third derivative* of a function $f(x)$ is the derivative of the second derivative $f''(x)$ and is denoted by $f'''(x)$. Compute $f'''(x)$ for the following functions:

(a) $f(x) = x^5 - x^4 + 3x$ (b) $f(x) = 4x^{5/2}$

34. Compute the third derivatives of the following functions:

(a) $f(t) = t^{10}$ (b) $f(z) = \dfrac{1}{z + 5}$

35. If $s = Tx^2 + 3xP + T^2$, find:

(a) $\dfrac{ds}{dx}$ (b) $\dfrac{ds}{dP}$ (c) $\dfrac{ds}{dT}$

36. If $s = 7x^2y\sqrt{z}$, find:

(a) $\dfrac{d^2s}{dx^2}$ (b) $\dfrac{d^2s}{dy^2}$ (c) $\dfrac{ds}{dz}$

37. Let $C(x)$ be the cost (in dollars) of manufacturing x bicycles per day in a certain factory. Interpret $C(50) = 5000$ and $C'(50) = 45$.

38. Estimate the cost of manufacturing 51 bicycles per day in Exercise 37.

39. The revenue from producing (and selling) x units of a product is given by $R(x) = 3x - .01x^2$ dollars.

  (a) Find the marginal revenue at a production level of 20.

  (b) Find the production levels where the revenue is $200.

40. Let $P(x)$ be the profit from producing (and selling) x units of goods. Match each question with the proper solution.

Questions

  A. What is the profit from producing 1000 units of goods?

  B. At what level of production will the marginal profit be 1000 dollars?

  C. What is the marginal profit from producing 1000 units of goods?

  D. For what level of production will the profit be 1000 dollars?

Solutions

  a. Compute $P'(1000)$.

  b. Find a value of a for which $P'(a) = 1000$.

  c. Set $P(x) = 1000$ and solve for x.

  d. Compute $P(1000)$.

41. Let $R(x)$ denote the revenue (in thousands of dollars) generated from the production of x units of computer chips per day, where each unit consists of 100 chips.

  (a) Represent the following statement by equations involving R or R': When 1200 chips are produced per day, the revenue is $22,000 and the marginal revenue is $.75 per chip.

  (b) If the marginal cost of producing 1200 chips is $1.5 per chip, what is the marginal profit at this production level?

42. Refer to Exercise 41. Is it profitable to produce 1300 chips per day if the cost of producing 1200 chips per day is $14,000?

Technology Exercises

43. For the given function, simultaneously graph the functions $f(x)$, $f'(x)$, and $f''(x)$ with the specified window setting. *Note:* Since we have not yet learned how to differentiate the given function, you must use your graphing utility's differentiation command to define the derivatives.

$$f(x) = \frac{x}{1 + x^2}, \ [-4, 4] \text{ by } [-2, 2]$$

44. Consider the cost function of Example 5.

  (a) Graph $C(x)$ in the window $[0, 60]$ by $[-300, 1260]$.

  (b) For what level of production will the cost be $535?

  (c) For what level of production will the marginal cost be $14?

Solutions to Practice Problems 1.7

1. $f(t) = t + t^{-1}$

$f'(t) = 1 + (-1)t^{(-1)-1} = 1 - t^{-2}$

$f''(t) = -(-2)t^{(-2)-1} = 2t^{-3} = \dfrac{2}{t^3}$

Therefore,

$$f''(2) = \frac{2}{2^3} = \frac{1}{4}.$$

[*Note:* It is essential first to compute the function $f''(t)$ and *then* to evaluate the function at $t = 2$.]

2. The expression $2\pi rh$ contains two numbers, 2 and π, and two letters, r and h. The notation $g(r)$ tells us that the expression $2\pi rh$ is to be regarded as a function of r. Therefore, h, and hence $2\pi h$, is to be treated as a constant, and differentiation is done with respect to the variable r. That is,

$$g(r) = (2\pi h)r$$
$$g'(r) = 2\pi h.$$

1.8 The Derivative as a Rate of Change

As we saw in previous sections, an important interpretation of the slope of a function at a point is as a rate of change. In this section, we reexamine this interpretation and discuss some additional applications in which it proves useful. The first step is to understand what is meant by "average rate of change" of a function $f(x)$.

Consider a function $y = f(x)$ defined on the interval $a \leq x \leq b$. The average rate of change of $f(x)$ over this interval is the change in $f(x)$ divided by the length of the interval. That is,

$$\left[\begin{array}{c} \text{average rate of change of } f(x) \\ \text{over the interval } a \leq x \leq b \end{array} \right] = \frac{f(b) - f(a)}{b - a}.$$

EXAMPLE 1   If $f(x) = x^2$, calculate the average rate of change of $f(x)$ over the following intervals:

(a) $1 \leq x \leq 2$ **(b)** $1 \leq x \leq 1.1$ **(c)** $1 \leq x \leq 1.01$

Solution **(a)** The average rate of change over the interval $1 \leq x \leq 2$ is

$$\frac{2^2 - 1^2}{2 - 1} = \frac{3}{1} = 3.$$

(b) The average rate of change over the interval $1 \leq x \leq 1.1$ is

$$\frac{1.1^2 - 1^2}{1.1 - 1} = \frac{.21}{.1} = 2.1.$$

(c) The average rate of change over the interval $1 \leq x \leq 1.01$ is

$$\frac{1.01^2 - 1^2}{1.01 - 1} = \frac{.0201}{.01} = 2.01. \qquad \blacksquare$$

In the special case where b is $a + h$, the value of $b - a$ is $(a + h) - a$, or h, and the average rate of change of the function over the interval is the familiar difference quotient

$$\frac{f(a + h) - f(a)}{h}.$$

Geometrically, this quotient is the slope of the secant line in Fig. 1. Recall that as h approaches 0, the slope of the secant line approaches the slope of the tangent line. Thus the average rate of change approaches $f'(a)$. For this reason, $f'(a)$ is called

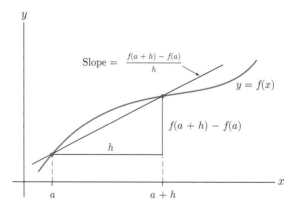

Figure 1. Average rate of change is the slope of the secant line.

the *(instantaneous) rate of change* of $f(x)$ exactly at the point where $x = a$. From now on, unless we explicitly use the word "average" when we refer to the rate of change of a function, we mean the "instantaneous" rate of change.

The derivative $f'(a)$ measures the rate of change of $f(x)$ at $x = a$.

EXAMPLE 2

Consider the function $f(x) = x^2$ of Example 1. Calculate the rate of change of $f(x)$ at $x = 1$.

Solution The rate of change of $f(x)$ at $x = 1$ is equal to $f'(1)$. We have

$$f'(x) = 2x$$
$$f'(1) = 2 \cdot 1 = 2.$$

That is, the rate of change is 2 units per unit change in x. Notice how the average rates of change in Example 1 approach the (instantaneous) rate of change as the intervals beginning at $x = 1$ shrink. ∎

EXAMPLE 3

The function $f(t)$ in Fig. 2 gives the population of the United States t years after the beginning of 1800. The figure also shows the tangent line through the point $(40, 17)$.

(a) What was the average rate of growth of the United States population from 1840 to 1870?

(b) How fast was the population growing in 1840?

(c) Was the population growing faster in 1810 or 1880?

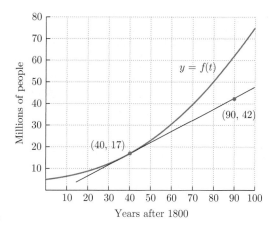

Figure 2. United States population from 1800 to 1900.

Solution (a) Since 1870 is 70 years after 1800 and 1840 is 40 years after 1800, the question asks for the average rate of change of $f(t)$ over the interval $40 \leq t \leq 70$. This value is

$$\frac{f(70) - f(40)}{70 - 40} = \frac{40 - 17}{30} = \frac{23}{30} \approx .77.$$

Therefore, from 1840 to 1870 the population grew at the average rate of about .77 million, or 770,000 people per year.

(b) The rate of growth of $f(t)$ at $t = 40$ is $f'(40)$, that is, the slope of the tangent line at $t = 40$. Since $(40, 17)$ and $(90, 42)$ are two points on the tangent line, the slope of the tangent line is

$$\frac{42 - 17}{90 - 40} = \frac{25}{50} = .5.$$

Therefore, in 1840 the population was growing at the rate of .5 million, or 500,000 people per year.

(c) The graph is clearly steeper in 1880 than in 1810. Therefore, the population was growing faster in 1880 than in 1810. ∎

Velocity and Acceleration   An everyday illustration of rate of change is given by the velocity of a moving object. Suppose that we are driving a car along a straight road and at each time t we let $s(t)$ be our position on the road, measured from some convenient reference point. See Fig. 3, where distances are positive to the right of the car. For the moment we shall assume that we are proceeding in only the positive direction along the road.

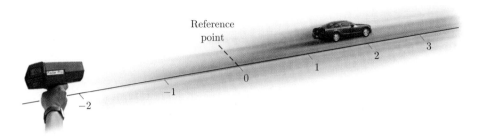

Figure 3.   Position of a car traveling on a straight road.

At any instant, the car's speedometer tells us how fast we are moving, that is, how fast our position $s(t)$ is changing. To show how the speedometer reading is related to our calculus concept of a derivative, let us examine what is happening at a specific time, say $t = 2$. Consider a short time interval of duration h from $t = 2$ to $t = 2 + h$. Our car will move from position $s(2)$ to position $s(2 + h)$, a distance of $s(2 + h) - s(2)$. Thus the *average velocity from $t = 2$ to $t = 2 + h$* is

$$\frac{[\text{distance traveled}]}{[\text{time elapsed}]} = \frac{s(2 + h) - s(2)}{h}. \tag{1}$$

If the car is traveling at a steady speed during this time period, the speedometer reading will equal the average velocity in formula (1).

From our discussion in Section 1.3, the ratio (1) approaches the derivative $s'(2)$ as h approaches zero. For this reason we call $s'(2)$ *the* (instantaneous) *velocity at* $t = 2$. This number will agree with the speedometer reading at $t = 2$ because, when h is very small, the car's speed will be nearly steady over the time interval from $t = 2$ to $t = 2 + h$; so the average velocity over this time interval will be nearly the same as the speedometer reading at $t = 2$.

The reasoning used for $t = 2$ holds for an arbitrary t as well. Thus the following definition makes sense:

> **DEFINITION**   If $s(t)$ denotes the position function of an object moving in a straight line, then the velocity $v(t)$ of the object at time t is given by
>
> $$v(t) = s'(t).$$

In our discussion we assumed that the car moved in the positive direction. If the car moves in the opposite direction, the ratio (1) and the limiting value $s'(2)$ will be negative. So we interpret negative velocity as movement in the negative direction along the road.

The derivative of the velocity function $v(t)$ is called the acceleration function and is often written as $a(t)$:

$$a(t) = v'(t).$$

Because $v'(t)$ measures the rate of change of the velocity $v(t)$, this use of the word "acceleration" agrees with our common usage in connection with automobiles. Note that since $v(t) = s'(t)$, the acceleration is actually the second derivative of the position function $s(t)$,

$$a(t) = s''(t).$$

EXAMPLE 4   When a ball is thrown straight up into the air, its position may be measured as the vertical distance from the ground. Regard "up" as the positive direction, and let $s(t)$ be the height of the ball in feet after t seconds. Suppose that $s(t) = -16t^2 + 128t + 5$. Figure 4 contains a graph of the function $s(t)$. *Note:* The graph does *not* show the path of the ball. The ball is rising *straight* up and then falling *straight* down.

(a) What is the velocity after 2 seconds?

(b) What is the acceleration after 2 seconds?

(c) At what time is the velocity -32 feet per second? (The negative sign indicates that the ball's height is decreasing; that is, the ball is falling.)

(d) When is the ball at a height of 117 feet?

Solution **(a)** The velocity is the rate of change of the height function, so

$$v(t) = s'(t) = -32t + 128.$$

The velocity when $t = 2$ is $v(2) = -32(2) + 128 = 64$ feet per second.

(b) $a(t) = v'(t) = -32$. The acceleration is -32 feet per second per second for all t. This constant acceleration is due to the downward (and therefore negative) force of gravity.

(c) Since the velocity is given and the time is unknown, we set $v(t) = -32$ and solve for t:

$$-32t + 128 = -32$$
$$-32t = -160$$
$$t = 5.$$

The velocity is -32 feet per second when t is 5 seconds.

(d) The question here involves the height function, not the velocity. Since the height is given and the time is unknown, we set $s(t) = 117$ and solve for t:

$$-16t^2 + 128t + 5 = 117$$
$$-16(t^2 - 8t + 7) = 0$$
$$-16(t - 1)(t - 7) = 0.$$

The ball is at a height of 117 feet once on the way up (when $t = 1$ second) and once on the way down (when $t = 7$ seconds). ∎

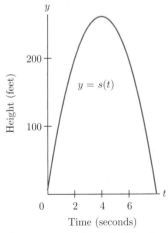

Figure 4. Height of ball in the air.

Approximating the Change in a Function   Consider the function $f(x)$ near $x = a$. As we just saw, the average rate of change of f over a small interval of length h is approximately equal to the instantaneous change at an endpoint of the interval. That is,

$$\frac{f(a+h) - f(a)}{h} \approx f'(a).$$

Multiplying both sides of this approximation by h, we have

$$f(a+h) - f(a) \approx f'(a) \cdot h. \tag{2}$$

If x changes from a to $a + h$, then the change in the value of the function $f(x)$ is approximately $f'(a)$ times the change h in the value of x. This result holds for both positive and negative values of h. In applications, the right side of formula (2) is calculated and used to estimate the left side.

Figure 5 contains a geometric interpretation of (2). Given the small change h in x, the quantity $f'(a) \cdot h$ gives the corresponding change in y *along the tangent line* at $(a, f(a))$. In contrast, the quantity $f(a+h) - f(a)$ gives the change in y *along the curve* $y = f(x)$. When h is small, $f'(a) \cdot h$ is a good approximation to the change in $f(x)$.

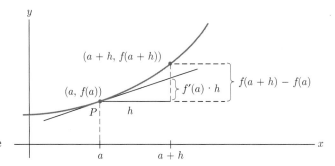

Figure 5.   Change in y along the tangent line and along the graph of $y = f(x)$.

Before we consider an application, it is instructive to present an alternative derivation of (2). From Section 1.3, the equation of the tangent line at P is

$$y - f(a) = f'(a)(x - a). \tag{3}$$

Now the tangent line approximates the graph of f near P, so if in formula (3) we replace y (the value of the tangent line) by $f(x)$, we get

$$f(x) - f(a) \approx f'(a)(x - a), \tag{4}$$

and (2) follows from (4) upon setting $h = x - a$ and $x = a + h$.

EXAMPLE 5   Let the production function $p(x)$ give the number of units of goods produced when employing x units of labor. If 5000 units of labor are currently employed, $p(5000) = 300$, and $p'(5000) = 2$.

(a) Interpret $p(5000) = 300$.

(b) Interpret $p'(5000) = 2$.

(c) Estimate the number of *additional* units of goods produced when x is increased from 5000 to $5000\frac{1}{2}$ units of labor.

(d) Estimate the *change* in the number of units of goods produced when x is decreased from 5000 to 4999 units of labor.

Solution   **(a)** When 5000 units of labor are employed, 300 units of goods will be produced.

(b) If 5000 units of labor are currently employed and we consider adding more labor, productivity will increase at approximately the rate of 2 units of goods for each additional unit of labor.

(c) Here $h = \frac{1}{2}$. By formula (2), the change in $p(x)$ will be approximately

$$p'(5000) \cdot \tfrac{1}{2} = 2 \cdot \tfrac{1}{2} = 1.$$

About 1 additional unit will be produced. Therefore, about 301 units of goods will be produced when $5000\frac{1}{2}$ units of labor are employed.

(d) Here $h = -1$, since the amount of labor is reduced. The change in $p(x)$ will be approximately

$$p'(5000) \cdot (-1) = 2 \cdot (-1) = -2.$$

About two fewer units (that is, 298) of goods will be produced.   ■

To simplify the use of formula (2) in marginal cost analysis, let us rewrite it with a cost function, as follows:

$$C(a + h) - C(a) \approx C'(a) \cdot h. \tag{5}$$

Taking $h = 1$, we obtain the familiar formula (1) of the previous section,

$$C(a + 1) - C(a) \approx C'(a). \tag{6}$$

Thus for a whole unit increase in production ($h = 1$), the change in cost is approximately equal to the marginal cost, but for an increase by h units, we must modify the marginal cost proportionally.

EXAMPLE 6

Marginal cost   The total cost in thousands of dollars of producing x units of a certain commodity is $C(x) = 6x^2 + 2x + 10$.

(a) Find the marginal cost function.

(b) Find the cost and marginal cost when 10 units are produced.

(c) Use marginal cost to approximate the cost of producing the 11th unit.

(d) Use marginal cost to estimate the additional cost incurred if the production level is increased from 10 to 10.5 units.

Solution   **(a)** The marginal cost function is the derivative of the cost function:

$$C'(x) = 12x + 2 \text{ (thousand dollars per unit)}.$$

(b) When 10 units are produced, the cost is

$$C(10) = 6(10^2) + 2(10) + 10 = 630 \text{ thousand dollars},$$

and the marginal cost is $C'(10) = 12(10) + 2 = 122$ thousand dollars per unit.

(c) The cost of producing the 11th unit is the difference in cost as x varies from 10 to 11 or $C(11) - C(10)$. By formula (6), this difference can be approximated by the marginal cost $C'(10) = 122$. Hence to produce the 11th unit it will cost approximately 122 thousand dollars.

(d) The additional cost when we raise the production level by half a unit from 10 to 10.5 is $C(10.5) - C(10)$. According to formula (5), with $a = 10$ and $h = .5$,

$$C(10.5) - C(10) \approx C'(10) \cdot (.5) = (122)(.5) = 61 \text{ thousand dollars.}$$ ■

Units for Rate of Change   Table 1 shows the units appearing in several examples from this section. In general,

$$[\text{unit of measure for } f'(x)]$$
$$= [\text{unit of measure for } f(x)] \text{ per } [\text{unit of measure for } x].$$

TABLE 1   Units of Measure

Example	Unit for $f(t)$ or $f(x)$	Unit for t or x	Unit for $f'(t)$ or $f'(x)$
U.S. population	millions of people	year	millions of people per year
Ball in air	feet	second	feet per second
Cost function	dollars	item	dollars per item

Practice Problems 1.8

Let $f(t)$ be the temperature (in degrees Celsius) of a liquid at time t (in hours). The rate of temperature change at time a has the value $f'(a)$. Listed next are typical questions about $f(t)$ and its slope at various points. Match each question with the proper method of solution.

Questions

1. What is the temperature of the liquid after 6 hours (that is, when $t = 6$)?

2. When is the temperature rising at the rate of 6 degrees per hour?

3. By how many degrees did the temperature rise during the first 6 hours?

4. When is the liquid's temperature only 6 degrees?

5. How fast is the temperature of the liquid changing after 6 hours?

6. What is the average rate of increase in the temperature during the first 6 hours?

Methods of Solution

(a) Compute $f(6)$.

(b) Set $f(t) = 6$ and solve for t.

(c) Compute $[f(6) - f(0)]/6$.

(d) Compute $f'(6)$.

(e) Find a value of a for which $f'(a) = 6$.

(f) Compute $f(6) - f(0)$.

EXERCISES 1.8

1. Suppose that $f(x) = 4x^2$.
   (a) What is the average rate of change of $f(x)$ over each of the intervals 1 to 2, 1 to 1.5, and 1 to 1.1?
   (b) What is the (instantaneous) rate of change of $f(x)$ when $x = 1$?

2. Suppose that $f(x) = -6/x$.
   (a) What is the average rate of change of $f(x)$ over each of the intervals 1 to 2, 1 to 1.5, and 1 to 1.2?
   (b) What is the (instantaneous) rate of change of $f(x)$ when $x = 1$?

3. Suppose that $f(t) = t^2 + 3t - 7$.
   (a) What is the average rate of change of $f(t)$ over the interval 5 to 6?
   (b) What is the (instantaneous) rate of change of $f(t)$ when $t = 5$?

4. Suppose that $f(t) = 3t + 2 - \dfrac{12}{t}$.
   (a) What is the average rate of change of $f(t)$ over the interval 2 to 3?
   (b) What is the (instantaneous) rate of change of $f(t)$ when $t = 2$?

5. **U.S. T-bill**   Refer to Fig. 6, where $f(t)$ is the percentage yield (interest rate) on a 3-month T-bill (U.S. Treasury bill) t years after January 1, 1980.
   (a) What was the average rate of change in the yield from January 1, 1981, to January 1, 1985?
   (b) How fast was the percentage yield rising on January 1, 1989?
   (c) Was the percentage yield rising faster on January 1, 1980, or January 1, 1989?

Figure 6.   Percentage yield on 3-month T-bill (1980–1992).

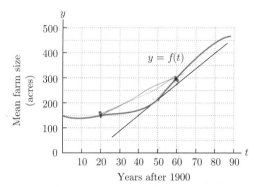

Figure 7.   Mean size of farms from 1900 to 1990.

6. Refer to Fig. 7, where $f(t)$ is the mean (average) size of a farm (in acres) t years after January 1, 1900.

   (a) What was the average rate of growth in the mean size of farms from 1920 to 1960?

   (b) How fast was the mean farm size rising on January 1, 1950?

   (c) Was the mean farm size rising faster on January 1, 1960, or January 1, 1980?

7. An object moving in a straight line travels $s(t)$ kilometers in t hours, where $s(t) = 2t^2 + 4t$.

   (a) What is the object's velocity when $t = 6$?

   (b) How far has the object traveled in 6 hours?

   (c) When is the object traveling at the rate of 6 kilometers per hour?

8. After an advertising campaign, the sales of a product often increase and then decrease. Suppose that t days after the end of the advertising, the daily sales are $f(t) = -3t^2 + 32t + 100$ units. What is the average rate of growth in sales during the fourth day, that is, from time $t = 3$ to $t = 4$? At what (instantaneous) rate are the sales changing when $t = 2$?

9. An analysis of the daily output of a factory assembly line shows that about $60t + t^2 - \frac{1}{12}t^3$ units are produced after t hours of work, $0 \le t \le 8$. What is the rate of production (in units per hour) when $t = 2$?

10. Liquid is pouring into a large vat. After t hours, there are $5t + \sqrt{t}$ gallons in the vat. At what rate is the liquid flowing into the vat (in gallons per hour) when $t = 4$?

11. The position of a particle moving on a line is given by $s(t) = 2t^3 - 21t^2 + 60t$, $t \ge 0$, where t is measured in seconds and s in feet.

   (a) What is the velocity after 3 seconds and after 6 seconds?

   (b) When is the particle moving in the positive direction?

   (c) Find the total distance traveled by the particle during the first 7 seconds.

12. Refer to Fig. 8, where $s(t)$ represents the position of a car moving in a straight line.

   (a) Was the car going faster at A or at B?

   (b) Was the car's acceleration positive or negative at B?

   (c) What happened to the car's velocity at C?

   (d) In which direction was the car moving at D?

   (e) What happened at E?

   (f) What happened after F?

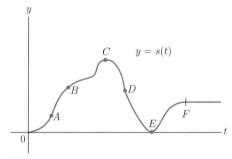

Figure 8

13. A toy rocket fired straight up into the air has height $s(t) = 160t - 16t^2$ feet after t seconds.

   (a) What is the rocket's initial velocity (when $t = 0$)?

   (b) What is the velocity after 2 seconds?

   (c) What is the acceleration when $t = 3$?

   (d) At what time will the rocket hit the ground?

   (e) At what velocity will the rocket be traveling just as it smashes into the ground?

14. A helicopter is rising straight up in the air. Its distance from the ground t seconds after takeoff is $s(t)$ feet, where $s(t) = t^2 + t$.

    (a) How long will it take for the helicopter to rise 20 feet?

    (b) Find the velocity and the acceleration of the helicopter when it is 20 feet above the ground.

15. Let $s(t)$ be the height (in feet) after t seconds of a ball thrown straight up into the air. Match each question with the proper solution.

Questions

    A. What is the velocity of the ball after 3 seconds?

    B. When is the velocity 3 feet per second?

    C. What is the average velocity during the first 3 seconds?

    D. When is the ball 3 feet above the ground?

    E. When does the ball hit the ground?

    F. How high is the ball after 3 seconds?

    G. How far did the ball travel during the first 3 seconds?

Solutions

    a. Set $s(t) = 0$ and solve for t.

    b. Compute $s'(3)$.

    c. Compute $s(3)$.

    d. Set $s'(t) = 3$ and solve for t.

    e. Find a value of a for which $s(a) = 3$.

    f. Compute $[s(3) - s(0)]/3$.

    g. Compute $s(3) - s(0)$.

16. Table 2 gives a car's trip-meter reading (in miles) at 1 hour into a trip and at several nearby times. What is the average speed during the time interval from 1 to 1.05 hours? Estimate the speed at time 1 hour into the trip.

TABLE 2 Trip-Meter Readings at Several Times

Time	Trip Meter
.96	43.2
.97	43.7
.98	44.2
.99	44.6
1.00	45.0
1.01	45.4
1.02	45.8
1.03	46.3
1.04	46.8
1.05	47.4

17. A particle is moving in a straight line in such a way that its position at time t (in seconds) is $s(t) = t^2 + 3t + 2$ feet to the right of a reference point, for $t \geq 0$.

    (a) What is the velocity of the object when the time is 6 seconds?

    (b) Is the object moving toward the reference point when $t = 6$? Explain your answer.

    (c) What is the object's velocity when the object is 6 feet from the reference point?

18. A car is traveling from New York to Boston and is partway between the two cities. Let $s(t)$ be the distance from New York during the next minute. Match each behavior with the corresponding graph of $s(t)$ in Fig. 9.

    (a) The car travels at a steady speed.

    (b) The car is stopped.

    (c) The car is backing up.

    (d) The car is accelerating.

    (e) The car is decelerating.

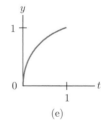

Figure 9. Possible graphs of $s(t)$.

19. If $f(100) = 5000$ and $f'(100) = 10$, estimate each of the following:

    (a) $f(101)$ **(b)** $f(100.5)$ **(c)** $f(99)$

    (d) $f(98)$ **(e)** $f(99.75)$

20. If $f(25) = 10$ and $f'(25) = -2$, estimate each of the following:

    (a) $f(27)$ **(b)** $f(26)$ **(c)** $f(25.25)$

    (d) $f(24)$ **(e)** $f(23.5)$

21. Let $f(t)$ be the temperature of a cup of coffee t minutes after it has been poured. Interpret $f(4) = 120$ and $f'(4) = -5$. Estimate the temperature of the coffee after 4 minutes and 6 seconds, that is, after 4.1 minutes.

22. Suppose that 5 mg of a drug is injected into the bloodstream. Let $f(t)$ be the amount present in the bloodstream after t hours. Interpret $f(3) = 2$ and $f'(3) = -.5$. Estimate the number of milligrams of the drug in the bloodstream after $3\frac{1}{2}$ hours.

23. Price affects sales Let $f(p)$ be the number of cars sold when the price is p dollars per car. Interpret the statements $f(10,000) = 200,000$ and $f'(10,000) = -3$.

24. Advertising affects sales Let $f(x)$ be the number of toys sold when x dollars are spent on advertising. Interpret the statements $f(100,000) = 3,000,000$ and $f'(100,000) = 30$.

25. Let $f(x)$ be the number (in thousands) of computers sold when the price is x hundred dollars per computer. Interpret the statements $f(12) = 60$ and $f'(12) = -2$. Then estimate the number of computers sold if the price is set at \$1250 per computer.

26. Let $C(x)$ be the cost (in dollars) of manufacturing x radios. Interpret the statements $C(2000) = 50,000$ and $C'(2000) = 10$. Estimate the cost of manufacturing 1998 radios.

27. Let $P(x)$ be the profit (in dollars) from manufacturing and selling x luxury cars. Interpret $P(100) = 90,000$ and $P'(100) = 1200$. Estimate the profit from manufacturing and selling 99 cars.

28. Let $f(x)$ be the value in dollars of one share of a company x days since the company went public.

 (a) Interpret the statements $f(100) = 16$ and $f'(100) = .25$.

 (b) Estimate the value of one share on the 101st day since the company went public.

29. Marginal Cost Analysis Consider the cost function $C(x) = 6x^2 + 14x + 18$ (thousand dollars).

 (a) What is the marginal cost at production level $x = 5$?

 (b) Estimate the cost of raising the production level from $x = 5$ to $x = 5.25$.

 (c) Let $R(x) = -x^2 + 37x + 38$ denote the revenue in thousands of dollars generated from the production of x units. What is the break-even point? (Recall that the break-even point is when revenue is equal to cost.)

 (d) Compute and compare the marginal revenue and marginal cost at the break-even point. Should the company increase production beyond the break-even point? Justify your answer using marginals.

30. Estimate how much the function

$$f(x) = \frac{1}{1 + x^2}$$

will change if x decreases from 1 to .9.

31. National health expenditures (in billions of dollars) from 1980 to 1998 are given by the function $f(t)$ in Fig. 10.

 (a) How much money was spent in 1987?

 (b) Approximately how fast were expenditures rising in 1987?

 (c) When did expenditures reach 1 trillion dollars?

 (d) When were expenditures rising at the rate of \$100 billion per year?

Figure 10

32. In an 8-second test run, a vehicle accelerates for several seconds and then decelerates. The function $s(t)$ gives the number of feet traveled after t seconds and is graphed in Fig. 11.

 (a) How far has the vehicle traveled after 3.5 seconds?

 (b) What is the velocity after 2 seconds?

 (c) What is the acceleration after 1 second?

 (d) When will the vehicle have traveled 120 feet?

 (e) When, during the second part of the test run, will the vehicle be traveling at the rate of 20 feet per second?

 (f) What is the greatest velocity? At what time is this greatest velocity reached? How far has the vehicle traveled at this time?

$y = s(t)$

$y = s'(t)$

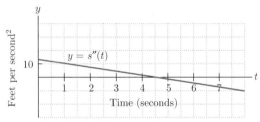

$y = s''(t)$

Figure 11

Technology Exercises

33. In a psychology experiment,[†] people improved their ability to recognize common verbal and semantic information with practice. Their judgment time after t days of practice was $f(t) = .36 + .77(t - .5)^{-.36}$ seconds.

    (a) Display the graphs of $f(t)$ and $f'(t)$ in the window $[.5, 6]$ by $[-3, 3]$. Use these graphs to answer the following questions.

    (b) What was the judgment time after 4 days of practice?

    (c) After how many days of practice was the judgment time about .8 second?

    (d) After 4 days of practice, at what rate was judgment time changing with respect to days of practice?

    (e) After how many days was judgment time changing at the rate of $-.08$ second per day of practice?

34. A ball thrown straight up into the air has height $s(t) = 102t - 16t^2$ feet after t seconds.

    (a) Display the graphs of $s(t)$ and $s'(t)$ in the window $[0, 7]$ by $[-100, 200]$. Use these graphs to answer the remaining questions.

    (b) How high is the ball after 2 seconds?

    (c) When, during descent, is the height 110 feet?

    (d) What is the velocity after 6 seconds?

    (e) When is the velocity 70 feet per second?

    (f) How fast is the ball traveling when it hits the ground?

Solutions to Practice Problems 1.8

1. Method (a). The question involves $f(t)$, the temperature at time t. Since the time is given, compute $f(6)$.

2. Method (e). The question involves $f'(t)$, the rate of change of temperature. The interrogative "when" indicates that the time is unknown. Set $f'(t) = 6$ and solve for t.

3. Method (f). The question asks for the change in the value of the function from time 0 to time 6, $f(6) - f(0)$.

4. Method (b). The question involves $f(t)$, and the time is unknown. Set $f(t) = 6$ and solve for t.

5. Method (d). The question involves $f'(t)$, and the time is given. Compute $f'(6)$.

6. Method (c). The question asks for the average rate of change of the function during the time interval 0 to 6, $[f(6) - f(0)]/6$.

[†]John R. Anderson, "Automaticity and the ACT Theory," *American Journal of Psychology*, 105:2 (Summer 1992), 165–180.

REVIEW OF FUNDAMENTAL CONCEPTS

1. Define the slope of a line and give a physical description.

2. What is the point–slope form of the equation of a line?

3. Describe how to find an equation for a line when you know the coordinates of two points on the graph of a line.

4. What can you say about the slopes of parallel lines? Perpendicular lines?

5. Give a physical description of what is meant by the slope of $f(x)$ at the point $(2, f(2))$.

6. What does $f'(2)$ represent?

7. Explain why the derivative of a constant function is 0.

8. State the power rule, constant-multiple rule, and sum rule, and give an example of each.

9. Explain how to calculate $f'(2)$ as the limit of slopes of secant lines through the point $(2, f(2))$.

10. In your own words, explain the meaning of $\lim_{x \to 2} f(x) = 3$. Give an example of a function with this property.

11. Give the limit definition of $f'(2)$, that is, the slope of $f(x)$ at the point $(2, f(2))$.

12. In your own words, explain the meaning of $\lim_{x \to \infty} f(x) = 3$. Give an example of such a function $f(x)$. Do the same for $\lim_{x \to -\infty} f(x) = 3$.

13. In your own words, explain the meaning of "$f(x)$ is continuous at $x = 2$." Give an example of a function $f(x)$ that is not continuous at $x = 2$.

14. In your own words, explain the meaning of "$f(x)$ is differentiable at $x = 2$." Give an example of a function $f(x)$ that is not differentiable at $x = 2$.

15. State the general power rule and give an example.

16. Give two different notations for the first derivative of $f(x)$ at $x = 2$. Do the same for the second derivative.

17. What is meant by the average rate of change of a function over an interval?

18. How is an (instantaneous) rate of change related to average rates of change?

19. Explain the relationship between derivatives and velocity and acceleration.

20. What expression involving a derivative gives an approximation to $f(a + h) - f(a)$?

21. Describe marginal cost in your own words.

22. How do you determine the proper units for a rate of change? Give an example.

SUPPLEMENTARY EXERCISES

Find the equation and sketch the graph of the following lines.

1. With slope -2, y-intercept $(0, 3)$

2. With slope $\frac{3}{4}$, y-intercept $(0, -1)$

3. Through $(2, 0)$, with slope 5

4. Through $(1, 4)$, with slope $-\frac{1}{3}$

5. Parallel to $y = -2x$, passing through $(3, 5)$

6. Parallel to $-2x + 3y = 6$, passing through $(0, 1)$

7. Through $(-1, 4)$ and $(3, 7)$

8. Through $(2, 1)$ and $(5, 1)$

9. Perpendicular to $y = 3x + 4$, passing through $(1, 2)$

10. Perpendicular to $3x + 4y = 5$, passing through $(6, 7)$

11. Horizontal with height 3 units above the x-axis

12. Vertical and 4 units to the right of the y-axis

13. The y-axis

14. The x-axis

Differentiate.

15. $y = x^7 + x^3$

16. $y = 5x^8$

17. $y = 6\sqrt{x}$

18. $y = x^7 + 3x^5 + 1$

19. $y = \dfrac{3}{x}$

20. $y = x^4 - \dfrac{4}{x}$

21. $y = (3x^2 - 1)^8$

22. $y = \frac{3}{4}x^{4/3} + \frac{4}{3}x^{3/4}$

23. $y = \dfrac{1}{5x - 1}$

24. $y = (x^3 + x^2 + 1)^5$

25. $y = \sqrt{x^2 + 1}$

26. $y = \dfrac{5}{7x^2 + 1}$

27. $f(x) = 1/\sqrt[4]{x}$

28. $f(x) = (2x + 1)^3$

29. $f(x) = 5$

30. $f(x) = \dfrac{5x}{2} - \dfrac{2}{5x}$

31. $f(x) = \left[x^5 - (x - 1)^5\right]^{10}$

32. $f(t) = t^{10} - 10t^9$

33. $g(t) = 3\sqrt{t} - \dfrac{3}{\sqrt{t}}$

34. $h(t) = 3\sqrt{2}$

35. $f(t) = \dfrac{2}{t - 3t^3}$

36. $g(P) = 4P^{.7}$

37. $h(x) = \frac{3}{2}x^{3/2} - 6x^{2/3}$

38. $f(x) = \sqrt{x + \sqrt{x}}$

39. If $f(t) = 3t^3 - 2t^2$, find $f'(2)$.

40. If $V(r) = 15\pi r^2$, find $V'\left(\frac{1}{3}\right)$.

41. If $g(u) = 3u - 1$, find $g(5)$ and $g'(5)$.

42. If $h(x) = -\frac{1}{2}$, find $h(-2)$ and $h'(-2)$.

43. If $f(x) = x^{5/2}$, what is $f''(4)$?

44. If $g(t) = \frac{1}{4}(2t - 7)^4$, what is $g''(3)$?

45. Find the slope of the graph of $y = (3x-1)^3 - 4(3x-1)^2$ at $x = 0$.

46. Find the slope of the graph of $y = (4-x)^5$ at $x = 5$.

Compute.

47. $\dfrac{d}{dx}(x^4 - 2x^2)$

48. $\dfrac{d}{dt}(t^{5/2} + 2t^{3/2} - t^{1/2})$

49. $\dfrac{d}{dP}(\sqrt{1 - 3P})$

50. $\dfrac{d}{dn}(n^{-5})$

51. $\dfrac{d}{dz}(z^3 - 4z^2 + z - 3)\Big|_{z=-2}$

52. $\dfrac{d}{dx}(4x - 10)^5\Big|_{x=3}$

53. $\dfrac{d^2}{dx^2}(5x + 1)^4$

54. $\dfrac{d^2}{dt^2}(2\sqrt{t})$

55. $\dfrac{d^2}{dt^2}(t^3 + 2t^2 - t)\Big|_{t=-1}$

56. $\dfrac{d^2}{dP^2}(3P + 2)\Big|_{P=4}$

57. $\dfrac{d^2y}{dx^2}$, where $y = 4x^{3/2}$

58. $\dfrac{d}{dt}\left(\dfrac{dy}{dt}\right)$, where $y = \dfrac{1}{3t}$

59. What is the slope of the graph of $f(x) = x^3 - 4x^2 + 6$ at $x = 2$? Write the equation of the line tangent to the graph of $f(x)$ at $x = 2$.

60. What is the slope of the curve $y = 1/(3x - 5)$ at $x = 1$? Write the equation of the line tangent to this curve at $x = 1$.

61. Find the equation of the tangent line to the curve $y = x^2$ at the point $\left(\frac{3}{2}, \frac{9}{4}\right)$. Sketch the graph of $y = x^2$ and sketch the tangent line at $\left(\frac{3}{2}, \frac{9}{4}\right)$.

62. Find the equation of the tangent line to the curve $y = x^2$ at the point $(-2, 4)$. Sketch the graph of $y = x^2$ and sketch the tangent line at $(-2, 4)$.

63. Determine the equation of the tangent line to the curve $y = 3x^3 - 5x^2 + x + 3$ at $x = 1$.

64. Determine the equation of the tangent line to the curve $y = (2x^2 - 3x)^3$ at $x = 2$.

65. In Fig. 1, the straight line has slope -1 and is tangent to the graph of $f(x)$. Find $f(2)$ and $f'(2)$.

$y = -x + 5$
$f(2) = 3$ $f'(2)$
$(5,0)$
$(2,5)$
$y - 2 = -1(x - 3)$
$y - 2 = -x + 5$
$y = -x + 7$
$f(2) = 5$ $f'(2) = -1$

$m = -1$

$y = f(x)$

2 5

Figure 1

66. In Fig. 2, the straight line is tangent to the graph of $f(x) = x^3$. Find the value of a.

$(0, 2)$

a

Figure 2

$-32t + 32$
$= -32t + 8$ $-32t \cdot t/8$

67. A helicopter is rising at a rate of 32 feet per second. At a height of 128 feet the pilot drops a pair of binoculars. After t seconds, the binoculars have height $s(t) = -16t^2 + 32t + 128$ feet from the ground. How fast will they be falling when they hit the ground?

68. Each day the total output of a coal mine after t hours of operation is approximately $40t + t^2 - \frac{1}{15}t^3$ tons, $0 \le t \le 12$. What is the rate of output (in tons of coal per hour) at $t = 5$ hours?

Exercises 69–72 refer to Fig. 3, where $s(t)$ is the number of feet traveled by a person after t seconds of walking along a straight path.

$y = s(t)$

Figure 3.   Walker's progress.

69. How far has the person traveled after 6 seconds?

70. What is the person's average velocity from time $t = 1$ to $t = 4$?

71. What is the person's velocity at time $t = 3$?

72. Without calculating velocities, determine whether the person is traveling faster at $t = 5$ or at $t = 6$.

73. A manufacturer estimates that the hourly cost of producing x units of a product on an assembly line is $C(x) = .1x^3 - 6x^2 + 136x + 200$ dollars.

   (a) Compute $C(21) - C(20)$, the extra cost of raising the production from 20 to 21 units.

   (b) Find the marginal cost when the production level is 20 units.

74. The number of people riding the subway daily from Silver Spring to Metro Center is a function $f(x)$ of the fare, x cents. If $f(235) = 4600$ and $f'(235) = -100$, approximate the daily number of riders for each of the following costs:

   (a) 237 cents   **(b)** 234 cents

   (c) 240 cents   **(d)** 232 cents

75. Let $h(t)$ be a boy's height (in inches) after t years. If $h'(12) = 1.5$, how much will his height increase (approximately) between ages 12 and $12\frac{1}{2}$?

76. If you deposit \$100 into a savings account at the end of each month for 2 years, the balance will be a function $f(r)$ of the interest rate, $r\%$. At 7% interest (compounded monthly), $f(7) = 2568.10$ and $f'(7) = 25.06$. Approximately how much additional money would you earn if the bank paid $7\frac{1}{2}\%$ interest?

Determine whether the following limits exist. If so, compute the limit.

77. $\displaystyle\lim_{x \to 2} \frac{x^2 - 4}{x - 2}$ **78.** $\displaystyle\lim_{x \to 3} \frac{1}{x^2 - 4x + 3}$

79. $\displaystyle\lim_{x \to 4} \frac{x - 4}{x^2 - 8x + 16}$ **80.** $\displaystyle\lim_{x \to 5} \frac{x - 5}{x^2 - 7x + 2}$

Use limits to compute the following derivatives.

81. $f'(5)$, where $f(x) = 1/(2x)$

82. $f'(3)$, where $f(x) = x^2 - 2x + 1$

83. What geometric interpretation may be given to

$$\frac{(3 + h)^2 - 3^2}{h}$$

in connection with the graph of $f(x) = x^2$?

84. As h approaches 0, what value is approached by

$$\frac{\dfrac{1}{2 + h} - \dfrac{1}{2}}{h}?$$

APPLICATIONS OF THE DERIVATIVE

Calculus techniques can be applied to a wide variety of problems in real life, such as the description of the growth of a yeast culture in Chapter 1. We consider many examples in this chapter. In each case we construct a function as a "mathematical model" of some problem and then analyze the function and its derivatives to gain information about the original problem. Our principal method for analyzing a function will be to sketch its graph. For this reason we devote the first part of the chapter to curve sketching.

2.1 Describing Graphs of Functions

Figure 1. An increasing function.

Let's examine the graph of a typical function, such as the one shown in Fig. 1, and introduce some terminology to describe its behavior. First observe that the graph is either rising or falling, depending on whether we look at it from left to right or from right to left. To avoid confusion, we shall always follow the accepted practice of reading a graph from left to right.

Let's now examine the behavior of a function $f(x)$ in an interval throughout which it is defined.

DEFINITION **Increasing Functions**   We say that a function $f(x)$ is *increasing in the interval* if the graph continuously rises as x goes from from left to right through the interval. That is, whenever x_1 and x_2 are in the interval with $x_1 < x_2$, we have $f(x_1) < f(x_2)$. We say that $f(x)$ is increasing at $x = c$ provided that $f(x)$ is increasing in some open interval on the x-axis that contains the point c.

DEFINITION **Decreasing Functions**   We say that a function $f(x)$ is *decreasing in an interval* provided that the graph continuously falls as x goes from left to right through the interval. That is, whenever x_1 and x_2 are in the interval with $x_1 < x_2$, we have $f(x_1) > f(x_2)$. We say that $f(x)$ is decreasing at $x = c$ provided that $f(x)$ is decreasing in some open interval that contains the point c.

Figure 2 shows graphs that are increasing and decreasing at $x = c$. Observe in Fig. 2(d) that when $f(c)$ is negative and $f(x)$ is decreasing, the values of $f(x)$ become *more* negative. When $f(c)$ is negative and $f(x)$ is increasing, as in Fig. 2(e), the values of $f(x)$ become *less* negative.

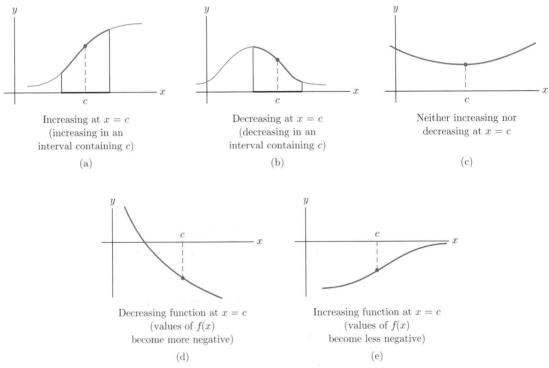

Figure 2.   Increasing and decreasing functions at $x = c$.

Extreme Points   A *relative extreme point* or an *extremum* of a function is a point at which its graph changes from increasing to decreasing, or vice versa. We distinguish the two possibilities in an obvious way.

DEFINITION **Relative Maximum and Minimum**   A *relative maximum point* is a point at which the graph changes from increasing to decreasing; a *relative minimum point* is a point at which the graph changes from decreasing to increasing. (See Fig. 3.)

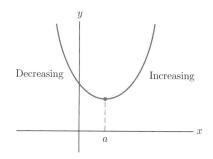

Relative maximum at $x = a$ Relative minimum at $x = a$

Figure 3. Relative extreme points.

The adjective "relative" in these definitions indicates that a point is maximal or minimal only relative to nearby points on the graph. The adjective "local" is also used in place of "relative."

> **DEFINITION Absolute Maximum and Minimum**   The *maximum value* (or *absolute maximum value*) of a function is the largest value that the function assumes on its domain. The *minimum value* (or *absolute minimum value*) of a function is the smallest value that the function assumes on its domain.

Functions may or may not have maximum or minimum values. (See Fig. 4.) However, it can be shown that a continuous function whose domain is an interval of the form $a \leq x \leq b$ has both a maximum and a minimum value.

Figure 4

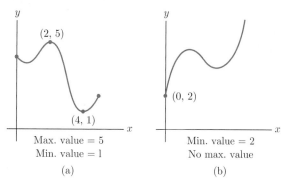

Max. value = 5   Min. value = 2
Min. value = 1   No max. value
(a)   (b)

Maximum values and minimum values of functions usually occur at relative maximum points and relative minimum points, as in Fig. 4(a). However, they can occur at endpoints of the domain, as in Fig. 4(b). If so, we say that the function has an *endpoint extreme value* (or *endpoint extremum*).

Relative maximum points and endpoint maximum points are higher than any nearby points. The maximum value of a function is the y-coordinate of the highest point on its graph. (The highest point is called the *absolute maximum point*.) Similar considerations apply to minima.

EXAMPLE 1   When a drug is injected intramuscularly (into a muscle), the concentration of the drug in the veins has the time–concentration curve shown in Fig. 5. Describe this graph using the terms introduced above.

Solution   Initially (when $t = 0$), there is no drug in the veins. When the drug is injected into the muscle, it begins to diffuse into the bloodstream. The concentration in the

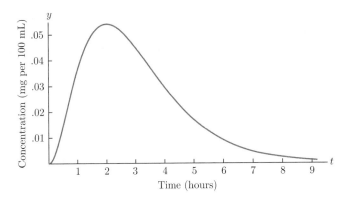

Figure 5. A drug time–concentration curve.

veins increases until it reaches its maximum value at $t = 2$. After this time the concentration begins to decrease as the body's metabolic processes remove the drug from the blood. Eventually, the drug concentration decreases to a level so small that, for all practical purposes, it is zero.   ■

Changing Slope   An important but subtle feature of a graph is the way the graph's slope *changes* (as we look from left to right). The graphs in Fig. 6 are both increasing, but there is a fundamental difference in the way they are increasing. Graph I, which describes the U.S. gross federal debt per person, is steeper for 1990 than for 1960. That is, the *slope* of graph I *increases* as we move from left to right. A newspaper description of graph I might read,

> U.S. gross federal debt per capita rose at an increasing rate during the years from 1960 to 1990.

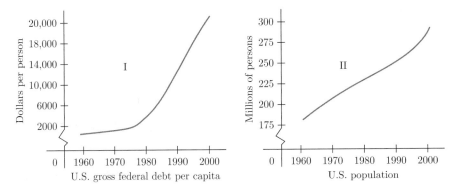

Figure 6.   Increasing and decreasing slopes.

In contrast, the *slope* of graph II *decreases* as we move from left to right from 1960 to 1990. Although the U.S. population is rising each year, the rate of increase declines throughout the years from 1960 to 1990. That is, the slope becomes less positive. The media might say,

> During the 1960s, '70s, and '80s, U.S. population rose at a decreasing rate.

EXAMPLE 2   The daily number of hours of sunlight in Washington, D.C., increased from 9.45 hours on December 21 to 12 hours on March 21 and then increased to 14.9 hours on June 21. From December 22 to March 21, the daily increase was greater than the previous daily increase, and from March 22 to June 21, the daily increase was less than the previous daily increase. Draw a possible graph of the number of hours of daylight as a function of time.

Solution Let $f(t)$ be the number of hours of daylight t months after December 21. See Fig. 7. The first part of the graph, December 21 to March 21, is increasing at an increasing rate. The second part of the graph, March 21 to June 21, is increasing at a decreasing rate. ∎

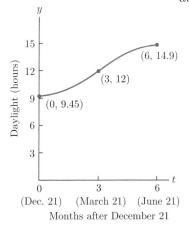

Figure 7. Hours of daylight in Washington, D.C.

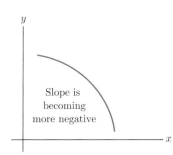

Figure 8. Slope is decreasing.

Figure 9. Slope is increasing.

WARNING ▶ Recall that when a negative quantity decreases, it becomes more negative. (Think about the temperature outside when it is below zero and the temperature is falling.) So if the slope of a graph is negative and the slope is decreasing, the slope is becoming more negative, as in Fig. 8. This technical use of the term "decreasing" runs counter to our intuition, because in popular discourse, "decrease" often means to become smaller in size. ∎

It is true that the curve in Fig. 9 is becoming "less steep" in a nontechnical sense (since steepness, if it were defined, would probably refer to the magnitude of the slope). However, the slope of the curve in Fig. 9 is increasing because it is becoming less negative. The popular press would probably describe the curve in Fig. 9 as decreasing at a decreasing rate, since the rate of fall tends to taper off. Since this terminology is potentially confusing, we shall not use it.

Concavity The U.S. debt and population graphs in Fig. 6 may also be described in geometric terms: Between 1960 and 1990, graph I opens up and lies above its tangent line at each point, whereas graph II opens down and lies below its tangent line at each point between 1960 and 1990. (See Fig. 10.)

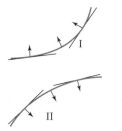

Figure 10. Relationship between concavity and tangent lines.

> **DEFINITION Concavity** We say that a function $f(x)$ is *concave up* at $x = a$ if there is an open interval on the x-axis containing a throughout which the graph of $f(x)$ lies above its tangent line. Equivalently, $f(x)$ is concave up at $x = a$ if the slope of the graph increases as we move from left to right through $(a, f(a))$.

The graph in Fig. 10(I) is an example of a function that is concave up at each point.

Similarly, we say that a function $f(x)$ is *concave down* at $x = a$ if there is an open interval on the x-axis containing a throughout which the graph of $f(x)$ lies below its tangent line. Equivalently, $f(x)$ is concave down at $x = a$ if the slope of the graph decreases as we move from left to right through $(a, f(a))$. The graph in Fig. 10(II) is concave down at each point.

Figure 11. Inflection points.

DEFINITION Inflection Point An *inflection point* is a point on the graph of a function at which the function is continuous and at which the graph changes from concave up to concave down, or vice versa.

At such a point, the graph crosses its tangent line. (See Fig. 11.) The continuity condition means that the graph cannot break at an inflection point.

EXAMPLE 3　　Use the terms defined earlier to describe the graph shown in Fig. 12.

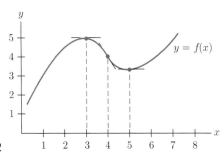

Figure 12

Solution　(a) For $x < 3$, $f(x)$ is increasing and concave down.

(b) Relative maximum point at $x = 3$.

(c) For $3 < x < 4$, $f(x)$ is decreasing and concave down.

(d) Inflection point at $x = 4$.

(e) For $4 < x < 5$, $f(x)$ is decreasing and concave up.

(f) Relative minimum point at $x = 5$.

(g) For $x > 5$, $f(x)$ is increasing and concave up. ■

Intercepts, Undefined Points, and Asymptotes　A point at which a graph crosses the y-axis is called a *y-intercept*, and a point at which it crosses the x-axis is called an *x-intercept*. A function can have at most one y-intercept. Otherwise its graph would violate the vertical line test for a function. Note, however, that a function can have any number of x-intercepts (possibly none). The x-coordinate of an x-intercept is sometimes called a "zero" of the function, since the function has the value zero there. (See Fig. 13.)

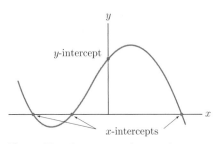

Figure 13. Intercepts of a graph.

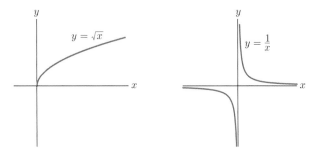

Figure 14. Graphs with undefined points.

Some functions are not defined for all values of x. For instance, $f(x) = 1/x$ is not defined for $x = 0$, and $f(x) = \sqrt{x}$ is not defined for $x < 0$. (See Fig. 14.) Many functions that arise in applications are defined only for $x \geq 0$. A properly drawn graph should leave no doubt as to the values of x for which the function is defined.

Graphs in applied problems sometimes straighten out and approach some straight line as x gets large (Fig. 15). Such a straight line is called an *asymptote* of the curve. The most common asymptotes are horizontal, as in (a) and (b) of Fig. 15. In Example 1, the t-axis is an asymptote of the drug time–concentration curve.

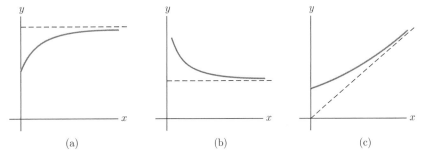

Figure 15. Graphs that approach asymptotes as x gets large.

(a)   (b)   (c)

The horizontal asymptotes of a graph may be determined by calculating the limits

$$\lim_{x \to \infty} f(x) \quad \text{and} \quad \lim_{x \to -\infty} f(x).$$

If either limit exists, the value of the limit determines a horizontal asymptote.

Occasionally, a graph will approach a vertical line as x approaches some fixed value, as in Fig. 16. Such a line is a *vertical asymptote*. Most often, we expect a vertical asymptote at a value x that would result in division by zero in the definition of $f(x)$. For example, $f(x) = 1/(x-3)$ has a vertical asymptote $x = 3$.

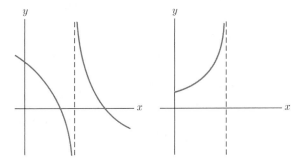

Figure 16. Examples of vertical asymptotes.

We now have six categories for describing the graph of a function.

Describing a Graph

1. Intervals in which the function is increasing (or decreasing), relative maximum points, relative minimum points

2. Maximum value, minimum value

3. Intervals in which the function is concave up (or concave down), inflection points

4. x-intercepts, y-intercept

5. Undefined points

6. Asymptotes

For us, the first three categories will be the most important. However, the last three categories should not be forgotten.

Practice Problems 2.1

1. Does the slope of the curve in Fig. 17 increase or decrease as x increases?

Figure 17

2. At what value of x is the slope of the curve in Fig. 18 minimized?

Figure 18

EXERCISES 2.1

Exercises 1–4 refer to graphs (a)–(f) in Fig. 19.

(a)

(b)

(c)

(d)

(e)

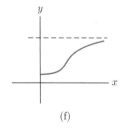

(f)

Figure 19

1. Which functions are increasing for all x?

2. Which functions are decreasing for all x?

3. Which functions have the property that the slope always increases as x increases?

4. Which functions have the property that the slope always decreases as x increases?

Describe each of the following graphs. Your descriptions should include each of the six categories mentioned previously.

5.

6.

7.

8.

9.

10.

11.

12.

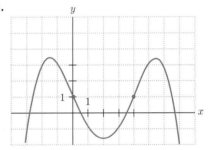

13. Describe the way the *slope* changes as you move along the graph (from left to right) in Exercise 5.

14. Describe the way the *slope* changes on the graph in Exercise 6.

15. Describe the way the *slope* changes on the graph in Exercise 8.

16. Describe the way the *slope* changes on the graph in Exercise 10.

Exercises 17 and 18 refer to the graph in Fig. 20.

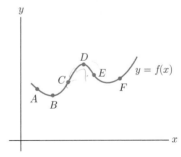

Figure 20

17. **(a)** At which labeled points is the function increasing?

(b) At which labeled points is the graph concave up?

(c) Which labeled point has the most positive slope?

18. **(a)** At which labeled points is the function decreasing?

(b) At which labeled points is the graph concave down?

(c) Which labeled point has the most negative slope (that is, negative and with the greatest magnitude)?

In Exercises 19–22, draw the graph of a function $y = f(x)$ with the stated properties.

19. Both the function and the slope increase as x increases.

20. The function increases and the slope decreases as x increases.

21. The function decreases and the slope increases as x increases. [*Note*: The slope is negative but becomes less negative.]

22. Both the function and the slope decrease as x increases. [*Note*: The slope is negative and becomes more negative.]

23. The annual world consumption of oil rises each year. Furthermore, the amount of the annual *increase* in oil consumption is also rising each year. Sketch a graph that could represent the annual world consumption of oil.

24. In certain professions the average annual income has been rising at an increasing rate. Let $f(T)$ denote the average annual income at year T for persons in one of these professions and sketch a graph that could represent $f(T)$.

25. At noon a child's temperature is $101°F$ and is rising at an increasing rate. At 1 P.M. the child is given medicine. After 2 P.M. the temperature is still increasing but at a decreasing rate. The temperature reaches a peak of $103°$ at 3 P.M. and decreases to $100°$ by 5 P.M. Draw a possible graph of the function $T(t)$, the child's temperature at time t.

26. The number of parking tickets given out each year in the District of Columbia increased from 114,000 in 1950 to 1,500,000 in 1990. Also, each year's increase was greater than the increase for the previous year. Draw a possible graph of the annual number of parking tickets as a function of time t. [*Note*: If $f(t)$ is the yearly rate t years after 1950, $f(0) = 114$ and $f(40) = 1500$ thousand tickets.]

27. Let $C(x)$ denote the total cost of manufacturing x units of some product. Then $C(x)$ is an increasing function for all x. For small values of x, the rate of increase of $C(x)$ decreases. (This is because of the savings that are possible with "mass production.") Eventually, however, for large values of x, the cost $C(x)$ increases at an increasing rate. (This happens when production facilities are strained and become less efficient.) Sketch a graph that could represent $C(x)$.

28. One method of determining the level of blood flow through the brain requires the person to inhale air containing a fixed concentration of N_2O, nitrous oxide. During the first minute, the concentration of N_2O in the jugular vein grows at an increasing rate to a level of .25%. Thereafter it grows at a decreasing rate and reaches a concentration of about 4% after 10 minutes. Draw a possible graph of the concentration of N_2O in the vein as a function of time.

29. Suppose that some organic waste products are dumped into a lake at time $t = 0$ and that the oxygen content of the lake at time t is given by the graph in Fig. 21. Describe the graph in physical terms. Indicate the significance of the inflection point at $t = b$.

30. Figure 22 gives the U.S. electrical energy production in trillion kilowatt-hours from 1935 ($t = 0$) to 1995 ($t = 60$) with projections. In what year was the level of production growing at the greatest rate?

31. Figure 23 gives the number of U.S. farms in millions from 1920 ($t = 20$) to 2000 ($t = 100$). In what year was the number of farms decreasing most rapidly?

32. Figure 24 shows the graph of the consumer price index for the years 1983 ($t = 0$) through 2002 ($t = 19$). This index measures how much a basket of commodities that cost $100 in the beginning of 1983 would cost at any given time. In what year was the rate of increase of the index greatest? The least?

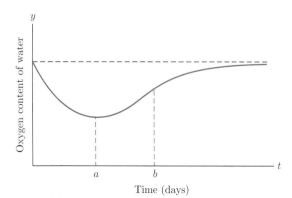

Figure 21.   A lake's recovery from pollution.

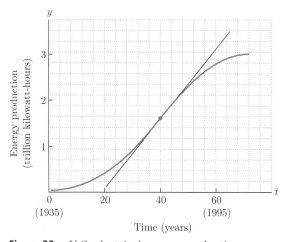

Figure 22.   U.S. electrical energy production.

Figure 23. Number of U.S. farms.

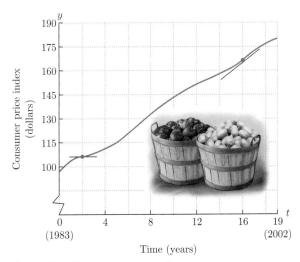

Figure 24. Consumer price index.

33. Let $s(t)$ be the distance (in feet) traveled by a parachutist after t seconds from the time of opening the chute, and suppose that $s(t)$ has the line $y = -15t + 10$ as an asymptote. What does this imply about the velocity of the parachutist? [*Note:* Distance traveled downward is given a negative value.]

34. Let $P(t)$ be the population of a bacteria culture after t days and suppose that $P(t)$ has the line $y = 25,000,000$ as an asymptote. What does this imply about the size of the population?

In Exercises 35–38, sketch the graph of a function having the given properties.

35. Defined for $0 \leq x \leq 10$; relative maximum point at $x = 3$; absolute maximum value at $x = 10$

36. Relative maximum points at $x = 1$ and $x = 5$; relative minimum point at $x = 3$; inflection points at $x = 2$ and $x = 4$

37. Defined and increasing for all $x \geq 0$; inflection point at $x = 5$; asymptotic to the line $y = (3/4)x + 5$

38. Defined for $x \geq 0$; absolute minimum value at $x = 0$; relative maximum point at $x = 4$; asymptotic to the line $y = (x/2) + 1$

39. Consider a smooth curve with no undefined points.

(a) If it has two relative maximum points, must it have a relative minimum point?

(b) If it has two relative extreme points, must it have an inflection point?

40. If the function $f(x)$ has a relative minimum at $x = a$ and a relative maximum at $x = b$, must $f(a)$ be less than $f(b)$?

The difference between the pressure inside the lungs and the pressure surrounding the lungs is called the transmural pressure *(or transmural pressure gradient). Figure 25 shows for*

three different persons how the volume of the lungs is related to the transmural pressure, based on static measurements taken while there is no air flowing through the mouth. (The functional reserve capacity mentioned on the vertical axis is the volume of air in the lungs at the end of a normal expiration.) The rate of change in lung volume with respect to transmural pressure is called the lung compliance.

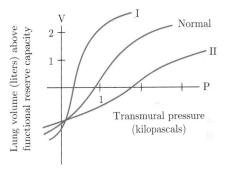

Figure 25. Lung pressure–volume curves.

41. If the lungs are less flexible than normal, an increase in pressure will cause a smaller change in lung volume than in a normal lung. In this case, is the compliance relatively high or low?

42. Most lung diseases cause a decrease in lung compliance. However, the compliance of a person with emphysema is higher than normal. Which curve (I or II) in Fig. 25 could correspond to a person with emphysema?

Technology Exercises

43. Graph the function

$$f(x) = \frac{1}{x^3 - 2x^2 + x - 2}$$

in the window $[0, 4]$ by $[-15, 15]$. For what value of x does $f(x)$ have a vertical asymptote?

44. The graph of the function

$$f(x) = \frac{2x^2 - 1}{.5x^2 + 6}$$

has a horizontal asymptote of the form $y = c$. Estimate the value of c by graphing $f(x)$ in the window $[0, 50]$ by $[-1, 6]$.

45. Simultaneously graph the functions

$$y = \frac{1}{x} + x \quad \text{and} \quad y = x$$

in the window $[-6, 6]$ by $[-6, 6]$. Convince yourself that the straight line is an asymptote of the first curve. How far apart are the two graphs when $x = 6$?

46. Simultaneously graph the functions

$$y = \frac{1}{x} + x + 2 \quad \text{and} \quad y = x + 2$$

in the window $[-6, 6]$ by $[-4, 8]$. Convince yourself that the straight line is an asymptote of the first curve. How far apart are the two graphs when $x = 6$?

Solutions to Practice Problems 2.1

1. The curve is concave up, so the slope increases. Even though the curve itself is decreasing, the slope becomes less negative as we move from left to right.

2. At $x = 3$. We have drawn in tangent lines at three points in Fig. 26. Note that as we move from left to right, the slopes decrease steadily until the point $(3, 2)$, at which time they start to increase. This is consistent with the fact that the graph is concave down (hence slopes are decreasing) to the left of $(3, 2)$ and concave up (hence, slopes are increasing) to the right of $(3, 2)$. Extreme values of slopes always occur at inflection points.

Figure 26

2.2 The First and Second Derivative Rules

We shall now show how properties of the graph of a function $f(x)$ are determined by properties of the derivatives, $f'(x)$ and $f''(x)$. These relationships will provide the key to the curve sketching and optimization discussed in the rest of the chapter.*

We begin with a discussion of the first derivative of a function $f(x)$. Suppose that for some value of x, say $x = a$, the derivative $f'(a)$ is positive. Then the tangent line at $(a, f(a))$ has positive slope and is a rising line (moving from left to right, of course). Since the graph of $f(x)$ near $(a, f(a))$ resembles its tangent line, the function must be increasing at $x = a$. Similarly, when $f'(a) < 0$, the function is decreasing at $x = a$. (See Fig. 1.)

Thus we have the following useful result.

First Derivative Rule   If $f'(a) > 0$, then $f(x)$ is increasing at $x = a$. If $f'(a) < 0$, then $f(x)$ is decreasing at $x = a$.

In other words, a function is increasing whenever the value of its derivative is positive; a function is decreasing whenever the value of its derivative is negative. The

---

*Throughout this chapter, we assume that we are dealing with functions that are not "too badly behaved." More precisely, it suffices to assume that all our functions have continuous first and second derivatives in the interval(s) (in x) where we are considering their graphs.

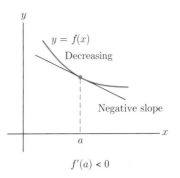

Figure 1. Illustration of the first derivative rule.

first derivative rule says nothing about the case when the derivative of a function is zero. If $f'(a) = 0$, the function might be increasing or decreasing or have a relative extreme point at $x = a$.

EXAMPLE 1   Sketch the graph of a function $f(x)$ that has all the following properties.

(a) $f(3) = 4$

(b) $f'(x) > 0$ for $x < 3$, $f'(3) = 0$, and $f'(x) < 0$ for $x > 3$

Solution   The only specific point on the graph is $(3, 4)$ [property (a)]. We plot this point and then use the fact that $f'(3) = 0$ to sketch the tangent line at $x = 3$. (See Fig. 2.)

From property (b) and the first derivative rule, we know that $f(x)$ must be increasing for x less than 3 and decreasing for x greater than 3. A graph with these properties might look like the curve in Fig. 3. ■

Figure 2

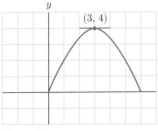

Figure 3

The second derivative of a function $f(x)$ gives useful information about the concavity of the graph of $f(x)$. Suppose that $f''(a)$ is negative. Then, since $f''(x)$ is the derivative of $f'(x)$, we conclude that $f'(x)$ has a negative derivative at $x = a$. In this case, $f'(x)$ must be a decreasing function at $x = a$; that is, the slope of the graph of $f(x)$ is decreasing as we move from left to right on the graph near $(a, f(a))$. (See Fig. 4.) This means that the graph of $f(x)$ is concave down at $x = a$. A similar analysis shows that if $f''(a)$ is positive, then $f(x)$ is concave up at $x = a$. Thus we have the following rule.

Second Derivative Rule   If $f''(a) > 0$, then $f(x)$ is concave up at $x = a$. If $f''(a) < 0$, then $f(x)$ is concave down at $x = a$.

When $f''(a) = 0$, the second derivative rule gives no information. In this case, the function might be concave up, concave down, or neither at $x = a$.

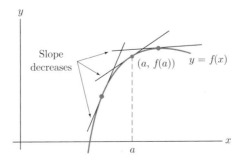

Figure 4. Illustration of the second derivative rule.

The following chart shows how a graph may combine the properties of increasing, decreasing, concave up, and concave down.

Conditions on the Derivatives	Description of $f(x)$ at $x = a$	Graph of $y = f(x)$ near $x = a$
1. $f'(a)$ positive $f''(a)$ positive	$f(x)$ increasing $f(x)$ concave up	
2. $f'(a)$ positive $f''(a)$ negative	$f(x)$ increasing $f(x)$ concave down	
3. $f'(a)$ negative $f''(a)$ positive	$f(x)$ decreasing $f(x)$ concave up	
4. $f'(a)$ negative $f''(a)$ negative	$f(x)$ decreasing $f(x)$ concave down	

EXAMPLE 2　Sketch the graph of a function $f(x)$ with all the following properties.

(a) $(2,3)$, $(4,5)$, and $(6,7)$ are on the graph.

(b) $f'(6) = 0$ and $f'(2) = 0$.

(c) $f''(x) > 0$ for $x < 4$, $f''(4) = 0$, and $f''(x) < 0$ for $x > 4$.

Solution　First we plot the three points from property (a) and then sketch two tangent lines, using the information from property (b). (See Fig. 5.) From property (c) and the second derivative rule, we know that $f(x)$ is concave up for $x < 4$. In particular, $f(x)$ is concave up at $(2,3)$. Also, $f(x)$ is concave down for $x > 4$, in particular, at $(6,7)$. Note that $f(x)$ must have an inflection point at $x = 4$ because the concavity changes there. We now sketch small portions of the curve near $(2,3)$ and $(6,7)$. (See Fig. 6.) We can now complete the sketch (Fig. 7), taking care to make the curve concave up for $x < 4$ and concave down for $x > 4$. ■

Figure 5

Figure 6

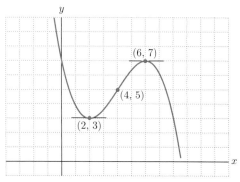

Figure 7

Connections between the Graphs of _f(x)_ and _f'(x)_ Think of the derivative of $f(x)$ as a "slope function" for $f(x)$. The "y-values" on the graph of $y = f'(x)$ are the *slopes* of the corresponding points on the original graph $y = f(x)$. This important connection is illustrated in the next three examples.

EXAMPLE 3   The function $f(x) = 8x - x^2$ is graphed in Fig. 8 along with the slope at several points. How is the slope changing on the graph? Compare the slopes on the graph with the y-coordinates of the points on the graph of $f'(x)$ in Fig. 9.

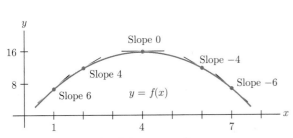

Figure 8. Graph of $f(x) = 8x - x^2$.

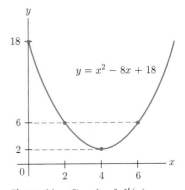

Figure 9. Graph of the derivative of the function in Fig. 8.

Solution   The slopes are decreasing (as we move from left to right). That is, $f'(x)$ is a decreasing function. Observe that the y-values of $f'(x)$ decrease to zero at $x = 4$ and then continue to decrease for x-values greater than 4.   ■

The graph in Fig. 8 is the shape of a typical revenue curve for a manufacturer. In this case, the graph of $f'(x)$ in Fig. 9 would be the *marginal revenue curve*. The graph in the next example has the shape of a typical cost curve. Its derivative produces a *marginal cost curve*.

EXAMPLE 4   The function $f(x) = \frac{1}{3}x^3 - 4x^2 + 18x + 10$ is graphed in Fig. 10. The slope decreases at first and then increases. Use the graph of $f'(x)$ to verify that the slope in Fig. 10 is minimum at the inflection point, where $x = 4$.

Solution   Several slopes are marked on the graph of $f(x)$. These values are y-coordinates of points on the graph of the derivative, $f'(x) = x^2 - 8x + 18$. Observe in Fig. 11 that the y-values on the graph of $f'(x)$ decrease at first and then begin to increase. The minimum value of $f'(x)$ occurs at $x = 4$.   ■

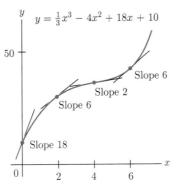

Figure 10. Graph of $f(x)$.

Figure 11. Graph of $f'(x)$.

EXAMPLE 5   Figure 12 shows the graph of $y = f'(x)$, the derivative of a function $f(x)$.

(a) What is the slope of the graph of $f(x)$ when $x = 1$?

(b) Describe how the values of $f'(x)$ change on the interval $1 \leq x \leq 2$.

(c) Describe the shape of the graph of $f(x)$ on the interval $1 \leq x \leq 2$.

(d) For what values of x does the graph of $f(x)$ have a horizontal tangent line?

(e) Explain why $f(x)$ has a relative maximum at $x = 3$.

Solution   (a) Since $f'(1)$ is 2, $f(x)$ has slope 2 when $x = 1$.

(b) The values of $f'(x)$ are positive and decreasing as x increases from 1 to 2.

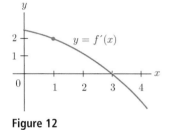

(c) On the interval $1 \leq x \leq 2$, the slope of the graph of $f(x)$ is positive and is decreasing as x increases. Therefore, the graph of $f(x)$ is increasing and concave down.

(d) The graph of $f(x)$ has a horizontal tangent line when the slope is 0, that is, when $f'(x)$ is 0. This occurs at $x = 3$.

(e) Since $f'(x)$ is positive to the left of $x = 3$ and negative to the right of $x = 3$, the graph of $f(x)$ changes from increasing to decreasing at $x = 3$. Therefore, $f(x)$ has a relative maximum at $x = 3$. ∎

Figure 12

Figure 13 shows the graph of a function $f(x)$ whose derivative has the shape shown in Fig. 12. Reread Example 5 and its solution while referring to Fig. 13.

Figure 13

INCORPORATING TECHNOLOGY

Graphing derivatives   Once we have defined a function Y_1 in our calculator, we can graph both the function and its derivative by setting $Y_2 = \mathbf{nDeriv}(Y_1, X, X)$. In Fig. 14 we do this for the function $f(x) = 8x - x^2$ from Example 3. ∎

(a)

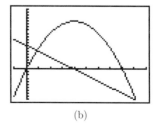

(b)

Figure 14

Practice Problems 2.2

1. Make a good sketch of the function $f(x)$ near the point where $x = 2$, given that $f(2) = 5$, $f'(2) = 1$, and $f''(2) = -3$.

2. The graph of $f(x) = x^3$ is shown in Fig. 15.

Figure 15

(a) Is the function increasing at $x = 0$?

(b) Compute $f'(0)$.

(c) Reconcile your answers to parts (a) and (b) with the first derivative rule.

3. The graph of $y = f'(x)$ is shown in Fig. 16. Explain why $f(x)$ must have a relative minimum point at $x = 3$.

Figure 16

EXERCISES 2.2

Exercises 1–4 refer to the functions whose graphs are given in Fig. 17.

1. Which functions have a positive first derivative for all x?

2. Which functions have a negative first derivative for all x?

3. Which functions have a positive second derivative for all x?

4. Which functions have a negative second derivative for all x?

5. Which one of the graphs in Fig. 18 could represent a function $f(x)$ for which $f(a) > 0$, $f'(a) = 0$, and $f''(a) < 0$?

6. Which one of the graphs in Fig. 18 could represent a function $f(x)$ for which $f(a) = 0$, $f'(a) < 0$, and $f''(a) > 0$?

In Exercises 7–12, sketch the graph of a function that has the properties described.

7. $f(2) = 1$; $f'(2) = 0$; concave up for all x

8. $f(-1) = 0$; $f'(x) < 0$ for $x < -1$, $f'(-1) = 0$ and $f'(x) > 0$ for $x > -1$

Figure 17

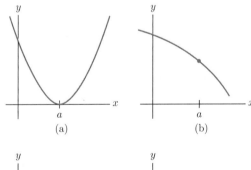

Figure 18

9. $f(3) = 5$; $f'(x) > 0$ for $x < 3$, $f'(3) = 0$ and $f'(x) > 0$ for $x > 3$

10. $(-2, -1)$ and $(2, 5)$ are on the graph; $f'(-2) = 0$ and $f'(2) = 0$; $f''(x) > 0$ for $x < 0$, $f''(0) = 0$, $f''(x) < 0$ for $x > 0$.

11. $(0, 6)$, $(2, 3)$, and $(4, 0)$ are on the graph; $f'(0) = 0$ and $f'(4) = 0$; $f''(x) < 0$ for $x < 2$, $f''(2) = 0$, $f''(x) > 0$ for $x > 2$.

12. $f(x)$ defined only for $x \geq 0$; $(0, 0)$ and $(5, 6)$ are on the graph; $f'(x) > 0$ for $x \geq 0$; $f''(x) < 0$ for $x < 5$, $f''(5) = 0$, $f''(x) > 0$ for $x > 5$.

In Exercises 13–18, use the given information to make a good sketch of the function $f(x)$ near $x = 3$.

13. $f(3) = 4$, $f'(3) = -\frac{1}{2}$, $f''(3) = 5$

14. $f(3) = -2$, $f'(3) = 0$, $f''(3) = 1$

15. $f(3) = 1$, $f'(3) = 0$, inflection point at $x = 3$, $f'(x) > 0$ for $x > 3$

16. $f(3) = 4$, $f'(3) = -\frac{3}{2}$, $f''(3) = -2$

17. $f(3) = -2$, $f'(3) = 2$, $f''(3) = 3$

18. $f(3) = 3$, $f'(3) = 1$, inflection point at $x = 3$, $f''(x) < 0$ for $x > 3$

19. Refer to the graph in Fig. 19. Fill in each entry of the grid with POS, NEG, or 0.

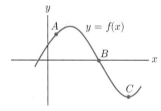

Figure 19

20. The first and second derivatives of the function $f(x)$ have the values given in Table 1.

   (a) Find the x-coordinates of all relative extreme points.

   (b) Find the x-coordinates of all inflection points.

TABLE 1 Values of the First Two Derivatives of a Function

x	$f'(x)$	$f''(x)$
$0 \leq x < 2$	Positive	Negative
2	0	Negative
$2 < x < 3$	Negative	Negative
3	Negative	0
$3 < x < 4$	Negative	Positive
4	0	0
$4 < x \leq 6$	Negative	Negative

21. Suppose that Fig. 20 contains the graph of $y = s(t)$, the distance traveled by a car after t hours. Is the car going faster at $t = 1$ or $t = 2$?

22. Suppose that Fig. 20 contains the graph of $y = v(t)$, the velocity of a car after t hours. Is the car going faster at $t = 1$ or $t = 2$?

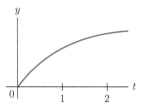

Figure 20

23. Refer to Fig. 21.

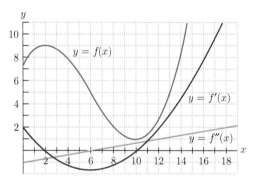

Figure 21

   (a) Looking at the graph of $f'(x)$, determine whether $f(x)$ is increasing or decreasing at $x = 9$. Look at the graph of $f(x)$ to confirm your answer.

   (b) Looking at the values of $f'(x)$ for $1 \leq x < 2$ and $2 < x \leq 3$, explain why the graph of $f(x)$ must have a relative maximum at $x = 2$. What are the coordinates of the relative maximum point?

   (c) Looking at the values of $f'(x)$ for x close to 10, explain why the graph of $f(x)$ has a relative minimum at $x = 10$.

   (d) Looking at the graph of $f''(x)$, determine whether $f(x)$ is concave up or concave down at $x = 2$. Look at the graph of $f(x)$ to confirm your answer.

   (e) Looking at the graph of $f''(x)$, determine where $f(x)$ has an inflection point. Look at the graph of $f(x)$ to confirm your answer. What are the coordinates of the inflection point?

   (f) Find the x-coordinate of the point on the graph of $f(x)$ at which $f(x)$ is increasing at the rate of 6 units per unit change in x.

24. In Fig. 22, the t-axis represents time in minutes.

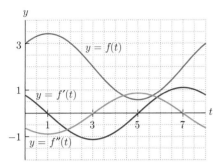

Figure 22

(a) What is $f(2)$?   **(b)** Solve $f(t) = 1$.

(c) When does $f(t)$ attain its greatest value?

(d) When does $f(t)$ attain its least value?

(e) What is the rate of change of $f(t)$ at $t = 7.5$?

(f) When is $f(t)$ decreasing at the rate of 1 unit per minute? That is, when is the rate of change equal to -1?

(g) When is $f(t)$ decreasing at the greatest rate?

(h) When is $f(t)$ increasing at the greatest rate?

Exercises 25–36 refer to Fig. 23, which contains the graph of $f'(x)$, the derivative of the function $f(x)$.

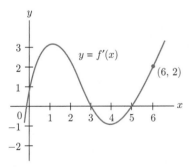

Figure 23

25. Explain why $f(x)$ must be increasing at $x = 6$.

26. Explain why $f(x)$ must be decreasing at $x = 4$.

27. Explain why $f(x)$ has a relative maximum at $x = 3$.

28. Explain why $f(x)$ has a relative minimum at $x = 5$.

29. Explain why $f(x)$ must be concave up at $x = 0$.

30. Explain why $f(x)$ must be concave down at $x = 2$.

31. Explain why $f(x)$ has an inflection point at $x = 1$.

32. Explain why $f(x)$ has an inflection point at $x = 4$.

33. If $f(6) = 3$, what is the equation of the tangent line to the graph of $y = f(x)$ at $x = 6$?

34. If $f(6) = 8$, what is an approximate value of $f(6.5)$?

35. If $f(0) = 3$, what is an approximate value of $f(.25)$?

36. If $f(0) = 3$, what is the equation of the tangent line to the graph of $y = f(x)$ at $x = 0$?

37. Melting snow causes a river to overflow its banks, and $h(t)$ is the number of inches of water on Main Street t hours after the melting begins.

(a) If $h'(100) = \frac{1}{3}$, by approximately how much will the water level change during the next half-hour?

(b) Which of the following two conditions is the better news?

(i) $h(100) = 3$, $h'(100) = 2$, $h''(100) = -5$

(ii) $h(100) = 3$, $h'(100) = -2$, $h''(100) = 5$

38. $T(t)$ is the temperature on a hot summer day at time t hours.

(a) If $T'(10) = 4$, by approximately how much will the temperature rise from 10:00 to 10:45?

(b) Which of the following two conditions is the better news?

(i) $T(10) = 95$, $T'(10) = 4$, $T''(10) = -3$

(ii) $T(10) = 95$, $T'(10) = -4$, $T''(10) = 3$

39. By looking at the first derivative, decide which of the curves in Fig. 24 could *not* be the graph of $f(x) = (3x^2 + 1)^4$ for $x \geq 0$.

Figure 24

40. By looking at the first derivative, decide which of the curves in Fig. 24 could *not* be the graph of $f(x) = x^3 - 9x^2 + 24x + 1$ for $x \geq 0$. [*Hint:* Factor the formula for $f'(x)$.]

41. By looking at the second derivative, decide which of the curves in Fig. 25 could be the graph of $f(x) = x^{5/2}$.

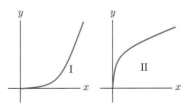

Figure 25

42. Match each observation (a)–(e) with a conclusion (A)–(E).

Observations

(a) The point $(3, 4)$ is on the graph of $f'(x)$.

(b) The point $(3, 4)$ is on the graph of $f(x)$.

(c) The point $(3, 4)$ is on the graph of $f''(x)$.

(d) The point $(3, 0)$ is on the graph of $f'(x)$ and the point $(3, 4)$ is on the graph of $f''(x)$.

(e) The point $(3, 0)$ is on the graph of $f'(x)$ and the point $(3, -4)$ is on the graph of $f''(x)$.

Conclusions

(A) $f(x)$ has a relative minimum point at $x = 3$.

(B) When $x = 3$, the graph of $f(x)$ is concave up.

(C) When $x = 3$, the tangent line to the graph of $y = f(x)$ has slope 4.

(D) When $x = 3$, the value of $f(x)$ is 4.

(E) $f(x)$ has a relative maximum point at $x = 3$.

43. The number of farms in the United States t years after 1925 is $f(t)$ million, where f is the function graphed in Fig. 26(a). [The graphs of $f'(t)$ and $f''(t)$ are shown in Fig. 26(b).]

(a)

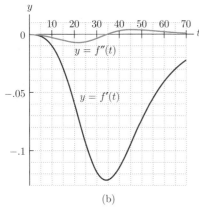

(b)

Figure 26

(a) Approximately how many farms were there in 1990?

(b) At what rate was the number of farms declining in 1990?

(c) In what year were there about 6 million farms?

(d) When was the number of farms declining at the rate of 60,000 farms per year?

(e) When was the number of farms declining fastest?

44. After a drug is taken orally, the amount of the drug in the bloodstream after t hours is $f(t)$ units. Figure 27 shows partial graphs of $f'(t)$ and $f''(t)$.

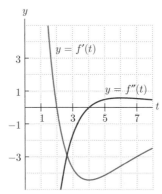

Figure 27

(a) Is the amount of the drug in the bloodstream increasing or decreasing at $t = 5$ hours?

(b) Is the graph of $f(t)$ concave up or concave down at $t = 5$ hours?

(c) When is the level of the drug in the bloodstream decreasing the fastest?

(d) At what time is the greatest level of drug in the bloodstream reached?

(e) When is the level of the drug in the bloodstream decreasing at the rate of 3 units per hour?

Technology Exercises

In Exercises 45 and 46, display the graph of the derivative of $f(x)$ in the specified window. Then use the graph of $f'(x)$ to determine the approximate values of x at which the graph of $f(x)$ has relative extreme points and inflection points. Then check your conclusions by displaying the graph of $f(x)$.

45. $3x^5 - 20x^3 - 120x$; $[-4, 4]$ by $[-325, 325]$

46. $x^4 - x^2$; $[-1.5, 1.5]$ by $[-.75, 1]$

Solutions to Practice Problems 2.2

1. Since $f(2) = 5$, the point $(2, 5)$ is on the graph. [See Fig. 28(a).] Since $f'(2) = 1$, the tangent line at the point $(2, 5)$ has slope 1. Draw in the tangent line. [See Fig. 28(b).] Near the point $(2, 5)$ the graph looks approximately like the tangent line. Since $f''(2) = -3$, a negative number, the graph is concave down at the point $(2, 5)$. Now we are ready to sketch the graph. [See Fig. 28(c).]

(Continued)

Figure 28

2. **(a)** Yes. The graph is steadily increasing as we pass through the point $(0, 0)$.

   (b) Since $f'(x) = 3x^2$, $f'(0) = 3 \cdot 0^2 = 0$.

   (c) There is no contradiction here. The first derivative rule says that if the derivative is positive the function is increasing. However, it does not say that this is the only condition under which a function is increasing. As we have just seen, sometimes we can have the first derivative zero and the function still increasing.

3. Since $f'(x)$, the derivative of $f(x)$, is negative to the left of $x = 3$ and positive to the right of $x = 3$, $f(x)$ is decreasing to the left of $x = 3$ and increasing to the right of $x = 3$. Therefore, by the definition of a relative minimum point, $f(x)$ has a relative minimum point at $x = 3$.

2.3 The First and Second Derivative Tests and Curve Sketching

In this section and the next we develop our ability to sketch the graphs of functions. There are two important reasons for doing so. First, a geometric "picture" of a function is often easier to comprehend than its abstract formula. Second, the material in this section will provide a foundation for the applications in Sections 2.5 through 2.7.

A "sketch" of the graph of a function $f(x)$ should convey the general shape of the graph; it should show where $f(x)$ is defined and where it is increasing and decreasing, and it should indicate, insofar as possible, where $f(x)$ is concave up and concave down. In addition, one or more key points should be accurately located on the graph. These points usually include extreme points, inflection points, and x- and y-intercepts. Other features of a graph may be important, too, but we shall discuss them as they arise in examples and applications.

Our general approach to curve sketching will involve four main steps:

1. Starting with $f(x)$, we compute $f'(x)$ and $f''(x)$.

2. Next, we locate all relative maximum and relative minimum points and make a partial sketch.

3. We study the concavity of $f(x)$ and locate all inflection points.

4. We consider other properties of the graph, such as the intercepts, and complete the sketch.

The first step was the main subject of Chapter 1. We discuss the second and third steps in this section and then present several completely worked examples that combine all four steps in the next section.

Locating Relative Extreme Points   The tangent line at a relative maximum or a relative minimum point of a function $f(x)$ has zero slope; that is, the derivative is zero there. Thus we may state the following useful rule.

> Look for possible relative extreme points of
> $f(x)$ by setting $f'(x) = 0$ and solving for x.
>
> (1)

A number a in the domain of f such that $f'(a) = 0$ is called a **critical number** or **critical value** of f. [We also call a value a in the domain of f a critical value if $f'(a)$ does not exist.] If a is a critical value, the point $(a, f(a))$ is called a **critical point**. Our rule states that if f has a relative extreme value at $x = a$, then a must be a critical value of f. But not every critical value of f yields a relative extreme point. The condition $f'(a) = 0$ tells us only that the tangent line is horizontal at a. In Fig. 1, we show four cases that can occur when $f'(a) = 0$. We can see from the figure that we have an extreme point when the first derivative, $f'(x)$, changes sign at $x = a$. But if the sign of the first derivative does not change, we do not have an extreme point. Based on these observations and the first derivative rule of the previous section, we obtain the following useful test for extreme points.

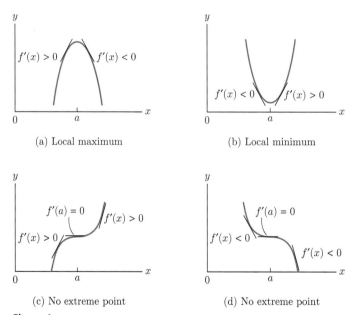

(a) Local maximum

(b) Local minimum

(c) No extreme point

(d) No extreme point

Figure 1

The First Derivative Test (for local extreme points)

Suppose that $f'(a) = 0$.

(a) If f' changes from positive to negative at $x = a$, then f has a local maximum at a. [See Fig. 1(a).]

(b) If f' changes from negative to positive at $x = a$, then f has a local minimum at a. [See Fig. 1(b).]

(c) If f' does not change sign at a (that is, f' is either positive on both sides of a [Fig. 1(c)] or negative on both sides of a [Fig. 1(d)], then f has no local extremum at a.

EXAMPLE 1

Applying the first derivative test Find the local maximum and minimum points of $f(x) = \frac{1}{3}x^3 - 2x^2 + 3x + 1$.

Solution First we find the critical values and critical points of f:

$$f'(x) = \frac{1}{3}(3)x^2 - 2(2)x + 3 = x^2 - 4x + 3$$
$$= (x - 1)(x - 3).$$

The first derivative $f'(x) = 0$ if $x - 1 = 0$ or $x - 3 = 0$. Thus the critical values are

$$x = 1 \quad \text{and} \quad x = 3.$$

Substitute the critical values into the expression of f:

$$f(1) = \frac{1}{3}(1)^3 - 2(1)^2 + 3(1) + 1 = \frac{1}{3} + 2 = \frac{7}{3};$$

$$f(3) = \frac{1}{3}(3)^3 - 2(3)^2 + 3(3) + 1 = 1.$$

Thus the critical points are $(1, \frac{7}{3})$ and $(3, 1)$. To determine whether we have a relative maximum, minimum, or neither at a critical point, we shall apply the first derivative test. This requires a careful study of the sign of $f'(x)$, which can be facilitated with the help of a chart. Here is how you can set up the chart.

- Divide the real line into intervals with the critical values as endpoints.
- Since the sign of f' depends on the signs of its two factors $x - 1$ and $x - 3$, determine the signs of the factors of f' over each interval. Usually this is done by testing the sign of a factor at points selected from each interval.
- In each interval, use a plus sign if the factor is positive or a minus sign if it is negative. Then determine the sign of f' over each interval by multiplying the signs of the factors and using

$$(+) \cdot (+) = +; \quad (+) \cdot (-) = -; \quad (-) \cdot (+) = -; \quad (-) \cdot (-) = +.$$

- A plus sign of f' corresponds to an increasing portion of the graph of f and a minus sign to a decreasing portion. Denote an increasing portion with an upward arrow and a decreasing portion with a downward arrow. The sequence of arrows should convey the general shape of the graph and, in particular, tell you whether or not your critical values correspond to extreme points. Here is the chart:

Critical Values		1		3	
	$x < 1$		$1 < x < 3$		$3 < x$
$x - 1$	$-$	0	$+$		$+$
$x - 3$	$-$		$-$	0	$+$
$f'(x)$	$+$	0	$-$	0	$+$
$f(x)$	Increasing on $(-\infty, 1)$	$7/3$	Decreasing on $(1, 3)$	1	Increasing on $(3, \infty)$

Local maximum $(1, 7/3)$ Local minimum $(3, 1)$

You can see from the chart that the sign of f' varies from positive to negative at $x = 1$. Thus, according to the first derivative test, f has a local maximum at $x = 1$. Also, the sign of f' varies from negative to positive at $x = 3$; so f has a local minimum at $x = 3$. You can confirm these assertions by following the direction of the arrows in the last row of the chart. In conclusion, f has a local maximum at $(1, \frac{7}{3})$ and a local minimum at $(3, 1)$. (See Fig. 2.)   ■

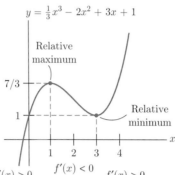

Figure 2 $f'(x) > 0$ $f'(x) < 0$ $f'(x) > 0$

The previous example illustrates cases (a) and (b) in Fig. 1, which correspond to local extreme points on the graph at each critical value. The following example illustrates cases (c) and (d) in Fig. 1.

EXAMPLE 2

Applying the first derivative test Find the local maximum and minimum points of $f(x) = (3x - 1)^3$.

Solution We begin by finding the critical values and critical points. Using the general power rule, we have

$$f'(x) = 3(3x - 1)^2(3) = 9(3x - 1)^2$$
$$f'(x) = 0 \Rightarrow 3x - 1 = 0$$
$$\Rightarrow x = \tfrac{1}{3}.$$

Plugging this value into the original expression for f, we find

$$f\left(\tfrac{1}{3}\right) = \left(3 \cdot \tfrac{1}{3} - 1\right)^3 = 0.$$

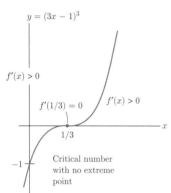

Figure 3

Thus $(\frac{1}{3}, 0)$ is the only critical point. To determine whether it is a local maximum, minimum, or neither, we use the first derivative test. Look at $f'(x) = 9(3x - 1)^2$. This expression is always nonnegative, because a square is always ≥ 0 (there is no need for a sign chart in this case). Consequently, f' is positive on both sides of $x = \frac{1}{3}$; so, since f' does not change sign at $x = \frac{1}{3}$, we conclude from part (c) of the first derivative test that there is no relative extreme point. Indeed, since $f'(x) \geq 0$ for all x, the graph is always increasing. (See Fig. 3.)   ■

Using Concavity to Determine Extreme Points Studying the variation of the graph of f and determining the extreme points can be tedious tasks. Any shortcut that helps us avoid excessive computations is always appreciated. Let us look back at Fig. 1 and give an alternative description of the extreme points using concavity. At $x = a$ the concavity is downward in Fig. 1(a), upward in Fig. 1(b), and neither in Figs. 1(c) and (d). These observations together with the second derivative rule from the previous section are the basis for the following important test.

> **The Second Derivative Test (for local extreme points)**
>
> (a) If $f'(a) = 0$ and $f''(a) < 0$, then f has a local maximum at a. [See Fig. 1(a).]
>
> (b) If $f'(a) = 0$ and $f''(a) > 0$, then f has a local minimum at a. [See Fig. 1(b).]

EXAMPLE 3   The graph of the quadratic function $f(x) = \frac{1}{4}x^2 - x + 2$ is a parabola and so has one relative extreme point. Find it and sketch the graph.

Solution   We begin by computing the first and second derivatives of $f(x)$.

$$f(x) = \tfrac{1}{4}x^2 - x + 2$$
$$f'(x) = \tfrac{1}{2}x - 1$$
$$f''(x) = \tfrac{1}{2}$$

Setting $f'(x) = 0$, we have $\frac{1}{2}x - 1 = 0$, so that $x = 2$ is the only critical value. Thus, $f'(2) = 0$. Geometrically, this means that the graph of $f(x)$ will have a horizontal tangent line at the point where $x = 2$. To plot this point, we substitute the value 2 for x in the original expression for $f(x)$.

$$f(2) = \tfrac{1}{4}(2)^2 - (2) + 2 = 1$$

Figure 4 shows the point $(2, 1)$ together with the horizontal tangent line. Is $(2, 1)$ a relative extreme point? To decide, we look at $f''(x)$. Since $f''(x) = \frac{1}{2}$, which is positive, the graph of $f(x)$ is concave up at $x = 2$, so $(2, 1)$ is a local minimum by the second derivative test. A partial sketch of the graph near $(2, 1)$ should look something like Fig. 5.

Figure 4

Figure 5

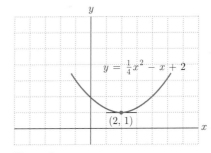

Figure 6

We see that $(2, 1)$ is a relative minimum point. In fact, it is the only relative extreme point, for there is no other place where the tangent line is horizontal. Since the graph has no other "turning points," it must be decreasing before it gets to $(2, 1)$ and then increasing to the right of $(2, 1)$. Note that since $f''(x)$ is positive (and equal to $\frac{1}{2}$) for all x, the graph is concave up at each point. A completed sketch is given in Fig. 6.   ■

EXAMPLE 4   **Applying the second derivative test**   Locate all possible relative extreme points on the graph of the function $f(x) = x^3 - 3x^2 + 5$. Check the concavity at these points and use this information to sketch the graph of $f(x)$.

Solution We have

$$f(x) = x^3 - 3x^2 + 5$$
$$f'(x) = 3x^2 - 6x$$
$$f''(x) = 6x - 6.$$

The easiest way to find the critical values is to factor the expression for $f'(x)$:

$$3x^2 - 6x = 3x(x - 2).$$

From this factorization it is clear that $f'(x)$ will be zero if and only if $x = 0$ or $x = 2$. In other words, the graph will have horizontal tangent lines when $x = 0$ and $x = 2$, and nowhere else.

To plot the points on the graph where $x = 0$ and $x = 2$, we substitute these values back into the original expression for $f(x)$. That is, we compute

$$f(0) = (0)^3 - 3(0)^2 + 5 = 5$$
$$f(2) = (2)^3 - 3(2)^2 + 5 = 1.$$

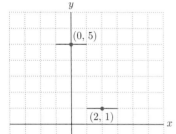

Figure 7

Figure 7 shows the points $(0, 5)$ and $(2, 1)$, along with the corresponding tangent lines.

Next, we check the sign of $f''(x)$ at $x = 0$ and $x = 2$ and apply the second derivative test:

$$f''(0) = 6(0) - 6 = -6 < 0 \quad \text{(local maximum)}$$
$$f''(2) = 6(2) - 6 = 6 > 0 \quad \text{(local minimum)}.$$

Since $f''(0)$ is negative, the graph is concave down at $x = 0$; since $f''(2)$ is positive, the graph is concave up at $x = 2$. A partial sketch of the graph is given in Fig. 8.

It is clear from Fig. 8 that $(0, 5)$ is a relative maximum point and $(2, 1)$ is a relative minimum point. Since they are the only turning points, the graph must be increasing before it gets to $(0, 5)$, decreasing from $(0, 5)$ to $(2, 1)$, and then increasing again to the right of $(2, 1)$. A sketch incorporating these properties appears in Fig. 9. ∎

Figure 8

Figure 9

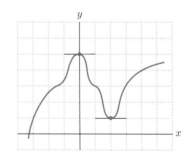

Figure 10

The facts that we used to sketch Fig. 9 could equally well be used to produce the graph in Fig. 10. Which graph really corresponds to $f(x) = x^3 - 3x^2 + 5$? The answer will be clear when we find the inflection points on the graph of $f(x)$.

Locating Inflection Points   An inflection point of a function $f(x)$ can occur only at a value of x for which $f''(x)$ is zero, because the curve is concave up where $f''(x)$ is positive and concave down where $f''(x)$ is negative. Thus we have the following rule.

> Look for possible inflection points by setting
> $f''(x) = 0$ and solving for x.   (2)

Once we have a value of x where the second derivative is zero, say at $x = b$, we must check the concavity of $f(x)$ at nearby points to see if the concavity really changes at $x = b$.

EXAMPLE 5   Find the inflection points of the function $f(x) = x^3 - 3x^2 + 5$ and explain why the graph in Fig. 9 has the correct shape.

Solution   From Example 4 we have $f''(x) = 6x - 6 = 6(x - 1)$. Clearly, $f''(x) = 0$ if and only if $x = 1$. We will want to plot the corresponding point on the graph, so we compute

$$f(1) = (1)^3 - 3(1)^2 + 5 = 3.$$

Figure 11

Therefore, the only possible inflection point is $(1, 3)$.

Now look back at Fig. 8, where we indicated the concavity of the graph at the relative extreme points. Since $f(x)$ is concave down at $(0, 5)$ and concave up at $(2, 1)$, the concavity must reverse somewhere between these points. Hence $(1, 3)$ must be an inflection point. Furthermore, since the concavity of $f(x)$ reverses nowhere else, the concavity at all points to the left of $(1, 3)$ must be the same (that is, concave down). Similarly, the concavity of all points to the right of $(1, 3)$ must be the same (concave up). Thus the graph in Fig. 9 has the correct shape. The graph in Fig. 10 has too many "wiggles," caused by frequent changes in concavity; that is, there are too many inflection points. A correct sketch showing the one inflection point at $(1, 3)$ is given in Fig. 11.   ∎

EXAMPLE 6   Sketch the graph of $y = -\frac{1}{3}x^3 + 3x^2 - 5x$.

Solution   Let

$$f(x) = -\frac{1}{3}x^3 + 3x^2 - 5x.$$

Then

$$f'(x) = -x^2 + 6x - 5$$
$$f''(x) = -2x + 6.$$

Step 1: Finding the critical points   We set $f'(x) = 0$ and solve for x:

$$-(x^2 - 6x + 5) = 0$$
$$-(x - 1)(x - 5) = 0$$
$$x = 1 \quad \text{or} \quad x = 5 \quad \text{(critical values)}.$$

Substituting these values of x back into $f(x)$, we find that

$$f(1) = -\frac{1}{3}(1)^3 + 3(1)^2 - 5(1) = -\frac{7}{3}$$
$$f(5) = -\frac{1}{3}(5)^3 + 3(5)^2 - 5(5) = \frac{25}{3}.$$

Step 2: Determining the The information we have so far is given in Fig. 12(a). The sketch in Fig. 12(b) is
extreme points obtained by computing

$$f''(1) = -2(1) + 6 = 4 > 0 \quad \text{(local minimum)}$$
$$f''(5) = -2(5) + 6 = -4 < 0 \quad \text{(local maximum)}.$$

The curve is concave up at $x = 1$ because $f''(1)$ is positive (local minimum), and
the curve is concave down at $x = 5$ because $f''(5)$ is negative (local maximum).

Step 3: Concavity and Since the concavity reverses somewhere between $x = 0$ and $x = 5$, there must be at
inflection points least one inflection point. If we set $f''(x) = 0$, we find that

$$-2x + 6 = 0$$
$$x = 3.$$

So the inflection point must occur at $x = 3$. To plot the inflection point, we compute

$$f(3) = -\tfrac{1}{3}(3)^3 + 3(3)^2 - 5(3) = 3.$$

The final sketch of the graph is given in Fig. 13. ■

(a)   (b)

Figure 12

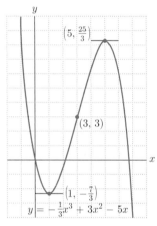

Figure 13

The argument in Example 6, that there must be an inflection point because
concavity reverses, is valid whenever $f(x)$ is a polynomial. However, it does not
always apply to a function whose graph has a break in it. For example, the function
$f(x) = 1/x$ is concave down at $x = -1$ and concave up at $x = 1$, but there is no
inflection point in between.

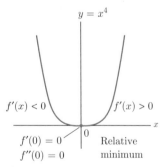

Figure 14

In Examples 1 and 2 we used the first derivative test to locate the extreme
points; but in the remaining examples we used the second derivative test. In general,
which test should we use? If $f''(x)$ is straightforward to compute (for example,
when $f(x)$ is a polynomial), you should try the second derivative test first. If the
second derivative is tedious to compute or if $f''(a) = 0$, use the first derivative test.
Remember that when $f''(a) = 0$, the second derivative test is inconclusive. Keep
the following examples in mind. The function $f(x) = x^3$ is always increasing and
has no local maximum or minimum, even though $f'(0) = 0$ and $f''(0) = 0$. The
function $f(x) = x^4$ is shown in Fig. 14. We have $f'(x) = 4x^3$ and $f''(x) = 12x^2$.
So $f'(0) = 0$ and $f''(0) = 0$, but, as you can see from Fig. 14, you have a local
minimum at $x = 0$. The reason for this follows from the first derivative test, since
f' changes sign at $x = 0$ and goes from negative to positive.

A summary of curve-sketching techniques appears at the end of the next section.
You may find steps 1, 2, and 3 helpful when working the exercises.

Practice Problems 2.3

1. Which of the curves in Fig. 15 could possibly be the graph of a function of the form $f(x) = ax^2 + bx + c$, where $a \neq 0$?

2. Which of the curves in Fig. 16 could possibly be the graph of a function of the form $f(x) = ax^3 + bx^2 + cx + d$, where $a \neq 0$?

(a)

(b)

(c)

(d)

Figure 15

(a)

(b)

(c)

(d)

Figure 16

EXERCISES 2.3

Each of the graphs of the functions in Exercises 1–8 has one relative maximum and one relative minimum point. Find these points using the first derivative test. Use a variation chart as in Example 1.

1. $f(x) = x^3 - 27x$

2. $f(x) = x^3 - 6x^2 + 1$

3. $f(x) = -x^3 + 6x^2 - 9x + 1$

4. $f(x) = -6x^3 - \frac{3}{2}x^2 + 3x - 3$

5. $f(x) = \frac{1}{3}x^3 - x^2 + 1$

6. $f(x) = \frac{4}{3}x^3 - x + 2$

7. $f(x) = -x^3 - 12x^2 - 2$

8. $f(x) = 2x^3 + 3x^2 - 3$

Each of the graphs of the functions in Exercises 9–16 has one relative extreme point. Plot this point and check the concavity there. Using only this information, sketch the graph. [As you work the problems, observe that if $f(x) = ax^2 + bx + c$, then $f(x)$ has a relative minimum point when $a > 0$ and a relative maximum point when $a < 0$.]

9. $f(x) = 2x^2 - 8$

10. $f(x) = x^2$

11. $f(x) = \frac{1}{2}x^2 + x - 4$

12. $f(x) = -3x^2 + 12x + 2$

13. $f(x) = 1 + 6x - x^2$

14. $f(x) = \frac{1}{2}x^2 + \frac{1}{2}$

15. $f(x) = -x^2 - 8x - 10$

16. $f(x) = -x^2 + 2x - 5$

Each of the graphs of the functions in Exercises 17–24 has one relative maximum and one relative minimum point. Plot these two points and check the concavity there. Using only this information, sketch the graph.

17. $f(x) = x^3 + 6x^2 + 9x$

18. $f(x) = \frac{1}{9}x^3 - x^2$

19. $f(x) = x^3 - 12x$

20. $f(x) = -\frac{1}{3}x^3 + 9x - 2$

21. $f(x) = -\frac{1}{9}x^3 + x^2 + 9x$

22. $f(x) = 2x^3 - 15x^2 + 36x - 24$

23. $f(x) = -\frac{1}{3}x^3 + 2x^2 - 12$

24. $f(x) = \frac{1}{3}x^3 + 2x^2 - 5x + \frac{8}{3}$

Sketch the following curves, indicating all relative extreme points and inflection points.

25. $y = x^3 - 3x + 2$

26. $y = x^3 - 6x^2 + 9x + 3$

27. $y = 1 + 3x^2 - x^3$

28. $y = -x^3 + 12x - 4$

29. $y = \frac{1}{3}x^3 - x^2 - 3x + 5$

30. $y = x^4 + \frac{1}{3}x^3 - 2x^2 - x + 1$ [*Hint:* $4x^3 + x^2 - 4x + 1 = (x^2 - 1)(4x - 1)$]

31. $y = 2x^3 - 3x^2 - 36x + 20$

32. $y = x^4 - \frac{4}{3}x^3$

33. Let a, b, c be fixed numbers with $a \neq 0$ and let $f(x) = ax^2 + bx + c$. Is it possible for the graph of $f(x)$ to have an inflection point? Explain your answer.

34. Let a, b, c, d be fixed numbers with $a \neq 0$ and let $f(x) = ax^3 + bx^2 + cx + d$. Is it possible for the graph of $f(x)$ to have more than one inflection point? Explain your answer.

The graph of each function in Exercises 35–40 has one relative extreme point. Find it (giving both x- and y-coordinates) and determine if it is a relative maximum or a relative minimum point. Do not include a sketch of the graph of the function.

35. $f(x) = \frac{1}{4}x^2 - 2x + 7$

36. $f(x) = 5 - 12x - 2x^2$

37. $g(x) = 3 + 4x - 2x^2$

38. $g(x) = x^2 + 10x + 10$

39. $f(x) = 5x^2 + x - 3$

40. $f(x) = 30x^2 - 1800x + 29{,}000$

In Exercises 41 and 42, determine which function is the derivative of the other.

41.

42.
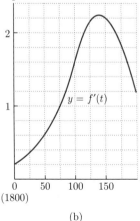

43. Consider the graph of $g(x)$ in Fig. 17.

    (a) If $g(x)$ is the first derivative of $f(x)$, what is the nature of $f(x)$ when $x = 2$?

    (b) If $g(x)$ is the second derivative of $f(x)$, what is the nature of $f(x)$ when $x = 2$?

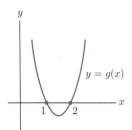

Figure 17

44. U.S. Population The population (in millions) of the United States (excluding Alaska and Hawaii) t years after 1800 is given by the function $f(t)$ in Fig. 18(a). The graphs of $f'(t)$ and $f''(t)$ are shown in Figs. 18(b) and 18(c).

    (a) What was the population in 1925?

    (b) Approximately when was the population 25 million?

    (c) How fast was the population growing in 1950?

    (d) When during the last 50 years was the population growing at the rate of 1.8 million people per year?

    (e) In what year was the population growing at the greatest rate?

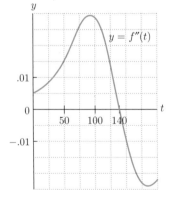

(a)　　　　　(b)　　　　　(c)

Figure 18. Population (in millions) of the United States from 1800 to 1998.

45. **Index-Fund Fees** When a mutual fund company charges a fee of 0.47% on its index funds, its assets in the fund are $41 billion. And when it charges a fee of 0.18%, its assets in the fund are $300 billion.*

    (a) Let x denote the fee that the company charges as a percentage of the index fund and $A(x)$ its assets in the fund. Express $A(x)$ as a linear function of x.

    (b) In September 2004, Fidelity Mutual lowered its fees on various index funds from an average of 0.3% to 0.10%. Let $R(x)$ denote the revenue of the company from fees when its index-fund fee is $x\%$. Compare the revenue of the company before and after lowering the fees. [*Hint*: Revenue is $x\%$ of assets.]

    (c) Find the fee that maximizes the revenue of the company and determine the maximum revenue.

46. **Index-Fund Fees** Suppose that the cost function in Exercise 45 is $C(x) = -2.5x+1$, where x is the index-fund fee. (The company has a fixed cost of $1 billion dollars, and the cost decreases as a function of the index-fund fee.) Find the value of x that maximizes profit. How well did Fidelity Mutual do before and after lowering the index-fund fees?

Technology Exercises

47. Draw the graph of $f(x) = \frac{1}{6}x^3 - x^2 + 3x + 3$ in the window $[-2, 6]$ by $[-10, 20]$. It has an inflection point

at $x = 2$, but no relative extreme points. Enlarge the window a few times to convince yourself that there are no relative extreme points anywhere. What does this tell you about $f'(x)$?

48. Draw the graph of $f(x) = \frac{1}{6}x^3 - \frac{5}{2}x^2 + 13x - 20$ in the window $[0, 10]$ by $[-20, 30]$. Algebraically determine the coordinates of the inflection point. Zoom in and zoom out to convince yourself that there are no relative extreme points anywhere.

49. Draw the graph of

$$f(x) = 2x + \frac{18}{x} - 10$$

in the window $[0, 16]$ by $[0, 16]$. In what ways is this graph like the graph of a parabola that opens upward? In what ways is it different?

50. Draw the graph of

$$f(x) = 3x + \frac{75}{x} - 25$$

in the window $[0, 25]$ by $[0, 50]$. Use the trace feature of the calculator or computer to estimate the coordinates of the relative minimum point. Then determine the coordinates algebraically. Convince yourself both graphically (with the calculator or computer) and algebraically that this function has no inflection points.

Solutions to Practice Problems 2.3

1. Answer: (a) and (d). Curve (b) has the shape of a parabola, but it is not the graph of any function, since vertical lines cross it twice. Curve (c) has two relative extreme points, but the derivative of $f(x)$ is a linear function, which could not be zero for two different values of x.

2. Answer: (a), (c), (d). Curve (b) has three relative extreme points, but the derivative of $f(x)$ is a quadratic function, which could not be zero for three different values of x.

2.4 Curve Sketching (Conclusion)

In Section 2.3 we discussed the main techniques for curve sketching. Here we add a few finishing touches and examine some slightly more complicated curves.

The more points we plot on a graph, the more accurate the graph becomes. This statement is true even for the simple quadratic and cubic curves in Section 2.3. Of course, the most important points on a curve are the relative extreme points and the inflection points. In addition, the x- and y-intercepts often have some intrinsic interest in an applied problem. The y-intercept is $(0, f(0))$. To find the x-intercepts on the graph of $f(x)$, we must find those values of x for which $f(x) = 0$. Since this can be a difficult (or impossible) problem, we shall find x-intercepts only when they are easy to find or when a problem specifically requires us to find them.

*Based on the article "Fidelity slashes index-fund fees," *Boston Globe*, September 1, 2004.

When $f(x)$ is a quadratic function, as in Example 1, we can easily compute the x-intercepts (if they exist) either by factoring the expression for $f(x)$ or by using the quadratic formula, which we now recall from Section 0.4:

The Quadratic Formula

$$x = \frac{-b \pm \sqrt{b^2 - 4ac}}{2a}$$

The sign $\pm$ tells us to form two expressions, one with $+$ and one with $-$. The equation has two distinct roots if $b^2 - 4ac > 0$; one double root if $b^2 - 4ac = 0$; and no (real) roots if $b^2 - 4ac < 0$.

EXAMPLE 1   **Applying the second derivative test**   Sketch the graph of $y = \frac{1}{2}x^2 - 4x + 7$.

Solution Let

$$f(x) = \tfrac{1}{2}x^2 - 4x + 7.$$

Then

$$f'(x) = x - 4$$
$$f''(x) = 1.$$

Since $f'(x) = 0$ only when $x = 4$ and since $f''(4)$ is positive, $f(x)$ must have a relative minimum point at $x = 4$ (the second derivative test). The relative minimum point is $(4, f(4)) = (4, -1)$.

The y-intercept is $(0, f(0)) = (0, 7)$. To find the x-intercepts, we set $f(x) = 0$ and solve for x:

$$\tfrac{1}{2}x^2 - 4x + 7 = 0.$$

The expression for $f(x)$ is not easily factored, so we use the quadratic formula to solve the equation:

$$x = \frac{-(-4) \pm \sqrt{(-4)^2 - 4(\tfrac{1}{2})(7)}}{2(\tfrac{1}{2})} = 4 \pm \sqrt{2}.$$

The x-intercepts are $(4 - \sqrt{2}, 0)$ and $(4 + \sqrt{2}, 0)$. To plot these points, we use the approximation $\sqrt{2} \approx 1.4$. (See Fig. 1.) ∎

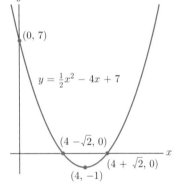

$y = \frac{1}{2}x^2 - 4x + 7$

$(0, 7)$

$(4 - \sqrt{2}, 0)$

$(4 + \sqrt{2}, 0)$

$(4, -1)$

Figure 1

The next example is interesting. The function that we consider has no critical points.

EXAMPLE 2   **A function with no critical points**   Sketch the graph of $f(x) = \frac{1}{6}x^3 - \frac{3}{2}x^2 + 5x + 1$.

Solution
$$f(x) = \tfrac{1}{6}x^3 - \tfrac{3}{2}x^2 + 5x + 1$$
$$f'(x) = \tfrac{1}{2}x^2 - 3x + 5$$
$$f''(x) = x - 3$$

Searching for critical points, we set $f'(x) = 0$ and try to solve for x:

$$\tfrac{1}{2}x^2 - 3x + 5 = 0. \tag{1}$$

If we apply the quadratic formula with $a = \frac{1}{2}$, $b = -3$, and $c = 5$, we see that $b^2 - 4ac$ is negative, and so there is no solution to (1). In other words, $f'(x)$ is never zero. Thus the graph cannot have relative extreme points. If we evaluate

$f'(x)$ at some x, say $x = 0$, we see that the first derivative is positive, and so $f(x)$ is increasing there. Since the graph of $f(x)$ is a smooth curve with no relative extreme points and no breaks, $f(x)$ must be increasing for all x. (If a function were increasing at $x = a$ and decreasing at $x = b$, it would have a relative extreme point between a and b.)

Now let us check the concavity.

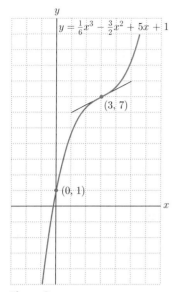

$y = \frac{1}{6}x^3 - \frac{3}{2}x^2 + 5x + 1$

(3, 7)

(0, 1)

Figure 2

$f''(x) = x - 3$		Graph of $f(x)$
$x < 3$	Negative	Concave down
$x = 3$	Zero	Concavity reverses
$x > 3$	Positive	Concave up

Since $f''(x)$ changes sign at $x = 3$, we have an inflection point at $(3, 7)$. The y-intercept is $(0, f(0)) = (0, 1)$. We omit the x-intercept because it is difficult to solve the cubic equation $\frac{1}{6}x^3 - \frac{3}{2}x^2 + 5x + 1 = 0$.

The quality of our sketch of the curve will be improved if we first sketch the tangent line at the inflection point. To do this, we need to know the slope of the graph at $(3, 7)$:

$$f'(3) = \frac{1}{2}(3)^2 - 3(3) + 5 = \frac{1}{2}.$$

We draw a line through $(3, 7)$ with slope $\frac{1}{2}$ and then complete the sketch as shown in Fig. 2. ∎

EXAMPLE 3

Using the first derivative test Sketch the graph of $f(x) = (x - 2)^4 - 1$.

Solution

$$f(x) = (x - 2)^4 - 1$$
$$f'(x) = 4(x - 2)^3$$
$$f''(x) = 12(x - 2)^2$$

Clearly, $f'(x) = 0$ only if $x = 2$. So the curve has a horizontal tangent at $(2, f(2)) = (2, -1)$. Since $f''(2) = 0$, the second derivative test is inconclusive. We shall apply the first derivative test. Note that

$$f'(x) = 4(x - 2)^3 \begin{cases} \text{negative} & \text{if } x < 2 \\ \text{positive} & \text{if } x > 2 \end{cases}$$

since the cube of a negative number is negative and the cube of a positive number is positive. Therefore, as x goes from left to right in the vicinity of 2, the first derivative changes sign and goes from negative to positive. By the first derivative test, the point $(2, -1)$ is a relative minimum.

The y-intercept is $(0, f(0)) = (0, 15)$. To find the x-intercepts, we set $f(x) = 0$ and solve for x:

$$(x - 2)^4 - 1 = 0$$
$$(x - 2)^4 = 1$$
$$x - 2 = 1 \quad \text{or} \quad x - 2 = -1$$
$$x = 3 \quad \text{or} \quad x = 1.$$

(See Fig. 3.) ∎

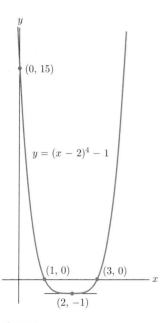

(0, 15)

$y = (x - 2)^4 - 1$

(1, 0) (3, 0)

(2, -1)

Figure 3

A Graph with Asymptotes Graphs similar to the one in the next example will arise in several applications later in this chapter.

EXAMPLE 4

Asymptotes Sketch the graph of $f(x) = x + (1/x)$, for $x > 0$.

Solution

$$f(x) = x + \frac{1}{x}$$

$$f'(x) = 1 - \frac{1}{x^2}$$

$$f''(x) = \frac{2}{x^3}$$

We set $f'(x) = 0$ and solve for x:

$$1 - \frac{1}{x^2} = 0$$

$$1 = \frac{1}{x^2}$$

$$x^2 = 1$$

$$x = 1.$$

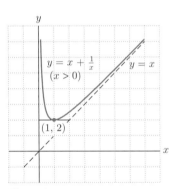

(We exclude the case $x = -1$ because we are only considering positive values of x.) The graph has a horizontal tangent at $(1, f(1)) = (1, 2)$. Since $f''(1) = 2 > 0$, the graph is concave up at $x = 1$ and, by the second derivative test, $(1, 2)$ is a relative minimum point. In fact, $f''(x) = (2/x^3) > 0$ for all positive x, and therefore the graph is concave up at all points.

Before sketching the graph, notice that as x approaches zero [a point at which $f(x)$ is not defined], the term $1/x$ in the formula for $f(x)$ is dominant. That is, this term becomes arbitrarily large, whereas the term x contributes a diminishing proportion to the function value as x approaches 0. Thus $f(x)$ has the y-axis as an asymptote. For large values of x, the term x is dominant. The value of $f(x)$ is only slightly larger than x since the term $1/x$ has decreasing significance as x becomes arbitrarily large; that is, the graph of $f(x)$ is slightly above the graph of $y = x$. As x increases, the graph of $f(x)$ has the line $y = x$ as an asymptote. (See Fig. 4.) ■

Figure 4

Summary of Curve-Sketching Techniques

1. Compute $f'(x)$ and $f''(x)$.

2. Find all relative extreme points.

(a) Find the critical values and critical points: Set $f'(x) = 0$ and solve for x. Suppose that $x = a$ is a solution (a critical value). Substitute $x = a$ into $f(x)$ to find $f(a)$, plot the critical point $(a, f(a))$, and draw a small horizontal tangent line through the point. Compute $f''(a)$.

The second derivative test

(i) If $f''(a) > 0$, draw a small concave up arc with $(a, f(a))$ as its lowest point. The curve has a relative minimum at $x = a$.

(ii) If $f''(a) < 0$, draw a small concave down arc with $(a, f(a))$ as its peak. The curve has a relative maximum at $x = a$.

The first derivative test

(iii) If $f''(a) = 0$, examine $f'(x)$ to the left and right of $x = a$ to determine if the function changes from increasing to decreasing, or vice versa. If a relative extreme point is indicated, draw an appropriate arc as in parts (i) and (ii).

(b) Repeat the preceding steps for each solution to $f'(x) = 0$.

3. Find all the inflection points of $f(x)$.

   (a) Set $f''(x) = 0$ and solve for x. Suppose that $x = b$ is a solution. Compute $f(b)$ and plot the point $(b, f(b))$.

   (b) Test the concavity of $f(x)$ to the right and left of b. If the concavity changes at $x = b$, then $(b, f(b))$ is an inflection point.

4. Consider other properties of the function and complete the sketch.

   (a) If $f(x)$ is defined at $x = 0$, the y-intercept is $(0, f(0))$.

   (b) Does the partial sketch suggest that there are x-intercepts? If so, they are found by setting $f(x) = 0$ and solving for x. (Solve only in easy cases or when a problem essentially requires you to calculate the x-intercepts.)

   (c) Observe where $f(x)$ is defined. Sometimes the function is given only for restricted values of x. Sometimes the formula for $f(x)$ is meaningless for certain values of x.

   (d) Look for possible asymptotes.

   (i) Examine the formula for $f(x)$. If some terms become insignificant as x gets large and if the rest of the formula gives the equation of a straight line, then that straight line is an asymptote.

   (ii) Suppose that there is some point a such that $f(x)$ is defined for x near a, but not at a (for example, $1/x$ at $x = 0$). If $f(x)$ gets arbitrarily large (in the positive or negative sense) as x approaches a, the vertical line $x = a$ is an asymptote for the graph.

   (e) Complete the sketch.

Practice Problems 2.4

Determine whether each of the following functions has an asymptote as x gets large. If so, give the equation of the straight line that is the asymptote.

1. $f(x) = \dfrac{3}{x} - 2x + 1$ **2.** $f(x) = \sqrt{x} + x$

3. $f(x) = \dfrac{1}{2x}$

EXERCISES 2.4

Find the x-intercepts of the given function.

1. $y = x^2 - 3x + 1$

2. $y = x^2 + 5x + 5$

3. $y = 2x^2 + 5x + 2$

4. $y = 4 - 2x - x^2$

5. $y = 4x - 4x^2 - 1$

6. $y = 3x^2 + 10x + 3$

7. Show that the function $f(x) = \frac{1}{3}x^3 - 2x^2 + 5x$ has no relative extreme points.

8. Show that the function $f(x) = -x^3 + 2x^2 - 6x + 3$ is always decreasing.

Sketch the graphs of the following functions.

9. $f(x) = x^3 - 6x^2 + 12x - 6$

10. $f(x) = -x^3$ **11.** $f(x) = x^3 + 3x + 1$

12. $f(x) = x^3 + 2x^2 + 4x$

13. $f(x) = 5 - 13x + 6x^2 - x^3$

14. $f(x) = 2x^3 + x - 2$ **15.** $f(x) = \frac{4}{3}x^3 - 2x^2 + x$

16. $f(x) = -3x^3 - 6x^2 - 9x - 6$

17. $f(x) = 1 - 3x + 3x^2 - x^3$

18. $f(x) = x^3 - 6x^2 + 12x - 5$

19. $f(x) = x^4 - 6x^2$ **20.** $f(x) = 3x^4 - 6x^2 + 3$

21. $f(x) = (x - 3)^4$ **22.** $f(x) = (x + 2)^4 - 1$

Sketch the graphs of the following functions for $x > 0$.

23. $y = \dfrac{1}{x} + \dfrac{1}{4}x$ **24.** $y = \dfrac{2}{x}$

25. $y = \dfrac{9}{x} + x + 1$ **26.** $y = \dfrac{12}{x} + 3x + 1$

27. $y = \dfrac{2}{x} + \dfrac{x}{2} + 2$

28. $y = \dfrac{1}{x^2} + \dfrac{x}{4} - \dfrac{5}{4}$ [*Hint*: $(1, 0)$ is an x-intercept.]

29. $y = 6\sqrt{x} - x$ **30.** $y = \dfrac{1}{\sqrt{x}} + \dfrac{x}{2}$

In Exercises 31 and 32, determine which function is the derivative of the other.

31.

32.

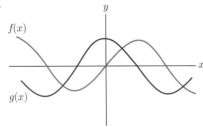

33. Find the quadratic function $f(x) = ax^2 + bx + c$ that goes through $(2, 0)$ and has a local maximum at $(0, 1)$.

34. Find the quadratic function $f(x) = ax^2 + bx + c$ that goes through $(0, 1)$ and has a local minimum at $(1, -1)$.

35. If $f'(a) = 0$ and $f'(x)$ is increasing at $x = a$, explain why $f(x)$ must have a local minimum at $x = a$. [*Hint:* Use the first derivative test.]

36. If $f'(a) = 0$ and $f'(x)$ is decreasing at $x = a$, explain why $f(x)$ must have a local maximum at $x = a$.

Technology Exercises

37. In a medical experiment,[*] the body weight of a baby rat in the control group after t days was $f(t) = 4.96 + .48t + .17t^2 - .0048t^3$ grams.

   (a) Graph $f(t)$ in the window $[0, 20]$ by $[-12, 50]$.

   (b) Approximately how much did the rat weigh after 7 days?

   (c) Approximately when did the rat's weight reach 27 grams?

   (d) Approximately how fast was the rat gaining weight after 4 days?

   (e) Approximately when was the rat gaining weight at the rate of 2 grams per day?

   (f) Approximately when was the rat gaining weight at the fastest rate?

38. The canopy height (in meters) of the tropical bunchgrass elephant millet[†] t days after mowing (for $t \geq 32$) is $f(t) = -3.14 + .142t - .0016t^2 + .0000079t^3 - .0000000133t^4$.

   (a) Graph $f(t)$ in the window $[32, 250]$ by $[-1.2, 4.5]$.

   (b) How tall was the canopy after 100 days?

   (c) When was the canopy 2 meters high?

   (d) How fast was the canopy growing after 80 days?

   (e) When was the canopy growing at the rate of .02 meters per day?

   (f) Approximately when was the canopy growing slowest?

   (g) Approximately when was the canopy growing fastest?

Solutions to Practice Problems 2.4

Functions with asymptotes as x gets large have the form $f(x) = g(x) + mx + b$, where $g(x)$ approaches zero as x gets large. The function $g(x)$ often looks like c/x or $c/(ax+d)$. The asymptote will be the straight line $y = mx + b$.

  1. Here $g(x)$ is $3/x$ and the asymptote is $y = -2x + 1$.

  2. This function has no asymptote as x gets large. Of

course, it can be written as $g(x) + mx + b$, where $m = 1$ and $b = 0$. However, $g(x) = \sqrt{x}$ does not approach zero as x gets large.

  3. Here $g(x)$ is $\dfrac{1}{2x}$ and the asymptote is $y = 0$. That is, the function has the x-axis as an asymptote.

---

[*]Johnson, Wogenrich, Hsi, Skipper, and Greenberg, "Growth Retardation During the Sucking Period in Expanded Litters of Rats; Observations of Growth Patterns and Protein Turnover," *Growth, Development and Aging*, 55(1991), 263–273.

[†]Woodward and Prine, "Crop Quality and Utilization," *Crop Science*, 33(1993), 818–824.

2.5 Optimization Problems

One of the most important applications of the derivative concept is to "optimization" problems, in which some quantity must be maximized or minimized. Examples of such problems abound in many areas of life. An airline must decide how many daily flights to schedule between two cities to maximize its profits. A doctor wants to find the minimum amount of a drug that will produce a desired response in one of her patients. A manufacturer needs to determine how often to replace certain equipment to minimize maintenance and replacement costs.

Our purpose in this section is to illustrate how calculus can be used to solve optimization problems. In each example we will find or construct a function that provides a "mathematical model" for the problem. Then, by sketching the graph of this function, we will be able to determine the answer to the original optimization problem by locating the highest or lowest point on the graph. The y-coordinate of this point will be the maximum value or minimum value of the function.

The first two examples are very simple because the functions to be studied are given explicitly.

EXAMPLE 1   Find the minimum value of the function $f(x) = 2x^3 - 15x^2 + 24x + 19$ for $x \geq 0$.

Solution   Using the curve-sketching techniques from Section 2.3, we obtain the graph in Fig. 1. As part of that process, we compute the derivatives:

$$f'(x) = 6x^2 - 30x + 24$$
$$f''(x) = 12x - 30.$$

The x-coordinate of the minimum point satisfies the equation

$$f'(x) = 0$$
$$6x^2 - 30x + 24 = 0$$
$$6(x - 4)(x - 1) = 0$$
$$x = 1, 4 \quad \text{(critical values)}.$$

The corresponding critical points on the curve are

$$(1, f(1)) = (1, 30)$$
$$(4, f(4)) = (4, 3).$$

Figure 1

Applying the second derivative test, we see that

$$f''(1) = 12 \cdot 1 - 30 = -18 < 0 \quad \text{(maximum)}$$
$$f''(4) = 12 \cdot 4 - 30 = 18 > 0 \quad \text{(minimum)}.$$

That is, the point $(1, 30)$ is a relative maximum and the point $(4, 3)$ is a relative minimum. The lowest point on the graph is $(4, 3)$. The minimum *value* of the function $f(x)$ is the y-coordinate of this point, 3.   ■

EXAMPLE 2   Suppose that a ball is thrown straight up into the air and its height after t seconds is $4 + 48t - 16t^2$ feet. Determine how long it will take for the ball to reach its maximum height and determine the maximum height.

Solution   Consider the function $f(t) = 4 + 48t - 16t^2$. For each value of t, $f(t)$ is the height of the ball at time t. We want to find the value of t for which $f(t)$ is the greatest. Using the techniques of Section 2.3, we sketch the graph of $f(t)$. (See Fig. 2.) Note that we may neglect the portions of the graph corresponding to points for which either $t < 0$ or $f(t) < 0$. [A negative value of $f(t)$ would correspond to the ball being underneath the ground.] The t-coordinate giving the maximum height is the solution of the equation:

$$f'(t) = 48 - 32t = 0$$
$$t = \tfrac{3}{2}.$$

Since

$$f''(t) = -32$$
$$f''(\tfrac{3}{2}) = -32 < 0 \quad (\text{maximum}),$$

we see from the second derivative test that $t = \tfrac{3}{2}$ is the location of a relative maximum. So $f(t)$ is greatest when $t = \tfrac{3}{2}$. At this value of t, the ball attains a height of 40 feet. Thus the ball reaches its maximum height of 40 feet in 1.5 seconds. [Note that the curve in Fig. 2 is the graph of $f(t)$, *not* a picture of the physical path of the ball.] ■

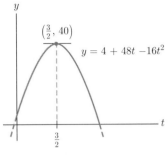

Figure 2

In the figure: $\left(\frac{3}{2}, 40\right)$; $y = 4 + 48t - 16t^2$; axis labeled y and t with $\frac{3}{2}$ marked.

EXAMPLE 3

Maximizing an area You want to plant a rectangular garden along one side of a house, with a picket fence on the other three sides of the garden. Find the dimensions of the largest garden that can be enclosed using 40 feet of fencing.

Solution Let us think about this problem before embarking on the solution. With 40 feet of fencing material, you can enclose a rectangular garden along your house in many different ways. Here are three illustrations: In Fig. 3(a), the enclosed area is $10 \times 15 = 150$ square feet; in Fig. 3(b), it is $16 \times 12 = 192$ square feet; and in Fig. 3(c), it is $32 \times 4 = 128$ square feet. Clearly, the enclosed area varies with your choice of dimensions. Which dimensions that total 40 feet yield the largest area? We next show how we can use calculus and optimization techniques to solve this problem.

(a) Area = $10 \times 15 = 150$ sq. ft. (b) Area = $16 \times 12 = 192$ sq. ft. (c) Area = $32 \times 4 = 128$ sq. ft.

Figure 3. Rectangular gardens enclosed with 40 feet of fencing.

Area = $w \cdot x$ sq. ft.
Fencing required = $2x + w$ ft.

Figure 4. General case.

Since we do not know the dimensions, the first step is to make a simple diagram that represents the general case and assign letters to the quantities that may vary. Let us denote the dimensions of the rectangular garden by w and x (Fig. 4). The phrase "largest garden" indicates that we must maximize the area, A, of the garden. In terms of the variables w and x,

$$A = wx. \tag{1}$$

The fencing on three sides must total 40 running feet; that is,

$$2x + w = 40. \tag{2}$$

We now solve equation (2) for w in terms of x:

$$w = 40 - 2x. \tag{3}$$

Substituting this expression for w into equation (1), we have

$$A = (40 - 2x)x = 40x - 2x^2. \tag{4}$$

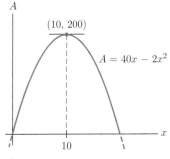

Figure 5

We now have a formula for the area A that depends on just one variable. From the statement of the problem, the value of $2x$ can be at most 40, so the domain of the function consists of x in the interval $(0, 20)$. Thus the function that we wish to maximize is

$$A(x) = 40x - 2x^2 \quad 0 \le x \le 20.$$

Its graph is a parabola looking downward (Fig. 5). To find the maximum point, we compute

$$A'(x) = 0$$
$$40 - 4x = 0$$
$$x = 10.$$

Since this value is in the domain of our function, we conclude that the absolute maximum of $A(x)$ occurs at $x = 10$. Alternatively, since $A''(x) = -4 < 0$, the concavity of the curve is always downward, and so the local maximum at $x = 10$ must be an absolute maximum. [The maximum area is $A(10) = 200$ square feet, but this fact is not needed for the problem.] From equation (3) we find that, when $x = 10$,

$$w = 40 - 2(10) = 20.$$

In conclusion, the rectangular garden with the largest area that can be enclosed using 40 feet of fencing has dimensions $x = 10$ feet and $w = 20$ feet. ∎

Equation (1) in Example 3 is called an *objective equation*. It expresses the quantity to be optimized (the area of the garden) in terms of the variables w and x. Equation (2) is called a *constraint equation* because it places a limit or constraint on the way x and w may vary.

EXAMPLE 4

Minimizing cost The manager of a department store wants to build a 600-square-foot rectangular enclosure on the store's parking lot to display some equipment. Three sides of the enclosure will be built of redwood fencing at a cost of $14 per running foot. The fourth side will be built of cement blocks, at a cost of $28 per running foot. Find the dimensions of the enclosure that will minimize the total cost of the building materials.

Solution Let x be the length of the side built out of cement blocks and let y be the length of an adjacent side, as shown in Fig. 6. The phrase "minimize the total cost" tells us that the objective equation should be a formula giving the total cost of the building materials.

$$[\text{cost of redwood}] = [\text{length of redwood fencing}] \times [\text{cost per foot}]$$
$$= (x + 2y) \cdot 14 = 14x + 28y$$
$$[\text{cost of cement blocks}] = [\text{length of cement wall}] \times [\text{cost per foot}]$$
$$= x \cdot 28$$

If C denotes the total cost of the materials, then

$$C = (14x + 28y) + 28x$$
$$C = 42x + 28y \quad \text{(objective equation)}. \tag{5}$$

Since the area of the enclosure must be 600 square feet, the constraint equation is

$$xy = 600. \tag{6}$$

Figure 6. Rectangular enclosure.

We simplify the objective equation by solving equation (6) for one of the variables, say y, and substituting into (5). Since $y = 600/x$,

$$C = 42x + 28 \left(\frac{600}{x} \right) = 42x + \frac{16,800}{x}.$$

We now have C as a function of the single variable x. From the context, we must have $x > 0$, since a length must be positive. However, to any positive value for x, there is a corresponding value for C. So the domain of C consists of all $x > 0$. We may now sketch the graph of C. (See Fig. 7.) (A similar curve was sketched in Example 4 of Section 2.4.) The x-coordinate of the minimum point is a solution of

$$C'(x) = 0$$

$$42 - \frac{16,800}{x^2} = 0$$

$$42x^2 = 16,800$$

$$x^2 = 400$$

$$x = 20.$$

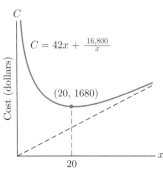

Figure 7

The corresponding value of C is

$$C(20) = 42(20) + \frac{16,800}{20} = \$1680.$$

That is, the minimum total cost of \$1680 occurs where $x = 20$. From equation (6) we find that the corresponding value of y is $\frac{600}{20} = 30$. To minimize the total cost of building the 600-square-foot rectangular enclosure, the manager must take the dimensions to be $x = 20$ feet and $y = 30$ feet. ■

EXAMPLE 5

Maximizing volume Parcel post regulations state that packages must have length plus girth of no more than 84 inches. Find the dimensions of the cylindrical package of greatest volume that is mailable by parcel post.

Solution Let l be the length of the package and let r be the radius of the circular end. (See Fig. 8.) The phrase "greatest volume" tells us that the objective equation should express the volume of the package in terms of the dimensions l and r. Let V denote the volume. Then

$$V = [\text{area of base}] \cdot [\text{length}]$$

$$V = \pi r^2 l \quad \text{(objective equation)}. \tag{7}$$

The girth equals the circumference of the end, that is, $2\pi r$. Since we want the package to be as large as possible, we must use the entire 84 inches allowable:

Figure 8. Cylindrical mailing package.

$$\text{length} + \text{girth} = 84$$

$$l + 2\pi r = 84 \quad \text{(constraint equation)}. \tag{8}$$

We now solve equation (8) for one of the variables, say $l = 84 - 2\pi r$. Substituting this expression into (7), we obtain

$$V = \pi r^2(84 - 2\pi r) = 84\pi r^2 - 2\pi^2 r^3. \tag{9}$$

Let $f(r) = 84\pi r^2 - 2\pi^2 r^3$. Then, for each value of r, $f(r)$ is the volume of the parcel with end radius r that meets the postal regulations. We want to find that value of r for which $f(r)$ is as large as possible.

Using curve-sketching techniques, we obtain the graph of $f(r)$ in Fig. 9. The domain excludes values of r that are negative and values of r for which the volume $f(r)$ is negative. Points corresponding to values of r not in the domain are shown with a dashed curve. We see that the volume is greatest when $r = 28/\pi$.

From equation (8) we find that the corresponding value of l is

$$l = 84 - 2\pi r = 84 - 2\pi\left(\frac{28}{\pi}\right) = 84 - 56 = 28.$$

The girth when $r = 28/\pi$ is

$$2\pi r = 2\pi\left(\frac{28}{\pi}\right) = 56.$$

Thus the dimensions of the cylindrical package of greatest volume are $l = 28$ inches and $r = 28/\pi$ inches. ∎

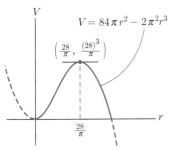

Figure 9

Suggestions for Solving an Optimization Problem

1. Draw a picture, if possible.

2. Decide what quantity Q is to be maximized or minimized.

3. Assign letters to other quantities that may vary.

4. Determine the "objective equation" that expresses Q as a function of the variables assigned in step 3.

5. Find the "constraint equation" that relates the variables to each other and to any constants that are given in the problem.

6. Use the constraint equation to simplify the objective equation in such a way that Q becomes a function of only one variable. Determine the domain of this function.

7. Sketch the graph of the function obtained in step 6 and use this graph to solve the optimization problem. Alternatively, you can use the second derivative test.

NOTE Optimization problems often involve geometric formulas. The most common formulas are shown in Fig. 10. ∎

Figure 10

Area $= xy$
Perimeter $= 2x + 2y$

Volume $= xyz$

Area $= \pi r^2$
Circumference $= 2\pi r$

Volume $= \pi r^2 h$

Practice Problems 2.5

1. *(Volume)* A canvas wind shelter for the beach has a back, two square sides, and a top (Fig. 11). If 96 square feet of canvas is to be used, find the dimensions of the shelter for which the space inside the shelter (the volume) is maximized.

2. In Practice Problem 1, what are the objective equation and the constraint equation?

Figure 11.   Wind shelter.

EXERCISES 2.5

1. For what x does the function $g(x) = 10 + 40x - x^2$ have its maximum value?

2. Find the maximum value of the function $f(x) = 12x - x^2$ and give the value of x where this maximum occurs.

3. Find the minimum value of $f(t) = t^3 - 6t^2 + 40$, $t \geq 0$, and give the value of t where this minimum occurs.

4. For what t does the function $f(t) = t^2 - 24t$ have its minimum value?

5. Find the maximum of $Q = xy$ if $x + y = 2$.

6. Find two positive numbers x and y that maximize $Q = x^2 y$ if $x + y = 2$.

7. Find the minimum of $Q = x^2 + y^2$ if $x + y = 6$.

8. In Exercise 7, can there be a maximum for $Q = x^2 + y^2$ if $x + y = 6$? Justify your answer.

9. Find the positive values of x and y that minimize $S = x + y$ if $xy = 36$, and find this minimum value.

10. Find the positive values of x, y, and z that maximize $Q = xyz$, if $x + y = 1$ and $y + z = 2$. What is this maximum value?

11. **Area** There are $320 available to fence in a rectangular garden. The fencing for the side of the garden facing the road costs $6 per foot and the fencing for the other three sides costs $2 per foot. [See Fig. 12(a).] Consider the problem of finding the dimensions of the largest possible garden.

    (a) Determine the objective and constraint equations.

    (b) Express the quantity to be maximized as a function of x.

    (c) Find the optimal values of x and y.

12. **Volume** Figure 12(b) shows an open rectangular box with a square base. Consider the problem of finding the values of x and h for which the volume is 32 cubic feet and the total surface area of the box is minimal.

(The surface area is the sum of the areas of the five faces of the box.)

(a) Determine the objective and constraint equations.

(b) Express the quantity to be minimized as a function of x.

(c) Find the optimal values of x and h.

(a)   (b)

Figure 12

13. **Volume** Postal requirements specify that parcels must have length plus girth of at most 84 inches. Consider the problem of finding the dimensions of the square-ended rectangular package of greatest volume that is mailable.

    (a) Draw a square-ended rectangular box. Label each edge of the square end with the letter x and label the remaining dimension of the box with the letter h.

    (b) Express the length plus the girth in terms of x and h.

    (c) Determine the objective and constraint equations.

    (d) Express the quantity to be maximized as a function of x.

    (e) Find the optimal values of x and h.

14. **Perimeter** Consider the problem of finding the dimensions of the rectangular garden of area 100 square meters for which the amount of fencing needed to surround the garden is as small as possible.

    (a) Draw a picture of a rectangle and select appropriate letters for the dimensions.

    (b) Determine the objective and constraint equations.

    (c) Find the optimal values for the dimensions.

15. **Cost** A rectangular garden of area 75 square feet is to be surrounded on three sides by a brick wall costing $10 per foot and on one side by a fence costing $5 per foot. Find the dimensions of the garden such that the cost of materials is minimized.

16. **Cost** A closed rectangular box with square base and a volume of 12 cubic feet is to be constructed using two different types of materials. The top is made of a metal costing $2 per square foot and the remainder of wood costing $1 per square foot. Find the dimension of the box for which the cost of materials is minimized.

17. **Surface Area** Find the dimensions of the closed rectangular box with square base and volume 8000 cubic centimeters that can be constructed with the least amount of material.

18. **Volume** A canvas wind shelter for the beach has a back, two square sides, and a top. Find the dimensions for which the volume will be 250 cubic feet and that requires the least possible amount of canvas.

19. **Area** A farmer has $1500 available to build an E-shaped fence along a straight river so as to create two identical rectangular pastures. (See Fig. 13.) The materials for the side parallel to the river cost $6 per foot, and the materials for the three sections perpendicular to the river cost $5 per foot. Find the dimensions for which the total area is as large as possible.

Figure 13. Rectangular pastures along a river.

20. **Area** Find the dimensions of the rectangular garden of greatest area that can be fenced off (all four sides) with 300 meters of fencing.

21. Find two positive numbers, x and y, whose sum is 100 and whose product is as large as possible.

22. Find two positive numbers, x and y, whose product is 100 and whose sum is as small as possible.

23. **Area** Figure 14(a) shows a Norman window, which consists of a rectangle capped by a semicircular region.

Find the value of x such that the perimeter of the window will be 14 feet and the area of the window will be as large as possible.

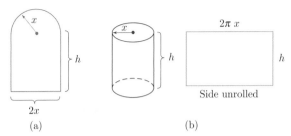

Side unrolled

(a)   (b)

Figure 14

24. **Surface Area** A large soup can is to be designed so that the can will hold 16π cubic inches (about 28 ounces) of soup. [See Fig. 14(b).] Find the values of x and h for which the amount of metal needed is as small as possible.

25. In Example 3 we can solve the constraint equation (2) for x instead of w to get $x = 20 - \frac{1}{2}w$. Substituting this for x in (1), we get

$$A = xw = \left(20 - \tfrac{1}{2}w\right)w.$$

Sketch the graph of the equation

$$A = 20w - \tfrac{1}{2}w^2$$

and show that the maximum occurs when $w = 20$ and $x = 10$.

26. **Cost** A ship uses $5x^2$ dollars of fuel per hour when traveling at a speed of x miles per hour. The other expenses of operating the ship amount to $2000 per hour. What speed minimizes the cost of a 500-mile trip? [*Hint:* Express cost in terms of speed and time. The constraint equation is *distance* = *speed* × *time*.]

27. **Cost** A cable is to be installed from one corner C of a rectangular factory to a machine M on the floor. The cable will run along one edge of the floor from C to a point P at a cost of $6 per meter, and then from P to M in a straight line buried under the floor at a cost of $10 per meter (see Fig. 15). Let x denote the distance from C to P. If M is 24 meters from the nearest point A on the edge of the floor on which P lies, and A is 20 meters from C, find the value of x that minimizes the cost of installing the cable and determine the minimum cost.

28. **Area** A rectangular page is to contain 50 square inches of print. The page has to have a 1-inch margin on top and at the bottom and a 1/2-inch margin on each side (see Fig. 16). Find the dimensions of the page that minimize the amount of paper used.

29. **Distance** Find the point on the graph of $y = \sqrt{x}$ that is closest to the point $(2, 0)$. See Fig. 17. [*Hint:* $\sqrt{(x-2)^2 + y^2}$ has its smallest value when $(x-2)^2 + y^2$ does. Therefore, just minimize the second expressi

Figure 15

Figure 16

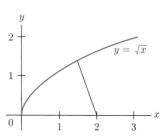

Figure 17. Shortest distance from a point to a curve.

30. Distance Developers from two towns, A and B, want to connect their towns to a major highway and plan to build rest stops and gas stations at the highway entrance. To minimize the cost of road construction, the developers must find the location for the highway entrance that minimizes the total distance, $d_1 + d_2$, from the towns A and B to the highway. Let x be as in Fig. 18. Find the value of x that solves the developers' problem, and find the minimum total distance.

31. Distance Find the point on the line $y = -2x + 5$ that is closest to the origin.

Technology Exercises

32. Find the value of x for which the rectangle inscribed in the semicircle of radius 3 in Fig. 19 has the greatest area. [*Note*: The equation of the semicircle is $y = \sqrt{9 - x^2}$.]

Figure 18

Figure 19

Solutions to Practice Problems 2.5

1. Since the sides of the wind shelter are square, we may let x represent the length of each side of the square. The remaining dimension of the wind shelter can be denoted by the letter h. (See Fig. 20.)

Figure 20

The volume of the shelter is x^2h, and this is to be maximized. Since we have learned to maximize only functions of a single variable, we must express h in terms of x. We must use the information that 96 feet of canvas are used; that is, $2x^2 + 2xh = 96$. (*Note*: The roof and the back each have an area xh, and each end has an area x^2.) We now solve this equation for h.

$$2x^2 + 2xh = 96$$
$$2xh = 96 - 2x^2$$
$$h = \frac{96}{2x} - \frac{2x^2}{2x} = \frac{48}{x} - x$$

The volume V is

$$x^2h = x^2\left(\frac{48}{x} - x\right) = 48x - x^3.$$

By sketching the graph of $V = 48x - x^3$, we see that V has a maximum value when $x = 4$. Then $h = \frac{48}{4} - 4 = 12 - 4 = 8$. So each end of the shelter should be a 4-foot by 4-foot square and the top should be 8 feet long.

2. The objective equation is $V = x^2h$, since it expresses the volume (the quantity to be maximized) in terms of the variables. The constraint equation is $2x^2 + 2xh = 96$, for it relates the variables to each other; that is, it can be used to express one of the variables in terms of the other.

2.6 Further Optimization Problems

In this section we apply the optimization techniques developed in Section 2.5 to some practical situations.

Inventory Control   When a firm regularly orders and stores supplies for later use or resale, it must decide on the size of each order. If it orders enough supplies to last an entire year, the business will incur heavy *carrying costs*. Such costs include insurance, storage costs, and cost of capital that is tied up in inventory. To reduce these carrying costs, the firm could order small quantities of the supplies at frequent intervals. However, such a policy increases the *ordering costs*. These might consist of minimum freight charges, the clerical costs of preparing the orders, and the costs of receiving and checking the orders when they arrive. Clearly, the firm must find an inventory ordering policy that lies between these two extremes.

To appreciate the inventory-control problem, let us first consider a simple problem that does not involve optimization.

EXAMPLE 1

A supermarket manager anticipates that 1200 cases of frozen orange juice will be sold at a steady rate during the next year. The manager plans on ordering these cases by placing equally spaced orders of the same size throughout the year. Given that it costs $8 to carry one case of orange juice in inventory for 1 year, find the carrying cost if:

(a) The manager places only one order during the year.

(b) The manager places two orders during the year.

(c) The manager places four orders during the year.

Carrying costs should be computed on the average inventory during the order-reorder period.

Solution **(a)** This plan calls for ordering 1200 cases all at once. The inventory of frozen orange juice in this case is shown in Fig. 1(a). It starts at 1200 and decreases steadily to 0 during this one order–reorder period. At any given time during the year, the inventory is between 1200 and 0. So there are $\frac{1200}{2}$, or 600 cases on average in inventory. Since carrying costs are computed on the average inventory, the carrying cost in this case is $C = 600 \times 8 = 4800$ dollars.

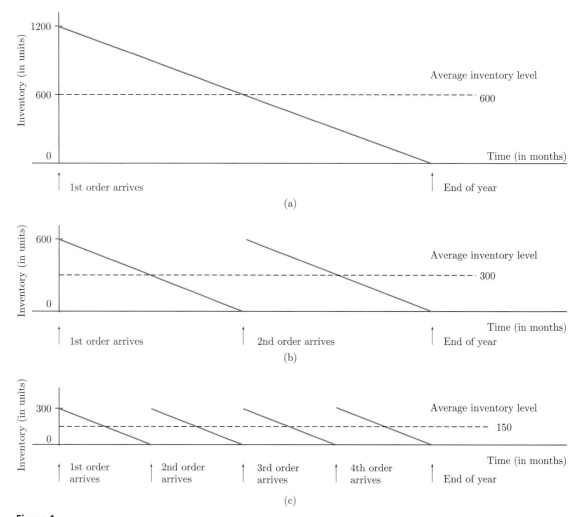

Figure 1

(b) This plan calls for ordering 1200 cases by placing two equally spaced orders of the same size. Thus the size of each order is equal to $\frac{1200}{2}$ or 600 cases. The inventory of frozen orange juice in this case is shown in Fig. 1(b). At the beginning of each order–reorder period, the inventory starts at 600 and decreases steadily to 0 at the end of the order–reorder period. At any given time during the order–reorder period, the inventory is between 600 and 0. So there are $\frac{600}{2}$, or 300 cases on average in inventory. Since the carrying costs are computed on the average inventory, the carrying cost in this case is $C = 300 \times 8 = 2400$ dollars.

(c) This plan calls for ordering 1200 cases by placing four equally spaced orders of the same size. The size of each order is $\frac{1200}{4} = 300$ cases. The inventory in this case is shown in Fig. 1(c). The average inventory during an order–reorder period is $\frac{300}{2}$, or 150 cases, and so the carrying cost in this case is $C = 150 \times 8 = 1200$ dollars. ∎

By increasing the number of orders in Example 1, the manager was able to reduce the carrying cost from $4800 down to $2400 down to $1200. Clearly, the manager can reduce the carrying cost even further by ordering more frequently. In reality, things are not so simple. There is a cost attached to each order, and by increasing the number of orders, the manager will increase the annual inventory cost, where

$$[\text{inventory cost}] = [\text{ordering cost}] + [\text{carrying cost}].$$

The inventory-control problem consists of finding the *economic order quantity*, commonly referred to in business as the EOQ,[*] that minimizes the inventory cost. We now show how we can use calculus to solve this problem.

EXAMPLE 2

Inventory control Suppose that the manager in Example 1 wants to establish an optimal inventory policy for frozen orange juice. Again, it is estimated that a total of 1200 cases will be sold at a steady rate during the next year. The manager plans to place several orders of the same size spaced equally throughout the year. Use the following data to determine the economic order quantity; that is, the order size that minimizes the total ordering and carrying cost.

1. The ordering cost for each delivery is $75.

2. It costs $8 to carry one case of orange juice in inventory for one year. (Carrying costs should be computed on the average inventory during the order–reorder period.)

Solution Since we are not given the size of each order or the number of orders that will be placed throughout the year, we let x be the order quantity and r the number of orders placed during the year. As in Example 1, the number of cases of orange juice in inventory declines steadily from x cases (each time a new order is filled) to 0 cases at the end of each order–reorder period. Figure 2 shows that the average number of cases in storage during the year is $x/2$. Since the carrying cost for one

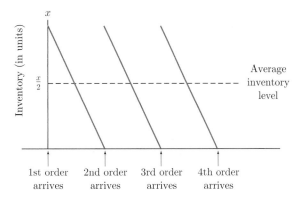

Figure 2. Average inventory level.

[*]James C. Van Horne, *Financial Management and Policy*, 6th ed. (Upper Saddle River, N.J.: Prentice-Hall, 1983), pp. 416–420.

case is \$8 per year, the cost for $x/2$ cases is $8 \cdot (x/2)$ dollars. Now

$$[\text{inventory cost}] = [\text{ordering cost}] + [\text{carrying cost}]$$

$$= 75r + 8 \cdot \frac{x}{2}$$

$$= 75r + 4x.$$

If C denotes the inventory cost, then the objective equation is

$$C = 75r + 4x.$$

Since there are r orders of x cases each, the total number of cases ordered during the year is $r \cdot x$. Therefore, the constraint equation is

$$r \cdot x = 1200.$$

The constraint equation says that $r = 1200/x$. Substitution into the objective equation yields

$$C = \frac{90{,}000}{x} + 4x.$$

Figure 3 is the graph of C as a function of x, for $x > 0$. The minimum point of the graph occurs where the first derivative equals zero. We find this point using the calculation

$$C'(x) = -\frac{90{,}000}{x^2} + 4 = 0$$

$$4x^2 = 90{,}000$$

$$x^2 = 22{,}500$$

$$x = 150 \quad (\text{ignore the negative root since } x > 0).$$

The total cost is at a minimum when $x = 150$. Therefore, the optimum inventory policy is to order 150 cases at a time and to place $1200/150 = 8$ orders during the year.

Figure 3. Cost function for inventory problem.

EXAMPLE 3 What should the inventory policy of Example 2 be if sales of frozen orange juice increase fourfold (i.e., 4800 cases are sold each year), but all other conditions are the same?

Solution The only change in our previous solution is in the constraint equation, which now becomes

$$r \cdot x = 4800.$$

The objective equation is, as before,

$$C = 75r + 4x.$$

Since $r = 4800/x$,

$$C = 75 \cdot \frac{4800}{x} + 4x = \frac{360{,}000}{x} + 4x.$$

Now

$$C' = -\frac{360{,}000}{x^2} + 4.$$

Setting $C' = 0$ yields

$$\frac{360{,}000}{x^2} = 4$$

$$90{,}000 = x^2$$

$$x = 300.$$

Therefore, the economic order quantity is 300 cases.

Notice that although the sales increased by a factor of 4, the economic order quantity increased by only a factor of 2 ($= \sqrt{4}$). In general, a store's inventory of an item should be proportional to the square root of the expected sales. (See Exercise 9 for a derivation of this result.) Many stores tend to keep their average inventories at a fixed percentage of sales. For example, each order may contain enough goods to last for 4 or 5 weeks. This policy is likely to create excessive inventories of high-volume items and uncomfortably low inventories of slower-moving items.

Manufacturers have an inventory-control problem similar to that of retailers. They have the carrying costs of storing finished products and the start-up costs of setting up each production run. The size of the production run that minimizes the sum of these two costs is called the *economic lot size*. See Exercises 6 and 7.

EXAMPLE 4

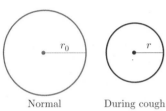

Figure 4. Contraction of the windpipe during coughing.

Lungs and trachea clearing Doctors studying coughing problems need to know the most effective condition for clearing the lungs and trachea (windpipe), that is, the radius of the trachea that gives the greatest air velocity. Using the following principles of fluid flow, determine this optimum value. (See Fig. 4.) Let

$$r_0 = \text{normal radius of the trachea}$$
$$r = \text{radius during a cough}$$
$$P = \text{increase in air pressure in the trachea during cough}$$
$$v = \text{velocity of air through trachea during cough.}$$

(1) $r_0 - r = aP$, for some positive constant a. (Experiment has shown that, during coughing, the decrease in the radius of the trachea is nearly proportional to the increase in the air pressure.)

(2) $v = b \cdot P \cdot \pi r^2$, for some positive constant b. (The theory of fluid flow requires that the velocity of the air forced through the trachea be proportional to the product of the increase in the air pressure and the area of a cross section of the trachea.)

Solution   In this problem the constraint equation (1) and the objective equation (2) are given directly. Solving equation (1) for P and substituting this result into equation (2), we have

$$v = b \left(\frac{r_0 - r}{a} \right) \pi r^2 = k(r_0 - r)r^2,$$

where $k = b\pi/a$. To find the radius at which the velocity v is a maximum, we first compute the derivatives with respect to r:

$$v = k(r_0 r^2 - r^3)$$

$$\frac{dv}{dr} = k(2r_0 r - 3r^2) = kr(2r_0 - 3r)$$

$$\frac{d^2 v}{dr^2} = k(2r_0 - 6r).$$

Figure 5. Graph for the windpipe problem.

We see that $\frac{dv}{dr} = 0$ when $r = 0$ or when $2r_0 - 3r = 0$; that is, when $r = \frac{2}{3}r_0$. It is easy to see that $\frac{d^2 v}{dr^2}$ is positive at $r = 0$ and negative at $r = \frac{2}{3}r_0$. The graph of v as a function of r is drawn in Fig. 5. The air velocity is maximized at $r = \frac{2}{3}r_0$. ∎

When solving optimization problems, we look for the maximum or minimum point on a graph. In our discussion of curve sketching, we saw that this point occurs either at a relative extreme point or at an endpoint of the domain of definition. In all our optimization problems so far, the maximum or minimum points were at relative extreme points. In the next example, the optimum point is an endpoint. This occurs often in applied problems.

EXAMPLE 5   **Area** A rancher has 204 meters of fencing from which to build two corrals: one square and the other rectangular with length that is twice the width. Find the dimensions that result in the greatest combined area.

Solution Let x be the width of the rectangular corral and h be the length of each side of the square corral. (See Fig. 6.) Let A be the combined area. Then

$$A = [\text{area of square}] + [\text{area of rectangle}] = h^2 + 2x^2.$$

The constraint equation is

$$204 = [\text{perimeter of square}] + [\text{perimeter of rectangle}] = 4h + 6x.$$

Since the perimeter of the rectangle cannot exceed 204, we must have $0 \leq 6x \leq 204$, or $0 \leq x \leq 34$. Solving the constraint equation for h and substituting into the objective equation leads to the function graphed in Fig. 7. The graph reveals that the area is minimized when $x = 18$. However, the problem asks for the *maximum* possible area. From Fig. 7 we see that this occurs at the endpoint where $x = 0$. Therefore, the rancher should build only the square corral, with $h = 204/4 = 51$ meters. In this example, the objective function has an endpoint extremum: The maximum value occurs at the endpoint where $x = 0$. ∎

Figure 6. Two corrals.

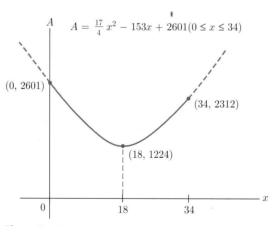

Figure 7. Combined area of the corrals.

Practice Problems 2.6

1. In the inventory problem of Example 2, suppose that the ordering cost is the same, but that it costs \$9 to carry one case of orange juice in inventory for 1 year, and carrying costs are computed on the *maximum* inventory during the order–reorder period. What is the new economic order quantity?

2. In the inventory problem of Example 2, suppose that the sales of frozen orange juice increase ninefold; that is, 10,800 cases are sold each year. What is the new economic order quantity?

EXERCISES 2.6

1. Figure 8 shows the inventory levels of dried Rainier cherries at a natural food store in Seattle and the order–reorder periods over 1 year. Refer to the figure to answer the following questions.

   (a) What is the average amount of cherries in inventory during one order–reorder period?

   (b) What is the maximum amount of cherries in inventory during one order–reorder period?

   (c) How many orders were placed during the year?

   (d) How many pounds of dried cherries were sold during 1 year?

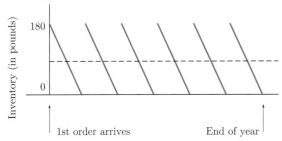

Figure 8

2. Refer to Fig. 8. Suppose that

   (i) the ordering cost for each delivery of dried cherries is $50, and

   (ii) it costs $7 to carry 1 pound of dried cherries in inventory for 1 year.

   (a) What is the inventory cost (carrying cost plus ordering cost) if carrying costs are computed on the average inventory during the order–reorder period?

   (b) What is the inventory cost if carrying costs are computed on the maximum inventory during the order–reorder period?

3. **Inventory Control** A pharmacist wants to establish an optimal inventory policy for a new antibiotic that requires refrigeration in storage. The pharmacist expects to sell 800 packages of this antibiotic at a steady rate during the next year. She plans to place several orders of the same size spaced equally throughout the year. The ordering cost for each delivery is $16, and carrying costs, based on the average number of packages in inventory, amount to $4 per year for one package.

   (a) Let x be the order quantity and r the number of orders placed during the year. Find the inventory cost (ordering cost plus carrying cost) in terms of x and r.

   (b) Find the constraint function.

   (c) Determine the economic order quantity that minimizes the inventory cost and then find the minimum inventory cost.

4. **Inventory Control** A furniture store expects to sell 640 sofas at a steady rate next year. The manager of the store plans to order these sofas from the manufacturer by placing several orders of the same size spaced equally throughout the year. The ordering cost for each delivery is $160, and carrying costs, based on the average number of sofas in inventory, amount to $32 per year for one sofa.

   (a) Let x be the order quantity and r the number of orders placed during the year. Find the inventory cost in terms of x and r.

   (b) Find the constraint function.

   (c) Determine the economic order quantity that minimizes the inventory cost and then find the minimum inventory cost.

5. **Inventory Control** A California distributor of sporting equipment expects to sell 10,000 cases of tennis balls during the coming year at a steady rate. Yearly carrying costs (to be computed on the average number of cases in stock during the year) are $10 per case, and the cost of placing an order with the manufacturer is $80.

   (a) Find the inventory cost incurred if the distributor orders 500 cases at a time during the year.

   (b) Determine the economic order quantity, that is, the order quantity that minimizes the inventory cost.

6. **Economic Lot Size** The Great American Tire Co. expects to sell 600,000 tires of a particular size and grade during the next year. Sales tend to be roughly the same from month to month. Setting up each production run costs the company $15,000. Carrying costs, based on the average number of tires in storage, amount to $5 per year for one tire.

   (a) Determine the costs incurred if there are 10 production runs during the year.

   (b) Find the economic lot size (that is, the production run size that minimizes the overall cost of producing the tires).

7. **Economic Lot Size** Foggy Optics, Inc. makes laboratory microscopes. Setting up each production run costs $2500. Insurance costs, based on the average number of microscopes in the warehouse, amount to $20 per microscope per year. Storage costs, based on the maximum number of microscopes in the warehouse, amount to $15 per microscope per year. If the company expects to sell 1600 microscopes at a fairly uniform rate throughout the year, determine the number of production runs that will minimize the company's overall expenses.

8. **Inventory Control** A bookstore is attempting to determine the most economical order quantity for a popular

book. The store sells 8000 copies of this book per year. The store figures that it costs \$40 to process each new order for books. The carrying cost (due primarily to interest payments) is \$2 per book, to be figured on the maximum inventory during an order–reorder period. How many times a year should orders be placed?

9. **Inventory Control** A store manager wants to establish an optimal inventory policy for an item. Sales are expected to be at a steady rate and should total Q items sold during the year. Each time an order is placed a cost of h dollars is incurred. Carrying costs for the year will be s dollars per item, to be figured on the average number of items in storage during the year. Show that the total inventory cost is minimized when each order calls for $\sqrt{2hQ/s}$ items.

10. Refer to the inventory problem of Example 2. If the distributor offers a discount of \$1 per case for orders of 600 or more cases, should the manager change the quantity ordered?

11. **Area** Starting with a 100-foot-long stone wall, a farmer would like to construct a rectangular enclosure by adding 400 feet of fencing, as shown in Fig. 9(a). Find the values of x and w that result in the greatest possible area.

(a)   (b)

Figure 9. Rectangular enclosures.

12. Rework Exercise 11 for the case where only 200 feet of fencing is added to the stone wall.

13. **Length** A rectangular corral of 54 square meters is to be fenced off and then divided by a fence into two sections, as shown in Fig. 9(b). Find the dimensions of the corral so that the amount of fencing required is minimized.

14. Refer to Exercise 13. If the cost of the fencing for the boundary is \$5 per meter and the dividing fence costs \$2 per meter, find the dimensions of the corral that minimize the cost of the fencing.

15. **Revenue** Shakespear's Pizza sells 1000 large vegi pizzas per week for \$18 a pizza. When the owner offers a \$5 discount, the weekly sales increase to 1500.

   (a) Assume a linear relation between the weekly sales $A(x)$ and the discount x. Find $A(x)$.

   (b) Find the value of x that maximizes the weekly revenue. [*Hint*: Revenue $= A(x) \cdot (\text{Price})$.]

   (c) Answer parts (a) and (b) if the price of one pizza is \$9 and all other data are unchanged.

16. **Surface Area** Design an open rectangular box with square ends, having volume 36 cubic inches, that minimizes the amount of material required for construction.

17. **Cost** A storage shed is to be built in the shape of a box with a square base. It is to have a volume of 150 cubic feet. The concrete for the base costs \$4 per square foot, the material for the roof costs \$2 per square foot, and the material for the sides costs \$2.50 per square foot. Find the dimensions of the most economical shed.

18. **Cost** A supermarket is to be designed as a rectangular building with a floor area of 12,000 square feet. The front of the building will be mostly glass and will cost \$70 per running foot for materials. The other three walls will be constructed of brick and cement block, at a cost of \$50 per running foot. Ignore all other costs (labor, cost of foundation and roof, and the like) and find the dimensions of the base of the building that will minimize the cost of the materials for the four walls of the building.

19. **Volume** A certain airline requires that rectangular packages carried on an airplane by passengers be such that the sum of the three dimensions is at most 120 centimeters. Find the dimensions of the square-ended rectangular package of greatest volume that meets this requirement.

20. **Area** An athletic field [Fig. 10(a)] consists of a rectangular region with a semicircular region at each end. The perimeter will be used for a 440-yard track. Find the value of x for which the area of the rectangular region is as large as possible.

21. **Volume** An open rectangular box is to be constructed by cutting square corners out of a 16- by 16-inch piece of cardboard and folding up the flaps. [See Fig. 10(b).] Find the value of x for which the volume of the box will be as large as possible.

(a)   (b)

Figure 10

22. **Volume** A closed rectangular box is to be constructed with a base that is twice as long as it is wide. If the total surface area must be 27 square feet, find the dimensions of the box that will maximize the volume.

23. Let $f(t)$ be the amount of oxygen (in suitable units) in a lake t days after sewage is dumped into the lake, and suppose that $f(t)$ is given approximately by

$$f(t) = 1 - \frac{10}{t + 10} + \frac{100}{(t + 10)^2}.$$

At what time is the oxygen content increasing the fastest?

24. The daily output of a coal mine after t hours of operation is approximately $40t + t^2 - \frac{1}{15}t^3$ tons, $0 \le t \le 12$. Find the maximum rate of output (in tons of coal per hour).

25. **Area** Consider a parabolic arch whose shape may be represented by the graph of $y = 9 - x^2$, where the base of the arch lies on the x-axis from $x = -3$ to $x = 3$. Find the dimensions of the rectangular window of maximum area that can be constructed inside the arch.

26. Advertising for a certain product is terminated, and t weeks later the weekly sales are $f(t)$ cases, where

$$f(t) = 1000(t + 8)^{-1} - 4000(t + 8)^{-2}.$$

At what time is the weekly sales amount falling the fastest?

27. **Surface Area** An open rectangular box of volume 400 cubic inches has a square base and a partition down the middle. See Fig. 11. Find the dimensions of the box for which the amount of material needed to construct the box is as small as possible.

Figure 11. Open rectangular box with dividing partition.

28. If $f(x)$ is defined on the interval $0 \le x \le 5$ and $f'(x)$ is negative for all x, for what value of x will $f(x)$ have its greatest value?

Technology Exercises

29. **Volume** A pizza box is formed from a 20-cm by 40-cm rectangular piece of cardboard by cutting out six squares of equal size, three from each long side of the rectangle, and then folding the cardboard in the obvious manner to create a box. See Fig. 12. Let x be the length of each side of the six squares. For what value of x will the box have greatest volume?

Figure 12

30. Coffee consumption in the United States is greater on a per capita basis than anywhere else in the world. However, due to price fluctuations of coffee beans and worries over the health effects of caffeine, coffee consumption has varied considerably over the years. According to data published in the *Wall Street Journal*, the number of cups $f(x)$ consumed daily per adult in year x (with 1955 corresponding to $x = 0$) is given by the mathematical model

$$f(x) = 2.77 + 0.0848x - 0.00832x^2 + 0.000144x^3.$$

(a) Graph $y = f(x)$ to show daily coffee consumption from 1955 through 1994.

(b) Use $f'(x)$ to determine the year in which coffee consumption was least during this period. What was the daily coffee consumption at that time?

(c) Use $f'(x)$ to determine the year in which coffee consumption was greatest during this period. What was the daily coffee consumption at that time?

(d) Use $f''(x)$ to determine the year in which coffee consumption was decreasing at the greatest rate.

Solutions to Practice Problems 2.6

1. Let x denote the order quantity and r the number of orders placed during the year. It is clear from Fig. 2 that the maximum inventory at any time during the order–reorder period is x cases. This is the number of cases in inventory at the moment an order arrives. Since the carrying cost for one case is $9, the carrying cost for x cases is $9 \cdot x$ dollars. This is the new carrying cost. The ordering cost is the same as in Example 2. So the objective function becomes

$$C = [\text{ordering cost}] + [\text{carrying cost}] = 75r + 9x.$$

The constraint equation is also the same as in Example 2, because the total number of cases ordered during the year has not changed. Thus $r \cdot x = 1200$.

(Continued)

From this point, the solution proceeds as in Example 2. We have

$$r = \frac{1200}{x}$$

$$C = 75 \cdot \frac{1200}{x} + 9x = \frac{90,000}{x} + 9x$$

$$C'(x) = -\frac{90,000}{x^2} + 9.$$

Solving $C'(x) = 0$, we find

$$-\frac{90,000}{x^2} + 9 = 0$$

$$9x^2 = 90,000$$

$$x^2 = 10,000$$

$$x = \sqrt{10,000} = 100 \quad \text{(ignore the negative root since } x > 0\text{).}$$

The total cost is at a minimum when $x = 100$. The optimum inventory policy is to order 100 cases at a time and to place $\frac{1200}{100}$ or 12 orders during the year.

2. This problem can be solved in the same manner that Example 3 was solved. However, the comment made at the end of Example 3 indicates that the economic order quantity should increase by a factor of 3, since $3 = \sqrt{9}$. Therefore, the economic order quantity is $3 \cdot 150 = 450$ cases.

2.7 Applications of Derivatives to Business and Economics

In recent years, economic decision making has become more and more mathematically oriented. Faced with huge masses of statistical data, depending on hundreds or even thousands of different variables, business analysts and economists have increasingly turned to mathematical methods to help them describe what is happening, predict the effects of various policy alternatives, and choose reasonable courses of action from the myriad possibilities. Among the mathematical methods employed is calculus. In this section we illustrate just a few of the many applications of calculus to business and economics. All our applications will center around what economists call the *theory of the firm*. In other words, we study the activity of a business (or possibly a whole industry) and restrict our analysis to a time period during which background conditions (such as supplies of raw materials, wage rates, and taxes) are fairly constant. We then show how derivatives can help the management of such a firm make vital production decisions.

Management, whether or not it knows calculus, utilizes many functions of the sort we have been considering. Examples of such functions are

$$C(x) = \text{cost of producing } x \text{ units of the product,}$$

$$R(x) = \text{revenue generated by selling } x \text{ units of the product,}$$

$$P(x) = R(x) - C(x) = \text{the profit (or loss) generated by producing and selling } x \text{ units of the product.}$$

Note that the functions $C(x)$, $R(x)$, and $P(x)$ are often defined only for nonnegative integers, that is, for $x = 0, 1, 2, 3, \ldots$. The reason is that it does not make sense to speak about the cost of producing -1 cars or the revenue generated by selling 3.62 refrigerators. Thus each function may give rise to a set of discrete points on a graph, as in Fig. 1(a). In studying these functions, however, economists usually draw a smooth curve through the points and assume that $C(x)$ is actually defined for all positive x. Of course, we must often interpret answers to problems in light of the fact that x is, in most cases, a nonnegative integer.

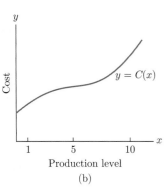

Figure 1. A cost function.

Cost Functions If we assume that a cost function $C(x)$ has a smooth graph as in Fig. 1(b), we can use the tools of calculus to study it. A typical cost function is analyzed in Example 1.

EXAMPLE 1

Marginal cost analysis Suppose that the cost function for a manufacturer is given by $C(x) = (10^{-6})x^3 - .003x^2 + 5x + 1000$ dollars.

(a) Describe the behavior of the marginal cost.

(b) Sketch the graph of $C(x)$.

Solution The first two derivatives of $C(x)$ are given by

$$C'(x) = (3 \cdot 10^{-6})x^2 - .006x + 5$$
$$C''(x) = (6 \cdot 10^{-6})x - .006.$$

Let us sketch the marginal cost $C'(x)$ first. From the behavior of $C'(x)$, we will be able to graph $C(x)$. The marginal cost function $y = (3 \cdot 10^{-6})x^2 - .006x + 5$ has as its graph a parabola that opens upward. Since $y' = C''(x) = .000006(x - 1000)$, we see that the parabola has a horizontal tangent at $x = 1000$. So the minimum value of $C'(x)$ occurs at $x = 1000$. The corresponding y-coordinate is

$$(3 \cdot 10^{-6})(1000)^2 - .006 \cdot (1000) + 5 = 3 - 6 + 5 = 2.$$

The graph of $y = C'(x)$ is shown in Fig. 2. Consequently, at first the marginal cost decreases. It reaches a minimum of 2 at production level 1000 and increases thereafter. This answers part (a). Let us now graph $C(x)$. Since the graph shown in Fig. 2 is the graph of the derivative of $C(x)$, we see that $C'(x)$ is never zero, so there are no relative extreme points. Since $C'(x)$ is always positive, $C(x)$ is always increasing (as any cost curve should be). Moreover, since $C'(x)$ decreases for x less than 1000 and increases for x greater than 1000, we see that $C(x)$ is concave down for x less than 1000, is concave up for x greater than 1000, and has an inflection point at $x = 1000$. The graph of $C(x)$ is drawn in Fig. 3. Note that the inflection point of $C(x)$ occurs at the value of x for which marginal cost is a minimum. ■

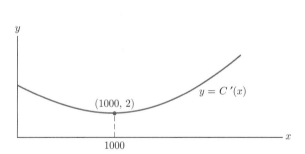

Figure 2. A marginal cost function.

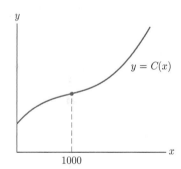

Figure 3. A cost function.

Actually, most marginal cost functions have the same general shape as the marginal cost curve of Example 1. For when x is small, production of additional units is subject to economies of production, which lowers unit costs. Thus, for x small, marginal cost decreases. However, increased production eventually leads to overtime, use of less efficient, older plants, and competition for scarce raw materials. As a result, the cost of additional units will increase for very large x. So we see that $C'(x)$ initially decreases and then increases.

Revenue Functions In general, a business is concerned not only with its costs, but also with its revenues. Recall that, if $R(x)$ is the revenue received from the sale of x units of some commodity, then the derivative $R'(x)$ is called the *marginal revenue*. Economists use this to measure the rate of increase in revenue per unit increase in sales.

If x units of a product are sold at a price p per unit, the total revenue $R(x)$ is given by

$$R(x) = x \cdot p.$$

If a firm is small and is in competition with many other companies, its sales have little effect on the market price. Then, since the price is constant as far as the one firm is concerned, the marginal revenue $R'(x)$ equals the price p [that is, $R'(x)$ is the amount that the firm receives from the sale of one additional unit]. In this case, the revenue function will have a graph as in Fig. 4.

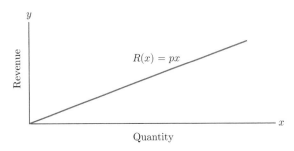

Figure 4.   A revenue curve.

An interesting problem arises when a single firm is the only supplier of a certain product or service, that is, when the firm has a monopoly. Consumers will buy large amounts of the commodity if the price per unit is low and less if the price is raised. For each quantity x, let $f(x)$ be the highest price per unit that can be set to sell all x units to customers. Since selling greater quantities requires a lowering of the price, $f(x)$ will be a decreasing function. Figure 5 shows a typical "demand curve" that relates the quantity demanded, x, to the price $p = f(x)$.

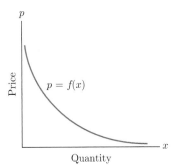

Figure 5.   A demand curve.

The *demand equation* $p = f(x)$ determines the total revenue function. If the firm wants to sell x units, the highest price it can set is $f(x)$ dollars per unit, and so the total revenue from the sale of x units is

$$R(x) = x \cdot p = x \cdot f(x). \tag{1}$$

The concept of a demand curve applies to an entire industry (with many producers) as well as to a single monopolistic firm. In this case, many producers offer the same product for sale. If x denotes the total output of the industry, $f(x)$ is the market price per unit of output and $x \cdot f(x)$ is the total revenue earned from the sale of the x units.

EXAMPLE 2

Maximizing revenue   The demand equation for a certain product is $p = 6 - \frac{1}{2}x$ dollars. Find the level of production that results in maximum revenue.

Solution   In this case, the revenue function $R(x)$ is

$$R(x) = x \cdot p = x \left(6 - \tfrac{1}{2}x\right) = 6x - \tfrac{1}{2}x^2$$

dollars. The marginal revenue is given by

$$R'(x) = 6 - x.$$

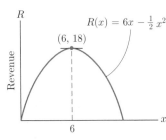

Figure 6.   Maximizing revenue.

The graph of $R(x)$ is a parabola that opens downward. (See Fig. 6.) It has a horizontal tangent precisely at those x for which $R'(x) = 0$—that is, for those x at which marginal revenue is 0. The only such x is $x = 6$. The corresponding value of revenue is

$$R(6) = 6 \cdot 6 - \tfrac{1}{2}(6)^2 = 18 \text{ dollars.}$$

Thus the rate of production resulting in maximum revenue is $x = 6$, which results in total revenue of 18 dollars.

EXAMPLE 3

Setting up a demand equation The WMA Bus Lines offers sightseeing tours of Washington, D.C. One tour, priced at $7 per person, had an average demand of about 1000 customers per week. When the price was lowered to $6, the weekly demand jumped to about 1200 customers. Assuming that the demand equation is linear, find the tour price that should be charged per person to maximize the total revenue each week.

Solution First we must find the demand equation. Let x be the number of customers per week and let p be the price of a tour ticket. Then $(x, p) = (1000, 7)$ and $(x, p) = (1200, 6)$ are on the demand curve. (See Fig. 7.) Using the point-slope formula for the line through these two points, we have

$$p - 7 = \frac{7 - 6}{1000 - 1200} \cdot (x - 1000) = -\frac{1}{200}(x - 1000) = -\frac{1}{200}x + 5,$$

so

$$p = 12 - \frac{1}{200}x. \qquad (2)$$

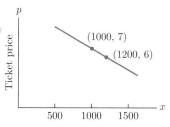

Figure 7. A demand curve.

From equation (1) we obtain the revenue function:

$$R(x) = x\left(12 - \frac{1}{200}x\right) = 12x - \frac{1}{200}x^2.$$

The marginal revenue function is

$$R'(x) = 12 - \frac{1}{100}x = -\frac{1}{100}(x - 1200).$$

Using $R(x)$ and $R'(x)$, we can sketch the graph of $R(x)$. (See Fig. 8.) The maximum revenue occurs when the marginal revenue is zero, that is, when $x = 1200$. The price corresponding to this number of customers is found from demand equation (2):

$$p = 12 - \frac{1}{200}(1200) = 6 \text{ dollars}.$$

Thus the price of $6 is most likely to bring the greatest revenue per week. ■

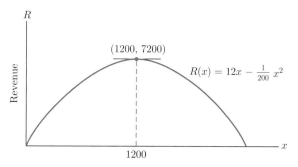

Figure 8. Maximizing revenue.

Profit Functions Once we know the cost function $C(x)$ and the revenue function $R(x)$, we can compute the profit function $P(x)$ from

$$P(x) = R(x) - C(x).$$

EXAMPLE 4

Maximizing profits Suppose that the demand equation for a monopolist is $p = 100 - .01x$ and the cost function is $C(x) = 50x + 10,000$. Find the value of x that maximizes the profit and determine the corresponding price and total profit for this level of production. (See Fig. 9.)

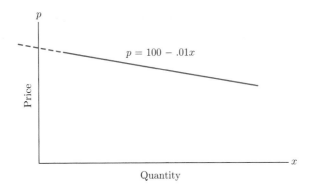

Figure 9. A demand curve.

Solution The total revenue function is

$$R(x) = x \cdot p = x(100 - .01x) = 100x - .01x^2.$$

Hence the profit function is

$$
\begin{aligned}
P(x) &= R(x) - C(x) \\
&= 100x - .01x^2 - (50x + 10{,}000) \\
&= -.01x^2 + 50x - 10{,}000.
\end{aligned}
$$

The graph of this function is a parabola that opens downward. (See Fig. 10.) Its highest point will be where the curve has zero slope, that is, where the marginal

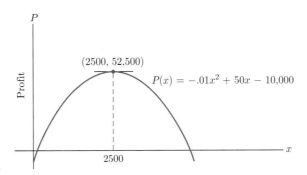

Figure 10. Maximizing profit.

profit $P'(x)$ is zero. Now

$$P'(x) = -.02x + 50 = -.02(x - 2500).$$

So $P'(x) = 0$ when $x = 2500$. The profit for this level of production is

$$P(2500) = -.01(2500)^2 + 50(2500) - 10{,}000 = 52{,}500 \text{ dollars.}$$

Finally, we return to the demand equation to find the highest price that can be charged per unit to sell all 2500 units:

$$p = 100 - .01(2500) = 100 - 25 = 75 \text{ dollars.}$$

Thus, to maximize the profit, produce 2500 units and sell them at $75 per unit. The profit will be $52,500. ■

EXAMPLE 5   Rework Example 4 under the condition that the government has imposed an excise tax of $10 per unit.

Solution For each unit sold, the manufacturer will have to pay $10 to the government. In other words, $10x$ dollars are added to the cost of producing and selling x units. The cost function is now

$$C(x) = (50x + 10{,}000) + 10x = 60x + 10{,}000.$$

The demand equation is unchanged by this tax, so the revenue is still

$$R(x) = 100x - .01x^2.$$

Proceeding as before, we have

$$P(x) = R(x) - C(x)$$
$$= 100x - .01x^2 - (60x + 10{,}000)$$
$$= -.01x^2 + 40x - 10{,}000.$$
$$P'(x) = -.02x + 40 = -.02(x - 2000).$$

The graph of $P(x)$ is still a parabola that opens downward, and the highest point is where $P'(x) = 0$, that is, where $x = 2000$. (See Fig. 11.) The corresponding profit is

$$P(2000) = -.01(2000)^2 + 40(2000) - 10{,}000 = 30{,}000 \text{ dollars.}$$

From the demand equation, $p = 100 - .01x$, we find the price that corresponds to $x = 2000$:

$$p = 100 - .01(2000) = 80 \text{ dollars.}$$

To maximize profit, produce 2000 units and sell them at $80 per unit. The profit will be $30,000. ■

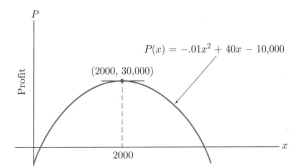

Figure 11. Profit after an excise tax.

Notice in Example 5 that the optimal price is raised from $75 to $80. If the monopolist wishes to maximize profits, he or she should pass only half the $10 tax on to the customer. The monopolist cannot avoid the fact that profits will be substantially lowered by the imposition of the tax. This is one reason why industries lobby against taxation.

Setting Production Levels Suppose that a firm has cost function $C(x)$ and revenue function $R(x)$. In a free-enterprise economy the firm will set production x in such a way as to maximize the profit function

$$P(x) = R(x) - C(x).$$

We have seen that if $P(x)$ has a maximum at $x = a$, then $P'(a) = 0$. In other words, since $P'(x) = R'(x) - C'(x)$,

$$R'(a) - C'(a) = 0$$
$$R'(a) = C'(a).$$

Thus profit is maximized at a production level for which marginal revenue equals marginal cost. (See Fig. 12.)

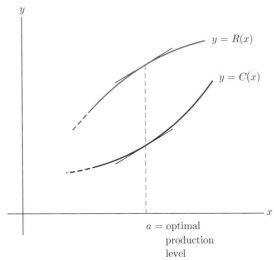

$y = R(x)$

$y = C(x)$

$a =$ optimal
production
level

Figure 12

Practice Problems 2.7

1. Rework Example 4 by finding the production level at which marginal revenue equals marginal cost.

2. Rework Example 4 under the condition that the fixed cost is increased from $10,000 to $15,000.

3. On a certain route, a regional airline carries 8000 passengers per month, each paying $50. The airline wants to increase the fare. However, the market research department estimates that for each $1 increase in fare the airline will lose 100 passengers. Determine the price that maximizes the airline's revenue.

EXERCISES 2.7

1. Given the cost function $C(x) = x^3 - 6x^2 + 13x + 15$, find the minimum marginal cost.

2. If a total cost function is $C(x) = .0001x^3 - .06x^2 + 12x + 100$, is the marginal cost increasing, decreasing, or not changing at $x = 100$? Find the minimum marginal cost.

3. The revenue function for a one-product firm is

$$R(x) = 200 - \frac{1600}{x + 8} - x.$$

Find the value of x that results in maximum revenue.

4. The revenue function for a particular product is $R(x) = x(4 - .0001x)$. Find the largest possible revenue.

5. **Cost and Profit** A one-product firm estimates that its daily total cost function (in suitable units) is $C(x) = x^3 - 6x^2 + 13x + 15$ and its total revenue function is $R(x) = 28x$. Find the value of x that maximizes the daily profit.

6. A small tie shop sells ties for $3.50 each. The daily cost function is estimated to be $C(x)$ dollars, where x is the number of ties sold on a typical day and $C(x) = .0006x^3 - .03x^2 + 2x + 20$. Find the value of x that will maximize the store's daily profit.

7. **Demand and Revenue** The demand equation for a certain commodity is

$$p = \tfrac{1}{12}x^2 - 10x + 300,$$

$0 \le x \le 60$. Find the value of x and the corresponding price p that maximize the revenue.

8. The demand equation for a product is $p = 2 - .001x$. Find the value of x and the corresponding price p that maximize the revenue.

9. **Profit** Some years ago it was estimated that the demand for steel approximately satisfied the equation $p = 256 - 50x$, and the total cost of producing x units of steel was $C(x) = 182 + 56x$. (The quantity x was measured in millions of tons and the price and total cost were measured in millions of dollars.) Determine the level of production and the corresponding price that maximize the profits.

10. Consider a rectangle in the xy-plane, with corners at $(0,0)$, $(a,0)$, $(0,b)$, and (a,b). If (a,b) lies on the graph of the equation $y = 30 - x$, find a and b such that the area of the rectangle is maximized. What economic interpretations can be given to your answer if the equation $y = 30 - x$ represents a demand curve and y is the price corresponding to the demand x?

11. **Demand, Revenue, and Profit**   Until recently hamburgers at the city sports arena cost $4 each. The food concessionaire sold an average of 10,000 hamburgers on a game night. When the price was raised to $4.40, hamburger sales dropped off to an average of 8000 per night.

   (a) Assuming a linear demand curve, find the price of a hamburger that will maximize the nightly hamburger revenue.

   (b) If the concessionaire has fixed costs of $1000 per night and the variable cost is $.60 per hamburger, find the price of a hamburger that will maximize the nightly hamburger profit.

12. **Demand and Revenue**   The average ticket price for a concert at the opera house was $50. The average attendance was 4000. When the ticket price was raised to $52, attendance declined to an average of 3800 persons per performance. What should the ticket price be to maximize the revenue for the opera house? (Assume a linear demand curve.)

13. **Demand and Revenue**   An artist is planning to sell signed prints of her latest work. If 50 prints are offered for sale, she can charge $400 each. However, if she makes more than 50 prints, she must lower the price of all the prints by $5 for each print in excess of the 50. How many prints should the artist make to maximize her revenue?

14. **Demand and Revenue**   A swimming club offers memberships at the rate of $200, provided that a minimum of 100 people join. For each member in excess of 100, the membership fee will be reduced $1 per person (for each member). At most 160 memberships will be sold. How many memberships should the club try to sell to maximize its revenue?

15. **Profit**   In the planning of a sidewalk café, it is estimated that for 12 tables the daily profit will be $10 per table. Because of overcrowding, for each additional table the profit per table (for every table in the cafe) will be reduced by $.50. How many tables should be provided to maximize the profit from the café?

16. **Demand and Revenue**   A certain toll road averages 36,000 cars per day when charging $1 per car. A survey concludes that increasing the toll will result in 300 fewer cars for each cent of increase. What toll should be charged to maximize the revenue?

17. **Price Setting**   The monthly demand equation for an electric utility company is estimated to be

$$p = 60 - (10^{-5})x,$$

where p is measured in dollars and x is measured in thousands of kilowatt-hours. The utility has fixed costs of $7,000,000 per month and variable costs of $30 per 1000 kilowatt-hours of electricity generated, so the cost function is

$$C(x) = 7 \cdot 10^6 + 30x.$$

   (a) Find the value of x and the corresponding price for 1000 kilowatt-hours that maximize the utility's profit.

   (b) Suppose that rising fuel costs increase the utility's variable costs from $30 to $40, so its new cost function is

$$C_1(x) = 7 \cdot 10^6 + 40x.$$

   Should the utility pass all this increase of $10 per thousand kilowatt-hours on to consumers? Explain your answer.

18. **Taxes, Profit, and Revenue**   The demand equation for a monopolist is $p = 200 - 3x$, and the cost function is

$$C(x) = 75 + 80x - x^2, \qquad 0 \le x \le 40.$$

   (a) Determine the value of x and the corresponding price that maximize the profit.

   (b) If the government imposes a tax on the monopolist of $4 per unit quantity produced, determine the new price that maximizes the profit.

   (c) The government imposes a tax of T dollars per unit quantity produced (where $0 \le T \le 120$), so the new cost function is

$$C(x) = 75 + (80 + T)x - x^2, \qquad 0 \le x \le 40.$$

   Determine the new value of x that maximizes the monopolist's profit as a function of T. Assuming that the monopolist cuts back production to this level, express the tax revenues received by the government as a function of T. Finally, determine the value of T that will maximize the tax revenue received by the government.

19. **Interest Rate**   A savings and loan association estimates that the amount of money on deposit will be 1,000,000 times the percentage rate of interest. For instance, a 4% interest rate will generate $4,000,000 in deposits. If the savings and loan association can loan all the money it takes in at 10% interest, what interest rate on deposits generates the greatest profit?

20. Let $P(x)$ be the annual profit for a certain product, where x is the amount of money spent on advertising. (See Fig. 13.)

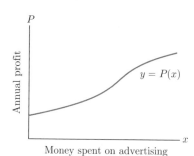

Figure 13.   Profit as a function of advertising.

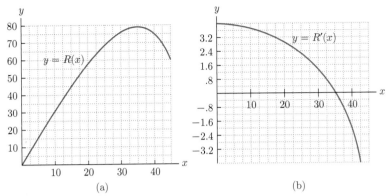

Figure 14. Revenue function and its first derivative.

(a) Interpret $P(0)$.

(b) Describe how the marginal profit changes as the amount of money spent on advertising increases.

(c) Explain the economic significance of the inflection point.

21. The revenue for a manufacturer is $R(x)$ thousand dollars, where x is the number of units of goods produced (and sold) and R and R' are the functions given in Figs. 14(a) and 14(b).

(a) What is the revenue from producing 40 units of goods?

(b) What is the marginal revenue when 17.5 units of goods are produced?

(c) At what level of production is the revenue $45,000?

(d) At what level(s) of production is the marginal revenue $800?

(e) At what level of production is the revenue greatest?

22. The cost function for a manufacturer is $C(x)$ dollars, where x is the number of units of goods produced and C, C', and C'' are the functions given in Fig. 15.

(a) What is the cost of manufacturing 60 units of goods?

(b) What is the marginal cost when 40 units of goods are manufactured?

(c) At what level of production is the cost $1200?

(d) At what level(s) of production is the marginal cost $22.50?

(e) At what level of production does the marginal cost have the least value? What is the marginal cost at this level of production?

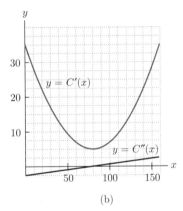

Figure 15. Cost function and its derivatives.

Solutions to Practice Problems 2.7

1. The revenue function is $R(x) = 100x - .01x^2$, so the marginal revenue function is $R'(x) = 100 - .02x$. The cost function is $C(x) = 50x + 10,000$, so the marginal cost function is $C'(x) = 50$. Let us now equate the two marginal functions and solve for x.

$$R'(x) = C'(x)$$
$$100 - .02x = 50$$
$$-.02x = -50$$
$$x = \frac{-50}{-.02} = \frac{5000}{2} = 2500$$

Of course, we obtain the same level of production as before.

2. If the fixed cost is increased from \$10,000 to \$15,000, the new cost function will be $C(x) = 50x + 15,000$, but the marginal cost function will still be $C'(x) = 50$. Therefore, the solution will be the same: 2500 units should be produced and sold at \$75 per unit. (Increases in fixed costs should not necessarily be passed on to the consumer if the objective is to maximize the profit.)

3. Let x denote the number of passengers per month and p the price per ticket. The number of passengers lost due to a fare increase is obtained by multiplying the number of dollars of fare increase, $p - 50$, by the number of passengers lost for each dollar of fare increase. So

$$x = 8000 - (p - 50)100 = -100p + 13,000.$$

Solving for p, we get the demand equation

$$p = -\frac{1}{100}x + 130.$$

From equation (1), the revenue function is

$$R(x) = x \cdot p = x\left(-\frac{1}{100}x + 130\right).$$

The graph is a parabola that opens downward, with x-intercepts at $x = 0$ and $x = 13,000$. (See Fig. 16.) Its maximum is located at the midpoint of the x-intercepts, or $x = 6500$. The price corresponding to this number of passengers is $p = -\frac{1}{100}(6500) + 130 = 65$ dollars. Thus the price of \$65 per ticket will bring the highest revenue to the airline company per month.

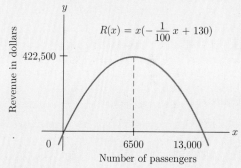

Figure 16

REVIEW OF FUNDAMENTAL CONCEPTS

1. State as many terms used to describe graphs of functions as you can recall.

2. What is the difference between having a relative maximum at $x = 2$ and having an absolute maximum at $x = 2$?

3. Give three characterizations of what it means for the graph of $f(x)$ to be concave up at $x = 2$. Concave down.

4. What does it mean to say that the graph of $f(x)$ has an inflection point at $x = 2$?

5. What is the difference between an x-intercept and a zero of a function?

6. How do you determine the y-intercept of a function?

7. What is an asymptote? Give an example.

8. State the first derivative rule. Second derivative rule.

9. Give two connections between the graphs of $f(x)$ and $f'(x)$.

10. Outline a method for locating the relative extreme points of a function.

11. Outline a method for locating the inflection points of a function.

12. Outline a procedure for sketching the graph of a function.

13. What is an objective equation?

14. What is a constraint equation?

15. Outline the procedure for solving an optimization problem.

16. How are the cost, revenue, and profit functions related?

SUPPLEMENTARY EXERCISES

1. Figure 1 contains the graph of $f'(x)$, the derivative of $f(x)$. Use the graph to answer the following questions about the graph of $f(x)$.

   (a) For what values of x is the graph of $f(x)$ increasing? Decreasing?

   (b) For what values of x is the graph of $f(x)$ concave up? Concave down?

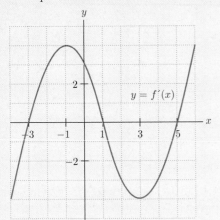

Figure 1

2. Figure 2 shows the graph of the function $f(x)$ and its tangent line at $x = 3$. Find $f(3)$, $f'(3)$, and $f''(3)$.

Figure 2

In Exercises 3–6, draw the graph of a function $f(x)$ for which the function and its first derivative have the stated property for all x.

3. $f(x)$ and $f'(x)$ increasing

4. $f(x)$ and $f'(x)$ decreasing

5. $f(x)$ increasing and $f'(x)$ decreasing

6. $f(x)$ decreasing and $f'(x)$ increasing

Exercises 7–12 refer to the graph in Fig. 3. List the labeled values of x at which the derivative has the stated property.

Figure 3

7. $f'(x)$ is positive.

8. $f'(x)$ is negative.

9. $f''(x)$ is positive.

10. $f''(x)$ is negative.

11. $f'(x)$ is maximized.

12. $f'(x)$ is minimized.

Properties of various functions are described next. In each case draw some conclusion about the graph of the function.

13. $f(1) = 2$, $f'(1) > 0$

14. $g(1) = 5$, $g'(1) = -1$

15. $h'(3) = 4$, $h''(3) = 1$

16. $F'(2) = -1$, $F''(2) < 0$

17. $G(10) = 2$, $G'(10) = 0$, $G''(10) > 0$

18. $f(4) = -2$, $f'(4) > 0$, $f''(4) = -1$

19. $g(5) = -1$, $g'(5) = -2$, $g''(5) = 0$

20. $H(0) = 0$, $H'(0) = 0$, $H''(0) = 1$

21. In Figs. 4(a) and 4(b), the t-axis represents time in hours.

    (a) When is $f(t) = 1$?

    (b) Find $f(5)$.

    (c) When is $f(t)$ changing at the rate of $-.08$ unit per hour?

    (d) How fast is $f(t)$ changing after 8 hours?

22. United States electrical energy production (in trillions of kilowatt-hours) in year t (with 1900 corresponding to $t = 0$) is given by $f(t)$, where f and its derivatives are graphed in Figs. 5(a) and 5(b).

    (a) How much electrical energy was produced in 1950?

    (b) How fast was energy production rising in 1950?

    (c) When did energy production reach 3000 trillion kilowatt-hours?

    (d) When was the level of energy production rising at the rate of 10 trillion kilowatt-hours per year?

(a)   (b)

Figure 4

(e) When was energy production growing at the greatest rate? What was the level of production at that time?

Sketch the following parabolas. Include their x- and y-intercepts.

23. $y = 3 - x^2$

24. $y = 7 + 6x - x^2$

25. $y = x^2 + 3x - 10$

26. $y = 4 + 3x - x^2$

27. $y = -2x^2 + 10x - 10$

28. $y = x^2 - 9x + 19$

29. $y = x^2 + 3x + 2$

30. $y = -x^2 + 8x - 13$

31. $y = -x^2 + 20x - 90$

32. $y = 2x^2 + x - 1$

Sketch the following curves.

33. $y = 2x^3 + 3x^2 + 1$

34. $y = x^3 - \frac{3}{2}x^2 - 6x$

35. $y = x^3 - 3x^2 + 3x - 2$

36. $y = 100 + 36x - 6x^2 - x^3$

37. $y = \frac{11}{3} + 3x - x^2 - \frac{1}{3}x^3$

38. $y = x^3 - 3x^2 - 9x + 7$

39. $y = -\frac{1}{3}x^3 - 2x^2 - 5x$

40. $y = x^3 - 6x^2 - 15x + 50$

41. $y = x^4 - 2x^2$

42. $y = x^4 - 4x^3$

43. $y = \frac{x}{5} + \frac{20}{x} + 3 \quad (x > 0)$

44. $y = \frac{1}{2x} + 2x + 1 \ (x > 0)$

45. Let $f(x) = (x^2 + 2)^{3/2}$. Show that the graph of $f(x)$ has a possible relative extreme point at $x = 0$.

46. Show that the function $f(x) = (2x^2 + 3)^{3/2}$ is decreasing for $x < 0$ and increasing for $x > 0$.

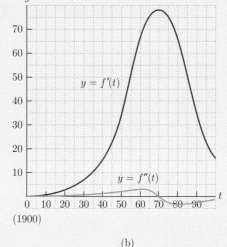

(a)   (b)

Figure 5. United States electrical energy production.

47. Let $f(x)$ be a function whose *derivative* is

$$f'(x) = \frac{1}{1 + x^2}.$$

Note that $f'(x)$ is always positive. Show that the graph of $f(x)$ definitely has an inflection point at $x = 0$.

48. Let $f(x)$ be a function whose *derivative* is

$$f'(x) = \sqrt{5x^2 + 1}.$$

Show that the graph of $f(x)$ definitely has an inflection point at $x = 0$.

49. A car is traveling on a straight road and $s(t)$ is the distance traveled after t hours. Match each set of information about $s(t)$ and its derivatives with the corresponding description of the car's motion.

Information

A. $s(t)$ is a constant function.

B. $s'(t)$ is a positive constant function.

C. $s'(t)$ is positive at $t = a$.

D. $s'(t)$ is negative at $t = a$.

E. $s'(t)$ and $s''(t)$ are positive at $t = a$.

F. $s'(t)$ is positive and $s''(t)$ is negative at $t = a$.

Descriptions

a. The car is moving forward and speeding up at time a.

b. The car is backing up at time a.

c. The car is standing still.

d. The car is moving forward, but slowing down, at time a.

e. The car is moving forward at a steady rate.

f. The car is moving forward at time a.

50. The water level in a reservoir varies during the year. Let $h(t)$ be the depth (in feet) of the water at time t days, where $t = 0$ at the beginning of the year. Match each set of information about $h(t)$ and its derivatives with the corresponding description of the reservoir's activity.

Information

A. $h(t)$ has the value 50 for $1 \le t \le 2$.

B. $h'(t)$ has the value .5 for $1 \le t \le 2$.

C. $h'(t)$ is positive at $t = a$.

D. $h'(t)$ is negative at $t = a$.

E. $h'(t)$ and $h''(t)$ are positive at $t = a$.

F. $h'(t)$ is positive and $h''(t)$ is negative at $t = a$.

Descriptions

a. The water level is rising at an increasing rate at time a.

b. The water level is receding at time a.

c. The water level is constant at 50 feet on January 2.

d. At time a the water level is rising, but the rate of increase is slowing down.

e. On January 2 the water rose steadily at a rate of .5 feet per day.

f. The water level is rising at time a.

51. Let $f(x)$ be the number of people living within x miles of the center of New York City.

(a) What does $f(10 + h) - f(10)$ represent?

(b) Explain why $f'(10)$ cannot be negative.

52. For what x does the function $f(x) = \frac{1}{4}x^2 - x + 2$, $0 \le x \le 8$, have its maximum value?

53. Find the maximum value of the function $f(x) = 2 - 6x - x^2$, $0 \le x \le 5$, and give the value of x where this maximum occurs.

54. Find the minimum value of the function $g(t) = t^2 - 6t + 9$, $1 \le t \le 6$.

55. **Surface Area** An open rectangular box is to be 4 feet long and have a volume of 200 cubic feet. Find the dimensions for which the amount of material needed to construct the box is as small as possible.

56. **Volume** A closed rectangular box with a square base is to be constructed using two different types of wood. The top is made of wood costing $3 per square foot and the remainder is made of wood costing $1 per square foot. If $48 is available to spend, find the dimensions of the box of greatest volume that can be constructed.

57. **Volume** A long rectangular sheet of metal 30 inches wide is to be made into a gutter by turning up strips vertically along the two sides. How many inches should be turned up on each side to maximize the amount of water that the gutter can carry?

58. **Maximizing the Total Yield** A small orchard yields 25 bushels of fruit per tree when planted with 40 trees. Because of overcrowding, the yield per tree (for each tree in the orchard) is reduced by $\frac{1}{2}$ bushel for each additional tree that is planted. How many trees should be planted to maximize the total yield of the orchard?

59. **Inventory Control** A publishing company sells 400,000 copies of a certain book each year. Ordering the entire amount printed at the beginning of the year ties up valuable storage space and capital. However, running off the copies in several partial runs throughout the year results in added costs for setting up each printing run, which costs $1000. The carrying costs, figured on the average number of books in storage, are 50¢ per book. Find the most economical lot size, that is, the production run size that minimizes the total setting up and carrying costs.

60. **Area** A poster is to have an area of 125 square inches. The printed material is to be surrounded by margins of 3 inches at the top and 2 inches at the bottom and sides. Find the dimensions of the poster that maximize the area of the printed material.

61. Profit If the demand equation for a monopolist is $p = 150 - .02x$ and the cost function is $C(x) = 10x + 300$, find the value of x that maximizes the profit.

62. Minimizing Time Jane wants to drive her tractor from point A on one side of her 5-mile-wide field to a point B on the opposite side of the field, as shown in Fig. 6. Jane could drive her tractor directly across the field to point C and then drive 15 miles down a road to B, or she could drive to some point P between B and C and then drive to B. If Jane drives her tractor 8 miles per hour across the field and 17 miles per hour on the road, find the point P where she should cut across the field to minimize the time it takes to reach B. [*Hint*: time $= \frac{\text{distance}}{\text{rate}}$.]

63. A travel agency offers a boat tour of several Caribbean islands for 3 days and 2 nights. For a group of 12 people, the cost per person is $800. For each additional person above the 12-person minimum, the cost per person is reduced by $20 for each person in the group. The maximum tour group size is 25. What tour group size produces the greatest revenue for the travel agency?

Figure 6

TECHNIQUES OF DIFFERENTIATION

3.1 The Product and Quotient Rules

3.2 The Chain Rule and the General Power Rule

3.3 Implicit Differentiation and Related Rates

We have seen that the derivative is useful in many applications. However, our ability to differentiate functions is somewhat limited. For example, we cannot yet readily differentiate the functions

$$f(x) = (x^2 - 1)^4 (x^2 + 1)^5, \qquad g(x) = \frac{x^3}{(x^2 + 1)^4}.$$

In this chapter we develop differentiation techniques that apply to functions like those just given. Two new rules are the *product rule* and the *quotient rule*. In Section 3.2 we extend the general power rule into a powerful formula called the *chain rule*.

3.1 The Product and Quotient Rules

We observed in our discussion of the sum rule for derivatives that the derivative of the sum of two differentiable functions is the sum of the derivatives. Unfortunately, however, the derivative of the product $f(x)g(x)$ is *not* the product of the derivatives. Rather, the derivative of a product is determined from the following rule:

Product Rule

$$\frac{d}{dx}[f(x)g(x)] = f(x)g'(x) + g(x)f'(x)$$

The derivative of the product of two functions is the first function times the derivative of the second plus the second function times the derivative of the first. At the end of this section we show why this statement is true.

EXAMPLE 1

Verifying the product rule Show that the product rule works for the case $f(x) = x^2$, $g(x) = x^3$.

Solution Since $x^2 \cdot x^3 = x^5$, we know that

$$\frac{d}{dx}(x^2 \cdot x^3) = \frac{d}{dx}(x^5) = 5x^4.$$

On the other hand, using the product rule,

$$\frac{d}{dx}(x^2 \cdot x^3) = x^2 \cdot \frac{d}{dx}(x^3) + x^3 \cdot \frac{d}{dx}(x^2)$$

$$= x^2(3x^2) + x^3(2x)$$

$$= 3x^4 + 2x^4 = 5x^4.$$

Thus the product rule gives the correct answer. ■

EXAMPLE 2

Finding the derivative of a product Differentiate $y = (2x^3 - 5x)(3x + 1)$.

Solution Let $f(x) = 2x^3 - 5x$ and $g(x) = 3x + 1$. Then

$$\frac{d}{dx}\left[(2x^3 - 5x)(3x + 1)\right] = (2x^3 - 5x) \cdot \frac{d}{dx}(3x + 1) + (3x + 1) \cdot \frac{d}{dx}(2x^3 - 5x)$$

$$= (2x^3 - 5x)(3) + (3x + 1)(6x^2 - 5)$$

$$= 6x^3 - 15x + 18x^3 - 15x + 6x^2 - 5$$

$$= 24x^3 + 6x^2 - 30x - 5.$$ ■

EXAMPLE 3

Apply the product rule to $y = g(x) \cdot g(x)$.

Solution

$$\frac{d}{dx}[g(x) \cdot g(x)] = g(x) \cdot g'(x) + g(x) \cdot g'(x)$$
$$= 2g(x)g'(x)$$

This answer is the same as that given by the general power rule:

$$\frac{d}{dx}[g(x) \cdot g(x)] = \frac{d}{dx}[g(x)]^2 = 2g(x)g'(x).$$ ■

EXAMPLE 4

Find

$$\frac{dy}{dx}, \quad \text{where } y = (x^2 - 1)^4(x^2 + 1)^5.$$

Solution Let $f(x) = (x^2 - 1)^4$, $g(x) = (x^2 + 1)^5$, and use the product rule. The general power rule is needed to compute $f'(x)$ and $g'(x)$.

$$\frac{dy}{dx} = (x^2 - 1)^4 \cdot \frac{d}{dx}(x^2 + 1)^5 + (x^2 + 1)^5 \cdot \frac{d}{dx}(x^2 - 1)^4$$

$$= (x^2 - 1)^4 \cdot 5(x^2 + 1)^4(2x) + (x^2 + 1)^5 \cdot 4(x^2 - 1)^3(2x) \qquad (1)$$

This form of $\frac{dy}{dx}$ is suitable for some purposes. For example, if we need to compute $\left.\frac{dy}{dx}\right|_{x=2}$, it is easier just to substitute 2 for x than to simplify and then substitute. However, it is often helpful to simplify the formula for $\frac{dy}{dx}$, such as in the case when we need to find x such that $\frac{dy}{dx} = 0$.

To simplify the answer in equation (1), we shall write $\frac{dy}{dx}$ as a single product, rather than as a sum of two products. The first step is to identify the common factors.

$$\frac{dy}{dx} = (x^2 - 1)^4 \cdot 5(x^2 + 1)^4 (2x) + (x^2 + 1)^5 \cdot 4(x^2 - 1)^3 (2x)$$

Both terms contain $2x$ and powers of $x^2 - 1$ and $x^2 + 1$. The most we can factor out of each term is $2x(x^2 - 1)^3(x^2 + 1)^4$. We obtain

$$\frac{dy}{dx} = 2x(x^2 - 1)^3(x^2 + 1)^4[5(x^2 - 1) + 4(x^2 + 1)].$$

Simplifying the right-most factor in this product, we have

$$\frac{dy}{dx} = 2x(x^2 - 1)^3(x^2 + 1)^4[9x^2 - 1]. \qquad (2)$$

(See Fig. 3 and Exercise 34 for a related application.) ∎

The answers to the exercises for this section appear in two forms, similar to those in (1) and (2). The unsimplified answers will allow you to check if you have mastered the differentiation rules. In each case you should make an effort to transform your original answer into the simplified version. Examples 4 and 6 show how to do this.

The Quotient Rule Another useful formula for differentiating functions is the quotient rule.

> **Quotient Rule**
>
> $$\frac{d}{dx}\left[\frac{f(x)}{g(x)}\right] = \frac{g(x)f'(x) - f(x)g'(x)}{[g(x)]^2}$$

We must be careful to remember the order of the terms in this formula because of the minus sign in the numerator.

EXAMPLE 5 Differentiate

$$y = \frac{x}{2x + 3}.$$

Solution Let $f(x) = x$ and $g(x) = 2x + 3$.

$$\frac{d}{dx}\left(\frac{x}{2x+3}\right) = \frac{(2x+3)\cdot\frac{d}{dx}(x) - (x)\cdot\frac{d}{dx}(2x+3)}{(2x+3)^2}$$

$$= \frac{(2x+3)\cdot 1 - x\cdot 2}{(2x+3)^2} = \frac{3}{(2x+3)^2} \qquad\blacksquare$$

EXAMPLE 6   **Simplifying after the quotient rule**   Find

$$\frac{dy}{dx}, \quad \text{where } y = \frac{x^3}{(x^2+1)^4}.$$

Solution Let $f(x) = x^3$ and $g(x) = (x^2 + 1)^4$.

$$\frac{d}{dx}\left(\frac{x^3}{(x^2+1)^4}\right) = \frac{(x^2+1)^4\cdot\frac{d}{dx}(x^3) - (x^3)\cdot\frac{d}{dx}(x^2+1)^4}{[(x^2+1)^4]^2}$$

$$= \frac{(x^2+1)^4\cdot 3x^2 - x^3\cdot 4(x^2+1)^3(2x)}{(x^2+1)^8}$$

If a simplified form of $\dfrac{dy}{dx}$ is desired, we can divide the numerator and the denominator by the common factor $(x^2+1)^3$.

$$\frac{dy}{dx} = \frac{(x^2+1)\cdot 3x^2 - x^3\cdot 4(2x)}{(x^2+1)^5}$$

$$= \frac{3x^4 + 3x^2 - 8x^4}{(x^2+1)^5}$$

$$= \frac{3x^2 - 5x^4}{(x^2+1)^5} = \frac{x^2(3-5x^2)}{(x^2+1)^5} \qquad\blacksquare$$

The derivative in Example 6 is a rational function. To simplify it, you can cancel common factors of the entire numerator and denominator, but terms in fractions should never be canceled. Here is another illustration of these rules.

EXAMPLE 7   Differentiate

$$y = \frac{x}{x + (x+1)^3}.$$

Solution Let $f(x) = x$ and $g(x) = x + (x+1)^3$. Then $f'(x) = 1$, $g'(x) = 1 + 3(x+1)^2$, and so

$$\frac{d}{dx}\left(\frac{x}{x+(x+1)^3}\right) = \frac{(x+(x+1)^3) - x\cdot(1+3(x+1)^2)}{(x+(x+1)^3)^2}.$$

Although there are several common terms in the denominator and numerator, none of them can be canceled because none of them are common factors of the entire numerator and denominator. You can continue simplifying as follows:

$$\frac{dy}{dx} = \frac{\cancel{x} + (x+1)^3 - \cancel{x} - 3x(x+1)^2}{(x+(x+1)^3)^2}$$

$$= \frac{\overbrace{(x+1)^3} - 3x \overbrace{(x+1)^2}}{(x+(x+1)^3)^2}$$

$$= \frac{(x+1)^2[(x+1)-3x]}{(x+(x+1)^3)^2} = \frac{(x+1)^2(-2x+1)}{(x+(x+1)^3)^2}. \quad \blacksquare$$

EXAMPLE 8

Minimizing the average cost   Suppose that the total cost of manufacturing x units of a certain product is given by the function $C(x)$. Then the *average cost per unit*, AC, is defined by

$$AC = \frac{C(x)}{x}.$$

Recall that the *marginal cost*, *MC*, is defined by

$$MC = C'(x).$$

Show that, at the level of production where the average cost is at a minimum, the average cost equals the marginal cost.

Solution   In practice, the marginal cost and average cost curves will have the general shapes shown in Fig. 1. So the minimum point on the average cost curve will occur when $\frac{d}{dx}(AC) = 0$. To compute the derivative, we need the quotient rule,

$$\frac{d}{dx}(AC) = \frac{d}{dx}\left(\frac{C(x)}{x}\right) = \frac{x \cdot C'(x) - C(x)}{x^2}.$$

Setting the derivative equal to zero and multiplying by x^2, we obtain

$$0 = x \cdot C'(x) - C(x)$$

$$C(x) = x \cdot C'(x)$$

$$\frac{C(x)}{x} = C'(x)$$

$$AC = MC.$$

Figure 1.   Marginal cost and average cost functions.

Thus when the output x is chosen so that the average cost is minimized, the average cost equals the marginal cost. $\blacksquare$

Verification of the Product and Quotient Rules

Verification of the Product Rule   From our discussion of limits, we compute the derivative of $f(x)g(x)$ at $x = a$ as the limit

$$\frac{d}{dx}[f(x)g(x)]\Big|_{x=a} = \lim_{h\to 0} \frac{f(a+h)g(a+h) - f(a)g(a)}{h}.$$

Let us add and subtract the quantity $f(a)g(a+h)$ in the numerator. After factoring and applying Limit Theorem III, Section 1.4, we obtain

$$\lim_{h\to 0} \frac{[f(a+h)g(a+h) - f(a)g(a+h)] + [f(a)g(a+h) - f(a)g(a)]}{h}$$

$$= \lim_{h\to 0} g(a+h) \cdot \frac{f(a+h) - f(a)}{h} + \lim_{h\to 0} f(a) \cdot \frac{g(a+h) - g(a)}{h}.$$

This expression may be rewritten by Limit Theorem V, Section 1.4, as

$$\lim_{h\to 0} g(a+h) \cdot \lim_{h\to 0} \frac{f(a+h) - f(a)}{h} + \lim_{h\to 0} f(a) \cdot \lim_{h\to 0} \frac{g(a+h) - g(a)}{h}.$$

Note, however, that since $g(x)$ is differentiable at $x = a$, it is continuous there, so $\lim_{h\to 0} g(a+h) = g(a)$. Therefore, the preceding expression equals

$$g(a)f'(a) + f(a)g'(a).$$

That is, we have proved that

$$\frac{d}{dx}[f(x)g(x)]\bigg|_{x=a} = g(a)f'(a) + f(a)g'(a),$$

which is the product rule. An alternative verification of the product rule not involving limit arguments is outlined in Exercise 66.

Verification of the Quotient Rule From the general power rule, we know that

$$\frac{d}{dx}\left[\frac{1}{g(x)}\right] = \frac{d}{dx}[g(x)]^{-1} = (-1)[g(x)]^{-2} \cdot g'(x).$$

We can now derive the quotient rule from the product rule.

$$\frac{d}{dx}\left[\frac{f(x)}{g(x)}\right] = \frac{d}{dx}\left[\frac{1}{g(x)} \cdot f(x)\right]$$

$$= \frac{1}{g(x)} \cdot f'(x) + f(x) \cdot \frac{d}{dx}\left[\frac{1}{g(x)}\right]$$

$$= \frac{g(x)f'(x)}{[g(x)]^2} + f(x) \cdot (-1)[g(x)]^{-2} \cdot g'(x)$$

$$= \frac{g(x)f'(x) - f(x)g'(x)}{[g(x)]^2}$$

INCORPORATING TECHNOLOGY

A quotient of continuous functions typically has a vertical asymptote at each zero of the denominator. Graphing calculators do not do well drawing graphs with vertical asymptotes. For example, the graph of

$$y = \frac{1}{x - 1}$$

has a vertical asymptote at $x = 1$, which is shown on the calculator screen as a spike. The problem is that the standard graphing mode on most calculators (**Connected** mode on TI calculators) connects consecutive dots along the graph. There is no point over $x = 1$ since the function is not defined there. When the calculator connects the dots on either side of the asymptote, a spike results. To display a graph more accurately with an asymptote, most calculators have a **Dot** mode in which consecutive dots are displayed, but not connected with line segments. On the TI-83, the **Dot** mode is selected from the **Mode** screen. The **Dot** mode setting is retained until you change it. ■

Practice Problems 3.1

1. Consider the function $y = (\sqrt{x} + 1)x$.
   (a) Differentiate y by the product rule.
   (b) First multiply out the expression for y and then differentiate.

2. Differentiate $y = \dfrac{5}{x^4 - x^3 + 1}$.

EXERCISES 3.1

Differentiate the functions in Exercises 1–28.

1. $y = (x+1)(x^3 + 5x + 2)$ **2.** $y = (-x^3 + 2)\left(\dfrac{x}{2} - 1\right)$

3. $y = (2x^4 - x + 1)(-x^5 + 1)$ **4.** $y = (x^2 + x + 1)^3(x - 1)^4$

5. $y = x(x^2 + 1)^4$ **6.** $y = x\sqrt{x}$

7. $y = (x^2 + 3)(x^2 - 3)^{10}$ **8.** $y = [(-2x^3 + x)(6x - 3)]^4$

9. $y = (5x + 1)(x^2 - 1) + \dfrac{2x + 1}{3}$

10. $y = x^7(3x^4 + 12x - 1)^2$

11. $y = \dfrac{x - 1}{x + 1}$ **12.** $y = \dfrac{1}{x^2 + x + 7}$

13. $y = \dfrac{x^2 - 1}{x^2 + 1}$ **14.** $y = \dfrac{x}{x + \frac{1}{x}}$

15. $y = \dfrac{x + 3}{(2x + 1)^2}$ **16.** $y = x^4 + 4\sqrt[4]{x}$

17. $y = \dfrac{1}{\pi} + \dfrac{2}{x^2 + 1}$ **18.** $y = \dfrac{ax + b}{cx + d}$

19. $y = \dfrac{3x^2 + 5x + 1}{3 - x^2}$ **20.** $y = \dfrac{x^2}{(x^2 + 1)^2}$

21. $y = [(3x^2 + 2x + 2)(x - 2)]^2$ **22.** $y = \dfrac{1}{\sqrt{x - 2}}$

23. $y = \dfrac{1}{\sqrt{x} + 1}$ **24.** $y = \dfrac{3}{\sqrt[3]{x} + 1}$

25. $y = \dfrac{x^4 - 4x^2 + 3}{x}$ **26.** $y = \dfrac{\sqrt{x} + 1}{\sqrt{x} - 1}$

27. $y = \sqrt{x + 2}(2x + 1)^2$ **28.** $y = \dfrac{\sqrt{3x - 1}}{x}$

29. Find the equation of the tangent line to the curve $y = (x - 2)^5(x + 1)^2$ at the point $(3, 16)$.

30. Find the equation of the tangent line to the curve $y = (x + 1)/(x - 1)$ at the point $(2, 3)$.

31. Find all x-coordinates of points (x, y) on the curve $y = (x - 2)^5/(x - 4)^3$ where the tangent line is horizontal.

32. Find the inflection points on the graph of $y = \dfrac{1}{x^2 + 1}$. (See Fig. 2.)

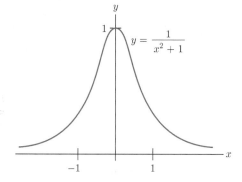

Figure 2

33. Find all x such that $\dfrac{dy}{dx} = 0$, where
$$y = (x^2 - 4)^3(2x^2 + 5)^5.$$

34. The graph of $y = (x^2 - 1)^4(x^2 + 1)^5$ is shown in Fig. 3. Find the coordinates of the local maxima and minima. [*Hint*: Use Example 4.]

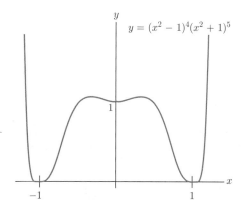

Figure 3

35. Find the point(s) on the graph of $y = (x^2 + 3x - 1)/x$ where the slope is 5.

36. Find the point(s) on the graph of $y = (2x^4 + 1)(x - 5)$ where the slope is 1.

Find $\dfrac{d^2y}{dx^2}$.

37. $y = (x^2 + 1)^4$ **38.** $y = \sqrt{x^2 + 1}$

39. $y = x\sqrt{x + 1}$ **40.** $y = \dfrac{2}{2 + x^2}$

In Exercises 41–44, a function $h(x)$ is defined in terms of a differentiable $f(x)$. Find an expression for $h'(x)$.

41. $h(x) = xf(x)$ **42.** $h(x) = (x^2 + 2x - 1)f(x)$

43. $h(x) = \dfrac{f(x)}{x^2 + 1}$ **44.** $h(x) = \left(\dfrac{f(x)}{x}\right)^2$

45. **Volume** An open rectangular box is 3 feet long and has a surface area of 16 square feet. Find the dimensions of the box for which the volume is as large as possible.

46. **Volume** A closed rectangular box is to be constructed with one side 1 meter long. The material for the top costs $20 per square meter, and the material for the sides and bottom costs $10 per square meter. Find the dimensions of the box with the largest possible volume that can be built at a cost of $240 for materials.

47. **Average Cost** A sugar refinery can produce x tons of sugar per week at a weekly cost of $.1x^2 + 5x + 2250$ dollars. Find the level of production for which the average cost is at a minimum and show that the average cost equals the marginal cost at that level of production.

48. Average Cost A cigar manufacturer produces x cases of cigars per day at a daily cost of $50x(x+200)/(x+100)$ dollars. Show that his cost increases and his average cost decreases as the output x increases.

49. Average Revenue Let $R(x)$ be the revenue received from the sale of x units of a product. The *average revenue per unit* is defined by $AR = R(x)/x$. Show that at the level of production where the average revenue is maximized, the average revenue equals the marginal revenue.

50. Average Velocity Let $s(t)$ be the number of miles a car travels in t hours. Then the average velocity during the first t hours is $\overline{v}(t) = s(t)/t$ miles per hour. If the average velocity is maximized at time t_0, show that at this time the average velocity $\overline{v}(t_0)$ equals the instantaneous velocity $s'(t_0)$. [*Hint:* Compute the derivative of $\overline{v}(t)$.]

51. Rate of Change The width of a rectangle is increasing at a rate of 3 inches per second and its length is increasing at the rate of 4 inches per second. At what rate is the area of the rectangle increasing when its width is 5 inches and its length is 6 inches? [*Hint:* Let $W(t)$ and $L(t)$ be the widths and lengths, respectively, at time t.]

52. Cost–Benefit of Emission Control A manufacturer plans to decrease the amount of sulfur dioxide escaping from its smokestacks. The estimated cost–benefit function is

$$f(x) = \frac{3x}{105 - x}, \qquad 0 \le x \le 100,$$

where $f(x)$ is the cost in millions of dollars for eliminating x percent of the total sulfur dioxide. (See Fig. 4.) Find the value of x at which the rate of increase of the cost–benefit function is 1.4 million dollars per unit. (Each unit is a 1 percentage point increase in pollutant removed.)

Figure 4. A cost–benefit function.

In Exercises 53 and 54, use the fact that at the beginning of 1998 the population of the United States was 268,924,000 people and growing at the rate of 1,856,000 people per year.

53. At the beginning of 1998, the annual per capita consumption of gasoline in the United States was 52.3 gallons and growing at the rate of .2 gallons per year. At what rate was the total annual consumption of gasoline in the United States increasing at that time? (*Hint:* [total annual consumption] = [population]·[annual per capita consumption].)

54. At the beginning of 1998, the annual consumption of ice cream in the United States was 12,582 million pints and growing at the rate of 212 million pints per year. At what rate was the annual per capita consumption of ice cream increasing at that time? (*Hint:* [annual per capita consumption] = $\frac{\text{[annual consumption]}}{\text{[population size]}}$.)

55. Figure 5 shows the graph of

$$y = \frac{10x}{1 + .25x^2}$$

for $x \ge 0$. Find the coordinates of the maximum point.

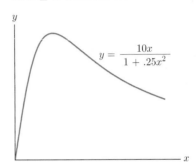

Figure 5

56. Figure 6 shows the graph of

$$y = \frac{1}{2} + \frac{x^2 - 2x + 1}{x^2 - 2x + 2}$$

for $0 \le x \le 2$. Find the coordinates of the minimum point.

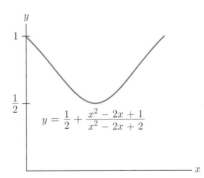

Figure 6

57. Show that the derivative of $\dfrac{x^4}{x^2 + 1}$ is not $\dfrac{4x^3}{2x}$.

58. Show that the derivative of $(5x^3)(2x^4)$ is not $(15x^2)(8x^3)$.

59. If $f(x)$ is a function whose derivative is $f'(x) = \dfrac{1}{1 + x^2}$, find the derivative of $\dfrac{f(x)}{1 + x^2}$.

60. If $f(x)$ and $g(x)$ are differentiable functions such that $f(1) = 2$, $f'(1) = 3$, $g(1) = 4$, and $g'(1) = 5$, find
$$\frac{d}{dx}[f(x)g(x)]\Big|_{x=1}.$$

61. Consider the functions of Exercise 60. Find
$$\frac{d}{dx}\left[\frac{f(x)}{g(x)}\right]\Big|_{x=1}.$$

62. If $f(x)$ is a function whose derivative is $f'(x) = 1/x$, find the derivative of $xf(x) - x$.

63. Let $f(x) = 1/x$ and $g(x) = x^3$.

    (a) Show that the product rule yields the correct derivative of $(1/x)x^3 = x^2$.

    (b) Compute the product $f'(x)g'(x)$ and note that it is *not* the derivative of $f(x)g(x)$.

64. The derivative of $(x^3 - 4x)/x$ is obviously $2x$ for $x \neq 0$, because $(x^3 - 4x)/x = x^2 - 4$ for $x \neq 0$. Verify that the quotient rule gives the same derivative.

65. Let $f(x)$, $g(x)$, and $h(x)$ be differentiable functions. Find a formula for the derivative of $f(x)g(x)h(x)$. (*Hint:* First differentiate $f(x)[g(x)h(x)]$.)

66. Alternative Verification of the Product Rule Apply the special case of the general power rule
$$\frac{d}{dx}[h(x)]^2 = 2h(x)h'(x)$$
and the identity
$$fg = \tfrac{1}{4}[(f + g)^2 - (f - g)^2]$$
to prove the product rule.

67. Body Mass Index The body mass index, or BMI, is a ratio of a person's weight divided by the square of his or her height. Let $b(t)$ denote the BMI; then
$$b(t) = \frac{w(t)}{[h(t)]^2},$$
where t is the age of the person, $w(t)$ the weight in kilograms, and $h(t)$ the height in meters. Find an expression for $b'(t)$.

68. The BMI is usually used as a guideline to determine whether a person is overweight or underweight. For example, according to the Centers for Disease Control, a 12-year-old boy is at risk of being overweight if his BMI is between 21 and 24 and is overweight if his BMI is above 24. A 13-year-old boy is at risk of being overweight if his BMI is between 22 and 25 and is overweight if his BMI is above 25.

    (a) At age 12, a boy's weight was 50 kilograms and his height was 1.55 meters. Find his BMI and determine whether he is overweight or at risk of being overweight.

    (b) Suppose that the boy's weight was increasing at the rate of 7 kilograms per year and his height was increasing at the rate of 5 centimeters per year. Find the rate of change of his BMI at age 12 or $b'(12)$.

    (c) Use the values $b(12)$ and $b'(12)$ to estimate $b(13)$, the BMI of the child at age 13. Is the child at risk of being overweight when he turns 13? [*Hint:* Use formula (2), Section 1.8, to approximate $b(13)$.]

Technology Exercises

69. Area of the Pupil The relationship between the area of the pupil of the eye and the intensity of light was analyzed by B. H. Crawford.[*] Crawford concluded that the area of the pupil is
$$f(x) = \frac{160x^{-.4} + 94.8}{4x^{-.4} + 15.8} \qquad (0 \leq x \leq 37)$$
square millimeters when x units of light are entering the eye per unit time.

    (a) Graph $f(x)$ and $f'(x)$ in the window $[0, 6]$ by $[-5, 20]$.

    (b) How large is the pupil when 3 units of light are entering the eye per unit time?

    (c) For what light intensity is the pupil size 11 square millimeters?

    (d) When 3 units of light are entering the eye per unit time, what is the rate of change of pupil size with respect to a unit change in light intensity?

---

[*]The dependence of pupil size on the external light stimulus under static and variable conditions; *Proceedings of the Royal Society Series B*, 121: 376–395, 1937.

Solutions to Practice Problems 3.1

1. (a) Apply the product rule to $y = (\sqrt{x} + 1)x$ with

$$f(x) = \sqrt{x} + 1 = x^{1/2} + 1$$
$$g(x) = x.$$

$$\frac{dy}{dx} = (x^{1/2} + 1) \cdot 1 + x \cdot \tfrac{1}{2}x^{-1/2}$$
$$= x^{1/2} + 1 + \tfrac{1}{2}x^{1/2}$$
$$= \tfrac{3}{2}\sqrt{x} + 1.$$

(b) $y = (\sqrt{x} + 1)x = (x^{1/2} + 1)x = x^{3/2} + x$

$$\frac{dy}{dx} = \tfrac{3}{2}x^{1/2} + 1.$$

Comparing parts (a) and (b), we note that it is helpful to simplify the function before differentiating.

2. We may apply the quotient rule to $y = \dfrac{5}{x^4 - x^3 + 1}$.

$$\frac{dy}{dx} = \frac{(x^4 - x^3 + 1) \cdot 0 - 5 \cdot (4x^3 - 3x^2)}{(x^4 - x^3 + 1)^2}$$
$$= \frac{-5x^2(4x - 3)}{(x^4 - x^3 + 1)^2}$$

However, it is slightly faster to use the general power rule, since $y = 5(x^4 - x^3 + 1)^{-1}$. Thus

$$\frac{dy}{dx} = -5(x^4 - x^3 + 1)^{-2}(4x^3 - 3x^2)$$
$$= -5x^2(x^4 - x^3 + 1)^{-2}(4x - 3).$$

3.2 The Chain Rule and the General Power Rule

In this section we show that the general power rule is a special case of a powerful differentiation technique called the chain rule. Applications of the chain rule appear throughout the text.

A useful way of combining functions $f(x)$ and $g(x)$ is to replace each occurrence of the variable x in $f(x)$ by the function $g(x)$. The resulting function is called the *composition* (or *composite*) of $f(x)$ and $g(x)$ and is denoted by $f(g(x))$.[†]

EXAMPLE 1   **Composition of functions** Let

$$f(x) = \frac{x - 1}{x + 1}, \quad g(x) = x^3.$$

What is $f(g(x))$?

Solution Replace each occurrence of x in $f(x)$ by $g(x)$ to obtain

$$f(g(x)) = \frac{g(x) - 1}{g(x) + 1} = \frac{x^3 - 1}{x^3 + 1}.$$

Given a composite function $f(g(x))$, we may think of $f(x)$ as the "outside" function that acts on the values of the "inside" function $g(x)$. This point of view often helps us to recognize a composite function.

EXAMPLE 2   **Decomposing composite functions** Write the following functions as composites of simpler functions.

(a) $h(x) = (x^5 + 9x + 3)^8$ **(b)** $k(x) = \sqrt{4x^2 + 1}$

Solution **(a)** $h(x) = f(g(x))$, where the outside function is the power function, $(\ldots)^8$, that is, $f(x) = x^8$. Inside this power function is $g(x) = x^5 + 9x + 3$.

(b) $k(x) = f(g(x))$, where the outside function is the square root function, $f(x) = \sqrt{x}$, and the inside function is $g(x) = 4x^2 + 1$.

[†]See Section 0.3 for additional information on function composition.

A function of the form $[g(x)]^r$ is a composite $f(g(x))$, where the outside function is $f(x) = x^r$. We have already given a rule for differentiating this function:

$$\frac{d}{dx}[g(x)]^r = r[g(x)]^{r-1}g'(x).$$

The *chain rule* has the same form, except that the outside function $f(x)$ can be *any* differentiable function.

> **The Chain Rule**   To differentiate $f(g(x))$, first differentiate the outside function $f(x)$ and substitute $g(x)$ for x in the result. Then multiply by the derivative of the inside function $g(x)$. Symbolically,
>
> $$\frac{d}{dx}f(g(x)) = f'(g(x))g'(x).$$

EXAMPLE 3

The general power rule as a chain rule   Use the chain rule to compute the derivative of $f(g(x))$, where $f(x) = x^8$ and $g(x) = x^5 + 9x + 3$.

Solution
$$f'(x) = 8x^7, \qquad g'(x) = 5x^4 + 9$$
$$f'(g(x)) = 8(x^5 + 9x + 3)^7$$

Finally, by the chain rule,

$$\frac{d}{dx}f(g(x)) = f'(g(x))g'(x) = 8(x^5 + 9x + 3)^7(5x^4 + 9).$$

Since in this example the outside function is a power function, the calculations are the same as when the general power rule is used to compute the derivative of $y = (x^5 + 9x + 3)^8$. However, the organization of the solution here emphasizes the notation of the chain rule. ∎

EXAMPLE 4

Derivative of a function that is not specified   If $h(x) = f(\sqrt{x})$, where f is a differentiable function that is not specified, find $h'(x)$ in terms of f'.

Solution   Let $g(x) = \sqrt{x}$; then h is the composition of $f(x)$ and $g(x)$: $h(x) = f(g(x))$. We have $g'(x) = \dfrac{1}{2\sqrt{x}}$ and, by the chain rule,

$$h'(x) = \frac{d}{dx}(f(g(x)) = f'(g(x))g'(x)$$
$$= f'(\sqrt{x}) \cdot \frac{1}{2\sqrt{x}} = \frac{f'(\sqrt{x})}{2\sqrt{x}}.$$ ∎

There is another way to write the chain rule. Given the function $y = f(g(x))$, set $u = g(x)$ so that $y = f(u)$. Then y may be regarded either as a function of u or, indirectly through u, as a function of x. With this notation, we have $\dfrac{du}{dx} = g'(x)$ and $\dfrac{dy}{du} = f'(u) = f'(g(x))$. Thus the chain rule says that

$$\frac{dy}{dx} = \frac{dy}{du}\frac{du}{dx}. \tag{1}$$

Although the derivative symbols in equation (1) are not really fractions, the apparent cancellation of the "du" symbols provides a mnemonic device for remembering this form of the chain rule.

Here is an illustration that makes (1) a plausible formula. Suppose that y, u, and x are three varying quantities, with y varying three times as fast as u and u varying two times as fast as x. It seems reasonable that y should vary six times as fast as x. That is, $\dfrac{dy}{dx} = \dfrac{dy}{du}\dfrac{du}{dx} = 3 \cdot 2 = 6$.

EXAMPLE 5   **The chain rule** Find

$$\frac{dy}{dx} \quad \text{if } y = u^5 - 2u^3 + 8 \text{ and } u = x^2 + 1.$$

Solution Since y is not given directly as a function of x, we cannot compute $\dfrac{dy}{dx}$ by differentiating y directly with respect to x. We can, however, differentiate, with respect to u, the relation $y = u^5 - 2u^3 + 8$, and get

$$\frac{dy}{du} = 5u^4 - 6u^2.$$

Similarly, we can differentiate with respect to x the relation $u = x^2 + 1$ and get

$$\frac{du}{dx} = 2x.$$

Applying the chain rule, as prescribed by (1), we obtain

$$\frac{dy}{dx} = \frac{dy}{du}\frac{du}{dx} = (5u^4 - 6u^2) \cdot (2x).$$

It is usually desirable to express $\dfrac{dy}{dx}$ as a function of x alone, so we substitute $x^2 + 1$ for u to obtain

$$\frac{dy}{dx} = [5(x^2 + 1)^4 - 6(x^2 + 1)^2] \cdot 2x. \qquad \blacksquare$$

You may have noticed another entirely mechanical way to work Example 5: First substitute $u = x^2 + 1$ into the original formula for y and obtain $y = (x^2 + 1)^5 - 2(x^2 + 1)^3 + 8$. Then $\dfrac{dy}{dx}$ is easily calculated by the sum rule, the general power rule, and the constant-multiple rule. The solution in Example 5, however, lays the foundation for applications of the chain rule in Section 3.3 and elsewhere.

In many situations involving composition of functions, the basic variable is time, t. It may happen that x is a function of t, say $x = g(t)$, and some other variable such as R is a function of x, say $R = f(x)$. Then $R = f(g(t))$, and the chain rule says that

$$\frac{dR}{dt} = \frac{dR}{dx}\frac{dx}{dt}.$$

EXAMPLE 6   **Marginal revenue and time rate of change** A store sells ties for $12 apiece. Let x be the number of ties sold in one day, and let R be the revenue received from the sale of x ties, so $R = 12x$. If daily sales are rising at the rate of four ties per day, how fast is the revenue rising?

Solution Clearly, revenue is rising at the rate of $48 per day, because each of the additional four ties brings in $12. This intuitive conclusion also follows from the chain rule,

$$\frac{dR}{dt} = \frac{dR}{dx} \cdot \frac{dx}{dt}$$

$$\begin{bmatrix} \text{rate of change} \\ \text{of revenue with} \\ \text{respect to time} \end{bmatrix} = \begin{bmatrix} \text{rate of change} \\ \text{of revenue with} \\ \text{respect to sales} \end{bmatrix} \cdot \begin{bmatrix} \text{rate of change} \\ \text{of sales with} \\ \text{respect to time} \end{bmatrix}$$

$$\begin{bmatrix} \text{\$48 increase} \\ \text{per day} \end{bmatrix} = \begin{bmatrix} \text{\$12 increase} \\ \text{per add'l tie} \end{bmatrix} \cdot \begin{bmatrix} \text{four additional} \\ \text{ties per day} \end{bmatrix}.$$

Notice that $\frac{dR}{dx}$ is actually the marginal revenue, studied earlier. This example shows that the time rate of change of revenue, $\frac{dR}{dt}$, is the marginal revenue multiplied by the time rate of change of sales. ∎

EXAMPLE 7

Time rate of change of revenue The demand equation for a certain brand of graphing calculator is $p = 86 - .002x$, where p is the price of one calculator (in dollars) and x is the number of calculators produced and sold. Determine the (time) rate of change of the total revenue if the company increases production by 200 calculators per day, when the production level is at 6000 calculators.

Solution The total revenue received from the production and sale of x calculators is $R(x) = x \cdot p = x(86 - .002x) = -.002x^2 + 86x$. We are told that, when $x = 6000$, the production will increase at the rate of 200 calculators per day; that is, $\frac{dx}{dt} = 200$ when $x = 6000$. We are asked to find the time rate of change of revenue, $\frac{dR}{dt}$, when $x = 6000$. Since R is not given directly as a function of t, to differentiate R with respect to t, we appeal to the chain rule. We have

$$\frac{dR}{dt} = \frac{dR}{dx} \cdot \frac{dx}{dt}.$$

(As in the previous example, the time rate of change of revenue, $\frac{dR}{dt}$, is the marginal revenue, $\frac{dR}{dx}$, multiplied by the time rate of change of sales, $\frac{dx}{dt}$.) Now

$$\frac{dR}{dx} = \frac{d}{dx}\overbrace{(-.002x^2 + 86x)}^{R(x)} = -.004x + 86.$$

So

$$\frac{dR}{dt} = \overbrace{(-.004x + 86)}^{dR/dx} \cdot \frac{dx}{dt}.$$

Substituting $x = 6000$ and $\frac{dx}{dt} = 200$, we find

$$\frac{dR}{dt} = [-.004(6000) + 86](200) = 12{,}400 \text{ dollars per day.}$$

Thus, if at production level $x = 6000$ the company increases production by 200 calculators per day, its total revenue will increase at the rate of $12,400 per day. ∎

Verification of the Chain Rule   Suppose that $f(x)$ and $g(x)$ are differentiable, and let $x = a$ be a number in the domain of $f(g(x))$. Since every differentiable function is continuous, we have

$$\lim_{h \to 0} g(a + h) = g(a),$$

which implies that

$$\lim_{h \to 0} [g(a + h) - g(a)] = 0. \tag{2}$$

Now $g(a)$ is a number in the domain of f, and the limit definition of the derivative gives us

$$f'(g(a)) = \lim_{k \to 0} \frac{f(g(a) + k) - f(g(a))}{k}. \tag{3}$$

Let $k = g(a + h) - g(a)$. By equation (2), k approaches zero as h approaches zero. Also, $g(a + h) = g(a) + k$. Therefore, (3) may be rewritten in the form

$$f'(g(a)) = \lim_{h \to 0} \frac{f(g(a + h)) - f(g(a))}{g(a + h) - g(a)}. \tag{4}$$

[Strictly speaking, we must assume that the denominator in (4) is never zero. This assumption may be avoided by a somewhat different and more technical argument that we omit.] Finally, we show that the function $f(g(x))$ has a derivative at $x = a$. We use the limit definition of the derivative, Limit Theorem V, and equation (4).

$$\frac{d}{dx} f(g(x)) \bigg|_{x=a} = \lim_{h \to 0} \frac{f(g(a + h)) - f(g(a))}{h}$$

$$= \lim_{h \to 0} \left[\frac{f(g(a + h)) - f(g(a))}{g(a + h) - g(a)} \cdot \frac{g(a + h) - g(a)}{h} \right]$$

$$= \lim_{h \to 0} \frac{f(g(a + h)) - f(g(a))}{g(a + h) - g(a)} \cdot \lim_{h \to 0} \frac{g(a + h) - g(a)}{h}$$

$$= f'(g(a)) \cdot g'(a)$$

Practice Problems 3.2

Consider the function $h(x) = (2x^3 - 5)^5 + (2x^3 - 5)^4$.

1. Write $h(x)$ as a composite function, $f(g(x))$.

2. Compute $f'(x)$ and $f'(g(x))$.

3. Use the chain rule to differentiate $h(x)$.

EXERCISES 3.2

Compute $f(g(x))$, where $f(x)$ and $g(x)$ are the following:

1. $f(x) = \dfrac{x}{x + 1}$, $g(x) = x^3$

2. $f(x) = x - 1$, $g(x) = \dfrac{1}{x + 1}$

3. $f(x) = x(x^2 + 1)$, $g(x) = \sqrt{x}$

4. $f(x) = \dfrac{x + 1}{x - 3}$, $g(x) = x + 3$

Each of the following functions may be viewed as a composite function $h(x) = f(g(x))$. Find $f(x)$ and $g(x)$.

5. $h(x) = (x^3 + 8x - 2)^5$
6. $h(x) = (9x^2 + 2x - 5)^7$

7. $h(x) = \sqrt{4 - x^2}$

8. $h(x) = (5x^2 + 1)^{-1/2}$

9. $h(x) = \dfrac{1}{x^3 - 5x^2 + 1}$

10. $h(x) = (4x - 3)^3 + \dfrac{1}{4x - 3}$

Differentiate the functions in Exercises 11–20 using one or more of the differentiation rules discussed thus far.

11. $y = (x^2 + 5)^{15}$

12. $y = (x^4 + x^2)^{10}$

13. $y = 6x^2(x - 1)^3$

14. $y = 5x^3(2 - x)^4$

15. $y = 2(x^3 - 1)(3x^2 + 1)^4$

16. $y = 2(2x - 1)^{5/4}(2x + 1)^{3/4}$

17. $y = \left(\dfrac{4}{1-x}\right)^3$

18. $y = \dfrac{4x^2 + x}{\sqrt{x}}$

19. $y = \left(\dfrac{4x-1}{3x+1}\right)^3$

20. $y = \left(\dfrac{x}{x^2+1}\right)^2$

In Exercises 21–26, a function $h(x)$ is defined in terms of a differentiable $f(x)$. Find an expression for $h'(x)$.

21. $h(x) = f(x^2)$

22. $h(x) = 2f(2x+1)$

23. $h(x) = -f(-x)$

24. $h(x) = f(f(x))$

25. $h(x) = \dfrac{f(x^2)}{x}$

26. $h(x) = \sqrt{f(x^2)}$

27. Sketch the graph of $y = 4x/(x+1)^2$, $x > -1$.

28. Sketch the graph of $y = 2/(1+x^2)$.

Compute $\dfrac{d}{dx} f(g(x))$, where $f(x)$ and $g(x)$ are the following:

29. $f(x) = x^5$, $g(x) = 6x - 1$

30. $f(x) = \sqrt{x}$, $g(x) = x^2 + 1$

31. $f(x) = \dfrac{1}{x}$, $g(x) = 1 - x^2$

32. $f(x) = \dfrac{1}{1+\sqrt{x}}$, $g(x) = \dfrac{1}{x}$

33. $f(x) = x^4 - x^2$, $g(x) = x^2 - 4$

34. $f(x) = \dfrac{4}{x} + x^2$, $g(x) = 1 - x^4$

35. $f(x) = (x^3 + 1)^2$, $g(x) = x^2 + 5$

36. $f(x) = x(x-2)^4$, $g(x) = x^3$

Compute $\dfrac{dy}{dx}$ using the chain rule in formula (1).

37. $y = u^{3/2}$, $u = 4x + 1$

38. $y = \sqrt{u+1}$, $u = 2x^2$

39. $y = \dfrac{u}{2} + \dfrac{2}{u}$, $u = x - x^2$

40. $y = \dfrac{u^2 + 2u}{u+1}$, $u = x(x+1)$

Compute $\dfrac{dy}{dt}\Big|_{t=t_0}$.

41. $y = x^2 - 3x$, $x = t^2 + 3$, $t_0 = 0$

42. $y = (x^2 - 2x + 4)^2$, $x = \dfrac{1}{t+1}$, $t_0 = 1$

43. $y = \dfrac{x+1}{x-1}$, $x = \dfrac{t^2}{4}$, $t_0 = 3$

44. $y = \sqrt{x+1}$, $x = \sqrt{t+1}$, $t_0 = 0$

45. Find the equation of the line tangent to the graph of $y = 2x(x-4)^6$ at the point $(5, 10)$.

46. Find the equation of the line tangent to the graph of $y = \dfrac{x}{\sqrt{2-x^2}}$ at the point $(1, 1)$.

47. Find the x-coordinates of all points on the curve $y = (-x^2 + 4x - 3)^3$ with a horizontal tangent line.

48. The function $f(x) = \sqrt{x^2 - 6x + 10}$ has one relative minimum point for $x \geq 0$. Find it.

49. The length, x, of the edge of a cube is increasing.

   (a) Write the chain rule for $\dfrac{dV}{dt}$, the time rate of change of the volume of the cube.

   (b) For what value of x is $\dfrac{dV}{dt}$ equal to 12 times the rate of increase of x?

50. **Allometric Equation** Many relations in biology are expressed by power functions, known as **allometric equations**, of the form $y = kx^a$, where k and a are constants. For example, the weight of a male hognose snake is approximately $446x^3$ grams, where x is its length in meters.[*] If a snake has length .4 meters and is growing at the rate of .2 meters per year, at what rate is the snake gaining weight?

51. Suppose that P, y, and t are variables, where P is a function of y and y is a function of t.

   (a) Write the derivative symbols for the following quantities: the rate of change of y with respect to t, the rate of change of P with respect to y, and the rate of change of P with respect to t. Select your answers from the following:
   $$\frac{dP}{dy}, \quad \frac{dy}{dP}, \quad \frac{dy}{dt}, \quad \frac{dP}{dt}, \quad \frac{dt}{dP}, \quad \text{and} \quad \frac{dt}{dy}.$$

   (b) Write the chain rule for $\dfrac{dP}{dt}$.

52. Suppose that Q, x, and y are variables, where Q is a function of x and x is a function of y. (Read this carefully.)

   (a) Write the derivative symbols for the following quantities: the rate of change of x with respect to y, the rate of change of Q with respect to y, and the rate of change of Q with respect to x. Select your answers from the following:
   $$\frac{dy}{dx}, \quad \frac{dx}{dy}, \quad \frac{dQ}{dx}, \quad \frac{dx}{dQ}, \quad \frac{dQ}{dy}, \quad \text{and} \quad \frac{dy}{dQ}.$$

   (b) Write the chain rule for $\dfrac{dQ}{dy}$.

53. **Marginal Profit and Time Rate of Change** When a company produces and sells x thousand units per week, its total weekly profit is P thousand dollars, where
$$P = \frac{200x}{100 + x^2}.$$

The production level at t weeks from the present is $x = 4 + 2t$.

---

[*]D. R. Platt, "Natural History of the Hognose Snakes *Heterodon Platyrhinos* and *Heterodon Nasicus*," University of Kansas Publications, *Museum of Natural History* 18:4 (1969), 253–420.

(a) Find the marginal profit, $\dfrac{dP}{dx}$.

(b) Find the time rate of change of profit, $\dfrac{dP}{dt}$.

(c) How fast (with respect to time) are profits changing when $t = 8$?

54. Marginal Cost and Time Rate of Change The cost of manufacturing x cases of cereal is C dollars, where $C = 3x + 4\sqrt{x} + 2$. Weekly production at t weeks from the present is estimated to be $x = 6200 + 100t$ cases.

(a) Find the marginal cost, $\dfrac{dC}{dx}$.

(b) Find the time rate of change of cost, $\dfrac{dC}{dt}$.

(c) How fast (with respect to time) are costs rising when $t = 2$?

55. A Model for Carbon Monoxide Levels Ecologists estimate that, when the population of a certain city is x thousand persons, the average level L of carbon monoxide in the air above the city will be L ppm (parts per million), where $L = 10 + .4x + .0001x^2$. The population of the city is estimated to be $x = 752 + 23t + .5t^2$ thousand persons t years from the present.

(a) Find the rate of change of carbon monoxide with respect to the population of the city.

(b) Find the time rate of change of the population when $t = 2$.

(c) How fast (with respect to time) is the carbon monoxide level changing at time $t = 2$?

56. Profit A manufacturer of microcomputers estimates that t months from now it will sell x thousand units of its main line of microcomputers per month, where $x = .05t^2 + 2t + 5$. Because of economies of scale, the profit P from manufacturing and selling x thousand units is estimated to be $P = .001x^2 + .1x - .25$ million dollars. Calculate the rate at which the profit will be increasing 5 months from now.

57. If $f(x)$ and $g(x)$ are differentiable functions, find $g(x)$ if you know that

$$\frac{d}{dx} f(g(x)) = 3x^2 \cdot f'(x^3 + 1).$$

58. If $f(x)$ and $g(x)$ are differentiable functions, find $g(x)$ if you know that $f'(x) = 1/x$ and

$$\frac{d}{dx} f(g(x)) = \frac{2x + 5}{x^2 + 5x - 4}.$$

59. If $f(x)$ and $g(x)$ are differentiable functions such that $f(1) = 2$, $f'(1) = 3$, $f'(5) = 4$, $g(1) = 5$, $g'(1) = 6$, $g'(2) = 7$, and $g'(5) = 8$, find $\dfrac{d}{dx} f(g(x)) \Big|_{x=1}$.

60. Consider the functions of Exercise 59. Find $\dfrac{d}{dx} g(f(x)) \Big|_{x=1}$.

61. Effect of Stocks on Total Assets of a Company After a computer software company went public, the price of one share of its stock on the stock market fluctuated according to the graph in Fig. 1(a). The total worth of the company depended on the value of one share and was estimated to be

$$W(x) = 10 \frac{12 + 8x}{3 + x},$$

where x is the value of one share (in dollars) and $W(x)$ is the total value of the company in millions of dollars. [See Fig. 1(b).]

Company went public.

(a)

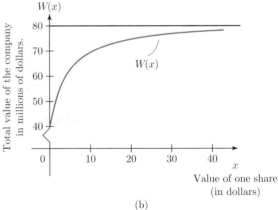

(b)

Figure 1

(a) Find the total value of the company when $t = 1.5$ and when $t = 3.5$.

(b) Find $\dfrac{dx}{dt}\Big|_{t=1.5}$ and $\dfrac{dx}{dt}\Big|_{t=3.5}$. Give an interpretation for these values.

62. Refer to Exercise 61. Use the chain rule to find $\dfrac{dW}{dt}\Big|_{t=1.5}$ and $\dfrac{dW}{dt}\Big|_{t=3.5}$. Give an interpretation for these values.

63. Refer to Exercise 61.

    (a) Find $\dfrac{dx}{dt}\Big|_{t=2.5}$ and $\dfrac{dx}{dt}\Big|_{t=4}$. Give an interpretation for these values.

    (b) Use the chain rule to find $\dfrac{dW}{dt}\Big|_{t=2.5}$ and $\dfrac{dW}{dt}\Big|_{t=4}$. Give an interpretation for these values.

64. Refer to Exercise 61.

    (a) What was the maximum value of the company during the first 6 months since it went public, and when was it attained?

    (b) Assuming that the value of one share will continue to increase at the rate that it did during the period following the sixth month, what is the limit of the total value of the company as t increases?

65. In an expression of the form $f(g(x))$, $f(x)$ is called the *outer* function and $g(x)$ is called the *inner* function. Give a written description of the chain rule using the words "inner" and "outer."

Solutions to Practice Problems 3.2

1. Let $f(x) = x^5 + x^4$ and $g(x) = 2x^3 - 5$.

2. $f'(x) = 5x^4 + 4x^3$, $f'(g(x)) = 5(2x^3 - 5)^4 + 4(2x^3 - 5)^3$.

3. We have $g'(x) = 6x^2$. Then, from the chain rule and

the result of Problem 2, we have

$$h'(x) = f'(g(x))g'(x)$$
$$= [5(2x^3 - 5)^4 + 4(2x^3 - 5)^3](6x^2).$$

3.3 Implicit Differentiation and Related Rates

This section presents two different applications of the chain rule. In each case we will need to differentiate one or more composite functions where the "inside" functions are not known explicitly.

Implicit Differentiation

In some applications, the variables are related by an equation rather than a function. In these cases, we can still determine the rate of change of one variable with respect to the other by the technique of implicit differentiation. As an illustration, consider the equation

$$x^2 + y^2 = 4. \tag{1}$$

The graph of this equation is the circle in Fig. 1. This graph obviously is not the graph of a function since, for instance, there are two points on the graph whose x-coordinate is 1. (Functions must satisfy the vertical line test. See Section 0.1.)

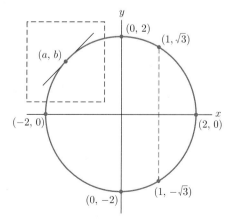

Figure 1. Graph of $x^2 + y^2 = 4$.

We denote the slope of the curve at the point $(1, \sqrt{3})$ by

$$\left.\frac{dy}{dx}\right|_{\substack{x=1 \\ y=\sqrt{3}}}.$$

In general, the slope at the point (a, b) is denoted by

$$\left.\frac{dy}{dx}\right|_{\substack{x=a \\ y=b}}.$$

In a small vicinity of the point (a, b), the curve looks like the graph of a function. [That is, on this part of the curve, $y = g(x)$ for some function $g(x)$.] We say that this function is defined *implicitly* by the equation.* We obtain a formula for $\dfrac{dy}{dx}$ by differentiating both sides of the equation with respect to x while treating y as a function of x.

EXAMPLE 1

Consider the graph of the equation $x^2 + y^2 = 4$.

(a) Use implicit differentiation to compute $\dfrac{dy}{dx}$.

(b) Find the slope of the graph at the points $(1, \sqrt{3})$ and $(1, -\sqrt{3})$.

Solution **(a)** The first term x^2 has derivative $2x$, as usual. We think of the second term y^2 as having the form $[g(x)]^2$. To differentiate, we use the chain rule (specifically, the general power rule):

$$\frac{d}{dx}[g(x)]^2 = 2[g(x)]g'(x)$$

or, equivalently,

$$\frac{d}{dx}y^2 = 2y\frac{dy}{dx}.$$

On the right side of the original equation, the derivative of the constant function 4 is zero. Thus implicit differentiation of $x^2 + y^2 = 4$ yields

$$2x + 2y\frac{dy}{dx} = 0.$$

Solving for $\dfrac{dy}{dx}$, we have

$$2y\frac{dy}{dx} = -2x.$$

If $y \neq 0$, then

$$\frac{dy}{dx} = \frac{-2x}{2y} = -\frac{x}{y}.$$

Notice that this slope formula involves y as well as x. This reflects the fact that the slope of the circle at a point depends on the y-coordinate of the point as well as the x-coordinate.

(b) At the point $(1, \sqrt{3})$ the slope is

$$\left.\frac{dy}{dx}\right|_{\substack{x=1 \\ y=\sqrt{3}}} = \left.-\frac{x}{y}\right|_{\substack{x=1 \\ y=\sqrt{3}}} = -\frac{1}{\sqrt{3}}.$$

---

*Of course, the tangent line at the point (a, b) must not be vertical. In this section, we assume that the given equations implicitly determine differentiable functions.

At the point $(1, -\sqrt{3})$ the slope is

$$\left.\frac{dy}{dx}\right|_{\substack{x=1\\y=-\sqrt{3}}} = \left.-\frac{x}{y}\right|_{\substack{x=1\\y=-\sqrt{3}}} = -\frac{1}{-\sqrt{3}} = \frac{1}{\sqrt{3}}.$$

(See Fig. 2.) The formula for $\dfrac{dy}{dx}$ gives the slope at every point on the graph of $x^2 + y^2 = 4$ except $(-2, 0)$ and $(2, 0)$. At these two points the tangent line is vertical and the slope of the curve is undefined. ■

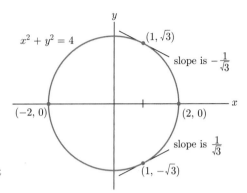

Figure 2. Slope of the tangent line.

The difficult step in Example 1(a) was to differentiate y^2 correctly. The derivative of y^2 *with respect to y* would be $2y$, by the ordinary power rule. But the derivative of y^2 *with respect to x* must be computed by the general power rule. To understand this distinction, think of y as a function of x, say, $y = f(x)$. Then

$$\frac{dy}{dx} = f'(x);$$
$$y^2 = [f(x)]^2;$$
$$\frac{d}{dx}y^2 = \frac{d}{dx}([f(x)]^2) = 2f(x)f'(x),$$

by the general power rule. Substituting $y = f(x)$ and $f'(x) = \dfrac{dy}{dx}$, we obtain

$$\frac{d}{dx}y^2 = 2y\frac{dy}{dx}.$$

Similarly, if a is a constant, then

$$\frac{d}{dx}(ay) = a\frac{dy}{dx};$$
$$\frac{d}{dx}(ay^2) = 2ay\frac{dy}{dx}.$$

Also, if r is a constant, the general power rule yields

$$\frac{d}{dx}y^r = ry^{r-1}\frac{dy}{dx}. \tag{2}$$

This rule is used to compute slope formulas in the next two examples.

EXAMPLE 2   Use implicit differentiation to calculate

$$\frac{dy}{dx}$$

for the equation $x^2y^6 = 1$.

Solution Differentiate each side of the equation $x^2y^6 = 1$ with respect to x. On the left side of the equation use the product rule and treat y as a function of x.

$$x^2 \frac{d}{dx}(y^6) + y^6 \frac{d}{dx}(x^2) = \frac{d}{dx}(1)$$

$$x^2 \cdot 6y^5 \frac{dy}{dx} + y^6 \cdot 2x = 0$$

Solve for $\frac{dy}{dx}$ by moving the term not involving $\frac{dy}{dx}$ to the right side and dividing by the factor that multiplies $\frac{dy}{dx}$.

$$6x^2y^5 \frac{dy}{dx} = -2xy^6$$

$$\frac{dy}{dx} = \frac{-2xy^6}{6x^2y^5} = -\frac{y}{3x} \qquad \blacksquare$$

EXAMPLE 3 Use implicit differentiation to calculate

$$\frac{dy}{dx}$$

when y is related to x by the equation $x^2y + xy^3 - 3x = 5$.

Solution Differentiate the equation term by term, taking care to differentiate x^2y and xy^3 by the product rule.

$$x^2 \frac{d}{dx}(y) + y \frac{d}{dx}(x^2) + x \frac{d}{dx}(y^3) + y^3 \frac{d}{dx}(x) - 3 = 0$$

$$x^2 \frac{dy}{dx} + y \cdot 2x + x \cdot 3y^2 \frac{dy}{dx} + y^3 \cdot 1 - 3 = 0$$

Solve for $\frac{dy}{dx}$ in terms of x and y following these steps:

Step 1 Keep all terms involving $\frac{dy}{dx}$ on the left side of the equation and move the other terms to the right side.

$$x^2 \frac{dy}{dx} + 3xy^2 \frac{dy}{dx} = 3 - y^3 - 2xy$$

Step 2 Factor out $\frac{dy}{dx}$ on the left side of the equation.

$$(x^2 + 3xy^2) \frac{dy}{dx} = 3 - y^3 - 2xy$$

Step 3 Divide both sides of the equation by the factor that multiplies $\frac{dy}{dx}$.

$$\frac{dy}{dx} = \frac{3 - y^3 - 2xy}{x^2 + 3xy^2} \qquad \blacksquare$$

REMINDER When a power of y is differentiated with respect to x, the result must include the factor $\frac{dy}{dx}$. When a power of x is differentiated, there is no factor $\frac{dy}{dx}$. $\blacksquare$

Here is the general procedure for implicit differentiation.

> **Finding $\dfrac{dy}{dx}$ by Implicit Differentiation**
>
> Differentiate each term of the equation *with respect to x*, treating y as a function of x.
>
> Move all terms involving $\dfrac{dy}{dx}$ to the left side of the equation and move the other terms to the right side.
>
> Factor out $\dfrac{dy}{dx}$ on the left side of the equation.
>
> Divide both sides of the equation by the factor that multiplies $\dfrac{dy}{dx}$.

Equations that implicitly define functions arise frequently in economic models. The economic background for the equation in the next example is discussed in Section 7.1.

EXAMPLE 4

Isoquants and marginal rate of substitution Suppose that x and y represent the amounts of two basic inputs into a production process and the equation

$$60x^{3/4}y^{1/4} = 3240$$

describes all input amounts (x, y) for which the output of the process is 3240 units. (The graph of this equation is called a *production isoquant* or *constant product curve*.) (See Fig. 3.) Use implicit differentiation to calculate the slope of the graph at the point on the curve where $x = 81$, $y = 16$.

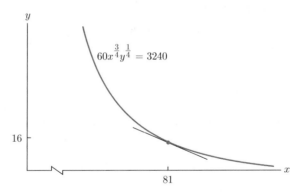

Figure 3. A production isoquant.

Solution We use the product rule and treat y as a function of x.

$$60x^{3/4}\frac{d}{dx}(y^{1/4}) + y^{1/4}\frac{d}{dx}(60x^{3/4}) = 0$$

$$60x^{3/4} \cdot \left(\frac{1}{4}\right)y^{-3/4}\frac{dy}{dx} + y^{1/4} \cdot 60\left(\frac{3}{4}\right)x^{-1/4} = 0$$

$$15x^{3/4}y^{-3/4}\frac{dy}{dx} = -45x^{-1/4}y^{1/4}$$

$$\frac{dy}{dx} = \frac{-45x^{-1/4}y^{1/4}}{15x^{3/4}y^{-3/4}} = \frac{-3y}{x}$$

When $x = 81$ and $y = 16$, we have

$$\left.\frac{dy}{dx}\right|_{\substack{x=81 \\ y=16}} = \frac{-3(16)}{81} = -\frac{16}{27}.$$

The number $-\frac{16}{27}$ is the slope of the production isoquant at the point $(81, 16)$. If the first input (corresponding to x) is increased by one small unit, the second input (corresponding to y) must decrease by approximately $\frac{16}{27}$ unit to keep the production output unchanged [that is, to keep (x, y) on the curve]. In economic terminology, the absolute value of $\dfrac{dy}{dx}$ is called the *marginal rate of substitution* of the first input for the second input.   ■

Related Rates   In implicit differentiation we differentiate an equation involving x and y, with y treated as a function of x. However, in some applications where x and y are related by an equation, both variables are functions of a third variable t (which may represent time). Often the formulas for x and y as functions of t are not known. When we differentiate such an equation with respect to t, we derive a relationship between the rates of change $\dfrac{dy}{dt}$ and $\dfrac{dx}{dt}$. We say that these derivatives are *related rates*. The equation relating the rates may be used to find one of the rates when the other is known.

EXAMPLE 5

A point moving along a graph   Suppose that x and y are both differentiable functions of t and are related by the equation

$$x^2 + 5y^2 = 36. \tag{3}$$

(a) Differentiate each term in the equation with respect to t and solve the resulting equation for $\dfrac{dy}{dt}$.

(b) Calculate $\dfrac{dy}{dt}$ at a time when $x = 4$, $y = 2$, and $\dfrac{dx}{dt} = 5$.

Solution (a) Since x is a function of t, the general power rule gives

$$\frac{d}{dt}(x^2) = 2x\frac{dx}{dt}.$$

A similar formula holds for the derivative of y^2. Differentiating each term in equation (3) with respect to t, we obtain

$$\frac{d}{dt}(x^2) + \frac{d}{dt}(5y^2) = \frac{d}{dt}(36)$$

$$2x\frac{dx}{dt} + 5 \cdot 2y\frac{dy}{dt} = 0$$

$$10y\frac{dy}{dt} = -2x\frac{dx}{dt}$$

$$\frac{dy}{dt} = -\frac{x}{5y}\frac{dx}{dt}.$$

(b) When $x = 4$, $y = 2$, and $\dfrac{dx}{dt} = 5$,

$$\frac{dy}{dt} = -\frac{4}{5(2)} \cdot (5) = -2.$$ ■

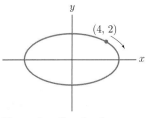

Figure 4. Graph of $x^2 + 5y^2 = 36$.

There is a helpful graphical interpretation of the calculations in Example 5. Imagine a point that is moving clockwise along the graph of the equation $x^2 + 5y^2 = 36$, which is an ellipse. (See Fig. 4.) Suppose that, when the point is at $(4, 2)$, the x-coordinate of the point is changing at the rate of 5 units per minute, so $\dfrac{dx}{dt} = 5$. In Example 5(b) we found that $\dfrac{dy}{dt} = -2$. This means that the y-coordinate of the point is decreasing at the rate of 2 units per minute when the moving point reaches $(4, 2)$.

EXAMPLE 6

Relating weekly sales to price   Suppose that x thousand units of a commodity can be sold weekly when the price is p dollars per unit and that x and p satisfy the demand equation

$$p + 2x + xp = 38.$$

(See Fig. 5.) How fast are weekly sales changing at a time when $x = 4$, $p = 6$, and the price is falling at the rate of $.40 per week?

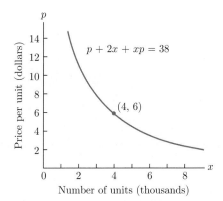

Figure 5. A demand curve.

Solution   Assume that p and x are differentiable functions of t, and differentiate the demand equation with respect to t.

$$\frac{d}{dt}(p) + \frac{d}{dt}(2x) + \frac{d}{dt}(xp) = \frac{d}{dt}(38)$$

$$\frac{dp}{dt} + 2\frac{dx}{dt} + x\frac{dp}{dt} + p\frac{dx}{dt} = 0 \qquad (4)$$

We want to know $\dfrac{dx}{dt}$ at a time when $x = 4$, $p = 6$, and $\dfrac{dp}{dt} = -.40$. (The derivative $\dfrac{dp}{dt}$ is negative because the price is decreasing.) We could solve equation (4) for $\dfrac{dx}{dt}$ and then substitute the given values, but since we do not need a general formula for $\dfrac{dx}{dt}$, it is easier to substitute first and then solve.

$$-.40 + 2\frac{dx}{dt} + 4(-.40) + 6\frac{dx}{dt} = 0$$

$$8\frac{dx}{dt} = 2$$

$$\frac{dx}{dt} = .25$$

Thus sales are rising at the rate of .25 thousand units (or 250 units) per week. ∎

Suggestions for Solving Related Rates Problems

1. Draw a picture, if possible.

2. Assign letters to quantities that vary, and identify one variable, say t, on which the other variables depend.

3. Find an equation that relates the variables to each other.

4. Differentiate the equation with respect to the independent variable t. Use the chain rule whenever appropriate.

5. Substitute all specified values for the variables and their derivatives.

6. Solve for the derivative that gives the unknown rate.

 INCORPORATING TECHNOLOGY   The graph of an equation in x and y can be easily obtained when y can be expressed as one or more functions of x. For instance, the graph of $x^2 + y^2 = 4$ can be plotted by simultaneously graphing $y = \sqrt{4 - x^2}$ and $y = -\sqrt{4 - x^2}$. ■

Practice Problems 3.3

Suppose that x and y are related by the equation $3y^2 - 3x^2 + y = 1$.

1. Use implicit differentiation to find a formula for the slope of the graph of the equation.

2. Suppose that x and y in the preceding equation are both functions of t. Differentiate both sides of the equation with respect to t and find a formula for $\dfrac{dy}{dt}$ in terms of x, y, and $\dfrac{dx}{dt}$.

EXERCISES 3.3

In Exercises 1–18, suppose that x and y are related by the given equation and use implicit differentiation to determine $\dfrac{dy}{dx}$.

1. $x^2 - y^2 = 1$

2. $x^3 + y^3 - 6 = 0$

3. $y^5 - 3x^2 = x$

4. $x^4 + (y + 3)^4 = x^2$

5. $y^4 - x^4 = y^2 - x^2$

6. $x^3 + y^3 = x^2 + y^2$

7. $2x^3 + y = 2y^3 + x$

8. $x^4 + 4y = x - 4y^3$

9. $xy = 5$

10. $xy^3 = 2$

11. $x(y + 2)^5 = 8$

12. $x^2 y^3 = 6$

13. $x^3 y^2 - 4x^2 = 1$

14. $(x + 1)^2 (y - 1)^2 = 1$

15. $x^3 + y^3 = x^3 y^3$

16. $x^2 + 4xy + 4y = 1$

17. $x^2 y + y^2 x = 3$

18. $x^3 y + xy^3 = 4$

Use implicit differentiation of the equations in Exercises 19–24 to determine the slope of the graph at the given point.

19. $4y^3 - x^2 = -5$; $x = 3$, $y = 1$

20. $y^2 = x^3 + 1$; $x = 2$, $y = -3$

21. $xy^3 = 2$; $x = -\frac{1}{4}$, $y = -2$

22. $\sqrt{x} + \sqrt{y} = 7$; $x = 9$, $y = 16$

23. $xy + y^3 = 14$; $x = 3$, $y = 2$

24. $y^2 = 3xy - 5$; $x = 2$, $y = 1$

25. Find the equation of the tangent line to the graph of $x^2 y^4 = 1$ at the point $(4, \frac{1}{2})$ and at the point $(4, -\frac{1}{2})$.

26. Find the equation of the tangent line to the graph of $x^4 y^2 = 144$ at the point $(2, 3)$ and at the point $(2, -3)$.

27. The graph of $x^4 + 2x^2 y^2 + y^4 = 4x^2 - 4y^2$ is the "lemniscate" in Fig. 6.

(a) Find $\dfrac{dy}{dx}$ by implicit differentiation.

(b) Find the slope of the tangent line to the lemniscate at $(\sqrt{6}/2, \sqrt{2}/2)$.

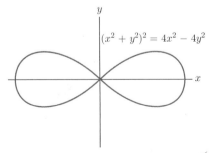

Figure 6. A lemniscate.

28. The graph of $x^4 + 2x^2y^2 + y^4 = 9x^2 - 9y^2$ is a lemniscate similar to that in Fig. 6.

(a) Find $\dfrac{dy}{dx}$ by implicit differentiation.

(b) Find the slope of the tangent line to the lemniscate at $(\sqrt{5}, -1)$.

29. Marginal Rate of Substitution Suppose that x and y represent the amounts of two basic inputs for a production process and that the equation

$$30x^{1/3}y^{2/3} = 1080$$

describes all input amounts where the output of the process is 1080 units.

(a) Find $\dfrac{dy}{dx}$.

(b) What is the marginal rate of substitution of x for y when $x = 16$ and $y = 54$?

30. Suppose that x and y represent the amounts of two basic inputs for a production process and

$$10x^{1/2}y^{1/2} = 600.$$

Find $\dfrac{dy}{dx}$ when $x = 50$ and $y = 72$.

In Exercises 31–36, suppose that x and y are both differentiable functions of t and are related by the given equation. Use implicit differentiation with respect to t to determine $\dfrac{dy}{dt}$ in terms of x, y, and $\dfrac{dx}{dt}$.

31. $x^4 + y^4 = 1$ **32.** $y^4 - x^2 = 1$

33. $3xy - 3x^2 = 4$ **34.** $y^2 = 8 + xy$

35. $x^2 + 2xy = y^3$ **36.** $x^2y^2 = 2y^3 + 1$

37. Point on a Curve A point is moving along the graph of $x^2 - 4y^2 = 9$. When the point is at $(5, -2)$, its x-coordinate is increasing at the rate of 3 units per second. How fast is the y-coordinate changing at that moment?

38. Point on a Curve A point is moving along the graph of $x^3y^2 = 200$. When the point is at $(2, 5)$, its x-coordinate is changing at the rate of -4 units per minute. How fast is the y-coordinate changing at that moment?

39. Demand Equation Suppose that the price p (in dollars) and the weekly sales x (in thousands of units) of a certain commodity satisfy the demand equation

$$2p^3 + x^2 = 4500.$$

Determine the rate at which sales are changing at a time when $x = 50$, $p = 10$, and the price is falling at the rate of $\$.50$ per week.

40. Suppose that the price p (in dollars) and the demand x (in thousands of units) of a commodity satisfy the demand equation

$$6p + x + xp = 94.$$

How fast is the demand changing at a time when $x = 4$, $p = 9$, and the price is rising at the rate of $\$2$ per week?

41. Advertising Affects Revenue The monthly advertising revenue A and the monthly circulation x of a magazine are related approximately by the equation

$$A = 6\sqrt{x^2 - 400}, \qquad x \geq 20,$$

where A is given in thousands of dollars and x is measured in thousands of copies sold. At what rate is the advertising revenue changing if the current circulation is $x = 25$ thousand copies and the circulation is growing at the rate of 2 thousand copies per month? $\left[\textit{Hint:}\text{ Use the chain rule }\dfrac{dA}{dt} = \dfrac{dA}{dx}\dfrac{dx}{dt}.\right]$

42. Suppose that in Boston the wholesale price p of oranges (in dollars per crate) and the daily supply x (in thousands of crates) are related by the equation $px + 7x + 8p = 328$. If there are 4 thousand crates available today at a price of $\$25$ per crate, and if the supply is changing at the rate of $-.3$ thousand crates per day, at what rate is the price changing?

43. Volume and Pressure of Gas Under certain conditions (called adiabatic expansion) the pressure P and volume V of a gas such as oxygen satisfy the equation $P^5V^7 = k$, where k is a constant. Suppose that at some moment the volume of the gas is 4 liters, the pressure is 200 units, and the pressure is increasing at the rate of 5 units per second. Find the (time) rate at which the volume is changing.

44. Spherical Volume The volume V of a spherical cancer tumor is given by $V = \pi x^3/6$, where x is the diameter of the tumor. A physician estimates that the diameter is growing at the rate of .4 millimeters per day, at a time when the diameter is already 10 millimeters. How fast is the volume of the tumor changing at that time?

45. Related Rates Figure 7 shows a 10-foot ladder leaning against a wall.

Figure 7

(a) Use the Pythagorean theorem to find an equation relating x and y.

(b) If the foot of the ladder is being pulled along the ground at the rate of 3 feet per second, how fast is the top end of the ladder sliding down the wall at the time when the foot of the ladder is 8 feet from the wall? That is, what is $\dfrac{dy}{dt}$ at the time when $\dfrac{dx}{dt} = 3$ and $x = 8$?

46. Related Rates An airplane flying 390 feet per second at an altitude of 5000 feet flew directly over an observer. Figure 8 shows the relationship of the airplane to the observer at a later time.

(a) Find an equation relating x and y.

(b) Find the value of x when y is 13,000.

(c) How fast is the distance from the observer to the airplane changing at the time when the airplane is 13,000 feet from the observer? That is, what is $\dfrac{dy}{dt}$ at the time when $\dfrac{dx}{dt} = 390$ and $y = 13,000$?

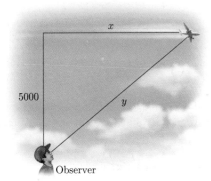

Figure 8

47. Related Rates A baseball diamond is a 90-foot by 90-foot square. (See Fig. 9.) A player runs from first to second base at the speed of 22 feet per second. How fast is the player's distance from third base changing when he is halfway between first and second base? [*Hint*: If x is the distance from the player to second base and y is his distance from third base, then $x^2 + 90^2 = y^2$.]

Figure 9. A baseball diamond.

48. A motorcyclist is driving over a ramp as shown in Fig. 10 at the speed of 80 miles per hour. How fast is he rising? [*Hint*: Use similar triangles to relate x and h and then compute dh/dt.]

Figure 10

Solutions to Practice Problems 3.3

1. $\dfrac{d}{dx}(3y^2) - \dfrac{d}{dx}(3x^2) + \dfrac{d}{dx}(y) = \dfrac{d}{dx}(1)$

$$6y\dfrac{dy}{dx} - 6x + \dfrac{dy}{dx} = 0$$

$$(6y + 1)\dfrac{dy}{dx} = 6x$$

$$\dfrac{dy}{dx} = \dfrac{6x}{6y + 1}$$

2. First, here is the solution without reference to Practice Problem 1.

$$\dfrac{d}{dt}(3y^2) - \dfrac{d}{dt}(3x^2) + \dfrac{d}{dt}(y) = \dfrac{d}{dt}(1)$$

$$6y\dfrac{dy}{dt} - 6x\dfrac{dx}{dt} + \dfrac{dy}{dt} = 0$$

$$(6y + 1)\dfrac{dy}{dt} = 6x\dfrac{dx}{dt}$$

$$\dfrac{dy}{dt} = \dfrac{6x}{6y + 1}\dfrac{dx}{dt}$$

Alternatively, we can use the chain rule and the formula for $\dfrac{dy}{dx}$ from Practice Problem 1.

$$\dfrac{dy}{dt} = \dfrac{dy}{dx}\dfrac{dx}{dt} = \dfrac{6x}{6y + 1}\dfrac{dx}{dt}$$

REVIEW OF FUNDAMENTAL CONCEPTS

1. State the product rule and quotient rule.
2. State the chain rule. Give an example.
3. What is the relationship between the chain rule and the general power rule?
4. What does it mean for a function to be defined implicitly by an equation?

5. State the formula for $\dfrac{d}{dx}y^r$, where y is defined implicitly as a function of x.

6. Outline the procedure for solving a related rates problem.

SUPPLEMENTARY EXERCISES

Differentiate the following functions.

1. $y = (4x - 1)(3x + 1)^4$
2. $y = 2(5 - x)^3(6x - 1)$
3. $y = x(x^5 - 1)^3$
4. $y = (2x + 1)^{5/2}(4x - 1)^{3/2}$
5. $y = 5(\sqrt{x} - 1)^4(\sqrt{x} - 2)^2$
6. $y = \dfrac{\sqrt{x}}{\sqrt{x} + 4}$
7. $y = 3(x^2 - 1)^3(x^2 + 1)^5$
8. $y = \dfrac{1}{(x^2 + 5x + 1)^6}$
9. $y = \dfrac{x^2 - 6x}{x - 2}$
10. $y = \dfrac{2x}{2 - 3x}$
11. $y = \left(\dfrac{3 - x^2}{x^3}\right)^2$
12. $y = \dfrac{x^3 + x}{x^2 - x}$

13. Let $f(x) = (3x + 1)^4(3 - x)^5$. Find all x such that $f'(x) = 0$.

14. Let $f(x) = (x^2 + 1)/(x^2 + 5)$. Find all x such that $f'(x) = 0$.

15. Find the equation of the line tangent to the graph of $y = (x^3 - 1)(x^2 + 1)^4$ at the point where $x = -1$.

16. Find the equation of the line tangent to the graph of $y = \dfrac{x - 3}{\sqrt{4 + x^2}}$ at the point where $x = 0$.

17. A botanical display is to be constructed as a rectangular region with a river as one side and a sidewalk 2 meters wide along the inside edges of the other three sides. (See Fig. 1.) The area for the plants must be 800 square meters. Find the outside dimensions of the region that minimizes the area of the sidewalk (and hence minimizes the amount of concrete needed for the sidewalk).

Figure 1. A botanical display.

18. Repeat Exercise 17, with the sidewalk on the inside of all four sides. In this case, the 800-square-meter planted region has dimensions $x - 4$ meters by $y - 4$ meters.

19. A store estimates that its cost when selling x lamps per day is C dollars, where $C = 40x + 30$ (the marginal cost per lamp is \$40). If daily sales are rising at the rate of three lamps per day, how fast are the costs rising? Explain your answer using the chain rule.

20. A company pays y dollars in taxes when its annual profit is P dollars. If y is some (differentiable) function of P and P is some function of time t, give a chain rule formula for the time rate of change of taxes $\dfrac{dy}{dt}$.

Exercises 21–26 refer to the graphs of the functions $f(x)$ and $g(x)$ in Fig. 2. Determine $h(1)$ and $h'(1)$.

21. $h(x) = 2f(x) - 3g(x)$
22. $h(x) = f(x) \cdot g(x)$
23. $h(x) = \dfrac{f(x)}{g(x)}$
24. $h(x) = [f(x)]^2$
25. $h(x) = f(g(x))$
26. $h(x) = g(f(x))$

In Exercises 27–29, find a formula for $\dfrac{d}{dx}f(g(x))$, where $f(x)$ is a function such that $f'(x) = 1/(x^2 + 1)$.

27. $g(x) = x^3$
28. $g(x) = \dfrac{1}{x}$
29. $g(x) = x^2 + 1$

In Exercises 30–32, find a formula for $\dfrac{d}{dx}f(g(x))$, where $f(x)$ is a function such that $f'(x) = x\sqrt{1 - x^2}$.

30. $g(x) = x^2$
31. $g(x) = \sqrt{x}$
32. $g(x) = x^{3/2}$

In Exercises 33–35, find $\dfrac{dy}{dx}$, where y is a function of u such that $\dfrac{dy}{du} = \dfrac{u}{u^2 + 1}$.

33. $u = x^{3/2}$
34. $u = x^2 + 1$
35. $u = \dfrac{5}{x}$

Figure 2

In Exercises 36–38, find $\dfrac{dy}{dx}$, where y is a function of u such that $\dfrac{dy}{du} = \dfrac{u}{\sqrt{1+u^4}}$.

36. $u = x^2$ **37.** $u = \sqrt{x}$ **38.** $u = \dfrac{2}{x}$

39. The revenue R that a company receives is a function of the weekly sales x. Also, the sales level x is a function of the weekly advertising expenditures A, and A in turn is a varying function of time.

(a) Write the derivative symbols for the following quantities: rate of change of revenue with respect to advertising expenditures, time rate of change of advertising expenditures, marginal revenue, and rate of change of sales with respect to advertising expenditures. Select your answers from the following:

$$\frac{dR}{dx}, \quad \frac{dR}{dt}, \quad \frac{dA}{dt}, \quad \frac{dA}{dR}, \quad \frac{dA}{dx}, \quad \frac{dx}{dA}, \quad \text{and} \quad \frac{dR}{dA}.$$

(b) Write a type of chain rule that expresses the time rate of change of revenue, $\dfrac{dR}{dt}$, in terms of three of the derivatives described in part (a).

40. The amount A of anesthetics that a certain hospital uses each week is a function of the number S of surgical operations performed each week. Also, S in turn

is a function of the population P of the area served by the hospital, while P is a function of time t.

(a) Write the derivative symbols for the following quantities: population growth rate, rate of change of anesthetic usage with respect to the population size, rate of change of surgical operations with respect to the population size, and rate of change of anesthetic usage with respect to the number of surgical operations. Select your answers from the following:

$$\frac{dS}{dP}, \quad \frac{dS}{dt}, \quad \frac{dP}{dS}, \quad \frac{dP}{dt}, \quad \frac{dA}{dS}, \quad \frac{dA}{dP}, \quad \text{and} \quad \frac{dS}{dA}.$$

(b) Write a type of chain rule that expresses the time rate of change of anesthetic usage, $\dfrac{dA}{dt}$, in terms of three of the derivatives described in part (a).

41. The graph of $x^{2/3} + y^{2/3} = 8$ is the *astroid* in Fig. 3.

(a) Find $\dfrac{dy}{dx}$ by implicit differentiation.

(b) Find the slope of the tangent line at $(8, -8)$.

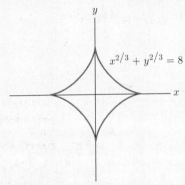

Figure 3. Astroid.

42. The graph of $x^3 + y^3 = 9xy$ is the folium of Descartes shown in Fig. 4.

(a) Find $\dfrac{dy}{dx}$ by implicit differentiation.

(b) Find the slope of the curve at $(2, 4)$.

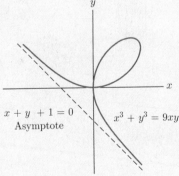

Figure 4. Folium of Descartes.

In Exercises 43–46, x and y are related by the given equation. Use implicit differentiation to calculate the value of $\frac{dy}{dx}$ for the given values of x and y.

43. $x^2 y^2 = 9$; $x = 1$, $y = 3$

44. $xy^4 = 48$; $x = 3$, $y = 2$

45. $x^2 - xy^3 = 20$; $x = 5$, $y = 1$

46. $xy^2 - x^3 = 10$; $x = 2$, $y = 3$

47. A factory's weekly production costs y and its weekly production quantity x are related by the equation $y^2 - 5x^3 = 4$, where y is in thousands of dollars and x is in thousands of units of output.

(a) Use implicit differentiation to find a formula for $\frac{dy}{dx}$, the marginal cost of production.

(b) Find the marginal cost of production when $x = 4$ and $y = 18$.

(c) Suppose that the factory begins to vary its weekly production level. Assuming that x and y are differentiable functions of time t, use the method of related rates to find a formula for $\frac{dy}{dt}$, the time rate of change of production costs.

(d) Compute $\frac{dy}{dt}$ when $x = 4$, $y = 18$, and the production level is rising at the rate of .3 thousand units per week (that is, when $\frac{dx}{dt} = .3$).

48. A town library estimates that when the population is x thousand persons approximately y thousand books will be checked out of the library during 1 year, where x and y are related by the equation $y^3 - 8000x^2 = 0$.

(a) Use implicit differentiation to find a formula for $\frac{dy}{dx}$, the rate of change of library circulation with respect to population size.

(b) Find the value of $\frac{dy}{dx}$ when $x = 27$ thousand persons and $y = 180$ thousand books per year.

(c) Assume that x and y are both differentiable functions of time t, and use the method of related rates to find a formula for $\frac{dy}{dt}$, the time rate of change of library circulation.

(d) Compute $\frac{dy}{dt}$ when $x = 27$, $y = 180$, and the population is rising at the rate of 1.8 thousand persons per year $\left(\frac{dx}{dt} = 1.8\right)$. Either use part (c) or use part (b) and the chain rule.

49. Suppose that the price p and quantity x of a certain commodity satisfy the demand equation $6p + 5x + xp = 50$ and that p and x are functions of time, t. Determine the rate at which the quantity x is changing when $x = 4$, $p = 3$, and $\frac{dp}{dt} = -2$.

50. An offshore oil well is leaking oil onto the ocean surface, forming a circular oil slick about .005 meter thick. If the radius of the slick is r meters, the volume of oil spilled is $V = .005\pi r^2$ cubic meters. If the oil is leaking at a constant rate of 20 cubic meters per hour so that $\frac{dV}{dt} = 20$, find the rate at which the radius of the oil slick is increasing at a time when the radius is 50 meters. $\left[\textit{Hint: Find a relation between } \frac{dV}{dt} \textit{ and } \frac{dr}{dt}.\right]$

51. Animal physiologists have determined experimentally that the weight W (in kilograms) and the surface area S (in square meters) of a typical horse are related by the empirical equation $S = 0.1W^{2/3}$. How fast is the surface area of a horse increasing at a time when the horse weighs 350 kg and is gaining weight at the rate of 200 kg per year? [*Hint:* Use the chain rule.]

52. Suppose that a kitchen appliance company's monthly sales and advertising expenses are approximately related by the equation $xy - 6x + 20y = 0$, where x is thousands of dollars spent on advertising and y is thousands of dishwashers sold. Currently, the company is spending 10 thousand dollars on advertising and is selling 2 thousand dishwashers each month. If the company plans to increase monthly advertising expenditures at the rate of \$1.5 thousand per month, how fast will sales rise? Use implicit differentiation to answer the question.

THE EXPONENTIAL AND NATURAL LOGARITHM FUNCTIONS

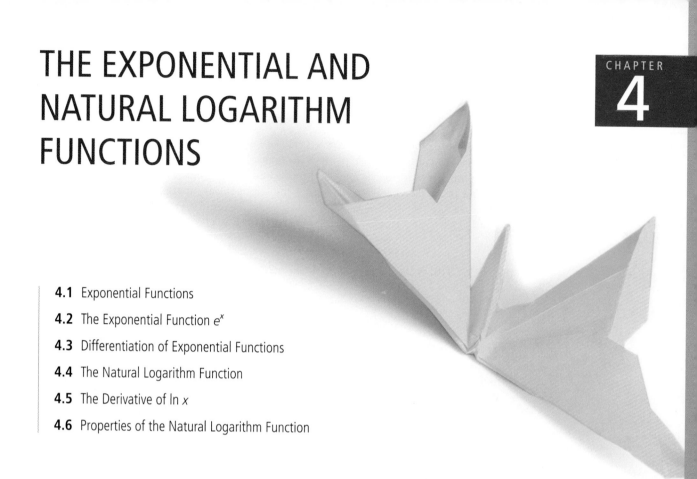

When an investment grows steadily at 15% per year, the rate of growth of the investment at any time is proportional to the value of the investment at that time. When a bacteria culture grows in a laboratory dish, the rate of growth of the culture at any moment is proportional to the total number of bacteria in the dish at that moment. These situations are examples of what is called *exponential growth*. A pile of radioactive uranium ^{235}U decays at a rate that at each moment is proportional to the amount of ^{235}U present. This decay of uranium (and of radioactive elements in general) is called *exponential decay*. Both exponential growth and exponential decay can be described and studied in terms of exponential functions and the natural logarithm function. The properties of these functions are investigated in this chapter. Subsequently, we shall explore a wide range of applications, in fields such as business, biology, archeology, public health, and psychology.

4.1 Exponential Functions

Throughout this section b will denote a positive number. The function

$$f(x) = b^x$$

is called an *exponential function*, because the variable x is in the exponent. The number b is called the *base* of the exponential function. In Section 0.5 we reviewed the definition of b^x for various values of b and x (although we used the letter r there instead of x). For instance, if $f(x)$ is the exponential function with base 2,

$$f(x) = 2^x,$$

then

$$f(0) = 2^0 = 1, \qquad f(1) = 2^1 = 2, \qquad f(4) = 2^4 = 2 \cdot 2 \cdot 2 \cdot 2 = 16,$$

and

$$f(-1) = 2^{-1} = \tfrac{1}{2}, \qquad f(\tfrac{1}{2}) = 2^{1/2} = \sqrt{2}, \qquad f(\tfrac{3}{5}) = (2^{1/5})^3 = (\sqrt[5]{2})^3.$$

Actually, in Section 0.5 we defined b^x only for rational (that is, integer or fractional) values of x. For other values of x (such as $\sqrt{3}$ or π), it is possible to define b^x by first approximating x with rational numbers and then applying a limiting process. We shall omit the details and simply assume henceforth that b^x can be defined for all numbers x in such a way that the usual laws of exponents remain valid.

Let us state the laws of exponents for reference.

(i) $b^x \cdot b^y = b^{x+y}$ (ii) $b^{-x} = \dfrac{1}{b^x}$

(iii) $\dfrac{b^x}{b^y} = b^x \cdot b^{-y} = b^{x-y}$ (iv) $(b^y)^x = b^{xy}$

(v) $a^x b^x = (ab)^x$ (vi) $\dfrac{a^x}{b^x} = \left(\dfrac{a}{b}\right)^x$

Property (iv) may be used to change the appearance of an exponential function. For instance, the function $f(x) = 8^x$ may also be written as $f(x) = (2^3)^x = 2^{3x}$, and $g(x) = (\tfrac{1}{9})^x$ may be written as $g(x) = (1/3^2)^x = (3^{-2})^x = 3^{-2x}$.

EXAMPLE 1   Use properties of exponents to write the following expressions in the form 2^{kx} for a suitable constant k.

(a) $4^{5x/2}$ (b) $(2^{4x} \cdot 2^{-x})^{1/2}$ (c) $8^{x/3} \cdot 16^{3x/4}$ (d) $\dfrac{10^x}{5^x}$

Solution (a) First express the base 4 as a power of 2, and then use Property (iv):

$$4^{5x/2} = (2^2)^{5x/2} = 2^{2(5x/2)} = 2^{5x}.$$

(b) First use Property (i) to simplify the quantity inside the parentheses, and then use Property (iv):

$$(2^{4x} \cdot 2^{-x})^{1/2} = (2^{4x-x})^{1/2} = (2^{3x})^{1/2} = 2^{(3/2)x}.$$

(c) First express the bases 8 and 16 as powers of 2, and then use (iv) and (i):

$$8^{x/3} \cdot 16^{3x/4} = (2^3)^{x/3} \cdot (2^4)^{3x/4} = 2^x \cdot 2^{3x} = 2^{4x}.$$

(d) Use Property (v) to change the numerator 10^x, and then cancel the common term 5^x:

$$\frac{10^x}{5^x} = \frac{(2 \cdot 5)^x}{5^x} = \frac{2^x \cdot 5^x}{5^x} = 2^x.$$

An alternative method is to use Property (vi):

$$\frac{10^x}{5^x} = \left(\frac{10}{5}\right)^x = 2^x. \qquad \blacksquare$$

Let us now study the graph of the exponential function $y = b^x$ for various values of b. We begin with the special case $b = 2$.

We have tabulated the values of 2^x for $x = 0, \pm 1, \pm 2, \pm 3$ and plotted these values in Fig. 1. Other intermediate values of 2^x for $x = \pm .1, \pm .2, \pm .3, \ldots$, may be obtained from a table or from a graphing calculator. [See Fig. 2(a).] By passing a smooth curve through these points, we obtain the graph of $y = 2^x$, Fig. 2(b).

x	2^x
-3	.125
-2	.25
-1	.5
0	1.0
1	2.0
2	4.0
3	8.0

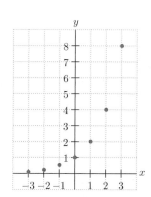

Figure 1. Values of 2^x.

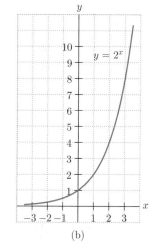

Figure 2. Graph of $y = 2^x$.

In the same manner, we have sketched the graph of $y = 3^x$ (Fig. 3). The graphs of $y = 2^x$ and $y = 3^x$ have the same basic shape. Also note that they both pass through the point $(0, 1)$ (because $2^0 = 1$, $3^0 = 1$).

In Fig. 4 we have sketched the graphs of several more exponential functions. Notice that the graph of $y = 5^x$ has a large slope at $x = 0$, since the graph at $x = 0$ is quite steep; however, the graph of $y = (1.1)^x$ is nearly horizontal at $x = 0$, and hence the slope is close to zero.

There is an important property of the function 3^x that is readily apparent from its graph. Since the graph is always increasing, the function 3^x never assumes the same y-value twice. That is, the only way 3^r can equal 3^s is to have $r = s$. This fact is useful when solving certain equations involving exponentials.

EXAMPLE 2   Let $f(x) = 3^{5x}$. Determine all x for which $f(x) = 27$.

Solution Since $27 = 3^3$, we must determine all x for which

$$3^{5x} = 3^3.$$

Equating exponents, we have

$$5x = 3$$

$$x = \tfrac{3}{5}. \qquad \blacksquare$$

Figure 3

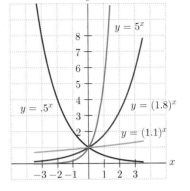

Figure 4

In general, for $b > 1$, the equation $b^r = b^s$ implies that $r = s$. This is because the graph of $y = b^x$ has the same basic shape as $y = 2^x$ and $y = 3^x$. Similarly, when $0 < b < 1$, the equation $b^r = b^s$ implies that $r = s$, because the graph of $y = b^x$ resembles the graph of $y = .5^x$ and is always decreasing.

There is no need at this point to become familiar with the graphs of the function b^x. We have shown a few graphs merely to make the reader more comfortable with the concept of an exponential function. The main purpose of this section has been to review properties of exponents in a context that is appropriate for our future work.

INCORPORATING TECHNOLOGY

Solving Equations and Intersection of Graphs   To determine the value of x at which $2^{2-4x} = 8$, we can proceed as follows. First, enter $Y_1 = 2^\wedge(2 - 4X)$ and $Y_2 = 8$. Next, press $\boxed{\text{2nd}}$ [CALC] $\boxed{5}$ for **Intersect**. Hit $\boxed{\text{ENTER}}$ twice to accept Y_1 as the first function and Y_2 as the second function. Next, enter a guess by either hitting $\boxed{\text{ENTER}}$ to accept the value given or entering your own value for X. Finally, hit $\boxed{\text{ENTER}}$ to have the calculator find the intersection point of the two graphs. From Fig. 5, we see that $2^{2-4x} = 8$ at $x = -0.25$.   ■

Figure 5

1. Can a function such as $f(x) = 5^{3x}$ be written in the form $f(x) = b^x$? If so, what is b?

2. Solve the equation $7 \cdot 2^{6-3x} = 28$.

EXERCISES 4.1

Write each expression in Exercises 1–14 in the form 2^{kx} or 3^{kx}, for a suitable constant k.

1. 4^x, $(\sqrt{3})^x$, $(\frac{1}{9})^x$

2. 27^x, $(\sqrt[3]{2})^x$, $(\frac{1}{8})^x$

3. $8^{2x/3}$, $9^{3x/2}$, $16^{-3x/4}$

4. $9^{-x/2}$, $8^{4x/3}$, $27^{-2x/3}$

5. $(\frac{1}{4})^{2x}$, $(\frac{1}{8})^{-3x}$, $(\frac{1}{81})^{x/2}$

6. $(\frac{1}{9})^{2x}$, $(\frac{1}{27})^{x/3}$, $(\frac{1}{16})^{-x/2}$

7. $6^x \cdot 3^{-x}$, $\dfrac{15^x}{5^x}$, $\dfrac{12^x}{2^{2x}}$

8. $7^{-x} \cdot 14^x$, $\dfrac{2^x}{6^x}$, $\dfrac{3^{2x}}{18^x}$

9. $\dfrac{3^{4x}}{3^{2x}}$, $\dfrac{2^{5x+1}}{2 \cdot 2^{-x}}$, $\dfrac{9^{-x}}{27^{-x/3}}$

10. $\dfrac{2^x}{6^x}$, $\dfrac{3^{-5x}}{3^{-2x}}$, $\dfrac{16^x}{8^{-x}}$

11. $2^{3x} \cdot 2^{-5x/2}$, $3^{2x} \cdot (\frac{1}{3})^{2x/3}$

12. $2^{5x/4} \cdot (\frac{1}{2})^x$, $3^{-2x} \cdot 3^{5x/2}$

13. $(2^{-3x} \cdot 2^{-2x})^{2/5}$, $(9^{1/2} \cdot 9^4)^{x/9}$

14. $(3^{-x} \cdot 3^{x/5})^5$, $(16^{1/4} \cdot 16^{-3/4})^{3x}$

15. Find a number b such that the function $f(x) = 3^{-2x}$ can be written in the form $f(x) = b^x$.

16. Find b so that $8^{-x/3} = b^x$ for all x.

Solve the following equations for x.

17. $5^{2x} = 5^2$

18. $10^{-x} = 10^2$

19. $(2.5)^{2x+1} = (2.5)^5$

20. $(3.2)^{x-3} = (3.2)^5$

21. $10^{1-x} = 100$

22. $2^{4-x} = 8$

23. $3(2.7)^{5x} = 8.1$

24. $4(2.7)^{2x-1} = 10.8$

25. $(2^{x+1} \cdot 2^{-3})^2 = 2$

26. $(3^{2x} \cdot 3^2)^4 = 3$

27. $2^{3x} = 4 \cdot 2^{5x}$

28. $3^{5x} \cdot 3^x - 3 = 0$

29. $(1+x)2^{-x} - 5 \cdot 2^{-x} = 0$

30. $(2-3x)5^x + 4 \cdot 5^x = 0$

31. $2^x - \dfrac{8}{2^{2x}} = 0$

32. $2^x - \dfrac{1}{2^x} = 0$

[Hint: In Exercises 33–36, let $X = 2^x$ or $X = 3^x$.]

33. $2^{2x} - 6 \cdot 2^x + 8 = 0$

34. $2^{2x+2} - 17 \cdot 2^x + 4 = 0$

35. $3^{2x} - 12 \cdot 3^x + 27 = 0$

36. $2^{2x} - 4 \cdot 2^x - 32 = 0$

The expressions in Exercises 37–42 may be factored as shown. Find the missing factors.

37. $2^{3+h} = 2^3(\quad)$

38. $5^{2+h} = 25(\quad)$

39. $2^{x+h} - 2^x = 2^x(\quad)$

40. $5^{x+h} + 5^x = 5^x(\quad)$

41. $3^{x/2} + 3^{-x/2} = 3^{-x/2}(\quad)$

42. $5^{7x/2} - 5^{x/2} = \sqrt{5^x}(\quad)$

Technology Exercises

43. Graph the function $f(x) = 2^x$ in the window $[-1, 2]$ by $[-1, 4]$, and estimate the slope of the graph at $x = 0$.

44. Graph the function $f(x) = 3^x$ in the window $[-1, 2]$ by $[-1, 8]$, and estimate the slope of the graph at $x = 0$.

45. By trial and error, find a number of the form $b = 2.\square$ (just one decimal place) with the property that the slope of the graph of $y = b^x$ at $x = 0$ is as close to 1 as possible.

Solutions to Practice Problems 4.1

1. If $5^{3x} = b^x$, then when $x = 1$, $5^{3(1)} = b^1$, which says that $b = 125$. This value of b certainly works, because
$$5^{3x} = (5^3)^x = 125^x.$$

2. Divide both sides of the equation by 7. We then obtain
$$2^{6-3x} = 4.$$
Now 4 can be written as 2^2. So we have
$$2^{6-3x} = 2^2.$$

Equate exponents to obtain

$$6 - 3x = 2$$
$$4 = 3x$$
$$x = \tfrac{4}{3}.$$

4.2 The Exponential Function e^x

We begin by examining the graphs of the exponential functions shown in Fig. 1. They all pass through $(0, 1)$, but with different slopes there. Notice that the graph of $y = 5^x$ is quite steep at $x = 0$, while the graph of $y = (1.1)^x$ is nearly horizontal at $x = 0$. It turns out that at $x = 0$ the graph of $y = 2^x$ has a slope of approximately .693, while the graph of 3^x has a slope of approximately 1.1.

Evidently, there is a particular value of the base b, between 2 and 3, where the graph of $y = b^x$ has slope *exactly* 1 at $x = 0$. We take this fact for granted and denote this special value of b by the letter e, and we call

$$f(x) = e^x$$

the exponential function. The number e is an important constant of nature that has been calculated to thousands of decimal places. To 10 significant digits, we have $e = 2.718281828$. For our purposes, it is usually sufficient to think of e as "approximately 2.7."

Our goal in this section is to find a formula for the derivative of $y = e^x$. It turns out that the calculations for e^x and 2^x are very similar. Since many people are more comfortable working with 2^x rather than e^x, we shall first analyze the graph of $y = 2^x$. Then we shall draw the appropriate conclusions about the graph of $y = e^x$.

Before computing the slope of $y = 2^x$ at an arbitrary x, let us consider the special case $x = 0$. Denote the slope at $x = 0$ by m. We shall use the secant-line approximation of the derivative to approximate m. We proceed by constructing the secant line in Fig. 2. The slope of the secant line through $(0, 1)$ and $(h, 2^h)$ is $\dfrac{2^h - 1}{h}$. As h approaches zero, the slope of the secant line approaches the slope of $y = 2^x$ at $x = 0$. That is,

$$m = \lim_{h \to 0} \frac{2^h - 1}{h}. \tag{1}$$

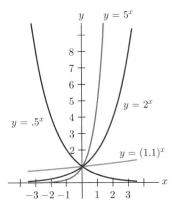

Figure 1. Several exponential functions.

Assuming that this limit exists, we can estimate the value of m by taking h smaller and smaller. Table 1 shows the values of the expressions for $h = .1, .01, \ldots, .0000001$. From the table it is reasonable to conclude that $m \approx .693$.

TABLE 1 $Y_1 = (2^\wedge X - 1)/X$

X	Y1	
.1	.71773	
.01	.69556	
.001	.69339	
1E-4	.69317	
1E-5	.69315	
1E-6	.69315	
1E-7	.69315	

$Y_1 = .693147$

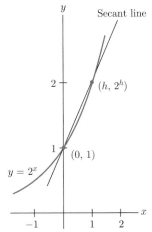

Figure 2. A secant line to the graph of $y = 2^x$.

Since m equals the slope of $y = 2^x$ at $x = 0$, we have

$$m = \frac{d}{dx}(2^x)\bigg|_{x=0} \approx .693. \tag{2}$$

Now that we have estimated the slope of $y = 2^x$ at $x = 0$, let us compute the slope for an arbitrary value of x. We construct a secant line through $(x, 2^x)$ and a nearby point $(x + h, 2^{x+h})$ on the graph. The slope of the secant line is

$$\frac{2^{x+h} - 2^x}{h}. \tag{3}$$

By a law of exponents, we have $2^{x+h} - 2^x = 2^x(2^h - 1)$ so that, by equation (1), we see that

$$\lim_{h \to 0} \frac{2^{x+h} - 2^x}{h} = \lim_{h \to 0} 2^x \frac{2^h - 1}{h} = 2^x \lim_{h \to 0} \frac{2^h - 1}{h} = m\, 2^x. \tag{4}$$

However, the slope of the secant (3) approaches the derivative of 2^x as h approaches zero. Consequently, we have

$$\frac{d}{dx}(2^x) = m\, 2^x, \quad \text{where } m = \frac{d}{dx}(2^x)\bigg|_{x=0}. \qquad (5)$$

EXAMPLE 1   Calculate

(a) $\dfrac{d}{dx}(2^x)\bigg|_{x=3}$ and   **(b)** $\dfrac{d}{dx}(2^x)\bigg|_{x=-1}.$

Solution (a) $\dfrac{d}{dx}(2^x)\bigg|_{x=3} = m \cdot 2^3 = 8m \approx 8(.693) = 5.544.$

(b) $\dfrac{d}{dx}(2^x)\bigg|_{x=-1} = m \cdot 2^{-1} = .5m \approx .5(.693) = .3465.$ ∎

The calculations just carried out for $y = 2^x$ can be carried out for $y = b^x$, where b is any positive number. Equation (5) will read exactly the same except that 2 will be replaced by b. Thus we have the following formula for the derivative of the function $f(x) = b^x$.

$$\frac{d}{dx}(b^x) = m\, b^x, \quad \text{where } m = \frac{d}{dx}(b^x)\bigg|_{x=0}. \qquad (6)$$

Our calculations showed that if $b = 2$ then $m \approx .693$. If $b = 3$, it turns out that $m \approx 1.1$. (See Exercise 1.) The derivative formula in (6) is simple when $m = 1$, that is, when the graph of $y = b^x$ has slope 1 at $x = 0$. As we said earlier, this special value of b is denoted by the letter e. Thus the number e has the property that

$$\frac{d}{dx}(e^x)\bigg|_{x=0} = 1 \qquad (7)$$

and

$$\frac{d}{dx}(e^x) = 1 \cdot e^x = e^x. \qquad (8)$$

The graphical interpretation of equation (7) is that the curve $y = e^x$ has slope 1 at $x = 0$. The graphical interpretation of (8) is that the slope of the curve $y = e^x$ at an arbitrary value of x is exactly equal to the value of the function e^x at that point. (See Fig. 3.)

The function e^x is the same type of function as 2^x and 3^x except that differentiating e^x is much easier. In fact, the next section shows that 2^x can be written as e^{kx} for a suitable constant k. The same is true for 3^x. For this reason, functions of the form e^{kx} are used in almost all applications that require an exponential type of function to describe a physical, economic, or biological phenomenon.

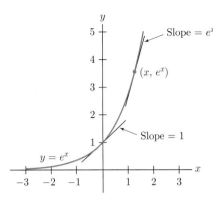

Figure 3. Fundamental properties of e^x.

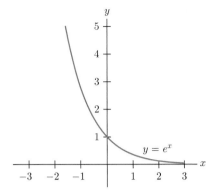

Figure 4

EXAMPLE 2

Sketch the graph of $y = e^{-x}$.

Solution Note that, since $e > 1$, we have $0 < 1/e < 1$. In fact, $1/e \approx .37$. Now

$$y = e^{-x} = \frac{1}{e^x} = \left(\frac{1}{e}\right)^x;$$

and so $f(x) = e^{-x}$ is an exponential function with base $b = \frac{1}{e}$, which is less than 1. Hence the graph of $y = e^{-x}$ is always decreasing and looks like the graph of the exponential function $y = .5^x$. (See Fig. 1.) With the help of a calculator, we form a table of values of $y = e^{-x}$ for selected values of x:

x	-3	-2	-1	0	1	2	3
e^{-x}	20.09	7.39	2.72	1	.37	.14	.05

Using these values, we sketch the graph of $y = e^{-x}$ in Fig. 4.   ◼

In the following section, we will compute the derivative of $y = e^{g(x)}$ for any differentiable function g and, in particular, $y = e^{kx}$. However, as we now illustrate, by relating e^{kx} to e^x, we can also differentiate $y = e^{kx}$ by using the generalized power rule for differentiation.

EXAMPLE 3

Find the derivative of $y = e^{-x}$.

Solution Let $f(x) = e^x$, then

$$e^{-x} = (e^x)^{-1} = (f(x))^{-1};$$

and so

$$\frac{d}{dx}[e^{-x}] = \frac{d}{dx}[(f(x))^{-1}]$$

$$= (-1)(f(x))^{-2}\frac{d}{dx}[f(x)] \quad \text{(the general power rule)}$$

$$= (-1)(e^x)^{-2}\frac{d}{dx}[e^x] \quad\quad (f(x) = e^x)$$

$$= (-1)e^{-2x}e^x \quad\quad\quad\quad \text{[by (8)]}$$

$$= -e^{-x} \quad\quad\quad\quad\quad\quad ◼$$

INCORPORATING TECHNOLOGY

All the techniques we have introduced in previous sections for analyzing functions with the TI-83/84 apply to exponential functions as well. To input an exponential function, press $\boxed{\text{2nd}}$ $[e^x]$. For example, in Fig. 5 we evaluate $e^{2/3}$, $e^{-2}/3$, and the derivative of e^x at $x = 0$. We emphasize the importance of using parentheses when evaluating e^x. ∎

```
e^(2/3)
            1.947734041
e^(-2)/3
            0.045111761
nDeriv(e^(X),X,0
)
            1.000000167
■
```

Figure 5

Practice Problems 4.2

In the following problems use the number 20 as the (approximate) value of e^3.

1. Find the equation of the tangent line to the graph of $y = e^x$ at $x = 3$.

2. Solve the following equation for x:

$$4e^{6x} = 80.$$

3. Differentiate $y = 3(x^2 + 1)e^x$.

EXERCISES 4.2

1. Show that

$$\frac{d}{dx}(3^x)\Big|_{x=0} \approx 1.1$$

by calculating the slope $\dfrac{3^h - 1}{h}$ of the secant line passing through the points $(0, 1)$ and $(h, 3^h)$. Take $h = .1$, $.01$, and $.001$.

2. Show that

$$\frac{d}{dx}(2.7^x)\Big|_{x=0} \approx .99$$

by calculating the slope $\dfrac{2.7^h - 1}{h}$ of the secant line passing through the points $(0, 1)$ and $(h, 2.7^h)$. Take $h = .1$, $.01$, and $.001$.

In Exercises 3–6, compute the given derivatives with the help of formulas (2), (5), and (6)–(8).

3. (a) $\dfrac{d}{dx}(2^x)\Big|_{x=1}$ **(b)** $\dfrac{d}{dx}(2^x)\Big|_{x=-2}$

4. (a) $\dfrac{d}{dx}(2^x)\Big|_{x=1/2}$ **(b)** $\dfrac{d}{dx}(2^x)\Big|_{x=2}$

5. (a) $\dfrac{d}{dx}(e^x)\Big|_{x=1}$ **(b)** $\dfrac{d}{dx}(e^x)\Big|_{x=-1}$

6. (a) $\dfrac{d}{dx}(e^x)\Big|_{x=e}$ **(b)** $\dfrac{d}{dx}(e^x)\Big|_{x=1/e}$

7. Estimate the slope of e^x at $x = 0$ by calculating the slope $\dfrac{e^h - 1}{h}$ of the secant line passing through the points $(0, 1)$ and (h, e^h). Take $h = .01$, $.001$, and $.0001$.

8. Use (8) and a familiar rule for differentiation to find

$$\frac{d}{dx}(5e^x).$$

9. Use (8) and a familiar rule for differentiation to find

$$\frac{d}{dx}(e^x)^{10}.$$

10. Use the fact that $e^{2+x} = e^2 \cdot e^x$ to find

$$\frac{d}{dx}(e^{2+x}).$$

[Remember that e^2 is just a constant—approximately $(2.7)^2$.]

11. (a) Use the fact that $e^{4x} = (e^x)^4$ to find $\dfrac{d}{dx}(e^{4x})$. Simplify the derivative as much as possible.

(b) Let k represent any constant. Generalize the calculations in part (a) to find a simple formula for $\dfrac{d}{dx}(e^{kx})$.

12. Find $\dfrac{d}{dx}(e^x + x^2)$.

Write each expression in the form e^{kx} for a suitable constant k.

13. $(e^2)^x$, $\left(\dfrac{1}{e}\right)^x$

14. $(e^3)^{x/5}$, $\left(\dfrac{1}{e^2}\right)^x$

15. $\left(\dfrac{1}{e^3}\right)^{2x}$, $e^{1-x} \cdot e^{3x-1}$

16. $\left(\dfrac{e^5}{e^3}\right)^x$, $e^{4x+2} \cdot e^{x-2}$

17. $(e^{4x} \cdot e^{6x})^{3/5}$, $\dfrac{1}{e^{-2x}}$

18. $\sqrt{e^{-x} \cdot e^{7x}}$, $\dfrac{e^{-3x}}{e^{-4x}}$

Solve each equation for x.

19. $e^{5x} = e^{20}$

20. $e^{1-x} = e^2$

21. $e^{x^2 - 2x} = e^8$

22. $e^{-x} = 1$

23. $e^x(x^2 - 1) = 0$

24. $4e^x(x^2 + 1) = 0$

Differentiate the following functions.

25. $y = 3e^x - 7x$

26. $y = \dfrac{2x + 4 - 5e^x}{4}$

27. $y = xe^x$

28. $y = (x^2 + x + 1)e^x$

29. $y = 8e^x(1 + 2e^x)^2$

30. $y = (1 + e^x)(1 - e^x)$

31. $y = \dfrac{e^x}{x + 1}$

32. $y = \dfrac{x + 1}{e^x}$

33. $y = \dfrac{e^x - 1}{e^x + 1}$

34. $y = \dfrac{1}{2e^x + 5}$

35. $y = \sqrt{2e^x + 1}$

36. $y = \sqrt[3]{1 + x + 2e^x}$

37. Find the point on the graph of $y = (1 + x^2)e^x$ where the tangent line is horizontal.

38. Show that the tangent line to the graph of $y = e^x$ at the point (a, e^a) is perpendicular to the tangent line to the graph of $y = e^{-x}$ at the point (a, e^{-a}).

39. Find the slope of the tangent line to the curve $y = xe^x$ at $(0, 0)$.

40. Find the slope of the tangent line to the curve $y = xe^x$ at $(1, e)$.

41. Find the equation of the tangent line to the curve $y = \dfrac{e^x}{1 + 2e^x}$ at $(0, \frac{1}{3})$.

42. Find the equation of the tangent line to the curve $y = \dfrac{e^x}{x + e^x}$ at $(0, 1)$.

Find the first and second derivatives.

43. $f(x) = e^x(1 + x)^2$

44. $f(x) = \dfrac{e^x}{x}$

45. The atmospheric pressure at an altitude of x kilometers is $f(x)$ g/cm^2 (grams per square centimeter), where $f(x) = 1035e^{-.12x}$. Give approximate answers to the following questions using the graphs of $f(x)$ and $f'(x)$ shown in Fig. 6.

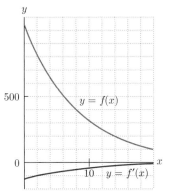

Figure 6

(a) What is the pressure at an altitude of 2 kilometers?

(b) At what altitude is the pressure 200 g/cm^2?

(c) At an altitude of 8 kilometers, at what rate is the atmospheric pressure changing (with respect to change in altitude)?

(d) At what altitude is atmospheric pressure falling at the rate of 100 g/cm^2 per kilometer?

46. Health Expenditures The national health expenditures (in billions of dollars) from 1960 to 1980 are given approximately by $f(t) = 27e^{.106t}$, with time in years measured from 1960. Give approximate answers to the following questions using the graphs of $f(t)$ and $f'(t)$ shown in Fig. 7.

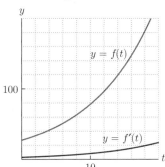

Figure 7

(a) How much money was spent in 1978?

(b) How fast were expenditures rising in 1972?

(c) When did expenditures reach 120 billion dollars?

(d) When were expenditures rising at the rate of 20 billion dollars per year?

Technology Exercises

47. Find the equation of the tangent line to the graph of $y = e^x$ at $x = 0$. Then graph the function and the tangent line together to confirm that your answer is correct.

48. (a) Graph $y = e^x$.

(b) Zoom in on the region near $x = 0$ until the curve appears as a straight line and estimate the slope of the line. This number is an estimate of $\dfrac{d}{dx}e^x$ at $x = 0$. Compare your answer with the actual slope, 1.

(c) Repeat parts (a) and (b) for $y = 2^x$. Observe that the slope at $x = 0$ is not 1.

49. Set $Y_1 = e^x$ and use your calculator's derivative command to specify Y_2 as the derivative of Y_1. Graph the two functions simultaneously in the window $[-1, 3]$ by $[-3, 20]$ and observe that the graphs overlap.

50. Demonstrate that e^x grows faster than any power function. For instance, graph $y = \dfrac{x^n}{e^x}$ for $n = 3$ with the

window $[0, 16]$ by $[0, 25]$ and observe that the function approaches zero as x gets large. Repeat for $n = 4$ and $n = 5$.

51. Calculate values of $\dfrac{10^x - 1}{x}$ for small values of x and use them to estimate $\dfrac{d}{dx}(10^x)\Big|_{x=0}$. What is the formula for $\dfrac{d}{dx}(10^x)$?

52. The graph of $y = e^{x-2}$ may be obtained by translating the graph of $y = e^x$ two units to the right. Find a constant k such that the graph of $y = ke^x$ is the same as the graph of $y = e^{x-2}$. Verify your result by graphing both functions.

Solutions to Practice Problems 4.2

1. When $x = 3$, $y = e^3 \approx 20$. So the point $(3, 20)$ is on the tangent line. Since $\dfrac{d}{dx}(e^x) = e^x$, the slope of the tangent line is e^3 or 20. Therefore, the equation of the tangent line in point–slope form is $y - 20 = 20(x-3)$.

2. This problem is similar to Practice Problem 2 of Section 4.1. First divide both sides of the equation by 4.

$$e^{6x} = 20.$$

The idea is to express 20 as a power of e and then equate exponents.

$$e^{6x} = e^3$$
$$6x = 3$$
$$x = \tfrac{1}{2}.$$

3. $\dfrac{d}{dx}[3(x^2 + 1)e^x] = 3\dfrac{d}{dx}[(x^2 + 1)e^x]$

(constant-multiple rule)

$$= 3(x^2 + 1)\dfrac{d}{dx}[e^x] + 3e^x\dfrac{d}{dx}[(x^2 + 1)]$$

(product rule)

$$= 3(x^2 + 1)e^x + 3e^x(2x)$$

[by (8)]

$$= 3e^x(x^2 + 2x + 1) = 3e^x(x + 1)^2$$

4.3 Differentiation of Exponential Functions

We have shown that $\dfrac{d}{dx}(e^x) = e^x$. Using this fact and the chain rule, we can differentiate functions of the form $y = e^{g(x)}$, where $g(x)$ is any differentiable function. This is because $e^{g(x)}$ is the composite of two functions. Indeed, if $f(x) = e^x$, then

$$e^{g(x)} = f(g(x)).$$

Thus, by the chain rule, we have

$$\dfrac{d}{dx}(e^{g(x)}) = f'(g(x))g'(x)$$
$$= f(g(x))g'(x) \qquad [\text{since } f'(x) = f(x)]$$
$$= e^{g(x)}g'(x).$$

So we have the following result:

> **Chain Rule for Exponential Functions** Let $g(x)$ be any differentiable function. Then
>
> $$\frac{d}{dx}(e^{g(x)}) = e^{g(x)}g'(x).$$
>
> (1)
>
> If we write $u = g(x)$, then equation (1) can be written in the form
>
> $$\frac{d}{dx}(e^u) = e^u \frac{du}{dx}.$$
>
> (1a)

EXAMPLE 1 Differentiate $y = e^{x^2+1}$.

Solution Here $g(x) = x^2 + 1$ and $g'(x) = 2x$, so

$$\frac{d}{dx}(e^{x^2+1}) = e^{x^2+1} \cdot 2x = 2xe^{x^2+1}.$$

■

EXAMPLE 2 Differentiate $y = e^{3x^2-(1/x)}$.

Solution $\dfrac{d}{dx}\left(e^{3x^2-(1/x)}\right) = e^{3x^2-(1/x)} \cdot \dfrac{d}{dx}\left(3x^2 - \dfrac{1}{x}\right) = e^{3x^2-(1/x)}\left(6x + \dfrac{1}{x^2}\right).$

■

EXAMPLE 3 Differentiate $y = e^{5x}$.

Solution $\dfrac{d}{dx}(e^{5x}) = e^{5x} \cdot \dfrac{d}{dx}(5x) = e^{5x} \cdot 5 = 5e^{5x}.$

■

Using a computation similar to that used in Example 3, we may differentiate $y = e^{kx}$ for any constant k. (In Example 3 we have $k = 5$.) The result is the following useful formula.

$$\frac{d}{dx}(e^{kx}) = ke^{kx}.$$

(2)

Many applications involve exponential functions of the form $y = Ce^{kx}$, where C and k are constants. In the next example we differentiate such functions.

EXAMPLE 4 Differentiate the following exponential functions.
(a) $y = 3e^{5x}$ (b) $y = 3e^{kx}$, where k is a constant
(c) $y = Ce^{kx}$, where C and k are constants

Solution (a) $\dfrac{d}{dx}(3e^{5x}) = 3\dfrac{d}{dx}(e^{5x}) = 3 \cdot 5e^{5x} = 15e^{5x}.$

(b) $\dfrac{d}{dx}(3e^{kx}) = 3\dfrac{d}{dx}(e^{kx})$
$\qquad = 3 \cdot ke^{kx}$ [by (2)]
$\qquad = 3ke^{kx}.$

(c) $\dfrac{d}{dx}(Ce^{kx}) = C\dfrac{d}{dx}(e^{kx}) = Cke^{kx}.$

■

If we let $y = Ce^{kx}$ then, by part (c) of Example 4, we have

$$y' = Cke^{kx} = k \cdot (Ce^{kx}) = ky.$$

In other words, the derivative of the function Ce^{kx} is k times the function itself. Let us record this fact.

Let C, k be any constants and let $y = Ce^{kx}$. Then y satisfies the equation

$$y' = ky.$$

The equation $y' = ky$ expresses a relationship between the function y and its derivative y'. Any equation expressing a relationship between a function y and one or more of its derivatives is called a *differential equation*.

Very often an applied problem will involve a function $y = f(x)$ that satisfies the differential equation $y' = ky$. It can be shown that y must then necessarily be an exponential function of the form $y = Ce^{kx}$. That is, we have the following result:

If $y = f(x)$ satisfies the differential equation

$$y' = ky, \tag{3}$$

then y is an exponential function of the form

$$y = Ce^{kx}, \qquad C \text{ a constant.}$$

We leave the verification of this result to Exercise 45.

EXAMPLE 5   Determine all functions $y = f(x)$ such that $y' = -.2y$.

Solution   The equation $y' = -.2y$ has the form $y' = ky$ with $k = -.2$. Therefore, any solution of the equation has the form

$$y = Ce^{-.2x},$$

where C is a constant.   ∎

EXAMPLE 6   Determine all functions $y = f(x)$ such that $y' = y/2$ and $f(0) = 4$.

Solution   The equation $y' = y/2$ has the form $y' = ky$ with $k = \frac{1}{2}$. Therefore,

$$f(x) = Ce^{(1/2)x}$$

for some constant C. We also require that $f(0) = 4$. That is,

$$4 = f(0) = Ce^{(1/2)0} = Ce^0 = C.$$

So $C = 4$ and

$$f(x) = 4e^{(1/2)x}.$$ ∎

The Functions $f(x) = e^{kx}$ Exponential functions of the form $f(x) = e^{kx}$ occur in many applications. Figure 1 shows the graphs of several functions of this type when k is a positive number. These curves $y = e^{kx}$, k positive, have several properties in common:

1. $(0, 1)$ is on the graph.
2. The graph lies strictly above the x-axis (e^{kx} is never zero).
3. The x-axis is an asymptote as x becomes large negatively.

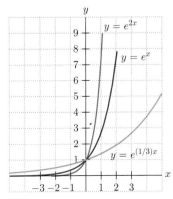

Figure 1

4. The graph is always increasing and concave up.

When k is negative, the graph of $y = e^{kx}$ is decreasing. (See Fig. 2.) Note the following properties of the curves $y = e^{kx}$, k negative:

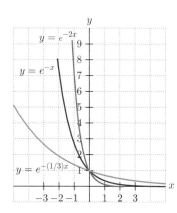

$y = e^{-2x}$
$y = e^{-x}$
$y = e^{-(1/3)x}$

Figure 2

1. $(0, 1)$ is on the graph.

2. The graph lies strictly above the x-axis.

3. The x-axis is an asymptote as x becomes large positively.

4. The graph is always decreasing and concave up.

The Functions $f(x) = b^x$ If b is a positive number, the function $f(x) = b^x$ may be written in the form $f(x) = e^{kx}$ for some k. For example, take $b = 2$. From Fig. 3 of the preceding section it is clear that there is some value of x such that $e^x = 2$. Call this value k, so $e^k = 2$. Then

$$2^x = (e^k)^x = e^{kx}$$

for all x. In general, if b is any positive number, there is a value of x, say $x = k$, such that $e^k = b$. In this case, $b^x = (e^k)^x = e^{kx}$. Thus all the curves $y = b^x$ discussed in Section 4.1 can be written in the form $y = e^{kx}$. This is one reason why we have focused on exponential functions with base e, instead of studying $y = 2^x$, $y = 3^x$, and so on.

Practice Problems 4.3

1. Differentiate $\left[e^{-3x}(1 + e^{6x})\right]^{12}$.

2. Determine all functions $y = f(x)$ such that $y' = -y/20$, $f(0) = 2$.

EXERCISES 4.3

Differentiate the following functions.

1. $f(x) = 4e^{2x}$

2. $f(x) = 3e^7$

3. $f(x) = 1 + 4x + e^{-2x}$

4. $f(x) = \dfrac{e^{3x}}{3}$

5. $f(t) = (1 + t)e^{t^2}$

6. $f(t) = te^{-t}$

7. $y = (e^x + e^{-x})^3$

8. $f(x) = 2e^{\sqrt{x}}$

9. $y = \dfrac{e^{\frac{t}{2}+1}}{4}$

10. $f(x) = \sqrt{e^{2x} - 2}$

11. $g(t) = e^{3/t}$

12. $f(x) = \sqrt{e^{x^2} + 1}$

13. $y = e^{x^2 - 5x + 4}$

14. $f(x) = \dfrac{e^{2x} - 1}{e^{2x} + 1}$

15. $y = \dfrac{x + e^{2x}}{e^x}$

16. $f(t) = (e^{2t})^{-4}$

17. $f(t) = \dfrac{e^{3t} + e^{-3t}}{e^t}$

18. $f(t) = e^t(e^{2t} - e^{-2t})$

[*Hint:* In Exercises 16–18, simplify $f(t)$ before differentiating.]

19. $f(t) = (t + 1)^2 e^{2t}$

20. $f(x) = \sqrt{e^{x/2} + 1}$

21. $f(x) = e^{e^x}$

22. $f(x) = e^x e^{e^x}$

23. $f(x) = e^{4x}/(4 + x)$

24. $f(x) = e^{e^x}$

25. $f(x) = \left(\dfrac{1}{x} + 3\right)e^{2x}$

26. $f(x) = \dfrac{4x^2}{x^2 + e^{2x}}$

In Exercises 27–32, find the values of x at which the function has a possible relative maximum or minimum point. (Recall that e^x is positive for all x.) Use the second derivative to determine the nature of the function at these points.

27. $f(x) = (1 + x)e^{-3x}$

28. $f(x) = (1 - x)e^{2x}$

29. $f(x) = \dfrac{3 - 4x}{e^{2x}}$

30. $f(x) = \dfrac{4x - 1}{e^{x/2}}$

31. $f(x) = (5x - 2)e^{1-2x}$

32. $f(x) = (2x - 5)e^{3x-1}$

33. Appreciation of Assets A painting purchased in 1998 for $100,000 is estimated to be worth $v(t) = 100,000e^{t/5}$ dollars after t years. At what rate will the painting be appreciating in 2003?

34. Depreciation of Assets The value of a computer t years after purchase is $v(t) = 2000e^{-.35t}$ dollars. At what rate is the computer's value falling after 3 years?

35. Velocity The velocity of a parachutist during free fall is

$$f(t) = 60(1 - e^{-.17t})$$

meters per second. Answer the following questions by reading the graph in Fig. 3. (Recall that acceleration is the derivative of velocity.)

(a) What is the velocity when $t = 8$ seconds?

(b) What is the acceleration when $t = 0$?

(c) When is the parachutist's velocity 30 m/sec?

(d) When is the acceleration 5 m/sec²?

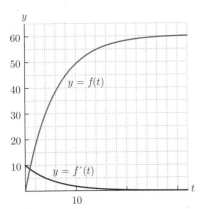

Figure 3

36. Suppose that the velocity of a parachutist is

$$v(t) = 65(1 - e^{-.16t})$$

meters per second. The graph of $v(t)$ is similar to that in Fig. 3. Calculate the parachutist's velocity and acceleration when $t = 9$ seconds.

37. Height of a Plant The height of a certain plant after t weeks is

$$f(t) = \frac{1}{.05 + e^{-.4t}}$$

inches. The graph of $f(t)$ resembles the graph in Fig. 4. Calculate the rate of growth of the plant after 7 weeks.

38. Height of a Plant The length of a certain weed after t weeks is

$$f(t) = \frac{6}{.2 + 5e^{-.5t}}$$

centimeters. Answer the following questions by reading the graph in Fig. 4.

(a) How fast is the weed growing after 10 weeks?

(b) When is the weed 10 centimeters long?

(c) When is the weed growing at the rate of 2 cm/week?

(d) What is the maximum rate of growth?

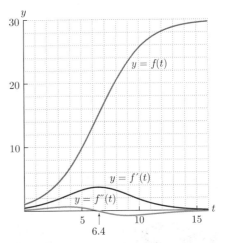

Figure 4

39. Gompertz Growth Curve Let a and b be positive numbers. A curve whose equation is $y = e^{-ae^{-bx}}$ is called a *Gompertz growth curve*. These curves are used in biology to describe certain types of population growth. Compute the derivative of $y = e^{-2e^{-.01x}}$.

40. Find $\dfrac{dy}{dx}$ if $y = e^{-(1/10)e^{-x/2}}$.

41. Determine all solutions of the differential equation $y' = -4y$.

42. Determine all solutions of the differential equation $y' = \frac{1}{3}y$.

43. Determine all functions $y = f(x)$ such that $y' = -.5y$ and $f(0) = 1$.

44. Determine all functions $y = f(x)$ such that $y' = 3y$ and $f(0) = \frac{1}{2}$.

45. Verify the result in equation (3). [*Hint:* Let $g(x) = f(x)e^{-kx}$. Show that $g'(x) = 0$.] You may assume that only a constant function has a zero derivative.

46. Let $f(x)$ be a function with the property that $f'(x) = 1/x$. Let $g(x) = f(e^x)$, and compute $g'(x)$.

47. As h approaches 0, what value is approached by the difference quotient $\dfrac{e^h - 1}{h}$? [*Hint:* $1 = e^0$.]

48. As h approaches 0, what value is approached by $\dfrac{e^{2h} - 1}{h}$? [*Hint:* $1 = e^0$.]

Technology Exercises

49. Size of a Tumor In a study, a cancerous tumor was found to have a volume of

$$f(t) = 1.825^3(1 - 1.6e^{-.4196t})^3$$

milliliters after t weeks, with $t > 1$.[*]

---

[*]Baker, Goddard, Clark, and Whimster, "Proportion of Necrosis in Transplanted Murine Adenocarcinomas and Its Relationship to Tumor Growth," *Growth, Development and Aging*, 54 (1990), 85–93.

(a) Sketch the graphs of $f(t)$ and $f'(t)$ for $1 \leq t \leq 15$. What do you notice about the tumor's volume?

(b) How large is the tumor after 5 weeks?

(c) When will the tumor have a volume of 5 milliliters?

(d) How fast is the tumor growing after 5 weeks?

(e) When is the tumor growing at the fastest rate?

(f) What is the fastest rate of growth of the tumor?

50. Height of a Plant Let $f(t)$ be the function from Exercise 37 that gives the height (inches) of a plant at time t (weeks).

(a) When is the plant 11 inches tall?

(b) When is the plant growing at the rate of 1 inch per week?

(c) What is the fastest rate of growth of the plant, and when does this occur?

Solutions to Practice Problems 4.3

1. We must use the general power rule. However, this is most easily done if we first use the laws of exponents to simplify the function inside the brackets.

$$e^{-3x}(1 + e^{6x}) = e^{-3x} + e^{-3x} \cdot e^{6x} = e^{-3x} + e^{3x}.$$

Now

$$\frac{d}{dx}[e^{-3x} + e^{3x}]^{12}$$

$$= 12 \cdot [e^{-3x} + e^{3x}]^{11} \cdot (-3e^{-3x} + 3e^{3x})$$

$$= 36 \cdot [e^{-3x} + e^{3x}]^{11} \cdot (-e^{-3x} + e^{3x}).$$

2. The differential equation $y' = -y/20$ is of the type $y' = ky$, where $k = -\frac{1}{20}$. Therefore, any solution has the form $f(x) = Ce^{-(1/20)x}$. Now

$$f(0) = Ce^{-(1/20)0} = Ce^0 = C,$$

so $C = 2$ when $f(0) = 2$. Therefore, the desired function is $f(x) = 2e^{-(1/20)x}$.

4.4 The Natural Logarithm Function

As a preparation for the definition of the natural logarithm, we shall make a geometric digression. In Fig. 1 we have plotted several pairs of points. Observe how they are related to the line $y = x$.

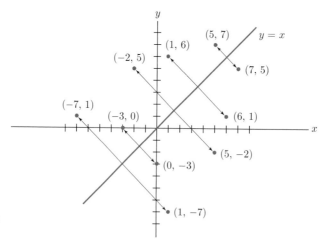

Figure 1. Reflections of points through the line $y = x$.

The points $(5, 7)$ and $(7, 5)$, for example, are the same distance from the line $y = x$. If we were to plot the point $(5, 7)$ with wet ink and then fold the page along the line $y = x$, the ink blot would produce a second blot at the point $(7, 5)$. If we think of the line $y = x$ as a mirror, then $(7, 5)$ is the mirror image of $(5, 7)$. We say that $(7, 5)$ is the *reflection* of $(5, 7)$ through the line $y = x$. Similarly, $(5, 7)$ is the reflection of $(7, 5)$ through the line $y = x$.

Now let us consider all points lying on the graph of the exponential function $y = e^x$. [See Fig. 2(a).] If we reflect each such point through the line $y = x$, we obtain a new graph. [See Fig. 2(b).] For each positive x, there is exactly one value of y such that (x, y) is on the new graph. We call this value of y the *natural logarithm of x*, denoted $\ln x$. Thus the reflection of the graph of $y = e^x$ through the line $y = x$ is the graph of the natural logarithm function $y = \ln x$.

From this we derive the following definition of the natural logarithm.

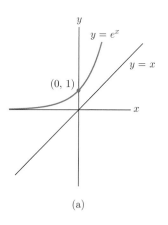

DEFINITION Natural Logarithm

$$y = \ln x \iff x = e^y$$

In other words, the natural logarithm function is the inverse function of the exponential function in base e.

We may deduce some properties of the natural logarithm function from an inspection of its graph.

1. The point $(1, 0)$ is on the graph of $y = \ln x$ [because $(0, 1)$ is on the graph of $y = e^x$]. In other words,

$$\ln 1 = 0. \tag{1}$$

2. $\ln x$ is defined only for positive values of x.

3. $\ln x$ is negative for x between 0 and 1.

4. $\ln x$ is positive for x greater than 1.

5. $y = \ln x$ is an increasing function and concave down.

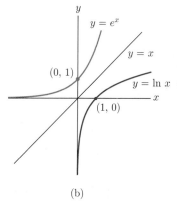

Figure 2. Obtaining the graph of $y = \ln x$ as a reflection of $y = e^x$.

Let us study the relationship between the natural logarithm and exponential functions more closely. From the way in which the graph of $y = \ln x$ was obtained, we know that (a, b) is on the graph of $y = \ln x$ if and only if (b, a) is on the graph of $y = e^x$. However, a typical point on the graph of $y = \ln x$ is of the form $(a, \ln a)$, $a > 0$. So for any positive value of a, the point $(\ln a, a)$ is on the graph of $y = e^x$. That is,

$$e^{\ln a} = a.$$

Since a was an arbitrary positive number, we have the following important relationship between the natural logarithm and exponential functions.

$$e^{\ln x} = x \quad \text{for } x > 0. \tag{2}$$

Equation (2) can be put into verbal form.

For each positive number x, $\ln x$ is the exponent to which we must raise e to get x.

If b is any number, e^b is positive and hence $\ln(e^b)$ makes sense. What is $\ln(e^b)$? Since (b, e^b) is on the graph of e^x, we know that (e^b, b) must be on the graph of $y = \ln x$. That is, $\ln(e^b) = b$. Thus we have shown that

$$\ln(e^x) = x \quad \text{for any } x. \tag{3}$$

The identities (2) and (3) express the fact that the natural logarithm function is the *inverse* of the exponential function (for $x > 0$). For instance, if we take a number x and compute e^x, we can, by (3), undo the effect of the exponential by taking the natural logarithm; that is, the logarithm of e^x equals the original number x. Similarly, if we take a positive number x and compute $\ln x$, we can, by (2), undo the effect of the logarithm by raising e to the $\ln x$ power; that is, $e^{\ln x}$ equals the original number x.

EXAMPLE 1   Simplify $e^{\ln 4 + 2 \ln 3}$.

Solution   Using properties of the exponential function, we have

$$e^{\ln 4 + 2 \ln 3} = e^{\ln 4} \cdot e^{2 \ln 3}$$

and

$$e^{2 \ln 3} = e^{(\ln 3) \cdot 2} = \left(e^{\ln 3}\right)^2.$$

By (2), we have $e^{\ln 4} = 4$ and $\left(e^{\ln 3}\right)^2 = 3^2$. So

$$e^{\ln 4 + 2 \ln 3} = e^{\ln 4} \cdot \left(e^{\ln 3}\right)^2 = 4 \cdot 3^2 = 36. \qquad \blacksquare$$

Scientific calculators have an LN key that will compute the natural logarithm of a number to as many as 10 significant figures. For instance, pressing the LN key and entering the number 2 into the calculator, we obtain $\ln 2 = .6931471806$ (to 10 significant figures).

The relationships (2) and (3) between e^x and $\ln x$ may be used to solve equations, as the next examples show.

EXAMPLE 2   Solve the equation $5e^{x-3} = 4$ for x.

Solution   First divide each side by 5,
$$e^{x-3} = .8.$$

Taking the logarithm of each side and using (3), we have

$$\ln(e^{x-3}) = \ln .8$$
$$x - 3 = \ln .8$$
$$x = 3 + \ln .8.$$

[If desired, the numerical value of x can be obtained by using a scientific calculator: $x = 3 - .22314 = 2.77686$ (to five decimal places).] $\blacksquare$

EXAMPLE 3   Solve the equation $2 \ln x + 7 = 0$ for x.

Solution
$$2 \ln x = -7$$
$$\ln x = -3.5$$
$$e^{\ln x} = e^{-3.5}$$
$$x = e^{-3.5} \qquad [\text{by (2)}]. \qquad \blacksquare$$

Other Exponential and Logarithm Functions   In our discussion of the exponential function, we mentioned that all exponential functions of the form $y = b^x$, where b is a fixed positive number, can be expressed in terms of *the* exponential function $y = e^x$. Now we can be quite explicit. For since $b = e^{\ln b}$, we see that

$$b^x = (e^{\ln b})^x = e^{(\ln b)x}.$$

Hence we have shown that

$$b^x = e^{kx}, \quad \text{where } k = \ln b.$$

The natural logarithm function is sometimes called the *logarithm to the base e*, for it is the inverse of the exponential function $y = e^x$. If we reflect the graph of the function $y = 2^x$ through the line $y = x$, we obtain the graph of a function called the *logarithm to the base 2*, denoted by $\log_2 x$. Similarly, if we reflect the graph of $y = 10^x$ through the line $y = x$, we obtain the graph of a function called the *logarithm to the base 10*, denoted by $\log_{10} x$. (See Fig. 3.)

Logarithms to the base 10 are sometimes called *common* logarithms. Common logarithms are usually introduced into algebra courses for the purpose of simplifying certain arithmetic calculations. However, with the advent of the modern computer and the widespread availability of graphing calculators, the need for common logarithms has diminished considerably. It can be shown that

$$\log_{10} x = \frac{1}{\ln 10} \cdot \ln x,$$

so $\log_{10} x$ is simply a constant multiple of $\ln x$. However, we shall not need this fact.

The natural logarithm function is used in calculus because differentiation and integration formulas are simpler than for $\log_{10} x$ or $\log_2 x$, and so on. (Recall that we prefer the function e^x over the functions 10^x and 2^x for the same reason.) Also, $\ln x$ arises "naturally" in the process of solving certain differential equations that describe various growth processes.

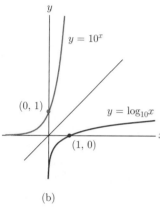

(a)

(b)

Figure 3.   Graphs of $y = \log_2 x$ and $y = \log_{10} x$ as reflections of $y = 2^x$ and $y = 10^x$.

Practice Problems 4.4

1. Find $\ln e$.

2. Solve $e^{-3x} = 2$ using the natural logarithm function.

EXERCISES 4.4

1. Find $\ln(\sqrt{e})$.

2. Find $\ln\left(\dfrac{1}{e^2}\right)$.

3. If $e^x = 5$, write x in terms of the natural logarithm.

4. If $e^{-x} = 3.2$, write x in terms of the natural logarithm.

5. If $\ln x = -1$, write x using the exponential function.

6. If $\ln x = 4.5$, write x using the exponential function.

Simplify the following expressions.

7. $\ln e^{-3}$

8. $e^{\ln 4.1}$

9. $e^{e^{\ln 1}}$

10. $\ln(e^{-2\ln e})$

11. $\ln(\ln e)$

12. $e^{4\ln 1}$

13. $e^{2\ln x}$

14. $e^{x \ln 2}$

15. $e^{-2\ln 7}$

16. $\ln(e^{-2}e^4)$

17. $e^{\ln x + \ln 2}$

18. $e^{\ln 3 - 2\ln x}$

Solve the following equations for x.

19. $e^{2x} = 5$

20. $e^{1-3x} = 4$

21. $\ln(4 - x) = \frac{1}{2}$

22. $\ln 3x = 2$

23. $\ln x^2 = 9$

24. $e^{x^2} = 25$

25. $6e^{-.00012x} = 3$

26. $4 - \ln x = 0$

27. $\ln 3x = \ln 5$

28. $\ln(x^2 - 5) = 0$

29. $\ln(\ln 3x) = 0$

30. $2\ln x = 7$

31. $2e^{x/3} - 9 = 0$

32. $e^{\sqrt{x}} = \sqrt{e^x}$

33. $5 \ln 2x = 8$

34. $750 e^{-.4x} = 375$

35. $(e^2)^x \cdot e^{\ln 1} = 4$

36. $e^{5x} \cdot e^{\ln 5} = 2$

37. $4e^x \cdot e^{-2x} = 6$

38. $(e^x)^2 \cdot e^{2-3x} = 4$

39. The graph of $f(x) = -5x + e^x$ is shown in Fig. 4. Find the coordinates of the minimum point.

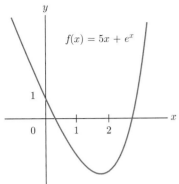

$f(x) = 5x + e^x$

Figure 4

40. The graph of $f(x) = -1 + (x-1)^2 e^x$ is shown in Fig. 5. Find the coordinates of the maximum and minimum points.

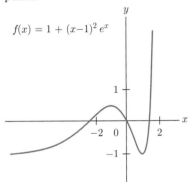

$f(x) = 1 + (x-1)^2 e^x$

Figure 5

41. (a) Find the point on the graph of $y = e^{-x}$ where the tangent line has slope -2.

(b) Plot the graphs of $y = e^{-x}$ and the tangent line in part (a).

42. Find the x-intercepts of $y = (x-1)^2 \ln(x+1)$, $x > -1$.

In Exercises 43 and 44, find the coordinates of each relative extreme point of the given function and determine if the point is a relative maximum point or a relative minimum point.

43. $f(x) = e^{-x} + 3x$

44. $f(x) = 5x - 2e^x$

45. Concentration of a Drug in the Blood When a drug or vitamin is administered intramuscularly (into a muscle),

the concentration in the blood at time t after injection can be approximated by a function of the form $f(t) = c(e^{-k_1 t} - e^{-k_2 t})$. The graph of $f(t) = 5(e^{-.01t} - e^{-.51t})$, for $t \geq 0$, is shown in Fig. 6. Find the value of t at which this function reaches its maximum value.

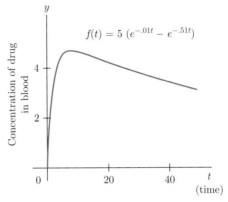

$f(t) = 5 (e^{-.01t} - e^{-.51t})$

Figure 6

46. Wind Velocity Under certain geographic conditions, the wind velocity v at a height x centimeters above the ground is given by $v = K \ln(x/x_0)$, where K is a positive constant (depending on the air density, average wind velocity, and the like) and x_0 is a roughness parameter (depending on the roughness of the vegetation on the ground).* Suppose that $x_0 = .7$ centimeter (a value that applies to lawn grass 3 centimeters high) and $K = 300$ centimeters per second.

(a) At what height above the ground is the wind velocity zero?

(b) At what height is the wind velocity 1200 centimeters per second?

47. Find k such that $2^x = e^{kx}$ for all x.

48. Find k such that $2^{-x/5} = e^{kx}$ for all x.

Technology Exercises

49. Graph the function $y = \ln(e^x)$ and use TRACE to convince yourself that it is the same as the function $y = x$. What do you observe about the graph of $y = e^{\ln x}$?

50. Graph $y = e^{2x}$ and $y = 5$ together and determine the x-coordinate of their point of intersection (to four decimal places). Express this number in terms of a logarithm.

51. Graph $y = \ln 5x$ and $y = 2$ together and determine the x-coordinate of their point of intersection (to four decimal places). Express this number in terms of a power of e.

---

*G. Cox, B. Collier, A. Johnson, and P. Miller, *Dynamic Ecology* (Englewood Cliffs, N.J.: Prentice-Hall, 1973), pp. 113–115.

Solutions to Practice Problems 4.4

1. Answer: 1. The number $\ln e$ is that exponent to which e must be raised to obtain e.

2. Take the logarithm of each side and use (3) to simplify the left side:

$$\ln e^{-3x} = \ln 2$$
$$-3x = \ln 2$$
$$x = -\frac{\ln 2}{3}.$$

4.5 The Derivative of ln *x*

Let us now compute the derivative of $y = \ln x$ for $x > 0$. Since $e^{\ln x} = x$, we have

$$\frac{d}{dx}(e^{\ln x}) = \frac{d}{dx}(x) = 1. \tag{1}$$

On the other hand, if we differentiate $y = e^{\ln x}$ by the chain rule, we find that

$$\frac{d}{dx}(e^{\ln x}) = e^{\ln x} \cdot \frac{d}{dx}(\ln x) = x \cdot \frac{d}{dx}(\ln x), \tag{2}$$

where the last equality used the fact that $e^{\ln x} = x$. By combining equations (1) and (2), we obtain

$$x \cdot \frac{d}{dx}(\ln x) = 1.$$

In other words,

$$\frac{d}{dx}(\ln x) = \frac{1}{x}, \qquad x > 0. \tag{3}$$

By combining this differentiation formula with the chain rule, product rule, and quotient rule, we can differentiate many functions involving $\ln x$.

EXAMPLE 1   Differentiate.

(a) $y = (\ln x)^5$ (b) $y = x \ln x$ (c) $y = \ln(x^3 + 5x^2 + 8)$

Solution   (a) By the general power rule,

$$\frac{d}{dx}(\ln x)^5 = 5(\ln x)^4 \cdot \frac{d}{dx}(\ln x) = 5(\ln x)^4 \cdot \frac{1}{x} = \frac{5(\ln x)^4}{x}.$$

(b) By the product rule,

$$\frac{d}{dx}(x \ln x) = x \cdot \frac{d}{dx}(\ln x) + (\ln x) \cdot 1 = x \cdot \frac{1}{x} + \ln x = 1 + \ln x.$$

(c) By the chain rule,

$$\frac{d}{dx}\ln(x^3 + 5x^2 + 8) = \frac{1}{x^3 + 5x^2 + 8} \cdot \frac{d}{dx}(x^3 + 5x^2 + 8) = \frac{3x^2 + 10x}{x^3 + 5x^2 + 8}. \qquad \blacksquare$$

Let $g(x)$ be any differentiable function. For any value of x for which $g(x)$ is positive, the function $y = \ln(g(x))$ is defined. For such a value of x, the derivative is given by the chain rule as

$$\frac{d}{dx}[\ln g(x)] = \frac{1}{g(x)} \cdot \frac{d}{dx} g(x) = \frac{g'(x)}{g(x)} \tag{4}$$

If $u = g(x)$, equation (4) can be written in the form

$$\frac{d}{dx}[\ln u] = \frac{1}{u}\frac{du}{dx}. \tag{4a}$$

Example 1(c) illustrates a special case of this formula.

EXAMPLE 2   The function $f(x) = (\ln x)/x$ has a relative extreme point for some $x > 0$. Find the point and determine whether it is a relative maximum or a relative minimum point.

Solution   By the quotient rule,

$$f'(x) = \frac{x \cdot \dfrac{1}{x} - (\ln x) \cdot 1}{x^2} = \frac{1 - \ln x}{x^2}$$

$$f''(x) = \frac{x^2 \cdot \left(-\dfrac{1}{x}\right) - (1 - \ln x)(2x)}{x^4} = \frac{2\ln x - 3}{x^3}.$$

If we set $f'(x) = 0$, then

$$1 - \ln x = 0$$
$$\ln x = 1$$
$$e^{\ln x} = e^1 = e$$
$$x = e.$$

Therefore, the only possible relative extreme point is at $x = e$. When $x = e$, $f(e) = (\ln e)/e = 1/e$. Furthermore,

$$f''(e) = \frac{2\ln e - 3}{e^3} = -\frac{1}{e^3} < 0,$$

which implies that the graph of $f(x)$ is concave down at $x = e$. Therefore, $(e, 1/e)$ is a relative maximum point of the graph of $f(x)$. (See Fig. 1.)   ∎

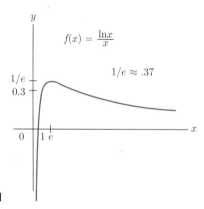

Figure 1

The next example introduces a function that will be needed later when we study integration.

EXAMPLE 3   The function $y = \ln |x|$ is defined for all nonzero values of x. Its graph is sketched in Fig. 2. Compute the derivative of $y = \ln |x|$.

Solution If x is positive, $|x| = x$, so

$$\frac{d}{dx} \ln |x| = \frac{d}{dx} \ln x = \frac{1}{x}.$$

If x is negative, $|x| = -x$ and, by the chain rule,

$$\frac{d}{dx} \ln |x| = \frac{d}{dx} \ln(-x) = \frac{1}{-x} \cdot \frac{d}{dx}(-x) = \frac{1}{-x} \cdot (-1) = \frac{1}{x}.$$ ■

Figure 2. Graph of $y = \ln |x|$.   Therefore, we have established the following useful fact.

$$\frac{d}{dx} \ln |x| = \frac{1}{x}, \qquad x \neq 0$$

Practice Problems 4.5

Differentiate.

1. $f(x) = \dfrac{1}{\ln(x^4 + 5)}$

2. $f(x) = \ln(\ln x)$

EXERCISES 4.5

Differentiate the following functions.

1. $y = \ln 2x$

2. $y = \dfrac{\ln x}{2}$

3. $y = \ln(x + 3) + \ln 5$

4. $y = \ln(6x^2 - 3x + 1)$

5. $y = \ln(e^x + e^{-x})$

6. $y = \ln\left(e^{x^2 + 2}\right)$

7. $y = e^{\ln x + x}$

8. $y = (1 + \ln x)^3$

9. $y = \ln x \cdot \ln 2x$

10. $y = \ln 2 \cdot \ln x$

11. $y = (\ln x)^2 + \ln(x^2)$

12. $y = \dfrac{\ln x}{\ln 2x}$

13. $y = \ln(kx)$, k constant

14. $y = \ln \dfrac{1}{x}$

15. $y = (x^2 + 1) \ln(x^2 + 1)$

16. $y = \dfrac{1}{\ln x}$

17. $y = \ln\left(\dfrac{x - 1}{x + 1}\right)$

18. $y = [\ln(e^{2x} + 1)]^2$

19. $y = \ln(e^{5x} + 1)$

20. $y = \sqrt{\ln 2x}$

Find.

21. $\dfrac{d^2}{dt^2}(t^2 \ln t)$

22. $\dfrac{d^2}{dt^2} \ln(\ln t)$

23. Determine the domain of definition of the given function.

   (a) $f(t) = \ln(\ln t)$ **(b)** $f(t) = \ln(\ln(\ln t))$

24. Find the equations of the tangent lines to the graph of $y = \ln |x|$ at $x = 1$ and $x = -1$.

25. Write the equation of the tangent line to the graph of $y = \ln(x^2 + e)$ at $x = 0$.

26. The function $f(x) = (\ln x + 1)/x$ has a relative extreme point for $x > 0$. Find the coordinates of the point. Is it a relative maximum point?

27. The graph of $f(x) = (\ln x)/\sqrt{x}$ is shown in Fig. 3. Find the coordinates of the maximum point.

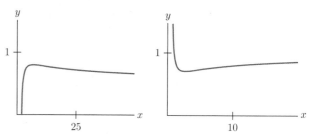

Figure 3   **Figure 4**

28. The graph of $f(x) = x/(\ln x + x)$ is shown in Fig. 4. Find the coordinates of the minimum point.

29. Find the coordinates of the relative extreme point of $y = x^2 \ln x$, $x > 0$. Then use the second derivative test to decide if the point is a relative maximum point or a relative minimum point.

30. Repeat the previous exercise with $y = \sqrt{x} \ln x$.

31. If a cost function is $C(x) = (100 \ln x)/(40 - 3x)$, find the marginal cost when $x = 10$.

32. If a cost function is $C(x) = (1000 \ln x)/\sqrt{100 - 3x}$, find the marginal cost when $x = 25$.

33. If the demand equation for a certain commodity is $p = 45/(\ln x)$, determine the marginal revenue function for this commodity, and compute the marginal revenue when $x = 20$.

34. Suppose that the total revenue function for a manufacturer is $R(x) = 300 \ln(x + 1)$, so the sale of x units of a product brings in about $R(x)$ dollars. Suppose also that the total cost of producing x units is $C(x)$ dollars, where $C(x) = 2x$. Find the value of x at which the profit function $R(x) - C(x)$ will be maximized. Show that the profit function has a relative maximum and not a relative minimum point at this value of x.

35. Evaluate $\lim\limits_{h \to 0} \dfrac{\ln(7 + h) - \ln 7}{h}$.

36. An Area Problem Find the maximum area of a rectangle in the first quadrant with one corner at the origin, an opposite corner on the graph of $y = -\ln x$, and two sides on the coordinate axes.

Technology Exercises

37. Graph the function $f(x) = \ln|x|$ in the window $[-5, 5]$ by $[-2, 2]$, and verify that the derivative formula gives the correct values of $f'(x)$ at $x = \pm 1, \pm 2$, and ± 4.

38. Analysis of the Effectiveness of an Insect Repellent Human hands covered with cotton fabrics impregnated with the insect repellent DEPA were inserted for 5 minutes into a test chamber containing 200 female mosquitoes.[*] The function $f(x) = 26.48 - 14.09 \ln x$ gives the number of mosquito bites received when the concentration was x percent. [*Note:* The answers to parts (b)–(e) can be obtained either algebraically or from the graphs. You might consider trying both methods.]

(a) Graph $f(x)$ and $f'(x)$ for $0 \le x \le 6$.

(b) How many bites were received when the concentration was 3.25%?

(c) What concentration resulted in 15 bites?

(d) At what rate is the number of bites changing with respect to concentration of DEPA when $x = 2.75$?

(e) For what concentration does the rate of change of bites with respect to concentration equal -10 bites per percentage increase in concentration?

Solutions to Practice Problems 4.5

1. Here $f(x) = [\ln(x^4 + 5)]^{-1}$. By the chain rule,

$$f'(x) = (-1) \cdot [\ln(x^4 + 5)]^{-2} \cdot \frac{d}{dx} \ln(x^4 + 5)$$

$$= -[\ln(x^4 + 5)]^{-2} \cdot \frac{4x^3}{x^4 + 5}.$$

2. $f'(x) = \dfrac{d}{dx} \ln(\ln x) = \dfrac{1}{\ln x} \cdot \dfrac{d}{dx} \ln x = \dfrac{1}{\ln x} \cdot \dfrac{1}{x} = \dfrac{1}{x \ln x}.$

4.6 Properties of the Natural Logarithm Function

The natural logarithm function $\ln x$ possesses many of the familiar properties of logarithms to base 10 (or common logarithms) that are encountered in algebra.

Let x and y be positive numbers, b any number.

LI $\ln(xy) = \ln x + \ln y$

LII $\ln\left(\dfrac{1}{x}\right) = -\ln x$

LIII $\ln\left(\dfrac{x}{y}\right) = \ln x - \ln y$

LIV $\ln(x^b) = b \ln x$

---

[*]Rao, K. M., Prakash, S., Kumar, S., Suryanarayana, M. V. S., Bhagwat, M. M., Gharia, M. M., and Bhavsar, R. B., "N-diethylphenylacetamide in Treated Fabrics, as a Repellent Against *Aedes aegypti* and *Culex quinquefasclatus* (Deptera: Culiciae)," *Journal of Medical Entomology*, 28 (January 1991), 1.

Verification of LI By equation (2) of Section 4.4, we have $e^{\ln(xy)} = xy$, $e^{\ln x} = x$, and $e^{\ln y} = y$. Therefore,

$$e^{\ln(xy)} = xy = e^{\ln x} \cdot e^{\ln y} = e^{\ln x + \ln y}.$$

By equating exponents, we get LI.

Verification of LII Since $e^{\ln(1/x)} = 1/x$, we have

$$e^{\ln(1/x)} = \frac{1}{x} = \frac{1}{e^{\ln x}} = e^{-\ln x}.$$

By equating exponents, we get LII.

Verification of LIII By LI and LII, we have

$$\ln\left(\frac{x}{y}\right) = \ln\left(x \cdot \frac{1}{y}\right) = \ln x + \ln\left(\frac{1}{y}\right) = \ln x - \ln y.$$

Verification of LIV Since $e^{\ln(x^b)} = x^b$, we have

$$e^{\ln(x^b)} = x^b = \left(e^{\ln x}\right)^b = e^{b \ln x}.$$

Equating exponents, we get LIV.

These properties of the natural logarithm should be learned thoroughly. You will find them useful in many calculations involving $\ln x$ and the exponential function.

EXAMPLE 1

Write $\ln 5 + 2 \ln 3$ as a single logarithm.

Solution

$$\ln 5 + 2 \ln 3 = \ln 5 + \ln 3^2 \qquad \text{(LIV)}$$
$$= \ln 5 + \ln 9$$
$$= \ln(5 \cdot 9) \qquad \text{(LI)}$$
$$= \ln 45. \qquad \blacksquare$$

EXAMPLE 2

Write $\frac{1}{2} \ln(4t) - \ln(t^2 + 1)$ as a single logarithm.

Solution

$$\tfrac{1}{2} \ln(4t) - \ln(t^2 + 1) = \ln\left[(4t)^{1/2}\right] - \ln(t^2 + 1) \qquad \text{(LIV)}$$
$$= \ln(2\sqrt{t}) - \ln(t^2 + 1)$$
$$= \ln\left(\frac{2\sqrt{t}}{t^2 + 1}\right). \qquad \text{(LIII)}$$
$$\blacksquare$$

EXAMPLE 3

Simplify $\ln x + \ln 3 + \ln y - \ln 5$.

Solution Use (LI) twice and (LIII) once.

$$(\ln x + \ln 3) + \ln y - \ln 5 = \ln 3x + \ln y - \ln 5$$
$$= \ln 3xy - \ln 5$$
$$= \ln\left(\frac{3xy}{5}\right). \qquad \blacksquare$$

EXAMPLE 4   Differentiate $f(x) = \ln[x(x + 1)(x + 2)]$.

Solution First rewrite $f(x)$ using (LI).

$$f(x) = \ln[x(x + 1)(x + 2)] = \ln x + \ln(x + 1) + \ln(x + 2).$$

Then $f'(x)$ is easily calculated:

$$f'(x) = \frac{1}{x} + \frac{1}{x + 1} + \frac{1}{x + 2}.$$ ∎

The natural logarithm function can be used to simplify the task of differentiating products. Suppose, for example, that we wish to differentiate the function

$$g(x) = x(x + 1)(x + 2).$$

As we showed in Example 4,

$$\frac{d}{dx} \ln g(x) = \frac{1}{x} + \frac{1}{x + 1} + \frac{1}{x + 2}.$$

However,

$$\frac{d}{dx} \ln g(x) = \frac{g'(x)}{g(x)}.$$

Therefore, equating the two expressions for $\dfrac{d}{dx} \ln g(x)$, we have

$$\frac{g'(x)}{g(x)} = \frac{1}{x} + \frac{1}{x + 1} + \frac{1}{x + 2}.$$

Finally, we solve for $g'(x)$:

$$g'(x) = g(x) \cdot \left(\frac{1}{x} + \frac{1}{x + 1} + \frac{1}{x + 2} \right)$$

$$= x(x + 1)(x + 2) \left(\frac{1}{x} + \frac{1}{x + 1} + \frac{1}{x + 2} \right).$$

In a similar way, we differentiate the product of any number of factors by first taking natural logarithms, then differentiating, and finally solving for the desired derivative. This procedure is called *logarithmic differentiation*.

EXAMPLE 5   Differentiate the function $g(x) = (x^2 + 1)(x^3 - 3)(2x + 5)$ using logarithmic differentiation.

Solution *Step 1*   Take the natural logarithm of both sides of the given equation:

$$\ln g(x) = \ln[(x^2 + 1)(x^3 - 3)(2x + 5)]$$

$$= \ln(x^2 + 1) + \ln(x^3 - 3) + \ln(2x + 5).$$

Step 2   Differentiate:

$$\frac{d}{dx}(\ln g(x)) = \frac{g'(x)}{g(x)} = \frac{2x}{x^2 + 1} + \frac{3x^2}{x^3 - 3} + \frac{2}{2x + 5}$$

Step 3   Solve for $g'(x)$:

$$g'(x) = g(x) \left(\frac{2x}{x^2 + 1} + \frac{3x^2}{x^3 - 3} + \frac{2}{2x + 5} \right)$$

$$= (x^2 + 1)(x^3 - 3)(2x + 5) \left(\frac{2x}{x^2 + 1} + \frac{3x^2}{x^3 - 3} + \frac{2}{2x + 5} \right).$$ ∎

Let us now use logarithmic differentiation to finally establish the power rule:

$$\frac{d}{dx}(x^r) = rx^{r-1}.$$

Verification of the Power Rule for x > 0   Let $f(x) = x^r$. Then

$$\ln f(x) = \ln x^r = r \ln x.$$

Differentiation of this equation yields

$$\frac{f'(x)}{f(x)} = r \cdot \frac{1}{x}$$

$$f'(x) = r \cdot \frac{1}{x} \cdot f(x) = r \cdot \frac{1}{x} \cdot x^r = rx^{r-1}.$$

Practice Problems 4.6

1. Differentiate $f(x) = \ln\left[\dfrac{e^x \sqrt{x}}{(x+1)^6}\right]$.

2. Use logarithmic differentiation to differentiate

$$f(x) = (x+1)^7(x+2)^8(x+3)^9.$$

EXERCISES 4.6

Simplify the following expressions.

1. $\ln 5 + \ln x$
2. $\ln x^5 - \ln x^3$
3. $\frac{1}{2}\ln 9$
4. $3\ln\frac{1}{2} + \ln 16$
5. $\ln 4 + \ln 6 - \ln 12$
6. $\ln 2 - \ln x + \ln 3$
7. $e^{2\ln x}$
8. $\frac{3}{2}\ln 4 - 5\ln 2$
9. $5\ln x - \frac{1}{2}\ln y + 3\ln z$
10. $e^{\ln x^2 + 3\ln y}$
11. $\ln x - \ln x^2 + \ln x^4$
12. $\frac{1}{2}\ln xy + \frac{3}{2}\ln\frac{x}{y}$

13. Which is larger, $2\ln 5$ or $3\ln 3$?
14. Which is larger, $\frac{1}{2}\ln 16$ or $\frac{1}{3}\ln 27$?

Evaluate the given expression. Use $\ln 2 = .69$ and $\ln 3 = 1.1$.

15. (a) $\ln 4$ (b) $\ln 6$ (c) $\ln 54$
16. (a) $\ln 12$ (b) $\ln 16$ (c) $\ln(9 \cdot 2^4)$
17. (a) $\ln\frac{1}{6}$ (b) $\ln\frac{2}{9}$ (c) $\ln\frac{1}{\sqrt{2}}$
18. (a) $\ln 100 - 2\ln 5$ (b) $\ln 10 + \ln\frac{1}{5}$ (c) $\ln\sqrt{108}$
19. Which of the following is the same as $4\ln 2x$?
    (a) $\ln 8x$ (b) $8\ln x$
    (c) $\ln 8 + \ln x$ (d) $\ln 16x^4$
20. Which of the following is the same as $\ln(9x) - \ln(3x)$?
    (a) $\ln 6x$ (b) $\ln(9x)/\ln(3x)$
    (c) $6 \cdot \ln(x)$ (d) $\ln 3$
21. Which of the following is the same as $\dfrac{\ln 8x^2}{\ln 2x}$?

(a) $\ln 4x$ (b) $4x$
(c) $\ln 8x^2 - \ln 2x$ (d) none of these
22. Which of the following is the same as $\ln 9x^2$?
    (a) $2 \cdot \ln 9x$ (b) $3x \cdot \ln 3x$
    (c) $2 \cdot \ln 3x$ (d) none of these

Solve the given equation for x.

23. $\ln x - \ln x^2 + \ln 3 = 0$ 24. $\ln\sqrt{x} - 2\ln 3 = 0$
25. $\ln x^4 - 2\ln x = 1$ 26. $\ln x^2 - \ln 2x + 1 = 0$
27. $(\ln x)^2 - 1 = 0$ 28. $3\ln x - \ln 3x = 0$
29. $\ln\sqrt{x} = \sqrt{\ln x}$ 30. $2(\ln x)^2 + \ln x - 1 = 0$
31. $\ln(x+1) - \ln(x-2) = 1$
32. $\ln[(x-3)(x+2)] - \ln(x+2)^2 - \ln 7 = 0$

Differentiate.

33. $y = \ln[(x+5)(2x-1)(4-x)]$
34. $y = \ln[(x+1)(2x+1)(3x+1)]$
35. $y = \ln\left[(1+x)^2(2+x)^3(3+x)^4\right]$
36. $y = \ln\left[e^{2x}(x^3+1)(x^4+5x)\right]$
37. $y = \ln\left[\sqrt{xe^{x^2+1}}\right]$
38. $y = \ln\dfrac{x+1}{x-1}$ 39. $y = \ln\dfrac{(x+1)^4}{e^{x-1}}$
40. $y = \ln\dfrac{(x+1)^4(x^3+2)}{x-1}$
41. $y = \ln(3x+1)\ln(5x+1)$ 42. $y = (\ln 4x)(\ln 2x)$

Use logarithmic differentiation to differentiate the following functions.

43. $f(x) = (x+1)^4(4x-1)^2$ **44.** $f(x) = e^x(3x-4)^8$

45. $f(x) = \dfrac{(x+1)(2x+1)(3x+1)}{\sqrt{4x+1}}$

46. $f(x) = \dfrac{(x-2)^3(x-3)^4}{(x+4)^5}$ **47.** $f(x) = 2^x$

48. $f(x) = \sqrt[x]{3}$ **49.** $f(x) = x^x$ **50.** $f(x) = \sqrt[x]{x}$

51. Allometric Equation   There are substantial empirical data to show that, if x and y measure the sizes of two organs of a particular animal, then x and y are related by an *allometric equation* of the form

$$\ln y - k \ln x = \ln c,$$

where k and c are positive constants that depend only on the type of parts or organs that are measured and are constant among animals belonging to the same species.* Solve this equation for y in terms of x, k, and c.

52. Epidemics Model   In the study of epidemics, we find the equation

$$\ln(1-y) - \ln y = C - rt,$$

where y is the fraction of the population that has a specific disease at time t. Solve the equation for y in terms of t and the constants C and r.

53. Determine the values of h and k for which the graph of $y = he^{kx}$ passes through the points $(1,6)$ and $(4,48)$.

54. Find values of k and r for which the graph of $y = kx^r$ passes through the points $(2,3)$ and $(4,15)$.

Solutions to Practice Problems 4.6

1. Use the properties of the natural logarithm to express $f(x)$ as a sum of simple functions before differentiating.

$$f(x) = \ln\left[\frac{e^x\sqrt{x}}{(x+1)^6}\right]$$

$$f(x) = \ln e^x + \ln\sqrt{x} - \ln(x+1)^6$$
$$= x + \tfrac{1}{2}\ln x - 6\ln(x+1)$$

$$f'(x) = 1 + \frac{1}{2x} - \frac{6}{x+1}.$$

2. $f(x) = (x+1)^7(x+2)^8(x+3)^9$
$$\ln f(x) = 7\ln(x+1) + 8\ln(x+2) + 9\ln(x+3).$$

Now we differentiate both sides of the equation.

$$\frac{f'(x)}{f(x)} = \frac{7}{x+1} + \frac{8}{x+2} + \frac{9}{x+3}$$

$$f'(x) = f(x)\left(\frac{7}{x+1} + \frac{8}{x+2} + \frac{9}{x+3}\right)$$
$$= (x+1)^7(x+2)^8(x+3)^9 \cdot$$
$$\left(\frac{7}{x+1} + \frac{8}{x+2} + \frac{9}{x+3}\right).$$

REVIEW OF FUNDAMENTAL CONCEPTS

1. State as many laws of exponents as you can recall.

2. What is e?

3. Write the differential equation satisfied by $y = Ce^{kt}$.

4. State the properties that graphs of the form $y = e^{kx}$ have in common when k is positive. When k is negative.

5. What are the coordinates of the reflection of the point (a, b) in the line $y = x$?

6. What is a logarithm?

7. What is the x-intercept of the graph of the natural logarithm function?

8. State the main features of the graph of $y = \ln x$.

9. State the two key equations giving the relationships between e^x and $\ln x$. [*Hint:* The right side of each equation is just x.]

10. What is the difference between a natural logarithm and a common logarithm?

11. Give the formula that converts a function of the form b^x to an exponential function with base e.

12. State the differentiation formula for each of the following functions:

(a) $f(x) = e^{kx}$ (b) $f(x) = e^{g(x)}$

(c) $f(x) = \ln g(x)$

13. State the four algebraic properties of the natural logarithm function.

14. Give an example of the use of logarithmic differentiation.

---

*E. Batschelet, *Introduction to Mathematics for Life Scientists* (New York: Springer-Verlag, 1971), pp. 305–307.

SUPPLEMENTARY EXERCISES

Calculate the following.

1. $27^{4/3}$ **2.** $4^{1.5}$ **3.** 5^{-2} **4.** $16^{-.25}$

5. $(2^{5/7})^{14/5}$ **6.** $8^{1/2}\cdot 2^{1/2}$ **7.** $\dfrac{9^{5/2}}{9^{3/2}}$ **8.** $4^{.2}\cdot 4^{.3}$

Simplify the following.

9. $\left(e^{x^2}\right)^3$ **10.** $e^{5x}\cdot e^{2x}$ **11.** $\dfrac{e^{3x}}{e^x}$

12. $2^x\cdot 3^x$ **13.** $(e^{8x}+7e^{-2x})e^{3x}$ **14.** $\dfrac{e^{5x/2}-e^{3x}}{\sqrt{e^x}}$

Solve the following equations for x.

15. $e^{-3x}=e^{-12}$ **16.** $e^{x^2-x}=e^2$

17. $(e^x\cdot e^2)^3=e^{-9}$ **18.** $e^{-5x}\cdot e^4=e$

Differentiate the following functions.

19. $y=10e^{7x}$ **20.** $y=e^{\sqrt{x}}$

21. $y=xe^{x^2}$ **22.** $y=\dfrac{e^x+1}{x-1}$

23. $y=e^{e^x}$ **24.** $y=(\sqrt{x}+1)e^{-2x}$

25. $y=\dfrac{x^2-x+5}{e^{3x}+3}$ **26.** $y=x^e$

27. Determine all solutions of the differential equation $y'=-y$.

28. Determine all functions $y=f(x)$ such that $y'=-1.5y$ and $f(0)=2000$.

29. Determine all solutions of the differential equation $y'=1.5y$ and $f(0)=2$.

30. Determine all solutions of the differential equation $y'=\frac{1}{3}y$.

Graph the following functions.

31. $y=e^{-x}+x$ **32.** $y=e^x-x$ **33.** $y=e^{-(1/2)x^2}$

34. $y=100(x-2)e^{-x}$ for $x\ge 2$

35. $y=e^{2x}+2e^{-x}$ **36.** $y=\dfrac{1}{e^x+e^{-x}}$

37. Find the equation of the tangent line to the graph of $y=\dfrac{e^x}{1+e^x}$ at $(0,.5)$.

38. Show that the tangent lines to the graph of $y=\dfrac{e^x-e^{-x}}{e^x+e^{-x}}$ at $x=1$ and $x=-1$ are parallel.

Simplify the following expressions.

39. $e^{(\ln 5)/2}$ **40.** $e^{\ln(x^2)}$ **41.** $\dfrac{\ln x^2}{\ln x^3}$

42. $e^{2\ln 2}$ **43.** $e^{-5\ln 1}$ **44.** $\left[e^{\ln x}\right]^2$

Solve the following equations for t.

45. $t^{\ln t}=e$ **46.** $\ln(\ln 3t)=0$

47. $3e^{2t}=15$ **48.** $3e^{t/2}-12=0$

49. $2\ln t=5$ **50.** $2e^{-.3t}=1$

Differentiate the following functions.

51. $y=\ln(x^6+3x^4+1)$ **52.** $y=\dfrac{x}{\ln x}$

53. $y=\ln(5x-7)$ **54.** $y=\ln(9x)$

55. $y=(\ln x)^2$ **56.** $y=(x\ln x)^3$

57. $y=\ln\left(\dfrac{xe^x}{\sqrt{1+x}}\right)$

58. $y=\ln\left[e^{6x}(x^2+3)^5(x^3+1)^{-4}\right]$

59. $y=x\ln x-x$ **60.** $y=e^{2\ln(x+1)}$

61. $y=\ln(\ln\sqrt{x})$ **62.** $y=\dfrac{1}{\ln x}$

63. $y=e^x\ln x$ **64.** $y=\ln(x^2+e^x)$

65. $y=\ln\sqrt{\dfrac{x^2+1}{2x+3}}$ **66.** $y=\ln|-2x+1|$

67. $y=\ln\left(\dfrac{e^{x^2}}{x}\right)$ **68.** $y=\ln\sqrt[3]{x^3+3x-2}$

69. $y=\ln(2^x)$ **70.** $y=\ln(3^{x+1})-\ln 3$

71. $y=\ln|x-1|$ **72.** $y=e^{2\ln(2x+1)}$

73. $y=\ln\left(\dfrac{1}{e^{\sqrt{x}}}\right)$ **74.** $y=\ln(e^x+3e^{-x})$

Use logarithmic differentiation to differentiate the following functions.

75. $f(x)=\sqrt[5]{\dfrac{x^5+1}{x^5+5x+1}}$ **76.** $f(x)=2^x$

77. $f(x)=x^{\sqrt{x}}$ **78.** $f(x)=b^x$, where $b>0$

79. $f(x)=(x^2+5)^6(x^3+7)^8(x^4+9)^{10}$

80. $f(x)=x^{1+x}$

81. $f(x)=10^x$ **82.** $f(x)=\sqrt{x^2+5}\,e^{x^2}$

83. $f(x)=\sqrt{\dfrac{xe^x}{x^3+3}}$

84. $f(x)=\dfrac{e^x\sqrt{x+1}(x^2+2x+3)^2}{4x^2}$

85. $f(x)=e^{x+1}(x^2+1)x$ **86.** $f(x)=e^xx^22^x$

Graph the following functions.

87. $y=\ln(1+e^x)$ **88.** $y=\ln(1+e^{-x})$

89. $y=\ln(e^x+e^{-x})$ **90.** $y=\ln(e^x+.5e^{-x})$

91. $y=(\ln x)^2$ **92.** $y=\ln(x^2+1)$

93. $y=x-\ln x$ **94.** $y=(\ln x)^3$

95. Find $\dfrac{dy}{dx}$ if $e^y-e^x=3$.

96. Find $\dfrac{dy}{dt}$ if $e^{t+y}=t$.

97. Find $\dfrac{dy}{dx}$ at the point $(1,0)$ if $e^{xy}=x$.

98. Find $\dfrac{dy}{dx}$ if $e^{2x}+e^{2y}=4$.

APPLICATIONS OF THE EXPONENTIAL AND NATURAL LOGARITHM FUNCTIONS

In Chapter 4 we introduced the exponential function $y = e^x$ and the natural logarithm function $y = \ln x$ and studied their most important properties. From the way we introduced these functions, it is by no means clear that they have any substantial connection with the physical world. However, as this chapter will demonstrate, the exponential and natural logarithm functions are involved in the study of many physical problems, often in a very curious and unexpected way.

Here the most significant fact that we require is that the exponential function be uniquely characterized by its differential equation. In other words, we will constantly make use of the following fact, stated previously:

The function[*] $y = Ce^{kt}$ satisfies the differential equation

$$y' = ky.$$

Conversely, if $y = f(t)$ satisfies the differential equation, then $y = Ce^{kt}$ for some constant C.

---

[*]Note that we use the independent variable t instead of x throughout this chapter. The reason is that, in most applications, the variable of our exponential function is time.

If $f(t) = Ce^{kt}$, then, by setting $t = 0$, we have

$$f(0) = Ce^0 = C.$$

Therefore, C is the value of $f(t)$ at $t = 0$.

5.1 Exponential Growth and Decay

In biology, chemistry, and economics it is often necessary to study the behavior of a quantity that is increasing as time passes. If, at every instant, the rate of increase of the quantity is proportional to the quantity at that instant, we say that the quantity is *growing exponentially* or is *exhibiting exponential growth*. A simple example of exponential growth is exhibited by the growth of bacteria in a culture. Under ideal laboratory conditions a bacteria culture grows at a rate proportional to the number of bacteria present. It does so because the growth of the culture is accounted for by the division of the bacteria. The more bacteria at a given instant, the greater the possibilities for division and hence the greater the rate of growth.

Let us study the growth of a bacteria culture as a typical example of exponential growth. Let $P(t)$ denotes the number of bacteria in a certain culture at time t. The rate of growth of the culture at time t is $P'(t)$. We assume that this rate of growth is proportional to the size of the culture at time t, so

$$P'(t) = kP(t), \tag{1}$$

where k is a positive constant of proportionality. If we let $y = P(t)$, then formula (1) can be written as

$$y' = ky.$$

Therefore, from our discussion at the beginning of this chapter we see that

$$y = P(t) = P_0 e^{kt}, \tag{2}$$

where P_0 is the number of bacteria in the culture at time $t = 0$. The number k is called the *growth constant*.

EXAMPLE 1

Exponential growth   A certain bacteria culture grows at a rate proportional to its size. At time $t = 0$, approximately 20,000 bacteria are present. In 5 hours there are 400,000 bacteria. Determine a function that expresses the size of the culture as a function of time, measured in hours.

Solution   Let $P(t)$ be the number of bacteria present at time t. By assumption, $P(t)$ satisfies a differential equation of the form $y' = ky$, so $P(t)$ has the form

$$P(t) = P_0 e^{kt},$$

where the constants P_0 and k must be determined. The value of P_0 and k can be obtained from the data that give the population size at two different times. We are told that

$$P(0) = 20{,}000, \qquad P(5) = 400{,}000. \tag{3}$$

The first condition immediately implies that $P_0 = 20{,}000$, so

$$P(t) = 20{,}000 e^{kt}.$$

Using the second condition in formula (3), we have

$$20{,}000e^{k(5)} = P(5) = 400{,}000$$
$$e^{5k} = 20$$
$$5k = \ln 20$$
$$k = \frac{\ln 20}{5} \approx .60. \tag{4}$$

So we may take

$$P(t) = 20{,}000e^{.6t}.$$

This function is a mathematical model of the growth of the bacteria culture. (See Fig. 1.) ∎

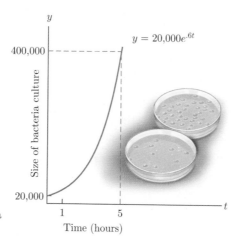

Figure 1. A model for bacteria growth.

EXAMPLE 2

Determining the growth constant A colony of fruit flies is growing according to the exponential law $P(t) = P_0 e^{kt}$, and the size of the colony doubles in 9 days. Determine the growth constant k.

Solution We do not know the initial size of the population at $t = 0$. However, we are told that $P(9) = 2P(0)$; that is,

$$P_0 e^{k(9)} = 2P_0$$
$$e^{9k} = 2$$
$$9k = \ln 2$$
$$k = \frac{\ln 2}{9} \approx .077. \quad ∎$$

The initial size P_0 of the population was not given in Example 2. But we were able to determine the growth constant because we were told the amount of time required for the colony to double in size. Thus the growth constant does not depend on the initial size of the population. This property is characteristic of exponential growth.

EXAMPLE 3   The initial size of the colony in Example 2 was 100.

(a) How large will the colony be after 41 days?

(b) How fast will the colony be growing at that time?

(c) At what time will the colony contain 800 fruit flies?

Solution (a) From Example 2 we have $P(t) = P_0 e^{.077t}$. Since $P(0) = 100$, we conclude that

$$P(t) = 100 e^{.077t}.$$

Therefore, after 41 days the size of the colony is

$$P(41) = 100 e^{.077(41)} = 100 e^{3.157} \approx 2350.$$

(b) Since the function satisfies the differential equation $y' = .077y$,

$$P'(t) = .077 P(t).$$

In particular, when $t = 41$,

$$P'(41) = .077 P(41) = .077 \cdot 2350 \approx 181.$$

Therefore, after 41 days the colony is growing at the rate of about 181 fruit flies per day.

(c) Use the formula for $P(t)$ and set $P(t) = 800$.

$$100 e^{.077t} = 800$$
$$e^{.077t} = 8$$
$$.077t = \ln 8$$
$$t = \frac{\ln 8}{.077} \approx 27 \text{ days.} \blacksquare$$

Exponential Decay An example of negative exponential growth, or *exponential decay*, is given by the disintegration of a radioactive element such as uranium 235. It is known that, at any instant, the rate at which a radioactive substance is decaying is proportional to the amount of the substance that has not yet disintegrated. If $P(t)$ is the quantity present at time t, then $P'(t)$ is the rate of decay. Of course, $P'(t)$ must be negative, since $P(t)$ is decreasing. Thus we may write $P'(t) = kP(t)$ for some negative constant k. To emphasize the fact that the constant is negative, k is often replaced by $-\lambda$, where λ is a positive constant.* Then $P(t)$ satisfies the differential equation

$$P'(t) = -\lambda P(t). \tag{5}$$

The general solution of (5) has the form

$$P(t) = P_0 e^{-\lambda t}$$

for some positive number P_0. We call such a function an *exponential decay function*. The constant λ is called the *decay constant*.

EXAMPLE 4   **Exponential decay** The decay constant for strontium 90 is $\lambda = .0244$, where time is measured in years. How long will it take for a quantity P_0 of strontium 90 to decay to one-half its original mass?

---

*λ is the Greek lowercase letter lambda.

Solution We have

$$P(t) = P_0 e^{-.0244t}.$$

Next, set $P(t)$ equal to $\frac{1}{2}P_0$ and solve for t:

$$P_0 e^{-.0244t} = \tfrac{1}{2}P_0$$
$$e^{-.0244t} = \tfrac{1}{2} = .5$$
$$-.0244t = \ln.5$$
$$t = \frac{\ln.5}{-.0244} \approx 28 \text{ years.} \qquad \blacksquare$$

The *half-life* of a radioactive element is the length of time required for a given quantity of that element to decay to one-half its original mass. Thus strontium 90 has a half-life of about 28 years. It takes 28 years for it to decay to half its original mass and another 28 years for it to decay to $\frac{1}{4}$ its original mass, another 28 years to decay to $\frac{1}{8}$, and so forth. (See Fig. 2.) Notice from Example 4 that the half-life does not depend on the initial amount P_0.

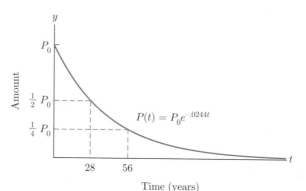

Figure 2. Half-life of radioactive strontium 90.

EXAMPLE 5

Half-life and decay constant Radioactive carbon 14 has a half-life of about 5730 years. Find its decay constant.

Solution If P_0 denotes the initial amount of carbon 14, the amount after t years will be

$$P(t) = P_0 e^{-\lambda t}.$$

After 5730 years, $P(t)$ will equal $\frac{1}{2}P_0$. That is,

$$P_0 e^{-\lambda(5730)} = P(5730) = \tfrac{1}{2}P_0 = .5P_0.$$

Solving for λ gives

$$e^{-5730\lambda} = .5$$
$$-5730\lambda = \ln.5$$
$$\lambda = \frac{\ln.5}{-5730} \approx .00012. \qquad \blacksquare$$

One problem connected with aboveground nuclear explosions is the radioactive debris, or fallout, that contaminates plants and grass, the food supply of animals. Strontium 90 is one of the most dangerous components of radioactive debris because it has a relatively long half-life. Also, it is chemically similar to calcium and is absorbed into the bone structure of animals (including humans) who eat contaminated food. Iodine 131 is also produced by nuclear explosions, but it presents less of a hazard because it has a half-life of 8 days.

EXAMPLE 6

Iodine level in dairy products If dairy cows eat hay containing too much iodine 131, their milk will be unfit to drink. If the hay contains 10 times the maximum allowable level of iodine 131, how many days should the hay be stored before it is fed to dairy cows?

Solution Let P_0 be the amount of iodine 131 present in the hay. Then the amount at time t is $P(t) = P_0 e^{-\lambda t}$ (t in days). The half-life of iodine 131 is 8 days, so

$$P_0 e^{-\lambda(8)} = .5 P_0$$
$$e^{-8\lambda} = .5$$
$$-8\lambda = \ln .5$$
$$\lambda = \frac{\ln .5}{-8} \approx .087$$

and

$$P(t) = P_0 e^{-.087t}.$$

Now that we have the formula for $P(t)$ we want to find t such that $P(t) = \frac{1}{10} P_0$. We have

$$P_0 e^{-.087t} = .1 P_0,$$

so

$$e^{-.087t} = .1$$
$$-.087t = \ln .1$$
$$t = \frac{\ln .1}{-.087} \approx 26 \text{ days.} \qquad \blacksquare$$

Radiocarbon Dating Knowledge about radioactive decay is valuable to archeologists and anthropologists who want to estimate the age of objects belonging to ancient civilizations. Several different substances are useful for radioactive-dating techniques; the most common is radiocarbon, ^{14}C. Carbon 14 is produced in the upper atmosphere when cosmic rays react with atmospheric nitrogen. Because the ^{14}C eventually decays, the concentration of ^{14}C cannot rise above certain levels. An equilibrium is reached where ^{14}C is produced at the same rate as it decays. Scientists usually assume that the total amount of ^{14}C in the biosphere has remained constant over the past 50,000 years. Consequently, it is assumed that the *ratio* of ^{14}C to ordinary nonradioactive carbon 12, ^{12}C, has been constant during this same period. (The ratio is about one part ^{14}C to 10^{12} parts of ^{12}C.) Both ^{14}C and ^{12}C are in the atmosphere as constituents of carbon dioxide. All living vegetation and most forms of animal life contain ^{14}C and ^{12}C in the same proportion as the atmosphere because plants absorb carbon dioxide through photosynthesis. The ^{14}C and ^{12}C in plants are distributed through the food chain to almost all animal life.

When an organism dies, it stops replacing its carbon; therefore the amount of ^{14}C begins to decrease through radioactive decay, but the ^{12}C in the dead organism remains constant. The ratio of ^{14}C to ^{12}C can be later measured to determine when the organism died. (See Fig. 3.)

EXAMPLE 7

Carbon dating A parchment fragment was discovered that had about 80% of the ^{14}C level found today in living matter. Estimate the age of the parchment.

Solution We assume that the original ^{14}C level in the parchment was the same as the level in living organisms today. Consequently, about eight-tenths of the original ^{14}C

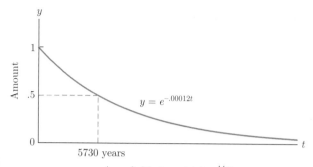

Figure 3. ^{14}C–^{12}C ratio compared to the ratio in living plants.

Age of object containing ^{14}C

remains. From Example 5 we obtain the formula for the amount of ^{14}C present t years after the parchment was made from an animal skin:

$$P(t) = P_0 e^{-.00012t},$$

where $P_0 = $ initial amount. We want to find t such that $P(t) = .8P_0$.

$$P_0 e^{-.00012t} = .8P_0$$
$$e^{-.00012t} = .8$$
$$-.00012t = \ln .8$$
$$t = \frac{\ln .8}{-.00012} \approx 1860 \text{ years old.} \qquad \blacksquare$$

A Sales Decay Curve If advertising and other promotions of a particular product are stopped and other market conditions remain fairly constant, marketing studies* have demonstrated that, at any time t, sales of the product will be declining at a rate proportional to the amount of current sales at time t. (See Fig. 4.) If S_0 is the number of sales in the last month during which advertising occurred and if $S(t)$ is the monthly level of sales t months after the cessation of promotional effort, a good mathematical model for $S(t)$ is

Figure 4. Exponential decay of sales.

$$S(t) = S_0 e^{-\lambda t},$$

where λ is a positive number called the *sales decay constant*. The value of λ depends on many factors, such as type of product, number of years of prior advertising, number of competing products, and other characteristics of the market.

The Time Constant Consider an exponential decay function $y = Ce^{-\lambda t}$. Figure 5 shows the tangent line to the decay curve when $t = 0$. The slope there is the initial rate of decay. If the decay process were to continue at this rate, the decay curve would follow the tangent line and y would be zero at some time T. This time is called the *time constant* of the decay curve. It can be shown (see Exercise 30) that $T = 1/\lambda$ for the curve $y = Ce^{-\lambda t}$. Thus $\lambda = 1/T$ and the decay curve can be written in the form

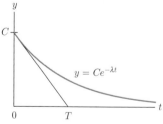

Figure 5. The time constant T in exponential decay: $T = 1/\lambda$.

$$y = Ce^{-t/T}.$$

If we have experimental data that tend to lie along an exponential decay curve, the numerical constants for the curve may be obtained from Fig. 5. First, sketch the curve and estimate the y-intercept, C. Then sketch an approximate tangent line

---

*M. Vidale and H. Wolfe, "An Operations-Research Study of Sales Response to Advertising," *Operations Research*, 5 (1957), 370–381. Reprinted in F. Bass et al., *Mathematical Models and Methods in Marketing* (Homewood, Ill.: Richard D. Irwin, Inc., 1961).

and from this estimate the time constant, T. This procedure is sometimes used in biology and medicine.

Practice Problems 5.1

1. **(a)** Solve the differential equation $P'(t) = -.6P(t)$, $P(0) = 50$.

   (b) Solve the differential equation $P'(t) = kP(t)$, $P(0) = 4000$, where k is some constant.

   (c) Find the value of k in part (b) for which $P(2) = 100P(0)$.

2. Under ideal conditions a colony of *Escherichia coli* bacteria can grow by a factor of 100 every 2 hours. If 4000 bacteria are present initially , how long will it take before there are 1,000,000 bacteria?

EXERCISES 5.1

1. **Population Growth** Let $P(t)$ be the population (in millions) of a certain city t years after 1990, and suppose that $P(t)$ satisfies the differential equation

   $$P'(t) = .02P(t), \qquad P(0) = 3.$$

   (a) Find the formula for $P(t)$.

   (b) What was the initial population, that is, the population in 1990?

   (c) What is the growth constant?

   (d) What was the population in 1998?

   (e) Use the differential equation to determine how fast the population is growing when it reaches 4 million people.

   (f) How large is the population when it is growing at the rate of 70,000 people per year?

2. **Growth of Bacteria** Approximately 10,000 bacteria are placed in a culture. Let $P(t)$ be the number of bacteria present in the culture after t hours, and suppose that $P(t)$ satisfies the differential equation

   $$P'(t) = .55P(t).$$

   (a) What is $P(0)$?

   (b) Find the formula for $P(t)$.

   (c) How many bacteria are there after 5 hours?

   (d) What is the growth constant?

   (e) Use the differential equation to determine how fast the bacteria culture is growing when it reaches 100,000.

   (f) What is the size of the bacteria culture when it is growing at a rate of 34,000 bacteria per hour?

3. **Growth of Cells** After t hours there are $P(t)$ cells present in a culture, where $P(t) = 5000e^{.2t}$.

   (a) How many cells were present initially?

   (b) Give a differential equation satisfied by $P(t)$.

   (c) When will the population double?

   (d) When will 20,000 cells be present?

4. **Insect Population** The size of a certain insect population is given by $P(t) = 300e^{.01t}$, where t is measured in days.

   (a) How many insects were present initially?

   (b) Give a differential equation satisfied by $P(t)$.

   (c) At what time will the population double?

   (d) At what time will the population equal 1200?

5. Determine the growth constant of a population that is growing at a rate proportional to its size, where the population doubles in size every 40 days.

6. Determine the growth constant of a population that is growing at a rate proportional to its size, where the population triples in size every 10 years.

7. A population is growing exponentially with growth constant .05. In how many years will the current population triple?

8. A population is growing exponentially with growth constant .04. In how many years will the current population double?

9. The rate of growth of a certain cell culture is proportional to its size. In 10 hours a population of 1 million cells grew to 9 million. How large will the cell culture be after 15 hours?

10. **World's Population** The world's population was 5.51 billion on January 1, 1993, and 5.88 billion on January 1, 1998. Assume that at any time the population grows at a rate proportional to the population at that time. In what year will the world's population reach 7 billion?

11. **Population of Mexico City** At the beginning of 1990, 20.2 million people lived in the metropolitan area of Mexico City, and the population was growing exponentially. The 1995 population was 23 million. (Part of the growth is due to immigration.) If this trend continues, how large will the population be in the year 2010?

12. The population (in millions) of a state t years after 1970 is given by the graph of the exponential function $y = P(t)$ with growth constant .025 in Fig. 6. [In parts (c) and (d) use the differential equation satisfied by $P(t)$.]

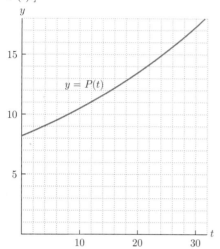

Figure 6

(a) What was the population in 1974?

(b) When was the population 10 million?

(c) How fast was the population growing in 1974?

(d) When is the population growing at the rate of 275,000 people per year?

13. Radioactive Decay A sample of 8 grams of radioactive material is placed in a vault. Let $P(t)$ be the amount remaining after t years, and let $P(t)$ satisfy the differential equation $P'(t) = -.021P(t)$.

(a) Find the formula for $P(t)$.

(b) What is $P(0)$?

(c) What is the decay constant?

(d) How much of the material will remain after 10 years?

(e) Use the differential equation to determine how fast the sample is disintegrating when just 1 gram remains.

(f) What amount of radioactive material remains when it is disintegrating at the rate of .105 gram per year?

(g) The radioactive material has a half-life of 33 years. How much will remain after 33 years? 66 years? 99 years?

14. Radioactive Decay Radium 226 is used in cancer radiotherapy, as a neutron source for some research purposes, and as a constituent of luminescent paints. Let $P(t)$ be the number of grams of radium 226 in a sample remaining after t years, and let $P(t)$ satisfy the differential equation

$$P'(t) = -.00043P(t), \qquad P(0) = 12.$$

(a) Find the formula for $P(t)$.

(b) What was the initial amount?

(c) What is the decay constant?

(d) Approximately how much of the radium will remain after 943 years?

(e) How fast is the sample disintegrating when just 1 gram remains? Use the differential equation.

(f) What is the weight of the sample when it is disintegrating at the rate of .004 gram per year?

(g) The radioactive material has a half-life of about 1612 years. How much will remain after 1612 years? 3224 years? 4836 years?

15. Decay of Penicillin in the Bloodstream A person is given an injection of 300 milligrams of penicillin at time $t = 0$. Let $f(t)$ be the amount (in milligrams) of penicillin present in the person's bloodstream t hours after the injection. Then the amount of penicillin decays exponentially and a typical formula is $f(t) = 300e^{-.6t}$.

(a) Give the differential equation satisfied by $f(t)$.

(b) How much will remain at time $t = 5$ hours?

(c) What is the biological half-life of the penicillin (that is, the time required for half of a given amount to decompose) in this case?

16. Radioactive Decay Ten grams of a radioactive substance with decay constant .04 is stored in a vault. Assume that time is measured in days, and let $P(t)$ be the amount remaining at time t.

(a) Give the formula for $P(t)$.

(b) Give the differential equation satisfied by $P(t)$.

(c) How much will remain after 5 days?

(d) What is the half-life of this radioactive substance?

17. The decay constant for the radioactive element cesium 137 is .023 when time is measured in years. Find its half-life.

18. Radioactive cobalt 60 has a half-life of 5.3 years. Find its decay constant.

19. A sample of radioactive material disintegrates from 5 to 2 grams in 100 days. After how many days will just 1 gram remain?

20. Ten grams of a radioactive material disintegrates to 3 grams in 5 years. What is the half-life of the radioactive material?

21. Decay of Sulfate in the Bloodstream In an animal hospital, 8 units of a sulfate were injected into a dog. After 50 minutes, only 4 units remained in the dog. Let $f(t)$ be the amount of sulfate present after t minutes. At any time, the rate of change of $f(t)$ is proportional to the value of $f(t)$. Find the formula for $f(t)$.

22. Forty grams of a certain radioactive material disintegrates to 16 grams in 220 years. How much of this material is left after 300 years?

23. Radioactive Decay A sample of radioactive material decays over time (measured in hours) with decay constant .2. The graph of the exponential function $y = P(t)$ in Fig. 7 gives the number of grams remaining after t hours. [*Hint*: In parts (c) and (d) use the differential equation satisfied by $P(t)$.]

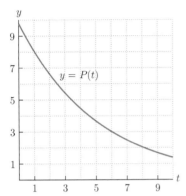

Figure 7

(a) How much was remaining after 1 hour?

(b) Approximate the half-life of the material.

(c) How fast was the sample decaying after 6 hours?

(d) When was the sample decaying at the rate of .4 gram per hour?

24. A sample of radioactive material has decay constant .25, where time is measured in hours. How fast will the sample be disintegrating when the sample size is 8 grams? For what sample size will the sample size be decreasing at the rate of 2 grams per day?

25. Carbon Dating In 1947, a cave with beautiful prehistoric wall paintings was discovered in Lascaux, France. Some charcoal found in the cave contained 20% of the ^{14}C expected in living trees. How old are the Lascaux cave paintings? (Recall that the decay constant for ^{14}C is .00012.)

26. King Arthur's Round Table According to legend, in the fifth century King Arthur and his knights sat at a huge round table. A round table alleged to have belonged to King Arthur was found at Winchester Castle in England. In 1976, carbon dating revealed the amount of radiocarbon in the table to be 91% of the radiocarbon present in living wood. Could the table possibly have belonged to King Arthur? Why? (Recall that the decay constant for ^{14}C is .00012.)

27. A 4500-year-old wooden chest was found in the tomb of the twenty-fifth century B.C. Chaldean king Meskalumdug of Ur. What percentage of the original ^{14}C would you expect to find in the wooden chest?

28. Population of the Pacific Northwest In 1938, sandals woven from strands of tree bark were found in Fort Rock Creek Cave in Oregon. The bark contained 34% of the level of ^{14}C found in living bark. Approximately how old were the sandals? [*Note*: This discovery by University of Oregon anthropologist Luther Cressman forced scientists to double their estimate of how long ago people came to the Pacific Northwest.]

29. Time of the Fourth Ice Age Many scientists believe there have been four ice ages in the past 1 million years. Before the technique of carbon dating was known, geologists erroneously believed that the retreat of the Fourth Ice Age began about 25,000 years ago. In 1950, logs from ancient spruce trees were found under glacial debris near Two Creeks, Wisconsin. Geologists determined that these trees had been crushed by the advance of ice during the Fourth Ice Age. Wood from the spruce trees contained 27% of the level of ^{14}C found in living trees. Approximately how long ago did the Fourth Ice Age actually occur?

30. Time Constant Let T be the time constant of the curve $y = Ce^{-\lambda t}$ as defined in Fig. 5. Show that $T = 1/\lambda$. [*Hint*: Express the slope of the tangent line in Fig. 5 in terms of C and T. Then set this slope equal to the slope of the curve $y = Ce^{-\lambda t}$ at $t = 0$.]

31. The amount in grams of a certain radioactive material present after t years is given by the function $P(t)$. Match each of the following answers with its corresponding question.

Answers

a. Solve $P(t) = .5P(0)$ for t.

b. Solve $P(t) = .5$ for t.

c. $P(.5)$

d. $P'(.5)$

e. $P(0)$

f. Solve $P'(t) = -.5$ for t.

g. $y' = ky$

h. $P_0 e^{kt}$, $k < 0$

Questions

A. Give a differential equation satisfied by $P(t)$.

B. How fast will the radioactive material be disintegrating in $\frac{1}{2}$ year?

C. Give the general form of the function $P(t)$.

D. Find the half-life of the radioactive material.

E. How many grams of the material will remain after $\frac{1}{2}$ year?

F. When will the radioactive material be disintegrating at the rate of $\frac{1}{2}$ gram per year?

G. When will there be $\frac{1}{2}$ gram remaining?

H. How much radioactive material was present initially?

32. **Time Constant and Half-life** Consider an exponential decay function $P(t) = P_0 e^{-\lambda t}$, and let T denote its time constant. Show that at $t = T$ the function $P(t)$ decays to about one-third of its initial size. Conclude that the time constant is always larger than the half-life.

33. Suppose that the function $P(t)$ satisfies the differential equation

$$y'(t) = -.5y(t), \quad y(0) = 10.$$

   (a) Find an equation of the tangent line to the graph

of $y = P(t)$ at $t = 0$. [*Hint*: What are $P'(0)$ and $P(0)$?]

   (b) Find $P(t)$.

   (c) What is the time constant of the decay curve $y = P(t)$?

34. **Time to Finish** Consider the exponential decay function $y = P_0 e^{-\lambda t}$ with time constant T. We define the time to finish to be the time it takes for the function to decay to about 1% of its initial value P_0. Show that the time to finish is about four times the time constant T.

Solutions to Practice Problems 5.1

1. (a) Answer: $P(t) = 50e^{-.6t}$. Differential equations of the type $y' = ky$ have as their solution $P(t) = Ce^{kt}$, where C is $P(0)$.

   (b) Answer: $P(t) = 4000e^{kt}$. This problem is like part (a) except that the constant is not specified. Additional information is needed if we want to determine a specific value for k.

   (c) Answer: $P(t) = 4000e^{2.3t}$. From the solution to part (b) we know that $P(t) = 4000e^{kt}$. We are given that $P(2) = 100P(0) = 100(4000) = 400,000$. So

$$P(2) = 4000e^{k(2)} = 400,000$$
$$e^{2k} = 100$$
$$2k = \ln 100$$
$$k = \frac{\ln 100}{2} \approx 2.3.$$

2. Let $P(t)$ be the number of bacteria present after t

hours. We must first find an expression for $P(t)$ and then determine the value of t for which $P(t) = 1,000,000$. From the discussion at the beginning of the section, we know that $P'(t) = k \cdot P(t)$. Also, we are given that $P(2)$ (the population after 2 hours) is $100P(0)$ (100 times the initial population). From Problem 1(c) we have an expression for $P(t)$:

$$P(t) = 4000e^{2.3t}.$$

Now we must solve $P(t) = 1,000,000$ for t.

$$4000e^{2.3t} = 1,000,000$$
$$e^{2.3t} = 250$$
$$2.3t = \ln 250$$
$$t = \frac{\ln 250}{2.3} \approx 2.4.$$

Therefore, after 2.4 hours there will be 1,000,000 bacteria.

5.2 Compound Interest

In Section 0.5, we introduced the notion of compound interest. Recall the fundamental formula developed there. If a principal amount P is compounded m times per year at an annual rate of interest r for t years, the compound amount A, the balance at the end of time t, is given by the formula

$$A = P\left(1 + \frac{r}{m}\right)^{mt}.$$

For instance, suppose that $1000 is invested at 6% interest for 1 year. In the formula, this corresponds to $P = \$1000$, $m = 1$, $r = .06$, and $t = 1$ year. If interest is compounded annually, the amount in the account at the end of the year is

$$A = P(1 + r)^1 = 1000(1 + .06) = \$1060.00.$$

If interest is compounded quarterly, the amount in the account at the end of the year is

$$A = P\left(1 + \frac{r}{4}\right)^{4t} = 1000\left(1 + \frac{.06}{4}\right)^4 \approx \$1061.36.$$

Table 1 contains these results along with those for monthly and daily compounding for 1 year.

TABLE 1 Effect of Increased Compounding Periods

Frequency of compounding	Annually	Quarterly	Monthly	Daily
m	1	4	12	360
Balance after 1 year ($)	1060.00	1061.36	1061.68	1061.83

Would the account balance in Table 1 be much more if the interest were compounded every hour? Every minute? It turns out that no matter how frequently interest is compounded the balance will not exceed 1061.84 (to the nearest cent). To understand why, we connect the concept of compound interest to the exponential function e^{rt}. Start from the formula for the compound amount and write it in the form

$$A = P\left(1 + \frac{r}{m}\right)^{mt} = P\left(1 + \frac{r}{m}\right)^{(m/r)\cdot rt}.$$

If we set $h = r/m$, then $1/h = m/r$, and

$$A = P(1 + h)^{(1/h)\cdot rt}.$$

As the frequency of compounding is increased, m gets large and $h = r/m$ approaches 0. To determine what happens to the compound amount, we must therefore examine the limit:

$$\lim_{h \to 0} P(1 + h)^{(1/h)\cdot rt}.$$

The following remarkable fact is proved in the appendix at the end of this section:

$$\lim_{h \to 0}(1 + h)^{1/h} = e.$$

Using this fact together with two limit theorems, we have

$$\lim_{h \to 0} P(1 + h)^{(1/h)rt} = P\left[\lim_{h \to 0}(1 + h)^{1/h}\right]^{rt} = Pe^{rt}.$$

These calculations show that the compound amount calculated from the formula

$$P\left(1 + \frac{r}{m}\right)^{mt}$$

gets closer to Pe^{rt} as the number m of interest periods per year is increased. When the formula

$$A = Pe^{rt} \tag{1}$$

is used to calculate the compound amount, we say that the interest is *compounded continuously*.

When $1000 is invested for 1 year at 6% interest compounded continuously, $P = 1000$, $r = .06$, and $t = 1$. The compound amount is

$$1000e^{.06(1)} \approx \$1061.84.$$

This is only 1 cent more than the result of the daily compounding shown in Table 1. Consequently, frequent compounding (such as every hour or every second) will produce at most 1 cent more.

In many computations, it is simpler to use the formula for interest compounded continuously than the formula for ordinary compound interest. In these instances, it is commonplace to use interest compounded continuously as an approximation to ordinary compound interest.

When interest is compounded continuously, the compound amount $A(t)$ is an exponential function of the number of years t that interest is earned: $A(t) = Pe^{rt}$. Hence $A(t)$ satisfies the differential equation

$$\frac{dA}{dt} = rA.$$

The rate of growth of the compound amount is proportional to the amount of money present. Since the growth comes from the interest, we conclude that under continuous compounding interest is earned continuously at a rate of growth proportional to the amount of money present.

The formula $A = Pe^{rt}$ contains four variables. (Remember that the letter e represents a specific constant, $e = 2.718\ldots$.) In a typical problem, we are given values for three of these variables and must solve for the remaining variable.

EXAMPLE 1

Compound interest One thousand dollars is invested at 5% interest compounded continuously.

(a) Give the formula for $A(t)$, the compound amount after t years.

(b) How much will be in the account after 6 years?

(c) After 6 years, at what rate will $A(t)$ be growing?

(d) How long is required for the initial investment to double?

Solution (a) $P = 1000$ and $r = .05$. By formula (1), $A(t) = 1000e^{.05t}$.

(b) $A(6) = 1000e^{.05(6)} = 1000e^{.3} \approx \$1349.86$.

(c) Rate of growth is different from interest rate. Interest rate is fixed at 5% and does not change with time. On the other hand, the rate of growth $A'(t)$ is always changing. Since $A(t) = 1000e^{.05t}$, $A'(t) = 1000 \cdot .05e^{.05t} = 50e^{.05t}$.

$$A'(6) = 50e^{.05(6)} = 50e^{.3} \approx \$67.49.$$

After 6 years the investment is growing at the rate of \$67.49 per year.

There is an easier way to answer part (c). Since $A(t)$ satisfies the differential equation $A'(t) = rA(t)$,

$$A'(6) = .05A(6) = .05 \cdot 1349.86 \approx \$67.49.$$

(d) We must find t such that $A(t) = \$2000$. So we set $1000e^{.05t} = 2000$ and solve for t.

$$1000e^{.05t} = 2000$$

$$e^{.05t} = 2$$

$$\ln e^{.05t} = \ln 2$$

$$.05t = \ln 2$$

$$t = \frac{\ln 2}{.05} \approx 13.86 \text{ years}$$

■

REMARK ▶ The calculations in Example 1(d) would be essentially unchanged after the first step if the initial amount of the investment were changed from \$1000 to any arbitrary amount P. When this investment doubles, the compound amount will be $2P$. So we set $2P = Pe^{.05t}$ and solve for t as we did previously to conclude that, at 5% interest compounded continuously, any amount doubles in about 13.86 years. ■

EXAMPLE 2

Appreciation of a painting Pablo Picasso's "The Dream" was purchased in 1941 for a war-distressed price of $7000. The painting was sold in 1997 for $48.4 million, the second highest price ever paid for a Picasso painting at auction. What rate of interest compounded continuously did this investment earn?

Solution Let Pe^{rt} be the value (in millions) of the painting t years after 1941. Since the initial value is .007 million, $P = .007$. Since the value after 56 years is 48.4 million dollars, $.007e^{r(56)} = 48.4$. Divide both sides of the equation by .007, take the logarithms of both sides, and then solve for r.

$$e^{r(56)} = \frac{48.4}{.007} \approx 6914.29$$

$$r(56) = \ln 6914.29$$

$$r = \frac{\ln 6914.29}{56} \approx .158.$$

Therefore, as an investment, the painting earned an interest rate of 15.8%. ■

If P dollars are invested today, the formula $A = Pe^{rt}$ gives the value of this investment after t years (assuming continuously compounded interest). We say that P is the *present value* of the amount A to be received in t years. If we solve for P in terms of A, we obtain

$$P = Ae^{-rt}. \qquad (2)$$

The concept of the present value of money is an important theoretical tool in business and economics. Problems involving depreciation of equipment, for example, may be analyzed by calculus techniques when the present value of money is computed from (2) using continuously compounded interest.

EXAMPLE 3

Find the present value of $5000 to be received in 2 years if money can be invested at 12% compounded continuously.

Solution Use formula (2) with $A = 5000$, $r = .12$, and $t = 2$.

$$P = 5000e^{-(.12)(2)} = 5000e^{-.24}$$

$$\approx \$3933.14.$$ ■

APPENDIX A Limit Formula for *e*

For $h \neq 0$, we have
$$\ln(1 + h)^{1/h} = (1/h)\ln(1 + h).$$

Taking the exponential of both sides, we find that
$$(1 + h)^{1/h} = e^{(1/h)\ln(1+h)}.$$

Since the exponential function is continuous,

$$\lim_{h \to 0}(1 + h)^{1/h} = e^{\left[\lim_{h \to 0}(1/h)\ln(1 + h)\right]}. \tag{3}$$

To examine the limit inside the exponential function, we note that $\ln 1 = 0$ and hence

$$\lim_{h \to 0}\left(\frac{1}{h}\right)\ln(1 + h) = \lim_{h \to 0}\frac{\ln(1 + h) - \ln 1}{h}.$$

The limit on the right is a difference quotient of the type used to compute a derivative. In fact,

$$\lim_{h \to 0}\frac{\ln(1 + h) - \ln 1}{h} = \frac{d}{dx}\ln x \bigg|_{x=1} = \frac{1}{x}\bigg|_{x=1} = 1.$$

Thus the limit inside the exponential function in (3) is 1. That is,

$$\lim_{h \to 0}(1 + h)^{1/h} = e^{[1]} = e.$$

Practice Problems 5.2

1. One thousand dollars is to be invested in a bank for 4 years. Would 8% interest compounded semiannually be better than $7\frac{3}{4}\%$ interest compounded continuously?

2. A building was bought for $150,000 and sold 10 years later for $400,000. What interest rate (compounded continuously) was earned on the investment?

EXERCISES 5.2

1. Savings Account Let $A(t) = 5000e^{.04t}$ be the balance in a savings account after t years.

(a) How much money was originally deposited?

(b) What is the interest rate?

(c) How much money will be in the account after 10 years?

(d) What differential equation is satisfied by $y = A(t)$?

(e) Use the results of parts (c) and (d) to determine how fast the balance is growing after 10 years.

(f) How large will the balance be when it is growing at the rate of $280 per year?

2. Savings Account Let $A(t)$ be the balance in a savings account after t years, and suppose that $A(t)$ satisfies the differential equation

$$A'(t) = .045A(t), \qquad A(0) = 3000.$$

(a) How much money was originally deposited into the account?

(b) What interest rate is being earned?

(c) Find the formula for $A(t)$.

(d) What is the balance after 5 years?

(e) Use part (d) and the differential equation to determine how fast the balance is growing after 5 years.

(f) How large will the balance be when it is growing at the rate of $270 per year?

3. Savings Account Four thousand dollars is deposited into a savings account at 3.5% interest compounded continuously.

(a) What is the formula for $A(t)$, the balance after t years?

(b) What differential equation is satisfied by $A(t)$, the balance after t years?

(c) How much money will be in the account after 2 years?

(d) When will the balance reach $5000?

(e) How fast is the balance growing when it reaches $5000?

4. Savings Account Ten thousand dollars is deposited into a savings account at 4.6% interest compounded continuously.

(a) What differential equation is satisfied by $A(t)$, the balance after t years?

(b) What is the formula for $A(t)$?

(c) How much money will be in the account after 3 years?

(d) When will the balance triple?

(e) How fast is the balance growing when it triples?

5. **Investment Analysis** An investment earns 4.2% interest compounded continuously. How fast is the investment growing when its value is $9000?

6. **Investment Analysis** An investment earns 5.1% interest compounded continuously and is currently growing at the rate of $765 per year. What is the current value of the investment?

7. **Continuous Compound** One thousand dollars is deposited in a savings account at 6% interest compounded continuously. How many years are required for the balance in the account to reach $2500?

8. **Continuous Compound** Ten thousand dollars is invested at 6.5% interest compounded continuously. When will the investment be worth $41,787?

9. **Technology Stock** One hundred shares of a technology stock were purchased on January 2, 1990, for $1200 and sold on January 2, 1998, for $12,500. What rate of interest compounded continuously did this investment earn?

10. **Appreciation of Art Work** Pablo Picasso's "Angel Fernandez de Soto" was acquired in 1946 for a postwar splurge of $22,220. The painting was sold in 1995 for $29.1 million. What rate of interest compounded continuously did this investment earn?

11. **Investment Analysis** How many years are required for an investment to double in value if it is appreciating at the rate of 4% compounded continuously?

12. What interest rate (compounded continuously) is earned by an investment that doubles in 10 years?

13. If an investment triples in 15 years, what interest rate (compounded continuously) does the investment earn?

14. **Real Estate Investment** If real estate in a certain city appreciates at the rate of 15% compounded continuously, when will a building purchased in 1998 triple in value?

15. **Real Estate Investment** In a certain town, property values tripled from 1980 to 1995. If this trend continues, when will property values be at five times their 1980 level? (Use an exponential model for the property value at time t.)

16. **Real Estate Investment** A lot purchased in 1980 for $5000 was appraised at $60,000 in 1998. If the lot continues to appreciate at the same rate, when will it be worth $100,000?

17. **Real Estate Investment** A farm purchased in 1985 for $1,000,000 was valued at $3,000,000 in 1995. If the farm continues to appreciate at the same rate (with continuous compounding), when will it be worth $10,000,000?

18. **Real Estate Investment** A parcel of land bought in 1990 for $10,000 was worth $16,000 in 1995. If the land continues to appreciate at this rate, in what year will it be worth $45,000?

19. **Present Value** Find the present value of $1000 payable at the end of 3 years, if money may be invested at 8% with interest compounded continuously.

20. **Present Value** Find the present value of $2000 to be received in 10 years, if money may be invested at 8% with interest compounded continuously.

21. **Present Value** How much money must you invest now at 4.5% interest compounded continuously to have $10,000 at the end of 5 years?

22. **Present Value** If the present value of $1000 to be received in 5 years is $559.90, what rate of interest, compounded continuously, was used to compute this present value?

23. Investment A is currently worth $70,200 and is growing at the rate of 13% per year compounded continuously. Investment B is currently worth $60,000 and is growing at the rate of 14% per year compounded continuously. After how many years will the two investments have the same value?

24. Ten thousand dollars is deposited into a money market fund paying 8% interest compounded continuously. How much interest will be earned during the second year of the investment?

25. A small amount of money is deposited in a savings account with interest compounded continuously. Let $A(t)$ be the balance in the account after t years. Match each of the following answers with its corresponding question.

Answers

a. Pe^{rt} **b.** $A(3)$ **c.** $A(0)$ **d.** $A'(3)$

e. Solve $A'(t) = 3$ for t.

f. Solve $A(t) = 3$ for t.

g. $y' = ry$

h. Solve $A(t) = 3A(0)$ for t.

Questions

A. How fast will the balance be growing in 3 years?

B. Give the general form of the function $A(t)$.

C. How long will it take for the initial deposit to triple?

D. Find the balance after 3 years.

E. When will the balance be 3 dollars?

F. When will the balance be growing at the rate of 3 dollars per year?

G. What was the principal amount?

H. Give a differential equation satisfied by $A(t)$.

26. The curve in Fig. 1 shows the growth of money in a savings account with interest compounded continuously.

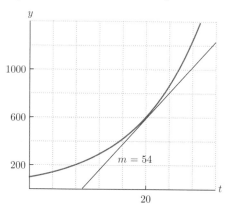

Figure 1. Growth of money in a savings account.

(a) What is the balance after 20 years?

(b) At what rate is the money growing after 20 years?

(c) Use the answers to parts (a) and (b) to determine the interest rate.

27. The function $A(t)$ in Fig. 2(a) gives the balance in a savings account after t years with interest compounded continuously. Figure 2(b) shows the derivative of $A(t)$.

(a) What is the balance after 20 years?

(b) How fast is the balance increasing after 20 years?

(c) Use the answers to parts (a) and (b) to determine the interest rate.

(d) When is the balance $300?

(e) When is the balance increasing at the rate of $12 per year?

(f) Why do the graphs of $A(t)$ and $A'(t)$ look the same?

28. When $1000 is invested at $r\%$ interest (compounded continuously) for 10 years, the balance is $f(r)$ dollars, where f is the function shown in Fig. 3.

(a) What will the balance be at 7% interest?

(b) For what interest rate will the balance be $3000?

(c) If the interest rate is 9%, what is the growth rate of the balance with respect to a unit increase in interest?

Technology Exercises

29. Convince yourself that $\lim_{h \to 0}(1+h)^{1/h} = e$ by graphing $f(x) = (1+x)^{1/x}$ near $x = 0$ and examining values of $f(x)$ for x near 0. For instance, use a window such as $[-.5, .5]$ by $[-.5, 4]$.

30. Convince yourself that daily compounding is nearly the same as continuous compounding by graphing $y = 100[1+(.05/360)]^{360x}$ together with $y = 100e^{.05x}$ in the window $[0, 64]$ by $[250, 2500]$. The two graphs should appear the same on the screen. Approximately how far apart are they when $x = 32$? When $x = 64$?

31. Internal Rate of Return An investment of $2000 yields payments of $1200 in 3 years, $800 in 4 years, and $500 in 5 years. Thereafter, the investment is worthless. What constant rate of return r would the investment need to produce to yield the payments specified? The number r is called the *internal rate of return* on the investment. We can consider the investment as consisting of three parts, each part yielding one payment. The sum of the present values of the three parts must total $2000. This yields the equation

$$2000 = 1200e^{-3r} + 800e^{-4r} + 500e^{-5r}.$$

Solve this equation to find the value of r.

Figure 2

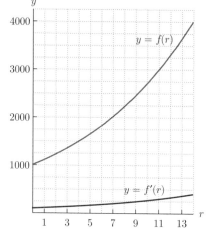

Figure 3. Effect of interest rate on balance.

1. Let us compute the balance after 4 years for each type of interest.

8% *compounded semiannually:* Use the formula at the beginning of this section. Here $P = 1000$, $r = .08$, $m = 2$ (semiannually means there are two interest periods per year), and $t = 4$. Therefore,

$$A = 1000 \left(1 + \frac{.08}{2}\right)^{2\cdot4} = 1000(1.04)^8 \approx \$1368.57.$$

$7\frac{3}{4}\%$ *compounded continuously:* Use the formula $A = Pe^{rt}$, where $P = 1000$, $r = .0775$, and $t = 4$. Then

$$A = 1000e^{(.0775)\cdot4} = 1000e^{.31} \approx \$1363.43.$$

Therefore, 8% compounded semiannually is better.

2. If the \$150,000 had been compounded continuously for 10 years at interest rate r, the balance would be $150{,}000e^{r\cdot10}$. The question is at what value of r will the balance be \$400,000? We need to solve an equation for r.

$$150{,}000e^{r\cdot10} = 400{,}000$$
$$e^{r\cdot10} \approx 2.67$$
$$r\cdot10 = \ln 2.67$$
$$r = \frac{\ln 2.67}{10} \approx .098.$$

Therefore, the investment earned 9.8% interest per year.

5.3 Applications of the Natural Logarithm Function to Economics

In this section we consider two applications of the natural logarithm to the field of economics. Our first application is concerned with relative rates of change and the second with elasticity of demand.

Relative Rates of Change   The *logarithmic derivative* of a function $f(t)$ is defined by the equation

$$\frac{d}{dt}\ln f(t) = \frac{f'(t)}{f(t)}. \tag{1}$$

The quantity on either side of equation (1) is often called the *relative rate of change of $f(t)$ per unit change of t*. Indeed, this quantity compares the rate of change of $f(t)$ [that is, $f'(t)$] with $f(t)$ itself. The *percentage rate of change* is the relative rate of change of $f(t)$ expressed as a percentage.

A simple example will illustrate these concepts. Suppose that $f(t)$ denotes the average price per pound of sirloin steak at time t and that $g(t)$ denotes the average price of a new car (of a given make and model) at time t, where $f(t)$ and $g(t)$ are given in dollars and time is measured in years. Then the ordinary derivatives $f'(t)$ and $g'(t)$ may be interpreted as the rate of change of the price of a pound of sirloin steak and of a new car, respectively, where both are measured in dollars per year. Suppose that, at a given time t_0, we have $f(t_0) = \$5.25$ and $g(t_0) = \$12{,}000$. Moreover, $f'(t_0) = \$.75$ and $g'(t_0) = \$1500$. Then, at time t_0, the price per pound of steak is increasing at a rate of \$.75 per year, while the price of a new car is increasing at a rate of \$1500 per year. Which price is increasing more quickly? It is not meaningful to say that the car price is increasing faster simply because \$1500 is larger than \$.75. We must take into account the vast difference between the actual cost of a car and the cost of steak. The usual basis of comparison of price increases is the percentage rate of increase. In other words, at $t = t_0$, the price of sirloin steak is increasing at the percentage rate

$$\frac{f'(t_0)}{f(t_0)} = \frac{.75}{5.25} \approx .143 = 14.3\%$$

per year, but at the same time the price of a new car is increasing at the percentage rate

$$\frac{g'(t_0)}{g(t_0)} = \frac{1500}{12,000} = .125 = 12.5\%$$

per year. Thus the price of sirloin steak is increasing at a faster percentage rate than the price of a new car.

Economists often use relative rates of change (or percentage rates of change) when discussing the growth of various economic quantities, such as national income or national debt, because such rates of change can be meaningfully compared.

EXAMPLE 1   **Gross domestic product**   A certain school of economists modeled the Gross Domestic Product of the United States at time t (measured in years from January 1, 1990) by the formula

$$f(t) = 3.4 + .04t + .13e^{-t},$$

where the Gross Domestic Product is measured in trillions of dollars. (See Fig. 1.) What was the predicted percentage rate of growth (or decline) of the economy at $t = 0$ and $t = 1$?

Solution   Since

$$f'(t) = .04 - .13e^{-t},$$

we see that

$$\frac{f'(0)}{f(0)} = \frac{.04 - .13}{3.4 + .13} = -\frac{.09}{3.53} \approx -2.5\%;$$

$$\frac{f'(1)}{f(1)} = \frac{.04 - .13e^{-1}}{3.4 + .04 + .13e^{-1}} \approx -\frac{.00782}{3.4878} \approx -.2\%.$$

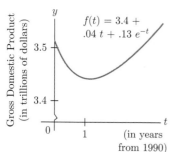

Figure 1

So on January 1, 1990, the economy is predicted to contract or decline at a relative rate of 2.5% per year; on January 1, 1991, the economy is predicted to be still contracting, but only at a relative rate of .2% per year.   ■

EXAMPLE 2   **Value of an investment**   The value in dollars of a certain business investment at time t may be approximated empirically by the function $f(t) = 750,000e^{.6\sqrt{t}}$. Use a logarithmic derivative to describe how fast the value of the investment is increasing when $t = 5$ years.

Solution   We have

$$\frac{f'(t)}{f(t)} = \frac{d}{dt}\ln f(t) = \frac{d}{dt}\left(\ln 750,000 + \ln e^{.6\sqrt{t}}\right)$$

$$= \frac{d}{dt}(\ln 750,000 + .6\sqrt{t})$$

$$= (.6)\left(\frac{1}{2}\right)t^{-1/2} = \frac{.3}{\sqrt{t}}.$$

When $t = 5$,

$$\frac{f'(5)}{f(5)} = \frac{.3}{\sqrt{5}} \approx .134 = 13.4\%.$$

Figure 2

Thus, when $t = 5$ years, the value of the investment is increasing at the relative rate of 13.4% per year. So when $t = 5$, we should expect the investment to grow by 13.4% in 1 year. (See Fig. 2.) Indeed, if we compute the increase in the investment from $t = 5$ to $t = 6$ and divide by its value at $t = 5$ to get the percent increase, we find

$$[\text{percentage rate of change}] = \frac{f(6) - f(5)}{f(5)} \approx \frac{32.6 - 28.7}{28.7} \approx .136$$

or 13.6%. This percentage is close to 13.4%, the relative rate of change of f at $t = 5$. ∎

In certain mathematical models, it is assumed that for a limited period of time the percentage rate of change of a particular function is constant. The following example shows that such a function must be an exponential function.

EXAMPLE 3   If the function $f(t)$ has a constant relative rate of change k, show that $f(t) = Ce^{kt}$ for some constant C.

Solution   We are given that

$$\frac{f'(t)}{f(t)} = k.$$

Hence $f'(t) = kf(t)$. But this is just the differential equation satisfied by the exponential function. Therefore, we must have $f(t) = Ce^{kt}$ for some constant C. ∎

Elasticity of Demand   In Section 2.7 we considered demand equations for monopolists and for entire industries. Recall that a demand equation expresses, for each quantity x to be produced, the market price that will generate a demand of exactly x. For instance, the demand equation

$$p = 150 - .01x \tag{2}$$

says that to sell x units the price must be set at $150 - .01x$ dollars. To be specific, to sell 6000 units the price must be set at $150 - .01(6000) = \$90$ per unit.

Equation (2) may be solved for x in terms of p to yield

$$x = 100(150 - p). \tag{3}$$

This last equation gives quantity in terms of price. If we let the letter q represent quantity, equation (3) becomes

$$q = 100(150 - p). \tag{3a}$$

This equation is of the form $q = f(p)$, where in this case $f(p)$ is the function $f(p) = 100(150 - p)$. In what follows it will be convenient to always write our demand functions so that the quantity q is expressed as a function $f(p)$ of the price p.

Usually, raising the price of a commodity lowers demand. Therefore, the typical demand function $q = f(p)$ is decreasing and has a negative slope everywhere. (See Fig. 3.)

Recall that the derivative $f'(p)$ compares the change in the quantity demanded with the change in price. By way of contrast, the concept of elasticity is designed to compare the *relative* rate of change of the quantity demanded with the *relative* rate of change of price.

Let us be more explicit. Consider a particular demand function $q = f(p)$. From our interpretation of the logarithmic derivative in (1), we know that the relative rate of change of the quantity demanded with respect to p is

$$\frac{(d/dp)f(p)}{f(p)} = \frac{f'(p)}{f(p)}.$$

Figure 3

Similarly, the relative rate of change of price with respect to p is

$$\frac{(d/dp)p}{p} = \frac{1}{p}.$$

Hence the ratio of the relative rate of change of the quantity demanded to the relative rate of change of price is

$$\frac{[\text{relative rate of change of quantity}]}{[\text{relative rate of change of price}]} = \frac{f'(p)/f(p)}{1/p} = \frac{pf'(p)}{f(p)}.$$

Since $f'(p)$ is always negative for a typical demand function, the quantity $pf'(p)/f(p)$ will be negative for all values of p. For convenience, economists prefer to work with positive numbers, and therefore the *elasticity of demand* is taken to be this quantity multiplied by -1.

> The elasticity of demand $E(p)$ at price p for the demand function $q = f(p)$ is defined to be
>
> $$E(p) = \frac{-pf'(p)}{f(p)}.$$

EXAMPLE 4   **Elasticity of demand**   The demand function for a certain metal is $q = 100 - 2p$, where p is the price per pound and q is the quantity demanded (in millions of pounds).

(a) What quantity can be sold at $30 per pound?

(b) Determine the function $E(p)$.

(c) Determine and interpret the elasticity of demand at $p = 30$.

(d) Determine and interpret the elasticity of demand at $p = 20$.

Solution   (a) In this case, $q = f(p)$, where $f(p) = 100 - 2p$. When $p = 30$, we have $q = f(30) = 100 - 2(30) = 40$. Therefore, 40 million pounds of the metal can be sold. We also say that the *demand* is 40 million pounds.

(b) $E(p) = \dfrac{-pf'(p)}{f(p)} = \dfrac{-p(-2)}{100 - 2p} = \dfrac{2p}{100 - 2p}.$

(c) The elasticity of demand at price $p = 30$ is $E(30)$.

$$E(30) = \frac{2(30)}{100 - 2(30)} = \frac{60}{40} = \frac{3}{2}$$

When the price is set at $30 per pound, a small increase in price will result in a relative rate of decrease in quantity demanded of about $\frac{3}{2}$ times the relative rate of increase in price. For example, if the price is increased from $30 by 1%, the quantity demanded will decrease by about 1.5%.

(d) When $p = 20$, we have

$$E(20) = \frac{2(20)}{100 - 2(20)} = \frac{40}{60} = \frac{2}{3}.$$

When the price is set at $20 per pound, a small increase in price will result in a relative rate of decrease in quantity demanded of only $\frac{2}{3}$ of the relative rate of increase of price. For example, if the price is increased from $20 by 1%, the quantity demanded will decrease by $\frac{2}{3}$ of 1%.   ■

Economists say that demand is *elastic* at price p_0 if $E(p_0) > 1$ and *inelastic* at price p_0 if $E(p_0) < 1$. In Example 4, the demand for the metal is elastic at $30 per pound and inelastic at $20 per pound.

The significance of the concept of elasticity may perhaps be best appreciated by studying how revenue, $R(p)$, responds to changes in price. Let us illustrate this response with a concrete example.

EXAMPLE 5

Figure 4 shows the elasticity of demand for the metal in Example 4.

$$E(p) = \frac{2p}{100 - 2p}$$

Figure 4

(a) For what values of p is demand elastic? Inelastic?

(b) Find and plot the revenue function for $0 < p < 50$.

(c) How does revenue respond to an increase in price when demand is elastic or respectively, inelastic?

Solution (a) In Example 4(b), we found the elasticity of demand to be

$$E(p) = \frac{2p}{100 - 2p}.$$

Let us solve $E(p) = 1$ for p.

$$\frac{2p}{100 - 2p} = 1$$
$$2p = 100 - 2p$$
$$4p = 100$$
$$p = 25.$$

By definition, demand is elastic at price p if $E(p) > 1$ and inelastic if $E(p) < 1$. From Figure 4, we see that demand is elastic if $25 < p < 50$ and inelastic if $0 < p < 25$.

(b) Recall that

$$[\text{revenue}] = [\text{quantity}] \cdot [\text{price per unit}].$$

Using the formula for demand (in millions of pounds) from Example 4, we obtain the revenue function

$$R = (100 - 2p) \cdot p = p(100 - 2p) \quad \text{(in millions of dollars)}.$$

This is a parabola opening down with p-intercepts at $p = 0$ and $p = 50$. Its maximum is located at the midpoint of the p-intercepts, or $p = 25$. (See Fig. 5.)

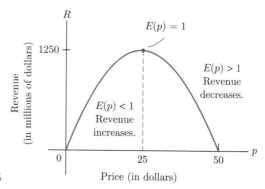

Figure 5

(c) In part (a), we determined that demand is elastic for $25 < p < 50$. For p in this price range, Figure 5 shows that an increase in price results in a decrease in revenue, and a decrease in price results in an increase in revenue. Hence we conclude that when demand is elastic the change of revenue is in the opposite direction of the change in price. Similarly, when demand is inelastic ($0 < p < 25$), Figure 5 shows that the change of revenue is in the same direction as the change in price. ∎

Example 5 illustrates an important application of elasticity as an indicator of the response of revenue to a change in price. As we now show, this response analysis can be applied in general.

Start by expressing revenue as a function of price:

$$R(p) = f(p) \cdot p,$$

where $f(p)$ is the demand function. If we differentiate $R(p)$ using the product rule, we find that

$$R'(p) = \frac{d}{dp}[f(p) \cdot p] = f(p) \cdot 1 + p \cdot f'(p)$$

$$= f(p)\left[1 + \frac{pf'(p)}{f(p)}\right]$$

$$= f(p)[1 - E(p)]. \tag{4}$$

Now suppose that demand is elastic at some price p_0. Then $E(p_0) > 1$ and $1 - E(p_0)$ is negative. Since $f(p)$ is always positive, we see from (4) that $R'(p_0)$ is negative. Therefore, by the first derivative rule, $R(p)$ is decreasing at p_0. So an increase in price will result in a decrease in revenue, and a decrease in price will result in an increase in revenue. On the other hand, if demand is inelastic at p_0, then $1 - E(p_0)$ will be positive and hence $R'(p_0)$ will be positive. In this case an increase in price will result in an increase in revenue, and a decrease in price will result in a decrease in revenue. We may summarize this as follows:

The change in revenue is in the opposite direction of the change in price when demand is elastic and in the same direction when demand is inelastic.

Practice Problems 5.3

The current toll for the use of a certain toll road is $2.50. A study conducted by the state highway department determined that with a toll of p dollars q cars will use the road each day, where $q = 60{,}000e^{-.5p}$.

1. Compute the elasticity of demand at $p = 2.5$.

2. Is demand elastic or inelastic at $p = 2.5$?

3. If the state increases the toll slightly, will the revenue increase or decrease?

EXERCISES 5.3

Determine the percentage rate of change of the functions at the points indicated.

1. $f(t) = t^2$ at $t = 10$ and $t = 50$

2. $f(t) = t^{10}$ at $t = 10$ and $t = 50$

3. $f(x) = e^{.3x}$ at $x = 10$ and $x = 20$

4. $f(x) = e^{-.05x}$ at $x = 1$ and $x = 10$

5. $f(t) = e^{.3t^2}$ at $t = 1$ and $t = 5$

6. $G(s) = e^{-.05s^2}$ at $s = 1$ and $s = 10$

7. $f(p) = 1/(p + 2)$ at $p = 2$ and $p = 8$

8. $g(p) = 5/(2p + 3)$ at $p = 1$ and $p = 11$

9. **Percentage Rate of Growth** The annual sales S (in dollars) of a company may be approximated empirically by the formula

$$S = 50{,}000\sqrt{e^{\sqrt{t}}},$$

where t is the number of years beyond some fixed reference date. Use a logarithmic derivative to determine the percentage rate of growth of sales at $t = 4$.

10. **Percentage Rate of Change** The price of wheat per bushel at time t (in months) is approximated by

$$f(t) = 4 + .001t + .01e^{-t}.$$

What is the percentage rate of change of $f(t)$ at $t = 0$? $t = 1$? $t = 2$?

11. An investment grows at a continuous 12% rate per year. In how many years will the value of the investment double?

12. The value of a piece of property is growing at a continuous $r\%$ rate per year and the value doubles in 3 years. Find r.

For each demand function, find $E(p)$ and determine if demand is elastic or inelastic (or neither) at the indicated price.

13. $q = 700 - 5p$, $p = 80$

14. $q = 600e^{-.2p}$, $p = 10$

15. $q = 400(116 - p^2)$, $p = 6$

16. $q = (77/p^2) + 3$, $p = 1$

17. $q = p^2 e^{-(p+3)}$, $p = 4$

18. $q = 700/(p + 5)$, $p = 15$

19. **Elasticity of Demand** Currently, 1800 people ride a certain commuter train each day and pay $4 for a ticket. The number of people q willing to ride the train at price p is $q = 600(5 - \sqrt{p})$. The railroad would like to increase its revenue.

   (a) Is demand elastic or inelastic at $p = 4$?

   (b) Should the price of a ticket be raised or lowered?

20. **Elasticity of Demand** An electronic store can sell $q = 10{,}000/(p + 50) - 30$ cellular phones at a price p dollars per phone. The current price is $150.

   (a) Is demand elastic or inelastic at $p = 150$?

   (b) If the price is lowered slightly, will revenue increase or decrease?

21. **Elasticity of Demand** A movie theater has a seating capacity of 3000 people. The number of people attending a show at price p dollars per ticket is $q = (18{,}000/p) - 1500$. Currently, the price is $6 per ticket.

   (a) Is demand elastic or inelastic at $p = 6$?

   (b) If the price is lowered, will revenue increase or decrease?

22. **Elasticity of Demand** A subway charges 65 cents per person and has 10,000 riders each day. The demand function for the subway is $q = 2000\sqrt{90 - p}$.

   (a) Is demand elastic or inelastic at $p = 65$?

   (b) Should the price of a ride be raised or lowered to increase the amount of money taken in by the subway?

23. **Elasticity of Demand** A country that is the major supplier of a certain commodity wishes to improve its balance of trade position by lowering the price of the commodity. The demand function is $q = 1000/p^2$.

   (a) Compute $E(p)$.

   (b) Will the country succeed in raising its revenue?

24. Show that any demand function of the form $q = a/p^m$ has constant elasticity m.

A cost function $C(x)$ gives the total cost of producing x units of a product. The elasticity of cost at quantity x, $E_c(x)$, is defined to be the ratio of the relative rate of change of cost (with respect to x) divided by the relative rate of change of quantity (with respect to x).

25. Show that $E_c(x) = x \cdot C'(x)/C(x)$.

26. Show that E_c is equal to the marginal cost divided by the average cost.

27. Let $C(x) = (1/10)x^2 + 5x + 300$. Show that $E_c(50) < 1$. (Hence when producing 50 units, a small relative increase in production results in an even smaller relative increase in total cost. Also, the average cost of producing 50 units is greater than the marginal cost at $x = 50$.)

28. Let $C(x) = 1000e^{.02x}$. Determine and simplify the formula for $E_c(x)$. Show that $E_c(60) > 1$ and interpret this result.

Technology Exercises

29. Consider the demand function $q = 60{,}000e^{-.5p}$ from Practice Problems 5.3.

(a) Determine the value of p for which the value of $E(p)$ is 1. For what values of p is demand inelastic?

(b) Graph the revenue function in the window $[0,4]$ by $[-5000, 50{,}000]$ and determine where its maximum value occurs. For what values of p is the revenue an increasing function?

Solutions to Practice Problems 5.3

1. The demand function is $f(p) = 60{,}000e^{-.5p}$.
$$f'(p) = -30{,}000e^{-.5p}$$
$$E(p) = \frac{-pf'(p)}{f(p)} = \frac{-p(-30{,}000)e^{-.5p}}{60{,}000e^{-.5p}} = \frac{p}{2}$$
$$E(2.5) = \frac{2.5}{2} = 1.25.$$

2. The demand is elastic, because $E(2.5) > 1$.

3. Since demand is elastic at \$2.50, a slight change in price causes revenue to change in the *opposite* direction. Hence revenue will decrease.

5.4 Further Exponential Models

After jumping out of an airplane, a skydiver falls at an increasing rate. However, the wind rushing past the skydiver's body creates an upward force that begins to counterbalance the downward force of gravity. This air friction finally becomes so great that the skydiver's velocity reaches a limiting speed called the *terminal velocity*. If we let $v(t)$ be the downward velocity of the skydiver after t seconds of free fall, a good mathematical model for $v(t)$ is given by

$$v(t) = M(1 - e^{-kt}), \tag{1}$$

where M is the terminal velocity and k is some positive constant. (See Fig. 1.) When t is close to zero, e^{-kt} is close to 1, and the velocity is small. As t increases, e^{-kt} becomes small and so $v(t)$ approaches M.

EXAMPLE 1   **Velocity of a skydiver** Show that the velocity given in equation (1) satisfies the equations
$$v'(t) = k[M - v(t)], \qquad v(0) = 0. \tag{2}$$

Solution From (1) we have $v(t) = M - Me^{-kt}$. Then
$$v'(t) = Mke^{-kt}.$$
However,
$$k[M - v(t)] = k[M - (M - Me^{-kt})] = kMe^{-kt},$$
so the differential equation $v'(t) = k[M - v(t)]$ holds. Also,
$$v(0) = M - Me^0 = M - M = 0. \qquad \blacksquare$$

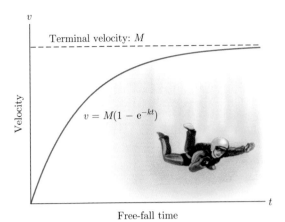

Figure 1. Velocity of a skydiver.

The differential equation (2) says that the rate of change in v is proportional to the difference between the terminal velocity M and the actual velocity v. It is not difficult to show that the only solution of (2) is given by the formula in (1).

The two equations (1) and (2) arise as mathematical models in a variety of situations. Some of these applications are described next.

The Learning Curve   Psychologists have found that in many learning situations a person's rate of learning is rapid at first and then slows down. Finally, as the task is mastered, the person's level of performance reaches a level above which it is almost impossible to rise. For example, within reasonable limits, each person seems to have a certain maximum capacity for memorizing a list of nonsense syllables. Suppose that a subject can memorize M syllables in a row if given sufficient time, say an hour, to study the list, but cannot memorize $M+1$ syllables in a row even if allowed several hours of study. By giving the subject different lists of syllables and varying lengths of time to study the lists, the psychologist can determine an empirical relationship between the number of nonsense syllables memorized accurately and the number of minutes of study time. It turns out that a good model for this situation is

$$f(t) = M(1 - e^{-kt})$$

for some appropriate positive constant k. (See Fig. 2.)

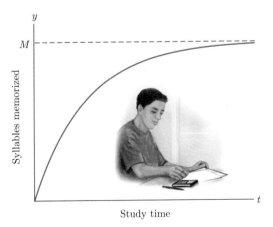

Figure 2. Learning curve, $f(t) = M(1 - e^{-kt})$.

The *slope* of this learning curve at time t is approximately the number of additional syllables that can be memorized if the subject is given 1 more minute of study time. Thus the slope is a measure of the *rate of learning*. The differential equation satisfied by the function $y = f(t)$ is

$$y' = k(M - y), \qquad f(0) = 0.$$

This equation says that, if the subject is given a list of M nonsense syllables, the rate of memorization is proportional to the number of syllables remaining to be memorized.

Diffusion of Information by Mass Media   Sociologists have found that the differential equation (2) provides a good model for the way information is spread (or "diffused") through a population when the information is being propagated constantly by mass media, such as television or magazines.* Given a fixed population P, let $f(t)$ be the number of people who have already heard a certain piece of information by time t. Then $P - f(t)$ is the number who have not yet heard the information by time t. Also, $f'(t)$ is the rate of increase of the number of people who have heard the news (the "rate of diffusion" of the information). If the information is being publicized often by some mass media, it is likely that the number of *newly informed* people per unit time is proportional to the number of people who have not yet heard the news. Therefore,

$$f'(t) = k[P - f(t)]. \tag{3}$$

Assume that $f(0) = 0$ (that is, there was a time $t = 0$ when nobody had heard the news). Then the remark following Example 1 shows that

$$f(t) = P(1 - e^{-kt}). \tag{4}$$

(See Fig. 3.)

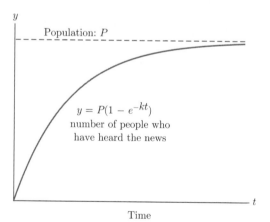

Figure 3.   Diffusion of information by mass media.

EXAMPLE 2   The news of the resignation of a public official is broadcast frequently by radio and television stations. Also, one-half of the residents of a city have heard the news within 4 hours of its initial release. Use the exponential model (4) to estimate when 90% of the residents will have heard the news.

Solution   We must find the value of k in (4). If P is the number of residents, the number who have heard the news in the first 4 hours is given by (4), with $t = 4$. By assumption,

---

*J. Coleman, *Introduction to Mathematical Sociology* (New York: The Free Press, 1964), p. 43.

this number is half the population. So

$$\tfrac{1}{2}P = P(1 - e^{-k\cdot 4})$$
$$.5 = 1 - e^{-4k}$$
$$e^{-4k} = 1 - .5 = .5.$$

Solving for k, we find that $k = \frac{\ln 2}{4} \approx .1733$. So the model for this particular situation is

$$f(t) = P(1 - e^{-kt}), \quad \text{where } k = \tfrac{\ln 2}{4}.$$

Now we want to find t such that $f(t) = .90P$. We solve for t:

$$.90P = P(1 - e^{-kt})$$
$$.90 = 1 - e^{-kt}$$
$$e^{-kt} = 1 - .90 = .10$$
$$-kt = \ln .10$$
$$t = \frac{\ln .10}{-k} = -4\frac{\ln .10}{\ln 2} \approx 13.29.$$

Therefore, 90% of the residents will hear the news within 13 hours of its initial release.

Intravenous Infusion of Glucose The human body both manufactures and uses glucose (blood sugar). Usually, there is a balance in these two processes, so the bloodstream has a certain equilibrium level of glucose. Suppose that a patient is given a single intravenous injection of glucose, and let $A(t)$ be the amount of glucose (in milligrams) above the equilibrium level. Then the body will start using up the excess glucose at a rate proportional to the amount of excess glucose; that is,

$$A'(t) = -\lambda A(t), \tag{5}$$

where λ is a positive constant called the *velocity constant of elimination*. This constant depends on how fast an individual patient's metabolic processes eliminate the excess glucose from the blood. Equation (5) describes a simple exponential decay process.

   Now suppose that, instead of a single shot, the patient receives a continuous intravenous infusion of glucose. A bottle of glucose solution is suspended above the patient, and a small tube carries the glucose down to a needle that runs into a vein. In this case, there are two influences on the amount of excess glucose in the blood: the glucose being added steadily from the bottle and the glucose being removed from the body by metabolic processes. Let r be the rate of infusion of glucose (often from 10 to 100 milligrams per minute). If the body did not remove any glucose, the excess glucose would increase at a constant rate of r milligrams per minute; that is,

$$A'(t) = r. \tag{6}$$

Taking into account the two influences on $A'(t)$ described by (5) and (6), we can write

$$A'(t) = r - \lambda A(t). \tag{7}$$

Define M to be r/λ and note that initially there is no excess glucose; then

$$A'(t) = \lambda(M - A(t)), \qquad A(0) = 0.$$

As stated in Example 1, a solution of this differential equation is given by

$$A(t) = M(1 - e^{-\lambda t}) = \frac{r}{\lambda}(1 - e^{-\lambda t}). \tag{8}$$

Note that M is the limiting value of the glucose level. Reasoning as in Example 1, we conclude that the amount of excess glucose rises until it reaches a stable level. (See Fig. 4.)

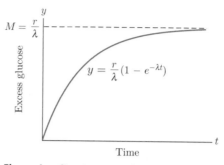

Figure 4. Continuous infusion of glucose. **Figure 5.** Logistic growth.

The Logistic Growth Curve The model for simple exponential growth discussed in Section 5.1 is adequate for describing the growth of many types of populations, but obviously a population cannot increase exponentially forever. The simple exponential growth model becomes inapplicable when the environment begins to inhibit the growth of the population. The logistic growth curve is an important exponential model that takes into account some of the effects of the environment on a population. (See Fig. 5.) For small values of t, the curve has the same basic shape as an exponential growth curve. Then, when the population begins to suffer from overcrowding or lack of food, the growth rate (the slope of the population curve) begins to slow down. Eventually, the growth rate tapers off to zero as the population reaches the maximum size that the environment will support. This latter part of the curve resembles the growth curves studied earlier in this section.

The equation for logistic growth has the general form

$$y = \frac{M}{1 + Be^{-Mkt}}, \tag{9}$$

where B, M, and k are positive constants. We can show that y satisfies the differential equation

$$y' = ky(M - y). \tag{10}$$

The factor y reflects the fact that the growth rate (y') depends in part on the size y of the population. The factor $M - y$ reflects the fact that the growth rate also depends on how close y is to the maximum level M.

The logistic curve is often used to fit experimental data that lie along an S-shaped curve. Examples are given by the growth of a fish population in a lake and the growth of a fruit fly population in a laboratory container. Also, certain enzyme reactions in animals follow a logistic law. One of the earliest applications of the logistic curve occurred in about 1840 when the Belgian sociologist P. Verhulst fit a logistic curve to six U.S. census figures, 1790 to 1840, and predicted the U.S. population for 1940. His prediction missed by less than 1 million persons (an error of about 1%).

EXAMPLE 3

Logistic growth A lake is stocked with 100 fish. After 3 months there are 250 fish. A study of the ecology of the lake predicts that the lake can support 1000 fish. Find a formula for the number $P(t)$ of fish in the lake t months after it has been stocked.

Solution The limiting population M is 1000. Therefore, we have

$$P(t) = \frac{1000}{1 + Be^{-1000kt}}.$$

At $t = 0$ there are 100 fish, so that

$$100 = P(0) = \frac{1000}{1 + Be^0} = \frac{1000}{1 + B}.$$

Thus $1 + B = 10$, or $B = 9$. Finally, since $P(3) = 250$, we have

$$250 = \frac{1000}{1 + 9e^{-3000k}}$$

$$1 + 9e^{-3000k} = 4$$

$$e^{-3000k} = \tfrac{1}{3}$$

$$-3000k = \ln \tfrac{1}{3}$$

$$k \approx .00037.$$

Therefore,

$$P(t) = \frac{1000}{1 + 9e^{-.37t}}. \qquad \blacksquare$$

Several theoretical justifications can be given for using (9) and (10) in situations where the environment prevents a population from exceeding a certain size. A discussion of this topic may be found in *Mathematical Models and Applications* by D. Maki and M. Thompson (Englewood Cliffs, N.J.: Prentice-Hall, Inc., 1973), pp. 312–317.

An Epidemic Model It will be instructive to actually "build" a mathematical model. Our example concerns the spread of a highly contagious disease. We begin by making several simplifying assumptions:

1. The population is a fixed number P and each member of the population is susceptible to the disease.
2. The duration of the disease is long, so that no cures occur during the time period under study.
3. All infected individuals are contagious and circulate freely among the population.
4. During each time period (such as 1 day or 1 week) each infected person makes c contacts, and each contact with an uninfected person results in transmission of the disease.

Consider a short period of time from t to $t + h$. Each infected person makes $c \cdot h$ contacts. How many of these contacts are with uninfected persons? If $f(t)$ is the number of infected persons at time t, then $P - f(t)$ is the number of uninfected persons, and $[P - f(t)]/P$ is the fraction of the population that is uninfected. Thus, of the $c \cdot h$ contacts made,

$$\left[\frac{P - f(t)}{P} \right] \cdot c \cdot h$$

will be with uninfected persons. This is the number of new infections produced by one infected person during the time period of length h. The total number of *new* infections during this period is

$$f(t) \left[\frac{P - f(t)}{P} \right] ch.$$

But this number must equal $f(t + h) - f(t)$, where $f(t + h)$ is the total number of infected persons at time $t + h$. So

$$f(t + h) - f(t) = f(t) \left[\frac{P - f(t)}{P} \right] ch.$$

Dividing by h, the length of the time period, we obtain the average number of new infections per unit time (during the small time period):

$$\frac{f(t + h) - f(t)}{h} = \frac{c}{P} f(t)[P - f(t)].$$

If we let h approach zero and let y stand for $f(t)$, the left-hand side approaches the rate of change in the number of infected persons and we derive the following equation:

$$\frac{dy}{dt} = \frac{c}{P} y(P - y). \tag{11}$$

This is the same type of equation as that used in (10) for logistic growth, although the two situations leading to this model appear to be quite dissimilar.

Comparing (11) with (10), we see that the number of infected individuals at time t is described by a logistic curve with $M = P$ and $k = c/P$. Therefore, by (9), we can write

$$f(t) = \frac{P}{1 + Be^{-ct}}.$$

B and c can be determined from the characteristics of the epidemic. (See Example 4.)

The logistic curve has an inflection point at that value of t for which $f(t) = P/2$. The position of this inflection point has great significance for applications of the logistic curve. From inspecting a graph of the logistic curve, we see that the inflection point is the point at which the curve has greatest slope. In other words, the inflection point corresponds to the instant of fastest growth of the logistic curve. This means, for example, that in the foregoing epidemic model the disease is spreading with the greatest rapidity precisely when half the population is infected. Any attempt at disease control (through immunization, for example) must strive to reduce the incidence of the disease to as low a point as possible, but in any case at least below the inflection point at $P/2$, at which point the epidemic is spreading fastest.

EXAMPLE 4

Spread of an epidemic The Public Health Service monitors the spread of an epidemic of a particularly long-lasting strain of flu in a city of 500,000 people. At the beginning of the first week of monitoring, 200 cases had been reported; during the first week, 300 new cases are reported. Estimate the number of infected individuals after 6 weeks.

Solution Here $P = 500,000$. If $f(t)$ denotes the number of cases at the end of t weeks, then

$$f(t) = \frac{P}{1 + Be^{-ct}} = \frac{500,000}{1 + Be^{-ct}}.$$

Moreover, $f(0) = 200$, so

$$200 = \frac{500,000}{1 + Be^0} = \frac{500,000}{1 + B},$$

and $B = 2499$. Consequently, since $f(1) = 300 + 200 = 500$, we have

$$500 = f(1) = \frac{500,000}{1 + 2499e^{-c}},$$

so $e^{-c} \approx .4$ and $c \approx .92$. Finally,

$$f(t) = \frac{500,000}{1 + 2499e^{-.92t}}$$

and

$$f(6) = \frac{500,000}{1 + 2499e^{-.92(6)}} \approx 45,000.$$

After 6 weeks, about 45,000 individuals are infected.   ∎

This epidemic model is used by sociologists (who still call it an epidemic model) to describe the spread of a rumor. In economics the model is used to describe the diffusion of knowledge about a product. An "infected person" represents an individual who possesses knowledge of the product. In both cases, it is assumed that the members of the population are themselves primarily responsible for the spread of the rumor or knowledge of the product. This situation is in contrast to the model described earlier, in which information was spread through a population by external sources, such as radio and television.

There are several limitations to this epidemic model. Each of the four simplifying assumptions made at the outset is unrealistic in varying degrees. More complicated models can be constructed that rectify one or more of these defects, but they require more advanced mathematical tools.

The Exponential Function in Lung Physiology

Let us conclude this section by deriving a useful model for the pressure in a person's lungs when the air is allowed to escape passively from the lungs with no use of the person's muscles. Let V be the volume of air in the lungs and let P be the relative pressure in the lungs when compared with the pressure in the mouth. The *total compliance* (of the respiratory system) is defined to be the derivative $\dfrac{dV}{dP}$. For normal values of V and P, we may assume that the total compliance is a positive constant, say C. That is,

$$\frac{dV}{dP} = C. \tag{12}$$

We assume that the airflow during the passive respiration is smooth and not turbulent. Then Poiseuille's law of fluid flow says that the rate of change of volume as a function of time (that is, the rate of airflow) satisfies

$$\frac{dV}{dt} = -\frac{P}{R}, \tag{13}$$

where R is a (positive) constant called the airway resistance. Under these conditions, we may derive a formula for P as a function of time. Since the volume is a function of pressure, and the pressure is in turn a function of time, we may use the chain rule to write

$$\frac{dV}{dt} = \frac{dV}{dP} \cdot \frac{dP}{dt}.$$

From (12) and (13),

$$-\frac{P}{R} = C \cdot \frac{dP}{dt},$$

so

$$\frac{dP}{dt} = -\frac{1}{RC} \cdot P.$$

From this differential equation, we conclude that P must be an exponential function of t. In fact,

$$P = P_0 e^{kt},$$

where P_0 is the initial pressure at time $t = 0$ and $k = -1/RC$. This relation between k and the product RC is useful to lung specialists, because they can experimentally compute k and the compliance C. They can then use the formula $k = -1/RC$ to determine the airway resistance R.

Practice Problems 5.4

1. A sociological study[*] was made to examine the process by which doctors decide to adopt a new drug. The doctors were divided into two groups. The doctors in group A had little interaction with other doctors and so received most of their information through mass media. The doctors in group B had extensive interaction with other doctors and so received most of their information through word of mouth. For each group, let $f(t)$ be the number who have learned about the new drug after t months. Examine the appropriate differential equations to explain why the two graphs were of the types shown in Fig. 6.

Group A

Group B

Figure 6. Results of a sociological study.

EXERCISES 5.4

1. Consider the function $f(x) = 5(1 - e^{-2x})$, $x \geq 0$.

   (a) Show that $f(x)$ is increasing and concave down for all $x \geq 0$.

   (b) Explain why $f(x)$ approaches 5 as x gets large.

   (c) Sketch the graph of $f(x)$, $x \geq 0$.

2. Consider the function $g(x) = 10 - 10e^{-.1x}$, $x \geq 0$.

   (a) Show that $g(x)$ is increasing and concave down for $x \geq 0$.

   (b) Explain why $g(x)$ approaches 10 as x gets large.

   (c) Sketch the graph of $g(x)$, $x \geq 0$.

3. If $y = 2(1 - e^{-x})$, compute y' and show that $y' = 2 - y$.

4. If $y = 5(1 - e^{-2x})$, compute y' and show that $y' = 10 - 2y$.

5. If $f(x) = 3(1 - e^{-10x})$, show that $y = f(x)$ satisfies the differential equation

$$y' = 10(3 - y), \qquad f(0) = 0.$$

6. Ebbinghaus Model for Forgetting A student learns a certain amount of material for some class. Let $f(t)$ denote the percentage of the material that the student can recall t weeks later. The psychologist Ebbinghaus found

that this percent retention can be modeled by a function of the form

$$f(t) = (100 - a)e^{-\lambda t} + a,$$

where λ and a are positive constants and $0 < a < 100$. Sketch the graph of the function $f(t) = 85e^{-.5t} + 15$, $t \geq 0$.

7. Spread of News When a grand jury indicted the mayor of a certain town for accepting bribes, the newspaper, radio, and television immediately began to publicize the news. Within an hour, one-quarter of the citizens heard about the indictment. Estimate when three-quarters of the town heard the news.

8. Examine formula (8) for the amount $A(t)$ of excess glucose in the bloodstream of a patient at time t. Describe what would happen if the rate r of infusion of glucose were doubled.

9. Glucose Elimination Describe an experiment that a doctor could perform to determine the velocity constant of elimination of glucose for a particular patient.

---

[*]James S. Coleman, Eliku Katz, and Herbert Menzel, "The Diffusion of an Innovation among Physicians," *Sociometry*, 20 (1957), 253–270.

10. **Concentration of Glucose in the Bloodstream** Physiologists usually describe the continuous intravenous infusion of glucose in terms of the excess *concentration* of glucose, $C(t) = A(t)/V$, where V is the total volume of blood in the patient. In this case, the rate of increase in the concentration of glucose due to the continuous injection is r/V. Find a differential equation that gives a model for the rate of change of the excess concentration of glucose.

11. **Spread of News** A news item is spread by word of mouth to a potential audience of 10,000 people. After t days,

$$f(t) = \frac{10{,}000}{1 + 50e^{-.4t}}$$

people will have heard the news. The graph of this function is shown in Fig. 7.

(a)

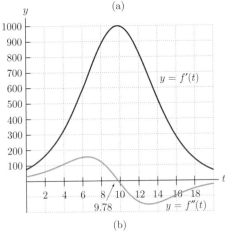

(b)

Figure 7

(a) Approximately how many people will have heard the news after 7 days?

(b) At approximately what rate will the news spread after 14 days?

(c) Approximately when will 7000 people have heard the news?

(d) Approximately when will the news spread at the rate of 600 people per day?

(e) When will the news spread at the greatest rate?

(f) Use equations (9) and (10) to determine the differential equation satisfied by $f(t)$.

(g) At what rate will the news spread when half of the potential audience has heard the news?

12. **Spread of News** A news item is broadcast by mass media to a potential audience of 50,000 people. After t days,

$$f(t) = 50{,}000(1 - e^{-.3t})$$

people will have heard the news. The graph of this function is shown in Fig. 8.

(a) How many people will have heard the news after 10 days?

(b) At what rate is the news spreading initially?

(c) When will 22,500 people have heard the news?

(d) Approximately when will the news spread at the rate of 2500 people per day?

(e) Use equations (3) and (4) to determine the differential equation satisfied by $f(t)$.

(f) At what rate will the news spread when half of the potential audience has heard the news?

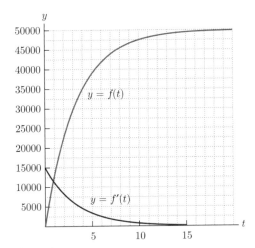

Figure 8

Technology Exercises

13. After a drug is taken orally, the amount of the drug in the bloodstream after t hours is $f(t) = 122(e^{-.2t} - e^{-t})$ units.

(a) Graph $f(t)$, $f'(t)$, and $f''(t)$ in the window $[0, 12]$ by $[-20, 75]$.

(b) How many units of the drug are in the bloodstream after 7 hours?

(c) At what rate is the level of drug in the bloodstream increasing after 1 hour?

(d) While the level is decreasing, when is the level of drug in the bloodstream 20 units?

(e) What is the greatest level of drug in the bloodstream and when is this level reached?

(f) When is the level of drug in the bloodstream decreasing the fastest?

14. A model incorporating growth restrictions for the number of bacteria in a culture after t days is given by $f(t) = 5000(20 + te^{-.04t})$.

(a) Graph $f'(t)$ and $f''(t)$ in the window $[0, 100]$ by $[-700, 300]$.

(b) How fast is the culture changing after 100 days?

(c) Approximately when is the culture growing at the rate of 76.6 bacteria per day?

(d) When is the size of the culture greatest?

(e) When is the size of the culture decreasing the fastest?

Solutions to Practice Problems 5.4

1. The difference between transmission of information by mass media and by word of mouth is that in the second case the rate of transmission depends not only on the number of people who have not yet received the information, but also on the number of people who know the information and therefore are capable of spreading it. Therefore, for group A, $f'(t) = k[P - f(t)]$, and for group B, $f'(t) = kf(t)[P - f(t)]$. Note that the spread of information by word of mouth follows the same pattern as the spread of an epidemic.

REVIEW OF FUNDAMENTAL CONCEPTS

1. What differential equation is key to solving exponential growth and decay problems? State a result about the solution to this differential equation.

2. What is a growth constant? Decay constant?

3. What is meant by the half-life of a radioactive element?

4. Explain how radiocarbon dating works.

5. State the formula for each of the following quantities:

(a) The compound amount of P dollars in t years at interest rate r compounded continuously

(b) The present value of A dollars in n years at interest rate r compounded continuously

6. What is the difference between a relative rate of change and a percentage rate of change?

7. Define the elasticity of demand, $E(p)$, for a demand function. How is $E(p)$ used?

8. Describe an application of the differential equation $y' = k(M - y)$.

9. Describe an application of the differential equation $y' = ky(M - y)$.

SUPPLEMENTARY EXERCISES

1. The atmospheric pressure $P(x)$ (measured in inches of mercury) at height x miles above sea level satisfies the differential equation $P'(x) = -.2P(x)$. Find the formula for $P(x)$ if the atmospheric pressure at sea level is 29.92.

2. The herring gull population in North America has been doubling every 13 years since 1900. Give a differential equation satisfied by $P(t)$, the population t years after 1900.

3. Find the present value of \$10,000 payable at the end of 5 years if money can be invested at 12% with interest compounded continuously.

4. One thousand dollars is deposited in a savings account at 10% interest compounded continuously. How many years are required for the balance in the account to reach \$3000?

5. The half-life of the radioactive element tritium is 12 years. Find its decay constant.

6. A piece of charcoal found at Stonehenge contained 63% of the level of ^{14}C found in living trees. Approximately how old is the charcoal?

7. From January 1, 1990, to January 1, 1997, the population of Texas grew from 17 million to 19.3 million.

(a) Give the formula for the population t years after 1990.

(b) If this growth continues, how large will the population be in 2000?

(c) In what year will the population reach 25 million?

8. A stock portfolio increased in value from \$100,000 to \$117,000 in 2 years. What rate of interest, compounded continuously, did this investment earn?

9. An investor initially invests $10,000 in a speculative venture. Suppose that the investment earns 20% interest compounded continuously for 5 years and then 6% interest compounded continuously for 5 years thereafter.

   (a) How much does the $10,000 grow to after 10 years?

   (b) The investor has the alternative of an investment paying 14% interest compounded continuously. Which investment is superior over a 10-year period, and by how much?

10. Two different bacteria colonies are growing near a pool of stagnant water. The first colony initially has 1000 bacteria and doubles every 21 minutes. The second colony has 710,000 bacteria and doubles every 33 minutes. How much time will elapse before the first colony becomes as large as the second?

11. The population of a city t years after 1990 satisfies the differential equation $y' = .02y$. What is the growth constant? How fast will the population be growing when the population reaches 3 million people? At what level of population will the population be growing at the rate of 100,000 people per year?

12. A colony of bacteria is growing exponentially with growth constant .4, with time measured in hours. Determine the size of the colony when the colony is growing at the rate of 200,000 bacteria per hour. Determine the rate at which the colony will be growing when its size is 1,000,000.

13. The population of a certain country is growing exponentially. The total population (in millions) in t years is given by the function $P(t)$. Match each of the following answers with its corresponding question.

Answers

   a. Solve $P(t) = 2$ for t.
   b. $P(2)$
   c. $P'(2)$
   d. Solve $P'(t) = 2$ for t.
   e. $y' = ky$
   f. Solve $P(t) = 2P(0)$ for t.
   g. $P_0 e^{kt}$, $k > 0$
   h. $P(0)$

Questions

   A. How fast will the population be growing in 2 years?
   B. Give the general form of the function $P(t)$.
   C. How long will it take for the current population to double?
   D. What will be the size of the population in 2 years?
   E. What is the initial size of the population?
   F. When will the size of the population be 2 million?
   G. When will the population be growing at the rate of 2 million people per year?
   H. Give a differential equation satisfied by $P(t)$.

14. You have 80 grams of a certain radioactive material, and the amount remaining after t years is given by the function $f(t)$ shown in Fig. 1.

   (a) How much will remain after 5 years?

   (b) When will 10 grams remain?

   (c) What is the half-life of this radioactive material?

   (d) At what rate will the radioactive material be disintegrating after 1 year?

   (e) After how many years will the radioactive material be disintegrating at the rate of about 5 grams per year?

Figure 1

15. A few years after money is deposited into a bank, the compound amount is $1000 and it is growing at the rate of $60 per year. What interest rate (compounded continuously) is the money earning?

16. The current balance in a savings account is $1230 and the interest rate is 4.5%. At what rate is the compound amount currently growing?

17. Find the percentage rate of change of the function $f(t) = 50e^{.2t^2}$ at $t = 10$.

18. Find $E(p)$ for the demand function $q = 4000 - 40p^2$, and determine if demand is elastic or inelastic at $p = 5$.

19. For a certain demand function, $E(8) = 1.5$. If the price is increased to $8.16, estimate the percentage decrease in the quantity demanded. Will the revenue increase or decrease?

20. Find the percentage rate of change of the function
$$f(p) = \frac{1}{3p + 1} \text{ at } p = 1.$$

21. A company can sell $q = 1000p^2 e^{-.02(p+5)}$ calculators at a price of p dollars per calculator. The current price is $200. If the price is decreased, will the revenue increase or decrease?

22. Consider a demand function of the form $q = ae^{-bp}$, where a and b are positive numbers. Find $E(p)$ and show that the elasticity equals 1 when $p = 1/b$.

23. Refer to Practice Problem 5.4. Out of 100 doctors in group A, none knew about the drug at time $t = 0$, but 66 of them were familiar with the drug after 13 months. Find the formula for $f(t)$.

24. The growth of the yellow nutsedge weed is described by a formula $f(t)$ of type (9) in Section 5.4. A typical weed has length 8 centimeters after 9 days, length 48 centimeters after 25 days, and reaches length 55 centimeters at maturity. Find the formula for $f(t)$.

25. When a rod of molten steel with a temperature of $1800°\mathrm{F}$ is placed in a large vat of water at temperature $60°\mathrm{F}$, the temperature of the rod after t seconds is

$$f(t) = 60(1 + 29e^{-.15t})°\mathrm{F}.$$

The graph of this function is shown in Fig. 2.

(a) What is the temperature of the rod after 11 seconds?

(b) At what rate is the temperature of the rod changing after 6 seconds?

(c) Approximately when is the temperature of the rod $200°\mathrm{F}$?

(d) Approximately when is the rod cooling at the rate of $200°\mathrm{F}$ per second?

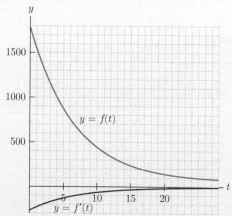

Figure 2

26. A certain bacteria culture grows at a rate proportional to its size. If 10,000 bacteria grow at the rate of 500 bacteria per day, how fast is the culture growing when it reaches 15,000 bacteria?

THE DEFINITE INTEGRAL

There are two fundamental problems of calculus: (1) to find the slope of a curve at a point and (2) to find the area of a region under a curve. These problems are quite simple when the curve is a straight line, as in Fig. 1. Both the slope of the line and the area of the shaded trapezoid can be calculated by geometric principles. When the graph consists of several line segments, as in Fig. 2, the slope of each line segment can be computed separately, and the area of the region can be found by adding the areas of the regions under each line segment.

Figure 1

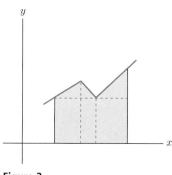

Figure 2

Calculus is needed when the curves are not straight lines. We have seen that the slope problem is resolved with the derivative of a function. In this chapter we describe how the area problem is connected with the notion of the "integral" of a function. Both the slope problem and the area problem were studied by the ancient Greeks and solved in special cases. But it was not until the development of calculus in the seventeenth century that the intimate connection between the two problems was discovered. In this chapter we will discuss this connection as stated in the fundamental theorem of calculus.

6.1 Antidifferentiation

We have developed several techniques for calculating the derivative $F'(x)$ of a function $F(x)$. In many applications, however, it is necessary to proceed in reverse. We are given the derivative $F'(x)$ and must determine the function $F(x)$. The process of determining $F(x)$ from $F'(x)$ is called *antidifferentiation*. The next example gives a typical application involving antidifferentiation.

EXAMPLE 1   **Federal health expenditures**   Based on data from the Centers for Medicare and Medicaid Services, during the late 1990s and early 2000s, the federal health expenditures grew at an exponential rate, with a growth constant of about .12. At the beginning of 2000, the rate was about 380 billion dollars per year. Let $R(t)$ denote the rate (in billions of dollars per year) of health expenditures at time t, where t is the number of years since the beginning of 2000. Then a reasonable model for $R(t)$ is given by

$$R(t) = 380e^{.12t}. \tag{1}$$

Assuming that the federal health expenditures will continue to grow at this rate, use the formula for $R(t)$ to determine the total amount of federal health expenditures from 2000 to 2010.

Solution   Let $T(t)$ denote the federal health expenditures from time 0 (2000) until time t. We wish to calculate $T(10)$, the health expenditures from 2000 to 2010. We do this by first determining a formula for $T(t)$. Since $T(t)$ is the health expenditures, the derivative, $T'(t)$, is the *rate* of health expenditures, or $R(t)$. Thus, although we do not yet have a formula for $T(t)$, we do know that

$$T'(t) = R(t) = 380e^{.12t}.$$

Thus the problem of determining a formula for $T(t)$ has been reduced to a problem of antidifferentiation: Find a function whose derivative is $R(t)$. We shall solve this particular problem after developing techniques for solving antidifferentiation problems in general.   ◼

Antiderivative   Suppose that $f(x)$ is a given function and $F(x)$ is a function having $f(x)$ as its derivative, that is, $F'(x) = f(x)$. We call $F(x)$ an *antiderivative* of $f(x)$.

EXAMPLE 2   Find an antiderivative of $f(x) = x^2$.

Solution   The derivative of x^3 is $3x^2$, which is almost the same as x^2 except for a factor of 3.

To make this factor 1 instead of 3, take $F(x) = \frac{1}{3}x^3$. Then

$$F'(x) = \underbrace{\frac{d}{dx}\left(\frac{1}{3}x^3\right) = \frac{1}{3}\left(\frac{d}{dx}x^3\right)}_{\text{constant multiple rule}} = \frac{1}{3}\cdot 3x^2 = x^2.$$

So $F(x)$ is an antiderivative of x^2. Another antiderivative is $\frac{1}{3}x^3 + 5$, because the derivative of a constant function is zero:

$$\frac{d}{dx}\left(\frac{1}{3}x^3 + 5\right) = \frac{d}{dx}\left(\frac{1}{3}x^3\right) + \frac{d}{dx}(5) = x^2 + 0 = x^2.$$

In fact, if C is any constant, the function $F(x) = \frac{1}{3}x^3 + C$ is also an antiderivative of x^2, since

$$\frac{d}{dx}\left(\frac{1}{3}x^3 + C\right) = \frac{1}{3}\cdot 3x^2 + 0 = x^2.$$

(The derivative of a constant function is zero.)   ■

EXAMPLE 3   Find an antiderivative of $f(x) = e^{-2x}$.

Solution   Recall that the derivative of e^{rx} is just a constant times e^{rx}. For an antiderivative of e^{-2x}, try a function of the form ke^{-2x}, where k is some constant to be determined. Then

$$\frac{d}{dx}ke^{-2x} = k\cdot(-2e^{-2x}) = -2ke^{-2x}.$$

Choose k to make $-2k = 1$; that is, choose $k = -\frac{1}{2}$. Then

$$\frac{d}{dx}\left(-\frac{1}{2}e^{-2x}\right) = \left(-\frac{1}{2}\right)(-2e^{-2x}) = 1\cdot e^{-2x} = e^{-2x}.$$

Thus $-\frac{1}{2}e^{-2x}$ is an antiderivative of e^{-2x}. Also, for any constant C, the function $-\frac{1}{2}e^{-2x} + C$ is an antiderivative of e^{-2x}, because

$$\frac{d}{dx}\left(-\frac{1}{2}e^{-2x} + C\right) = e^{-2x} + 0 = e^{-2x}.$$ ■

Examples 2 and 3 illustrate the fact that if $F(x)$ is an antiderivative of $f(x)$, so is $F(x) + C$, where C is any constant. [The derivative of the constant function C is zero.] The next theorem says that *all* antiderivatives of $f(x)$ can be produced in this way.

Theorem I   Suppose that $f(x)$ is a continuous function on an interval I. If $F_1(x)$ and $F_2(x)$ are two antiderivatives of $f(x)$, then $F_1(x)$ and $F_2(x)$ differ by a constant on I. In other words, there is a constant C such that

$$F_2(x) = F_1(x) + C \quad \text{for all } x \text{ in } I.$$

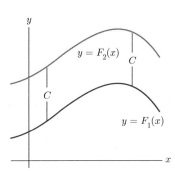

Figure 1.   Two antiderivatives of the same function.

Geometrically, the graph of any antiderivative $F_2(x)$ is obtained by shifting the graph of $F_1(x)$ vertically. (See Fig. 1.)

Our verification of Theorem I is based on the following fact, which is important in its own right.

Theorem II   If $F'(x) = 0$ for all x in an interval I, there is a constant C such that $F(x) = C$ for all x in I.

It is easy to see why Theorem II is reasonable. (A formal proof of the theorem requires an important theoretical result called the mean value theorem.) If $F'(x) = 0$ for all x, the curve $y = F(x)$ has slope equal to zero at every point. Thus the tangent line to $y = F(x)$ at any point is horizontal, which implies that the graph of $y = F(x)$ is a horizontal line. (Try to draw the graph of a function with a horizontal tangent everywhere. There is no choice but to keep your pencil moving on a constant, horizontal line!) If the horizontal line is $y = C$, then $F(x) = C$ for all x.

Verification of Theorem I If $F_1(x)$ and $F_2(x)$ are two antiderivatives of $f(x)$, the function $F(x) = F_2(x) - F_1(x)$ has the derivative

$$F'(x) = F_2'(x) - F_1'(x) = f(x) - f(x) = 0.$$

So, by Theorem II, we know that $F(x) = C$ for some constant C. In other words, $F_2(x) - F_1(x) = C$, so

$$F_2(x) = F_1(x) + C,$$

which is Theorem I.

Using Theorem I, we can find *all* antiderivatives of a given function once we know one antiderivative. For instance, since one antiderivative of x^2 is $\frac{1}{3}x^3$ (Example 2), all antiderivatives of x^2 have the form $\frac{1}{3}x^3 + C$, where C is a constant.

> **DEFINITION Indefinite Integral** Suppose that $f(x)$ is a function whose antiderivatives are $F(x) + C$. The standard way to express this fact is to write
>
> $$\int f(x)\, dx = F(x) + C.$$
>
> The symbol $\int$ is called an *integral sign*. The entire notation $\int f(x)\, dx$ is called an *indefinite integral* and stands for antidifferentiation of the function $f(x)$.

We always record the variable of interest by prefacing it by the letter d. For example, if the variable of interest is t rather than x, we would write $\int f(t)\, dt$ for the antiderivative.

EXAMPLE 4   Determine

(a) $\displaystyle\int x^r\, dx, \quad r$ a constant $\neq -1$ (b) $\displaystyle\int e^{kx}\, dx, \quad k$ a constant $\neq 0$

Solution (a) By the constant-multiple and power rules,

$$\frac{d}{dx}\left(\frac{1}{r+1}x^{r+1}\right) = \frac{1}{r+1} \cdot \frac{d}{dx}x^{r+1} = \frac{1}{r+1} \cdot (r+1)x^r = x^r.$$

Thus $x^{r+1}/(r+1)$ is an antiderivative of x^r. Letting C represent any constant, we have

$$\int x^r\, dx = \frac{1}{r+1}x^{r+1} + C, \qquad r \neq -1. \tag{2}$$

(b) An antiderivative of e^{kx} is e^{kx}/k, since

$$\frac{d}{dx}\left(\frac{1}{k}e^{kx}\right) = \frac{1}{k} \cdot \frac{d}{dx}e^{kx} = \frac{1}{k}(ke^{kx}) = e^{kx}.$$

Hence

$$\int e^{kx}\,dx = \frac{1}{k}e^{kx} + C, \qquad k \neq 0. \tag{3}$$

■

EXAMPLE 5   Determine

(a) $\displaystyle\int \sqrt{x}\,dx$ (b) $\displaystyle\int \frac{1}{x^2}\,dx$

Solution Both integrals follow by applying formula (2). For part (a), take $r = 1/2$ in (2); then

$$\int x^{\frac{1}{2}}\,dx = \frac{1}{\frac{1}{2}+1}x^{\frac{1}{2}+1} + C = \frac{1}{3/2}x^{\frac{3}{2}} + C = \frac{2}{3}x^{\frac{3}{2}} + C.$$

For part (b), take $r = -2$ in (2), then

$$\int x^{-2}\,dx = \frac{1}{-2+1}x^{-2+1} + C = -x^{-1} + C = -\frac{1}{x} + C.$$

■

There is a subtle point to be made concerning the integral of a discontinuous function. Consider the integral of $f(x) = \frac{1}{x^2}$ from Example 5(b). Since $f(x)$ is not defined at $x = 0$ and is continuous everywhere else, Theorem I tells us only that the antiderivatives of $f(x)$ are $(-1/x) + C$ on any interval that does not contain 0. So to express the antiderivatives of $f(x)$, we exclude 0 and write

$$\int \frac{1}{x^2}\,dx = \begin{cases} -\frac{1}{x} + C_1 & \text{if } x > 0, \\ -\frac{1}{x} + C_2 & \text{if } x < 0. \end{cases}$$

Note that, since we have two separate intervals ($x < 0$ and $x > 0$), we need two separate constants of integration, C_1 and C_2. For notational convenience, however, we will continue to use only one constant of integration, as we did in Example 5(b).

Formula (2) does not give an antiderivative of x^{-1} because $1/(r+1)$ is undefined for $r = -1$. However, we know that for $x \neq 0$ the derivative of $\ln|x|$ is $1/x$. Hence $\ln|x|$ is an antiderivative of $1/x$, and we have

$$\int \frac{1}{x}\,dx = \ln|x| + C, \qquad x \neq 0. \tag{4}$$

Formula (4) supplies the missing case ($r = -1$) of formula (2). Because $1/x$ is not defined at $x = 0$, as we explained following Example 5, we should distinguish two intervals for the validity of formula (4) and interpret it to mean

$$\int \frac{1}{x}\,dx = \begin{cases} \ln x + C_1 & \text{if } x > 0, \\ \ln(-x) + C_2 & \text{if } x < 0. \end{cases}$$

Formulas (2), (3), and (4) are each obtained by "reversing" a familiar differentiation rule. In a similar fashion, we may use the sum rule and constant-multiple rule for derivatives to obtain corresponding rules for antiderivatives:

$$\int [f(x) + g(x)]\,dx = \int f(x)\,dx + \int g(x)\,dx. \tag{5}$$

$$\int kf(x)\,dx = k\int f(x)\,dx, \qquad k \text{ a constant.} \tag{6}$$

In words, (5) says that a sum of functions may be antidifferentiated term by term, and (6) says that a constant multiple may be moved through the integral sign.

EXAMPLE 6   Compute

$$\int \left(x^{-3} + 7e^{5x} + \frac{4}{x} \right) dx.$$

Solution   Using the preceding rules, we have

$$\int \left(x^{-3} + 7e^{5x} + \frac{4}{x} \right) dx = \int x^{-3}\, dx + \int 7e^{5x}\, dx + \int \frac{4}{x}\, dx$$

$$= \int x^{-3}\, dx + 7 \int e^{5x}\, dx + 4 \int \frac{1}{x}\, dx$$

$$= \frac{1}{-2} x^{-2} + 7 \left(\frac{1}{5} e^{5x} \right) + 4 \ln |x| + C$$

$$= -\frac{1}{2} x^{-2} + \frac{7}{5} e^{5x} + 4 \ln |x| + C. \qquad \blacksquare$$

After some practice, most of the intermediate steps shown in the solution of Example 6 can be omitted.

A function has infinitely many different antiderivatives, corresponding to various choices of the constant C. In applications, it is often necessary to satisfy an additional condition, which then determines a specific value of C.

EXAMPLE 7   Find the function $f(x)$ for which $f'(x) = x^2 - 2$ and $f(1) = \frac{4}{3}$. [Equivalently, in this problem you are asked to solve the differential equation $y' = x^2 - 2$, $y(1) = \frac{4}{3}$.]

Solution   The unknown function $f(x)$ is an antiderivative of $x^2 - 2$. One antiderivative of $x^2 - 2$ is $\frac{1}{3}x^3 - 2x$. Therefore, by Theorem I,

$$f(x) = \tfrac{1}{3}x^3 - 2x + C, \quad C \text{ a constant.}$$

Figure 2 shows the graphs of $f(x)$ for several choices of C. We want the function whose graph passes through $(1, \frac{4}{3})$. To find the value of C that makes $f(1) = \frac{4}{3}$, we set

$$\tfrac{4}{3} = f(1) = \tfrac{1}{3}(1)^3 - 2(1) + C = -\tfrac{5}{3} + C$$

and find $C = \frac{4}{3} + \frac{5}{3} = 3$. Therefore, $f(x) = \frac{1}{3}x^3 - 2x + 3$. $\qquad \blacksquare$

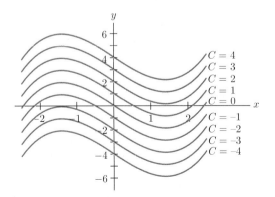

Figure 2.   Several antiderivatives of $x^2 - 2$.

Having introduced the basics of antidifferentiation, let us now solve the health expenditure problem in Example 1.

Solution of Example 1 (Continued) The rate of federal health expenditures at time t is $R(t) = 380e^{.12t}$ billion dollars per year. Moreover, we observed that the health expenditures $T(t)$, from time 0 to time t, is an antiderivative of $R(t)$. Using formulas (3) and (6), we have

$$T(t) = \int 380e^{.12t}\, dt = \frac{380}{.12}e^{.12t} + C = 3166.67e^{.12t} + C,$$

where C is a constant. However, in our particular example, $T(0) = 0$, since $T(0)$ represents the expenditures from time 0 to time 0. Therefore, the constant C must satisfy

$$0 = T(0) = 3166.67e^{.12(0)} + C = 3166.67 + C$$
$$C = -3166.67.$$

Thus,

$$T(t) = 3166.67e^{.12t} - 3166.67 = 3166.67(e^{.12t} - 1),$$

and so

$$T(10) = 3166.67(e^{.12(10)} - 1) \approx 7347 \text{ billion dollars.}$$

Hence the federal health expenditures from 2000 to 2010 will exceed 7 trillion dollars.

■

EXAMPLE 8   **Position of a rocket**   A rocket is fired vertically into the air. Its velocity at t seconds after lift-off is $v(t) = 6t + .5$ meters per second. Before launch, the top of the rocket is 8 meters above the launch pad. Find the height of the rocket (measured from the top of the rocket to the launch pad) at time t.

Solution   If $s(t)$ denotes the height of the rocket at time t, then $s'(t)$ is the rate at which the height is changing. That is, $s'(t) = v(t)$, and therefore $s(t)$ is an antiderivative of $v(t)$. Thus

$$s(t) = \int v(t)\, dt = \int (6t + .5)\, dt = 3t^2 + .5t + C,$$

where C is a constant. When $t = 0$, the rocket's height is 8 meters. That is, $s(0) = 8$ and

$$8 = s(0) = 3(0)^2 + .5(0) + C = C.$$

Thus $C = 8$ and

$$s(t) = 3t^2 + .5t + 8.$$

■

EXAMPLE 9   **Cost function**   A company's marginal cost function is $C'(x) = .015x^2 - 2x + 80$ dollars, where x denotes the number of units produced in 1 day. The company has fixed costs of \$1000 per day.

(a) Find the cost of producing x units per day.

(b) If the current production level is $x = 30$, determine the amount that costs will rise if the production level is raised to $x = 60$ units.

Solution   **(a)** Let $C(x)$ be the cost of producing x units in 1 day. The derivative $C'(x)$ is the marginal cost. In other words, $C(x)$ is an antiderivative of the marginal cost function. Thus

$$C(x) = \int (.015x^2 - 2x + 80)\, dx = .005x^3 - x^2 + 80x + C.$$

The $1000 fixed costs are the costs incurred when producing 0 units. That is, $C(0) = 1000$. So

$$1000 = C(0) = .005(0)^3 - (0)^2 + 80(0) + C.$$

Therefore, $C = 1000$, and

$$C(x) = .005x^3 - x^2 + 80x + 1000.$$

(b) The cost when $x = 30$ is $C(30)$, and the cost when $x = 60$ is $C(60)$. So the *increase* in cost when production is raised from $x = 30$ to $x = 60$ is $C(60) - C(30)$. We compute

$$C(60) = .005(60)^3 - (60)^2 + 80(60) + 1000 = 3280$$
$$C(30) = .005(30)^3 - (30)^2 + 80(30) + 1000 = 2635.$$

Thus the increase in cost is $3280 - 2635 = \$645.$ ■

INCORPORATING TECHNOLOGY

Graphing an Antiderivative and Solving a Differential Equation   To graph the solution of the differential equation in Example 7, proceed as follows. First, press $\boxed{\text{Y=}}$; then set $Y_1 = \mathbf{fnInt}(X^2 - 2, X, 1, X) + 4/3$. To display **fnInt** press $\boxed{\text{MATH}}$ $\boxed{9}$. Upon pressing $\boxed{\text{GRAPH}}$, you will notice that graphing Y_1 proceeds very slowly. This can be speeded up by setting xRes (in the $\boxed{\text{WINDOW}}$ screen) to a higher value. Figure 3 shows the graph of the solution to Example 7.   ■

Figure 3

Practice Problems 6.1

1. Determine the following:

(a) $\displaystyle\int t^{7/2}\, dt$ **(b)** $\displaystyle\int \left(\frac{x^3}{3} + \frac{3}{x^3} + \frac{3}{x}\right) dx$

2. Find the value of k that makes the following antidifferentiation formula true.

$$\int (1 - 2x)^3\, dx = k(1 - 2x)^4 + C.$$

EXERCISES 6.1

Find all antiderivatives of each following function:

1. $f(x) = x$ **2.** $f(x) = 9x^8$ **3.** $f(x) = e^{3x}$

4. $f(x) = e^{-3x}$ **5.** $f(x) = 3$ **6.** $f(x) = -4x$

Determine the following:

7. $\displaystyle\int 4x^3\, dx$ **8.** $\displaystyle\int \frac{x}{3}\, dx$ **9.** $\displaystyle\int 7\, dx$

10. $\displaystyle\int k^2\, dx$ (k a constant)

11. $\displaystyle\int \frac{x}{c}\, dx$ (c a constant $\neq 0$)

12. $\displaystyle\int x \cdot x^2\, dx$ **13.** $\displaystyle\int \left(\frac{2}{x} + \frac{x}{2}\right) dx$

14. $\displaystyle\int \frac{1}{7x}\, dx$ **15.** $\displaystyle\int x\sqrt{x}\, dx$

16. $\int \left(\dfrac{2}{\sqrt{x}} + 2\sqrt{x} \right) dx$ **17.** $\int \left(x - 2x^2 + \dfrac{1}{3x} \right) dx$

18. $\int \left(\dfrac{7}{2x^3} - \sqrt[3]{x} \right) dx$ **19.** $\int 3e^{-2x}\, dx$

20. $\int e^{-x}\, dx$ **21.** $\int e\, dx$

22. $\int \dfrac{7}{2e^{2x}}\, dx$ **23.** $\int -2(e^{2x} + 1)\, dx$

24. $\int \left(-3e^{-x} + 2x - \dfrac{e^{.5x}}{2} \right) dx$

In Exercises 25–36, find the value of k that makes the antidifferentiation formula true. [Note: You can check your answer without looking in the answer section. How?]

25. $\int 5e^{-2t}\, dt = ke^{-2t} + C$

26. $\int 3e^{t/10}\, dt = ke^{t/10} + C$

27. $\int 2e^{4x-1}\, dx = ke^{4x-1} + C$

28. $\int \dfrac{4}{e^{3x+1}}\, dx = \dfrac{k}{e^{3x+1}} + C$

29. $\int (5x - 7)^{-2}\, dx = k(5x - 7)^{-1} + C$

30. $\int \sqrt{x + 1}\, dx = k(x + 1)^{3/2} + C$

31. $\int (4 - x)^{-1}\, dx = k\ln|4 - x| + C$

32. $\int \dfrac{7}{(8 - x)^4}\, dx = \dfrac{k}{(8 - x)^3} + C$

33. $\int (3x + 2)^4\, dx = k(3x + 2)^5 + C$

34. $\int (2x - 1)^3\, dx = k(2x - 1)^4 + C$

35. $\int \dfrac{3}{2 + x}\, dx = k\ln|2 + x| + C$

36. $\int \dfrac{5}{2 - 3x}\, dx = k\ln|2 - 3x| + C$

Find all functions f(t) with the following property:

37. $f'(t) = t^{3/2}$ **38.** $f'(t) = \dfrac{4}{6 + t}$

39. $f'(t) = 0$ **40.** $f'(t) = t^2 - 5t - 7$

Find all functions f(x) with the following properties:

41. $f'(x) = .5e^{-.2x},\ f(0) = 0$

42. $f'(x) = 2x - e^{-x},\ f(0) = 1$

43. $f'(x) = x,\ f(0) = 3$

44. $f'(x) = 8x^{1/3},\ f(1) = 4$

45. $f'(x) = \sqrt{x} + 1,\ f(4) = 0$

46. $f'(x) = x^2 + \sqrt{x},\ f(1) = 3$

47. Figure 4 shows the graphs of several functions $f(x)$ for which $f'(x) = \dfrac{2}{x}$. Find the expression for the function $f(x)$ whose graph passes through $(1, 2)$.

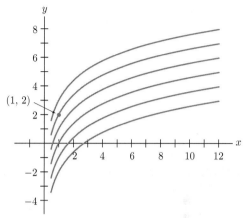

Figure 4

48. Figure 5 shows the graphs of several functions $f(x)$ for which $f'(x) = \frac{1}{3}$. Find the expression for the function $f(x)$ whose graph passes through $(6, 3)$.

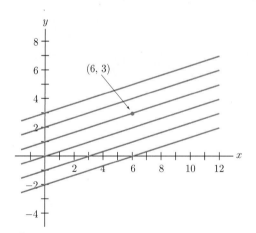

Figure 5

49. Which of the following is $\int \ln x\, dx$?

(a) $\dfrac{1}{x} + C$ (b) $x \cdot \ln x - x + C$

(c) $\dfrac{1}{2} \cdot (\ln x)^2 + C$

50. Which of the following is $\int x\sqrt{x + 1}\, dx$?

(a) $\frac{2}{5}(x + 1)^{5/2} - \frac{2}{3}(x + 1)^{3/2} + C$

(b) $\frac{1}{2}x^2 \cdot \frac{2}{3}(x + 1)^{3/2} + C$

51. Figure 6 contains the graph of a function $F(x)$. On the same coordinate system, draw the graph of the function $G(x)$ having the properties $G(0) = 0$ and $G'(x) = F'(x)$ for each x.

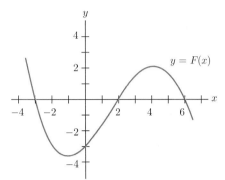

Figure 6

52. Figure 7 contains an antiderivative of the function $f(x)$. Draw the graph of another antiderivative of $f(x)$.

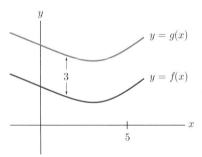

Figure 7

53. The function $g(x)$ in Fig. 8 was obtained by shifting the graph of $f(x)$ up 3 units. If $f'(5) = \frac{1}{4}$, what is $g'(5)$?

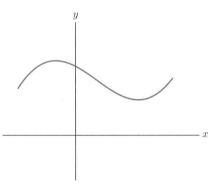

Figure 8

54. The function $g(x)$ in Fig. 9 was obtained by shifting the graph of $f(x)$ up 2 units. What is the derivative of $h(x) = g(x) - f(x)$?

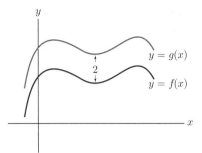

Figure 9

55. Position of a Ball A ball is thrown upward from a height of 256 feet above the ground, with an initial velocity of 96 feet per second. From physics it is known that the velocity at time t is $v(t) = 96 - 32t$ feet per second.

(a) Find $s(t)$, the function giving the height of the ball at time t.

(b) How long will the ball take to reach the ground?

(c) How high will the ball go?

56. A rock is dropped from the top of a 400-foot cliff. Its velocity at time t seconds is $v(t) = -32t$ feet per second.

(a) Find $s(t)$, the height of the rock above the ground at time t.

(b) How long will the rock take to reach the ground?

(c) What will be its velocity when it hits the ground?

57. Let $P(t)$ be the total output of a factory assembly line after t hours of work. If the rate of production at time t is $P'(t) = 60 + 2t - \frac{1}{4}t^2$ units per hour, find the formula for $P(t)$.

58. After t hours of operation, a coal mine is producing coal at the rate of $C'(t) = 40 + 2t - \frac{1}{5}t^2$ tons of coal per hour. Find a formula for the total output of the coal mine after t hours of operation.

59. Heat Diffusion A package of frozen strawberries is taken from a freezer at $-5°$C into a room at $20°$C. At time t the average temperature of the strawberries is increasing at the rate of $T'(t) = 10e^{-.4t}$ degrees Celsius per hour. Find the temperature of the strawberries at time t.

60. Epidemic A flu epidemic hits a town. Let $P(t)$ be the number of persons sick with the flu at time t, where time is measured in days from the beginning of the epidemic and $P(0) = 100$. After t days, if the flu is spreading at the rate of $P'(t) = 120t - 3t^2$ people per day, find the formula for $P(t)$.

61. Profit A small tie shop finds that at a sales level of x ties per day its marginal profit is $MP(x)$ dollars per tie, where $MP(x) = 1.30 + .06x - .0018x^2$. Also, the shop will lose \$95 per day at a sales level of $x = 0$. Find the profit from operating the shop at a sales level of x ties per day.

62. Cost A soap manufacturer estimates that its marginal cost of producing soap powder is $C'(x) = .2x + 1$ hundred dollars per ton at a production level of x tons per day. Fixed costs are \$200 per day. Find the cost of producing x tons of soap powder per day.

63. U.S. Consumption of Iron Ore The United States has been consuming iron ore at the rate of $R(t)$ million metric tons per year at time t, where $t = 0$ corresponds to 1980 and $R(t) = 94e^{.016t}$. Find a formula for the total U.S. consumption of iron ore from 1980 until time t.

64. U.S. Production of Natural Gas Since 1987, the rate of production of natural gas in the United States has been approximately $R(t)$ quadrillion British thermal units per year at time t, with $t = 0$ corresponding to 1987 and $R(t) = 17.04e^{.016t}$. Find a formula for the total U.S. production of natural gas from 1987 until time t.

65. Cost Drilling of an oil well has a fixed cost of \$10,000 and a marginal cost of $C'(x) = 1000 + 50x$ dollars per foot, where x is the depth in feet. Find the expression for $C(x)$, the total cost of drilling x feet. [*Note:* $C(0) = 10,000$.]

Technology Exercises

In Exercises 66 and 67, find an antiderivative of $f(x)$, call it $F(x)$, and compare the graphs of $F(x)$ and $f(x)$ in the given window to check that the expression for $F(x)$ is reasonable. [That is, determine whether the two graphs are consistent. When $F(x)$ has a relative extreme point, $f(x)$ should be zero; when $F(x)$ is increasing, $f(x)$ should be positive, and so on.]

66. $f(x) = 2x - e^{-.02x}$, $[-10, 10]$ by $[-20, 100]$

67. $f(x) = e^{2x} + e^{-x} + \frac{1}{2}x^2$, $[-2.4, 1.7]$ by $[-10, 10]$

68. Plot the graph of the differential equation $y' = e^{-x^2}$, $y(0) = 0$. Observe that the graph approaches the value $\sqrt{\pi}/2 \approx .9$ as x increases.

Solutions to Practice Problems 6.1

1. (a) $\displaystyle\int t^{7/2}\, dt = \frac{1}{\frac{9}{2}}t^{9/2} + C = \frac{2}{9}t^{9/2} + C$

(b) $\displaystyle\int \left(\frac{x^3}{3} + \frac{3}{x^3} + \frac{3}{x} \right) dx$

$\displaystyle = \int \left(\frac{1}{3}\cdot x^3 + 3x^{-3} + 3\cdot\frac{1}{x} \right) dx$

$\displaystyle = \frac{1}{3}\left(\frac{1}{4}x^4 \right) + 3\left(-\frac{1}{2}x^{-2} \right) + 3\ln|x| + C$

$\displaystyle = \frac{1}{12}x^4 - \frac{3}{2}x^{-2} + 3\ln|x| + C.$

2. Since we are told that the antiderivative has the general form $k(1 - 2x)^4$, all we have to do is determine the value of k. Differentiating, we obtain

$$4k(1 - 2x)^3(-2) \quad \text{or} \quad -8k(1 - 2x)^3,$$

which is supposed to equal $(1 - 2x)^3$. Therefore, $-8k = 1$, so $k = -\frac{1}{8}$.

6.2 Areas and Riemann Sums

This section and the next reveal the important connection between antiderivatives and areas of regions under curves. Although the full story will have to wait until the next section, we can give a hint now by describing how "area" is related to a problem solved in Section 6.1.

Example 9 in Section 6.1 concerned a company's marginal cost function, $f(x) = .015x^2 - 2x + 80$. A calculation with an antiderivative of $f(x)$ showed that if the production level is increased from $x = 30$ to $x = 60$ units per day, the change in total costs is \$645. As we will see, this change in cost exactly equals the area of the region under the graph of the marginal cost curve in Fig. 1 from $x = 30$ to $x = 60$. First, however, we need to learn how to find areas of such regions.

Area under a Graph If $f(x)$ is a continuous nonnegative function on the interval $a \le x \le b$, we refer to the area of the region shown in Fig. 2 as the *area under the graph of $f(x)$ from a to b.*

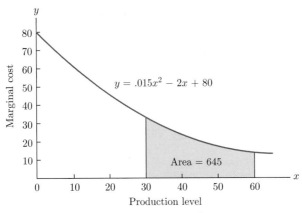

Figure 1.   Area under a marginal cost curve.

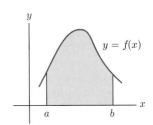

Figure 2.   Area under a graph.

The computation of the area in Fig. 2 is not a trivial matter when the top boundary of the region is curved. However, we can *estimate* the area to any desired degree of accuracy. The basic idea is to construct rectangles whose total area is approximately the same as the area to be computed. The area of each rectangle, of course, is easy to compute.

Figure 3 shows three rectangular approximations to the area under a graph. When the rectangles are thin, the mismatch between the rectangles and the region under the graph is quite small. In general, a rectangular approximation can be made as closely as desired to the exact area simply by making the width of the rectangles sufficiently small.

Given a continuous nonnegative function $f(x)$ on the interval $a \leq x \leq b$, divide the x-axis interval into n equal subintervals, where n represents some positive integer. Such a subdivision is called a *partition* of the interval from a to b. Since the entire interval is of width $b - a$, the width of each n subinterval is $(b - a)/n$. For brevity, denote this width by Δx. That is,

$$\Delta x = \frac{b - a}{n} \quad \text{(width of one subinterval)}.$$

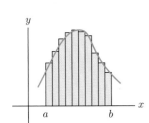

Figure 3.   Approximating a region with rectangles.

In each subinterval, select a point. (Any point in the subinterval will do.) Let x_1 be the point selected from the first subinterval, x_2 the point from the second subinterval, and so on. These points are used to form rectangles that approximate the region under the graph of $f(x)$. Construct the first rectangle with height $f(x_1)$ and the first subinterval as base, as in Fig. 4. The top of the rectangle touches the graph directly above x_1. Notice that

$$[\text{area of first rectangle}] = [\text{height}][\text{width}] = f(x_1)\,\Delta x.$$

The second rectangle rests on the second subinterval and has height $f(x_2)$. Thus

$$[\text{area of second rectangle}] = [\text{height}][\text{width}] = f(x_2)\,\Delta x.$$

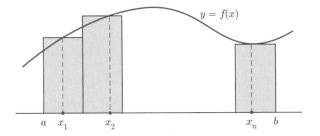

Figure 4.   Rectangles with heights $f(x_1), \ldots, f(x_n)$.

Continuing in this way, we construct n rectangles with a combined area of

$$f(x_1)\,\Delta x + f(x_2)\,\Delta x + \cdots + f(x_n)\,\Delta x. \tag{1}$$

A sum as in (1) is called a *Riemann sum*. It provides an approximation to the area under the graph of $f(x)$ when $f(x)$ is nonnegative and continuous. In fact, as the number of subintervals increases indefinitely, the Riemann sums in (1) approach a limiting value, the area under the graph.*

The value in (1) is easier to calculate when written as

$$[f(x_1) + f(x_2) + \cdots + f(x_n)]\,\Delta x.$$

This calculation requires only one multiplication.

EXAMPLE 1   Estimate the area under the graph of the marginal cost function $f(x) = .015x^2 - 2x + 80$ from $x = 30$ to $x = 60$. Use partitions of 5, 20, and 100 subintervals. Use the midpoints of the subintervals as $x_1, x_2, \ldots, x_n$ to construct the rectangles. (See Fig. 5.)

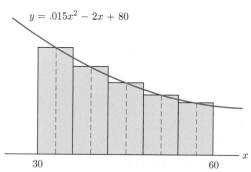

$y = .015x^2 - 2x + 80$

Figure 5. Estimating the area under a marginal cost curve.

Solution   The partition of $30 \le x \le 60$ with $n = 5$ is shown in Fig. 6. The length of each subinterval is

$$\Delta x = \frac{60 - 30}{5} = 6.$$

Figure 6. A partition of the interval $30 \le x \le 60$.

Observe that the first midpoint is $\Delta x/2$ units from the left endpoint, and the midpoints themselves are Δx units apart. The first midpoint is $x_1 = 30 + \Delta x/2 = 30 + 3 = 33$. Subsequent midpoints are found by successively adding $\Delta x = 6$:

$$\text{midpoints: } 33, 39, 45, 51, 57.$$

---

*Riemann sums are named after the nineteenth-century German mathematician G. B. Riemann (pronounced "Reemahn"), who used them extensively in his work on calculus. The concept of a Riemann sum has several uses: to approximate areas under curves, to construct mathematical models in applied problems, and to give a formal definition of area. In the next section, Riemann sums are used to define the definite integral of a function.

The corresponding estimate for the area under the graph of $f(x)$ is

$$f(33)\,\Delta x + f(39)\,\Delta x + f(45)\,\Delta x + f(51)\,\Delta x + f(57)\,\Delta x$$
$$= [f(33) + f(39) + f(45) + f(51) + f(57)]\,\Delta x$$
$$= [30.335 + 24.815 + 20.375 + 17.015 + 14.735]\cdot 6$$
$$= 107.275\cdot 6 = 643.65.$$

A similar calculation with 20 subintervals produces an area estimate of 644.916. With 100 subintervals the estimate is 644.997. ∎

The approximations in Example 1 seem to confirm the claim made at the beginning of the section that the area under the marginal cost curve equals the change in total cost, \$645. The verification of this fact will be given in the next section.

Although the midpoints of subintervals are often selected as the $x_1, x_2, \ldots, x_n$ in a Riemann sum, left endpoints and right endpoints may also be convenient.

EXAMPLE 2   Use a Riemann sum with $n = 4$ to estimate the area under the graph of $f(x) = x^2$ from 1 to 3. Select the right endpoints of the subintervals as x_1, x_2, x_3, x_4.

Solution   Here $\Delta x = (3-1)/4 = .5$. The right endpoint of the first subinterval is $1+\Delta x = 1.5$. Subsequent right endpoints are obtained by successively adding .5, as follows:

The corresponding Riemann sum is

$$f(x_1)\,\Delta x + f(x_2)\,\Delta x + f(x_3)\,\Delta x + f(x_4)\,\Delta x$$
$$= [f(x_1) + f(x_2) + f(x_3) + f(x_4)]\,\Delta x$$
$$= [(1.5)^2 + (2)^2 + (2.5)^2 + (3)^2](.5)$$
$$= [2.25 + 4 + 6.25 + 9](.5)$$
$$= 21.5\cdot(.5) = 10.75.$$

The rectangles used for this Riemann sum are shown in Fig. 7. The right endpoints here give an area estimate that is obviously greater than the exact area. Midpoints would work better. But if the rectangles are sufficiently narrow, even a Riemann sum using right endpoints will be close to the exact area. ∎

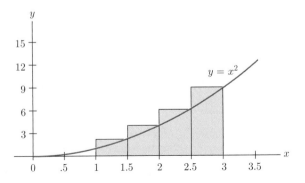

Figure 7. A Riemann sum using right endpoints.

EXAMPLE 3

To estimate the area of a 100-foot-wide waterfront lot, a surveyor measured the distance from the street to the waterline at 20-foot intervals, starting 10 feet from one corner of the lot. Use the data to construct a Riemann sum approximation to the area of the lot. (See Fig. 8.)

Figure 8. Survey of a waterfront property.

Solution Treat the street as the x-axis, and consider the waterline along the property as the graph of a function $f(x)$ over the interval from 0 to 100. The five "vertical" distances give $f(x_1), \ldots, f(x_5)$, where $x_1 = 10, \ldots, x_5 = 90$. Since there are five points $x_1, \ldots, x_5$ spread across the interval $0 \leq x \leq 100$, we partition the interval into five subintervals, with $\Delta x = \frac{100}{5} = 20$. Fortunately, each subinterval contains one x_i. (In fact, each x_i is the midpoint of a subinterval.) Thus the area of the lot is approximated by the Riemann sum

$$f(x_1)\,\Delta x + \cdots + f(x_5)\,\Delta x = [f(x_1) + \cdots + f(x_5)]\,\Delta x$$
$$= [39 + 46 + 44 + 40 + 41.5] \cdot 20$$
$$= 210.5 \cdot 20 = 4210 \text{ square feet.}$$

For a better estimate of the area, the surveyor will have to make more measurements from the street to the waterline. Note that we are able to approximate the area without ever knowing an analytic expression for the function $f(x)$. ∎

The final example shows how Riemann sums arise in an application. Since the variable is time, t, we write Δt in place of Δx.

EXAMPLE 4

The velocity of a rocket at time t is $v(t)$ feet per second. Construct a Riemann sum that estimates how far the rocket travels in the first 10 seconds. (*Note:* Don't calculate the Riemann sum. Just set it up.) What happens when the number of subintervals in the partition increases without bound?

Solution Partition the interval $0 \leq t \leq 10$ into n subintervals of width Δt, and select points $t_1, t_2, \ldots, t_n$ from these subintervals. Although the rocket's velocity is not constant, it does not change much during a small subinterval of time. So we may use $v(t_1)$ as an approximation of the rocket's velocity during the first subinterval. Over this time subinterval of length $\Delta t = (10 - 0)/n$,

$$[\text{distance traveled}] \approx [\text{velocity}] \cdot [\text{time}]$$
$$= v(t_1)\,\Delta t. \tag{2}$$

The distance traveled during the second time subinterval is approximately $v(t_2)\,\Delta t$, and so on. Thus an estimate for the total distance traveled is

$$v(t_1)\,\Delta t + v(t_2)\,\Delta t + \cdots + v(t_n)\,\Delta t. \tag{3}$$

The sum in (3) is a Riemann sum for the velocity function on the interval $0 \le t \le 10$. As n increases, such a Riemann sum approaches the area under the graph of the velocity function. However, from our derivation of (3), it seems reasonable that as n increases the sums become a better and better estimate of the total distance traveled. We conclude that

$$\begin{bmatrix} \text{total distance rocket travels} \\ \text{during the first 10 seconds} \end{bmatrix} = \begin{bmatrix} \text{area under graph of} \\ v(t) \text{ over } 0 \le t \le 10 \end{bmatrix}.$$

See Fig. 9. ∎

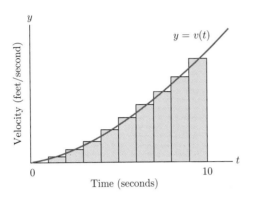

Figure 9. Area under a velocity curve.

It is helpful to look at Example 4 from a slightly different point of view. The height of the rocket is an increasing quantity, and $v(t)$ is the *rate* of change of height at time t. The area under the graph of $v(t)$ from time $a = 0$ to time $b = 10$ is the *amount* of increase in the height during the first 10 seconds.

This connection between the rate of change of a function and the amount of increase of the function generalizes to a variety of situations.

If a quantity is increasing, the area under the rate of change function from a to b is the amount of increase in the quantity from a to b.

Table 1 shows four instances of this principle. The result is justified in the next section. The first example in the table was discussed in Example 1.

TABLE 1 Interpretation of Areas

Function	a to b	Area under the Graph from a to b
Marginal cost at production level x	30 to 60	Additional cost when production is increased from 30 to 60 units
Rate of sulfur emissions from a power plant t years after 1990	1 to 3	Amount of sulfur released from 1991 to 1993
Birth rate t years after 1980 (in babies per year)	0 to 10	Number of babies born from 1980 to 1990
Rate of gas consumption t years after 1985	3 to 6	Amount of gas used from 1988 to 1991

INCORPORATING TECHNOLOGY

Riemann Sums When the points selected for a Riemann sum are all midpoints, all left endpoints, or all right endpoints, $[f(x_1) + f(x_2) + \cdots + f(x_n)]$ is the sum of the sequence of values $f(x_1), f(x_2), \ldots, f(x_n)$, where successive numbers $x_1, x_2, \ldots, x_n$ each differ by the value Δx. In such a case, this sum of a sequence can be evaluated on a graphing calculator. Figure 10 shows the computation of the Riemann sum asked for in Example 1. To do this, first set $Y_1 = 0.015X^2 - 2X + 80$. Return to the home screen and press [2nd] [LIST], and move the cursor right to MATH. Press [5] to display **sum(** and press [2nd] [LIST]; then move the cursor right to OPS. Press [5] to display **seq(**. Now complete the expression as in Fig. 10. The numbers 33, 57, and 6 are determined as explained in Example 1. ∎

Figure 10

Practice Problems 6.2

1. Determine Δx and the midpoints of the subintervals formed by partitioning the interval $-2 \leq x \leq 2$ into five subintervals.

2. The graph in Fig. 11 gives the rate at which new jobs were created (in millions of jobs per year) in the United States, where $t = 0$ corresponds to 1983. For instance, on January 1, 1986, jobs were being created at the rate of 2.4 million new jobs per year. Interpret the area of the shaded region.

Figure 11. Rate of job creation.

EXERCISES 6.2

Determine Δx and the midpoints of the subintervals formed by partitioning the given interval into n subintervals. [Hint: Decimals are sometimes easier to use than fractions.]

1. $0 \leq x \leq 2$; $n = 4$
2. $0 \leq x \leq 3$; $n = 6$
3. $1 \leq x \leq 4$; $n = 5$
4. $3 \leq x \leq 5$; $n = 5$

In Exercises 5–10, use a Riemann sum to approximate the area under the graph of $f(x)$ on the given interval, with selected points as specified.

5. $f(x) = x^2$; $1 \leq x \leq 3$, $n = 4$, midpoints of subintervals
6. $f(x) = x^2$; $-2 \leq x \leq 2$, $n = 4$, midpoints of subintervals
7. $f(x) = x^3$; $1 \leq x \leq 3$, $n = 5$, left endpoints
8. $f(x) = x^3$; $0 \leq x \leq 1$, $n = 5$, right endpoints
9. $f(x) = e^{-x}$; $2 \leq x \leq 3$, $n = 5$, right endpoints
10. $f(x) = \ln x$; $2 \leq x \leq 4$, $n = 5$, left endpoints

In Exercises 11–14, use a Riemann sum to approximate the area under the graph of $f(x)$ in Fig. 12 on the given interval, with selected points as specified. Draw the approximating rectangles.

11. $0 \leq x \leq 8$, $n = 4$, midpoints of subintervals

12. $3 \leq x \leq 7$, $n = 4$, left endpoints
13. $4 \leq x \leq 9$, $n = 5$, right endpoints
14. $1 \leq x \leq 7$, $n = 3$, midpoints of subintervals

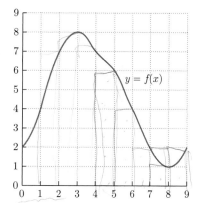

Figure 12. Graph for Exercises 11–14.

15. Use a Riemann sum with $n = 4$ and left endpoints to estimate the area under the graph of $f(x) = 4 - x$ on the interval $1 \leq x \leq 4$. Then repeat with $n = 4$ and

midpoints. Compare the answers with the exact answer, 4.5, which can be computed from the formula for the area of a triangle.

16. Use a Riemann sum with $n = 4$ and right endpoints to estimate the area under the graph of $f(x) = 2x - 4$ on the interval $2 \le x \le 3$. Then repeat with $n = 4$ and midpoints. Compare the answers with the exact answer, 1, which can be computed from the formula for the area of a triangle.

17. The graph of the function $f(x) = \sqrt{1 - x^2}$ on the interval $-1 \le x \le 1$ is a semicircle. The area under the graph is $\frac{1}{2}\pi(1)^2 = \pi/2 = 1.57080$, to five decimal places. Use a Riemann sum with $n = 5$ and midpoints to estimate the area under the graph. (See Fig. 13.) Carry out the calculations to five decimal places and compute the error (the difference between the estimate and 1.57080).

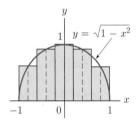

Figure 13

18. Use a Riemann sum with $n = 5$ and midpoints to estimate the area under the graph of $f(x) = \sqrt{1 - x^2}$ on the interval $0 \le x \le 1$. The graph is a quarter circle, and the area under the graph is .78540, to five decimal places. (See Fig. 14.) Carry out the calculations to five decimal places and compute the error. If you double the estimate from this exercise, will it be more accurate than the estimate computed in Exercise 17?

Figure 14

19. **Airflow through the Lungs** Lung physiologists measure the velocity of air passing through a patient's throat by having the patient breathe into a pneumotachograph. This machine produces a graph that plots airflow rate as a function of time. The graph in Fig. 15 shows the flow rate while a patient is breathing out. The area under the graph gives the total volume of air during exhalation. Estimate this volume with a Riemann sum. Use $n = 5$ and midpoints.

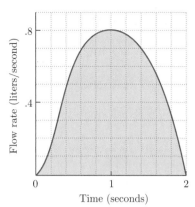

Figure 15. Data from a pneumotachograph.

20. Estimate the area (in square feet) of the residential property shown in Fig. 16.

Figure 16. Area of a residential property.

21. **Velocity** The velocity of a car (in feet per second) is recorded from the speedometer every 10 seconds, beginning 5 seconds after the car starts to move. (See Table 2.) Use a Riemann sum to estimate the distance the car travels during the first 60 seconds. [*Note:* Each velocity is given at the *middle* of a 10-second interval. The first interval extends from 0 to 10, and so on.]

TABLE 2	A Car's Velocity					
Time	5	15	25	35	45	55
Velocity	20	44	32	39	65	80

22. Table 3 shows the velocity (in feet per second) at the end of each second for a person starting a morning jog. Make three Riemann sum estimates of the total distance jogged during the time from $t = 2$ to $t = 8$.

(a) $n = 6$, left endpoints

(b) $n = 6$, right endpoints

(c) $n = 3$, midpoints

TABLE 3	A Jogger's Velocity								
Time	0	1	2	3	4	5	6	7	8
Velocity	0	2	3	5	5	6	6	7	8

TABLE 4 Interpretation of Areas

Function	a to b	Area under the Graph from a to b
Rate of growth of a population t years after 1900 (in millions of people per year)	10 to 50	
	5 to 7	Number of cigarettes smoked from 1990 to 1992
Marginal profit at production level x		Additional profit created by increasing production from 20 to 50 units

23. Complete the three missing entries in Table 4.

24. Interpret the area of the shaded region in Fig. 17.

25. Interpret the area of the shaded region in Fig. 18.

26. Interpret the area of the shaded region in Fig. 19.

In Exercises 27 and 28, let A be the area under the graph of $y = f(x)$ from $x = a$ to $x = b$.

27. If the function $f(x)$ is always increasing on the interval $a \le x \le b$, explain why [approximation given by the Riemann sum with left endpoints] $< A <$ [approximation given by the Riemann sum with right endpoints].

28. If the function $f(x)$ is always decreasing on the interval $a \le x \le b$, explain why [approximation given by the Riemann sum with left endpoints] $> A >$ [approximation given by the Riemann sum with right endpoints].

Technology Exercises

Evaluate a Riemann sum to approximate the area under the graph of $f(x)$ on the given interval, with points selected as specified.

29. $f(x) = x\sqrt{1 + x^2}$; $1 \le x \le 3$, $n = 20$, midpoints of subintervals

30. $f(x) = \sqrt{1 - x^2}$; $-1 \le x \le 1$, $n = 20$, midpoints of subintervals

31. $f(x) = e^{-x^2}$; $-2 \le x \le 2$, $n = 100$, left endpoints

32. $f(x) = x \ln x$; $2 \le x \le 6$, $n = 100$, right endpoints

33. $f(x) = (\ln x)^2$; $2 \le x \le 3$, $n = 200$, right endpoints

34. $f(x) = (1 + x)^2 e^{4x}$; $2 \le x \le 4$, $n = 200$, left endpoints

Figure 17. Marginal revenue.

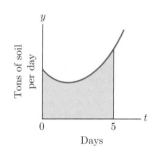

Figure 18. Rate of soil erosion.

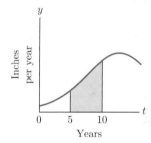

Figure 19. A child's growth rate.

Solutions to Practice Problems 6.2

1. Since $n = 5$, $\Delta x = \frac{2 - (-2)}{5} = \frac{4}{5} = .8$. The first midpoint is $x_1 = -2 + .8/2 = -1.6$. Subsequent midpoints are found by successively adding .8, to obtain

$x_2 = -.8$, $x_3 = 0$, $x_4 = .8$, and $x_5 = 1.6$.

2. The area under the curve from $t = 2$ to $t = 6$ is the number of new jobs created from 1985 to 1989.

6.3 Definite Integrals and the Fundamental Theorem

In Section 6.2 we saw that the area under the graph of a continuous nonnegative function $f(x)$ from a to b is the limiting value of Riemann sums of the form

$$f(x_1)\,\Delta x + f(x_2)\,\Delta x + \cdots + f(x_n)\,\Delta x$$

as n increases without bound or, equivalently, as Δx approaches zero. (Recall that $x_1, x_2, \ldots, x_n$ are selected points from a partition of $a \le x \le b$, and Δx is the width of each n subinterval.) It can be shown that even if $f(x)$ has negative values the Riemann sums still approach a limiting value as $\Delta x \to 0$. This number is called the *definite integral of $f(x)$ from a to b* and is denoted by

$$\int_a^b f(x)\, dx.$$

That is,

$$\int_a^b f(x)\, dx = \lim_{\Delta x \to 0} [f(x_1)\,\Delta x + f(x_2)\,\Delta x + \cdots + f(x_n)\,\Delta x]. \qquad (1)$$

If $f(x)$ is a nonnegative function, we know from Section 6.2 that the Riemann sum on the right side of (1) approaches the area under the graph of $f(x)$ from a to b. Thus, *the definite integral of a nonnegative function $f(x)$ equals the area under the graph of $f(x)$.*

EXAMPLE 1   Calculate

$$\int_1^4 \left(\frac{1}{3}x + \frac{2}{3}\right) dx.$$

Solution Figure 1 shows the graph of the function $f(x) = \frac{1}{3}x + \frac{2}{3}$. Since $f(x)$ is nonnegative for $1 \le x \le 4$, the definite integral of $f(x)$ equals the area of the shaded region in Fig. 1. The region consists of a rectangle and a triangle. By geometry,

$$[\text{area of rectangle}] = [\text{width}] \cdot [\text{height}] = 3 \cdot 1 = 3,$$

$$[\text{area of triangle}] = \tfrac{1}{2}[\text{width}] \cdot [\text{height}] = \tfrac{1}{2} \cdot 3 \cdot 1 = \tfrac{3}{2}.$$

Thus the area under the graph is $4\frac{1}{2}$, and hence

$$\int_1^4 \left(\tfrac{1}{3}x + \tfrac{2}{3}\right) dx = 4\tfrac{1}{2}. \qquad \blacksquare$$

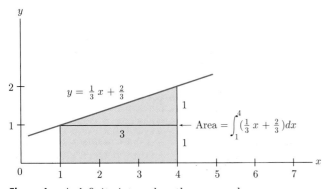

Figure 1. A definite integral as the area under a curve.

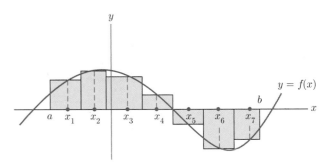

Figure 2

If $f(x)$ is negative at some points in the interval, we may also give a geometric interpretation of the definite integral. Consider the function $f(x)$ shown in Fig. 2. The figure shows a rectangular approximation of the region between the graph and the x-axis from a to b. Consider a typical rectangle located above or below the selected point x_i. If $f(x_i)$ is nonnegative, the area of the rectangle equals $f(x_i)\,\Delta x$. In case $f(x_i)$ is negative, the area of the rectangle equals $(-f(x_i))\,\Delta x$. So the expression $f(x_i)\,\Delta x$ equals either the area of the corresponding rectangle or the negative of the area, according to whether $f(x_i)$ is nonnegative or negative, respectively. In particular, the Riemann sum

$$f(x_1)\,\Delta x + f(x_2)\,\Delta x + \cdots + f(x_n)\,\Delta x$$

is equal to the area of the rectangles above the x-axis minus the area of the rectangles below the x-axis. Now take the limit as Δx approaches 0. On the one hand, the Riemann sum approaches the definite integral. On the other hand, the rectangular approximations approach the area bounded by the graph that is above the x-axis minus the area bounded by the graph that is below the x-axis. This gives us the following geometric interpretation of the definite integral.

> Suppose that $f(x)$ is continuous on the interval $a \leq x \leq b$. Then
>
> $$\int_a^b f(x)\,dx$$
>
> is equal to the area above the x-axis bounded by the graph of $y = f(x)$ from $x = a$ to $x = b$ minus the area below the x-axis. Referring to Fig. 3, we have
>
> $$\int_a^b f(x)\,dx = [\text{area of } B \text{ and } D] - [\text{area of } A \text{ and } C]$$

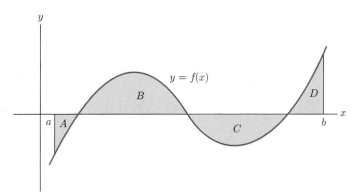

Figure 3.   Regions above and below the x-axis.

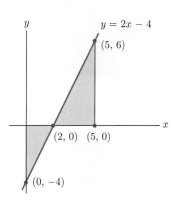

Figure 4

EXAMPLE 2   Calculate

$$\int_0^5 (2x - 4)\,dx.$$

Solution   Figure 4 shows the graph of the function $f(x) = 2x - 4$ on the interval $0 \leq x \leq 5$. The area of the triangle above the x-axis is 9 and the area of the triangle below the x-axis is 4. Therefore, from geometry, we find that

$$\int_0^5 (2x - 4)\,dx = 9 - 4 = 5. \qquad \blacksquare$$

The values of the definite integrals in Examples 1 and 2 follow from simple geometric formulas. For integrals of more complex functions, analogous area formulas are not available. Of course, Riemann sums may always be used to estimate the value of a definite integral to any desired degree of accuracy. However, for most of the integrals in this text, the following fundamental theorem of calculus will rescue us from such calculations.

Fundamental Theorem of Calculus   Suppose that $f(x)$ is continuous on the interval $a \leq x \leq b$, and let $F(x)$ be an antiderivative of $f(x)$. Then

$$\int_a^b f(x)\, dx = F(b) - F(a). \tag{2}$$

This theorem connects the two key concepts of calculus: the integral and the derivative. An explanation of why the theorem is true is given later in this section. First we show how to use the theorem to evaluate definite integrals.

EXAMPLE 3   Use the fundamental theorem of calculus to evaluate the following definite integrals.

(a) $\displaystyle\int_1^4 \left(\tfrac{1}{3}x + \tfrac{2}{3}\right) dx$ (b) $\displaystyle\int_0^5 (2x - 4)\, dx$

Solution (a) An antiderivative of $f(x) = \tfrac{1}{3}x + \tfrac{2}{3}$ is $F(x) = \tfrac{1}{6}x^2 + \tfrac{2}{3}x$. Therefore, by the fundamental theorem,

$$\int_1^4 \left(\tfrac{1}{3}x + \tfrac{2}{3}\right) dx = F(4) - F(1)$$

$$= \left[\tfrac{1}{6}(4)^2 + \tfrac{2}{3}(4)\right] - \left[\tfrac{1}{6}(1)^2 + \tfrac{2}{3}(1)\right]$$

$$= \left[\tfrac{16}{6} + \tfrac{8}{3}\right] - \left[\tfrac{1}{6} + \tfrac{2}{3}\right] = 4\tfrac{1}{2}.$$

This result is the same as in Example 1.

(b) An antiderivative of $f(x) = 2x - 4$ is $F(x) = x^2 - 4x$. Therefore,

$$\int_0^5 (2x - 4)\, dx = F(5) - F(0) = \left[5^2 - 4(5)\right] - \left[0^2 - 4(0)\right] = 5.$$

This result is the same as in Example 2.   ▪

EXAMPLE 4   Evaluate

$$\int_2^5 3x^2\, dx.$$

Solution An antiderivative of $f(x) = 3x^2$ is $F(x) = x^3 + C$, where C is any constant. Then

$$\int_2^5 3x^2\, dx = F(5) - F(2) = \left[5^3 + C\right] - \left[2^3 + C\right] = 5^3 + C - 2^3 - C = 117.$$

Notice how the C in $F(2)$ is subtracted from the C in $F(5)$. Thus the value of the definite integral does not depend on the choice of the constant C. For convenience, we may omit C when evaluating a definite integral.   ▪

The quantity $F(b) - F(a)$ is called the *net change of $F(x)$ from $x = a$ to $x = b$.* It is abbreviated by the symbol $F(x)\big|_a^b$. For instance, the net change of $F(x) = \tfrac{1}{3}e^{3x}$ from $x = 0$ to $x = 2$ is written as $\tfrac{1}{3}e^{3x}\big|_0^2$ and is evaluated as $F(2) - F(0)$.

EXAMPLE 5   Evaluate

$$\int_0^2 e^{3x}\, dx.$$

Solution An antiderivative of $f(x) = e^{3x}$ is $F(x) = \frac{1}{3}e^{3x}$. Therefore,

$$\int_0^2 e^{3x}\, dx = \frac{1}{3}e^{3x}\Big|_0^2 = \frac{1}{3}e^{3(2)} - \frac{1}{3}e^{3(0)} = \frac{1}{3}e^6 - \frac{1}{3}. \qquad \blacksquare$$

EXAMPLE 6

Compute the area under the curve $y = x^2 - 4x + 5$ from $x = -1$ to $x = 3$.

Solution The graph of $f(x) = x^2 - 4x + 5$ is shown in Fig. 5. Since $f(x)$ is nonnegative for $-1 \leq x \leq 3$, the area under the curve is given by the definite integral

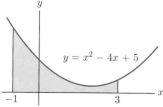

$y = x^2 - 4x + 5$

Figure 5

$$\int_{-1}^3 (x^2 - 4x + 5)\, dx = \left(\frac{x^3}{3} - 2x^2 + 5x \right)\Big|_{-1}^3$$

$$= \left[\frac{(3)^3}{3} - 2(3)^2 + 5(3) \right] - \left[\frac{(-1)^3}{3} - 2(-1)^2 + 5(-1) \right]$$

$$= \left[9 - 18 + 15 \right] - \left[-\tfrac{1}{3} - 2 - 5 \right]$$

$$= \left[6 \right] - \left[-\tfrac{22}{3} \right] = \tfrac{18}{3} + \tfrac{22}{3} = \tfrac{40}{3}. \qquad \blacksquare$$

Study this solution carefully. The calculations show how to use parentheses to avoid errors in arithmetic. It is particularly important to include the outside brackets around the value of the antiderivative at -1.

Areas in Applications At the end of Section 6.2, we described how the area under a graph can represent the amount of change (now called the "net change") in a quantity. This interpretation of area follows immediately from the fundamental theorem of calculus, which we may rewrite as

$$\int_a^b F'(x)\, dx = F(b) - F(a) \qquad (3)$$

because the $f(x)$ in (2) is the derivative of $F(x)$. If the derivative is nonnegative, the integral in (3) is the area under the graph of the "rate of change" function $F'(x)$. Equation (3) says that this area equals the net change in $F(x)$ over the interval from a to b.

EXAMPLE 7

Position of a rocket A rocket is fired vertically into the air. Its velocity at t seconds after lift-off is $v(t) = 6t + .5$ meters per second.

(a) Describe a region whose area represents the distance the rocket travels from time $t = 40$ to $t = 100$ seconds.

(b) Compute the distance in part (a).

Solution (a) Let $s(t)$ be the position of the rocket at time t measured from some reference point. (Example 8 in Section 6.1 measured distances from the launch pad, for example.) Then the required distance is $s(100) - s(40)$, the net change in position over the interval $40 \leq t \leq 100$. This distance is represented by the area under the velocity curve from 40 to 100. (See Fig. 6.)

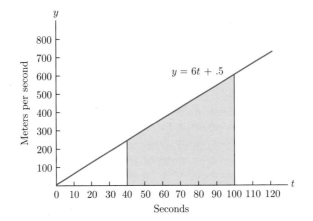

Figure 6. Area under the velocity curve is distance traveled.

(b) Now that we know the connection between antiderivatives and area, we can easily compute the area in part (a) corresponding to the distance traveled by the rocket:

$$s(100) - s(40) = \int_{40}^{100} s'(t)\,dt = \int_{40}^{100} v(t)\,dt$$

$$= \int_{40}^{100} (6t + .5)\,dt = \left(3t^2 + .5t\right)\Big|_{40}^{100}$$

$$= \left[3(100)^2 + .5(100)\right] - \left[3(40)^2 + .5(40)\right]$$

$$= 30{,}050 - 4820 = 25{,}230 \text{ meters.} \qquad \blacksquare$$

EXAMPLE 8

Worldwide oil consumption During the early 1970s, the annual worldwide rate of oil consumption was $R(t) = 16.1e^{.07t}$ billion barrels of oil per year, where t is the number of years since the beginning of 1970.

(a) Determine the amount of oil consumed from 1972 to 1974.

(b) Represent the answer to part (a) as an area.

Solution **(a)** We are interested in $T(4) - T(2)$, where $T(t)$ is the total consumption of oil since 1970. This difference is the net change in $T(t)$ over the time period from $t = 2$ (1972) to $t = 4$ (1974). Now, $T(t)$ is an antiderivative of the rate function $R(t)$. Hence

$$T(4) - T(2) = \int_2^4 R(t)\,dt = \int_2^4 16.1e^{.07t}\,dt = \frac{16.1}{.07}e^{.07t}\Big|_2^4$$

$$= 230e^{.07(4)} - 230e^{.07(2)} \approx 39.76 \text{ billion barrels of oil.}$$

(b) The net change in $T(t)$ is the area of the region under the graph of the rate function $T'(t) = R(t)$ from $t = 2$ to $t = 4$. (See Fig. 7.) $\qquad \blacksquare$

Verification of the Fundamental Theorem of Calculus We shall give two explanations of why the fundamental theorem of calculus is true. Both convey important ideas about definite integrals and the fundamental theorem.

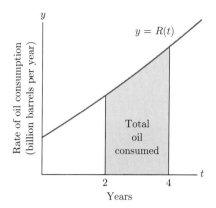

Figure 7. Total oil consumed.

The first explanation of the fundamental theorem is based directly on the Riemann sum definition of the integral. As we observed earlier, the fundamental theorem may be written in the form

$$\int_a^b F'(x)\,dx = F(b) - F(a), \tag{4}$$

where $F(x)$ is any function with a continuous derivative on $a \leq x \leq b$. There are three key ideas in an explanation of why (4) is true.

I. If the interval $a \leq x \leq b$ is partitioned into n subintervals of width $\Delta x = (b-a)/n$, the net change in $F(x)$ over $a \leq x \leq b$ is the sum of the net changes in $F(x)$ over each subinterval.

For example, partition $a \leq x \leq b$ into three subintervals and denote the left endpoints by x_1, x_2, x_3, as follows:

$$a = x_1 \qquad x_2 \qquad x_3 \qquad b$$

Then

$$[\text{change in } F(x) \text{ over 1st interval}] = F(x_2) - F(a),$$
$$[\text{change in } F(x) \text{ over 2nd interval}] = F(x_3) - F(x_2),$$
$$[\text{change in } F(x) \text{ over 3rd interval}] = F(b) - F(x_3).$$

When these changes are summed, the intermediate terms cancel:

$$F(b) - F(x_3) + F(x_3) - F(x_2) + F(x_2) - F(a) = F(b) - F(a).$$

II. If Δx is small, the change in $F(x)$ over the ith subinterval is approximately $F'(x_i)\,\Delta x$.

This is the approximation of the change in a function discussed in Section 1.8. Here x_i is the left endpoint of the ith subinterval, as follows:

$$\longleftarrow \Delta x \longrightarrow$$
$$x_i \qquad\qquad x_i + \Delta x$$

III. The sum of the approximations in (II) is a Riemann sum for the definite integral $\int_a^b F'(x)\,dx$.

That is,

$$F(b) - F(a)$$

$$= \left[\begin{array}{c} \text{change over} \\ \text{1st subinterval} \end{array}\right] + \left[\begin{array}{c} \text{change over} \\ \text{2nd subinterval} \end{array}\right] + \cdots + \left[\begin{array}{c} \text{change over} \\ n\text{th subinterval} \end{array}\right]$$

$$\approx F'(x_1)\,\Delta x + F'(x_2)\,\Delta x + \cdots + F'(x_n)\,\Delta x.$$

As $\Delta x \to 0$, these approximations improve. Thus, in the limit

$$F(b) - F(a) = \int_a^b F'(x)\,dx.$$

To summarize, the derivative of $F(x)$ determines the approximate change of $F(x)$ on small subintervals. The definite integral sums these approximate changes and, in the limit, gives the exact change of $F(x)$ over the entire interval $a \le x \le b$.

If you look back at the solution of Example 4 of Section 6.2, you will see essentially the same argument. The distance a rocket travels is the sum of the distances it travels over small intervals of time. On each subinterval, this distance is approximately the velocity (the derivative) times Δt.

An Area Function as an Antiderivative The second explanation of the fundamental theorem of calculus applies only when $f(x)$ is nonnegative. We begin with the following theorem, which describes the basic relationship between area and antiderivatives. In fact, this theorem is sometimes referred to as an alternative version of the fundamental theorem of calculus.

Theorem III Let $f(x)$ be a continuous nonnegative function for $a \le x \le b$. Let $A(x)$ be the area of the region under the graph of the function from a to the number x. (See Fig. 8.) Then $A(x)$ is an antiderivative of $f(x)$.

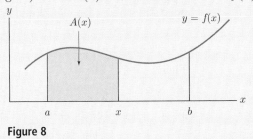

Figure 8

The fundamental theorem of calculus for a nonnegative function follows easily from Theorem III. Let $F(x)$ be an antiderivative of $f(x)$. Since the "area function" $A(x)$ is also an antiderivative of $f(x)$, by Theorem III, we have

$$A(x) = F(x) + C$$

for some constant C. Notice that $A(a)$ is 0 and $A(b)$ equals the area of the region under the graph of $f(x)$ for $a \le x \le b$. Therefore,

$$\int_a^b f(x)\,dx = A(b) = A(b) - A(a)$$

$$= [F(b) + C] - [F(a) + C]$$

$$= F(b) - F(a).$$

It is not difficult to explain why Theorem III is reasonable, although we shall not give a formal proof of the theorem. If h is a small positive number, $A(x+h) - A(x)$ is the area of the darker shaded region in Fig. 9. This shaded region is approximately a rectangle of width h, height $f(x)$, and area $h \cdot f(x)$. Thus

$$A(x + h) - A(x) \approx h \cdot f(x),$$

where the approximation becomes better as h approaches zero. Dividing by h, we have

$$\frac{A(x + h) - A(x)}{h} \approx f(x).$$

Since the approximation improves as h approaches zero, the quotient must approach $f(x)$. However, the limit definition of the derivative tells us that the quotient approaches $A'(x)$ as h approaches zero. Therefore, we have $A'(x) = f(x)$. Since x represented any number between a and b, this shows that $A(x)$ is an antiderivative of $f(x)$.

Figure 9

Figure 10

 INCORPORATING TECHNOLOGY

Definite Integral   The definite integral from Example 4 is evaluated in Figure 10. To do this, select MATH 9 to display **fnInt(**. Now complete the expression to read **fnInt**$(3X^2, X, 2, 5)$ and press ENTER. The result is shown in Fig. 10.   ■

Practice Problems 6.3

1. Find the area under the curve $y = e^{x/2}$ from $x = -3$ to $x = 2$.

2. Let $MR(x)$ be a company's marginal revenue at production level x. Give an economic interpretation of the number $\int_{75}^{80} MR(x)\, dx$.

EXERCISES 6.3

In Exercises 1–3, set up the definite integral that gives the area of the shaded region.

1.

2.

3.

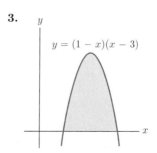

In Exercises 4–6, draw the region whose area is given by the definite integral.

4. $\displaystyle\int_2^4 x^2\,dx$

5. $\displaystyle\int_0^4 (8-2x)\,dx$

6. $\displaystyle\int_0^4 \sqrt{x}\,dx$

Calculate the following definite integrals.

7. $\displaystyle\int_{-1}^1 x\,dx$

8. $\displaystyle\int_0^1 \left(x+\frac{1}{2}\right)\,dx$

9. $\displaystyle\int_1^2 5\,dx$

10. $\displaystyle\int_1^4 x^2\sqrt{x}\,dx$

11. $\displaystyle\int_1^4 \frac{1}{2\sqrt{x}}\,dx$

12. $\displaystyle\int_0^1 (4x^3-1)\,dx$

13. $\displaystyle\int_0^1 4e^{-3x}\,dx$

14. $\displaystyle\int_0^1 2e^{2x}\,dx$

15. $\displaystyle\int_{-1}^1 \frac{e^x+e^{-x}}{2}\,dx$

16. $\displaystyle\int_0^1 \sqrt{e^x}\,dx$

17. $\displaystyle\int_0^{\ln 3} e^{-2t}\,dt$

18. $\displaystyle\int_{\ln 2}^{\ln 3} \frac{3}{e^{3t}}\,dt$

19. $\displaystyle\int_3^6 x^{-1}\,dx$

20. $\displaystyle\int_1^e \frac{1}{7x}\,dx$

21. $\displaystyle\int_{\frac{1}{5}}^5 \frac{2}{x}\,dx$

22. $\displaystyle\int_1^2 \left(\frac{x}{2}+\frac{2}{x}+\frac{1}{2x^2}\right)\,dx$

23. $\displaystyle\int_0^1 (3x^{1/3}-1-e^{.5x})\,dx$

24. $\displaystyle\int_{-1}^1 \frac{1-x-e^{-x}}{2}\,dx$

25. $\displaystyle\int_0^3 (x^3+x-7)\,dx$

26. $\displaystyle\int_0^1 \left(e^{x/3}-\frac{2x}{5}\right)\,dx$

27. $\displaystyle\int_2^3 (5-2t)^4\,dt$

28. $\displaystyle\int_4^9 \frac{3}{t-2}\,dt$

Find the area under each of the given curves.

29. $y=4x$; $x=2$ to $x=3$

30. $y=3x^2$; $x=-1$ to $x=1$

31. $y=e^{x/2}$; $x=0$ to $x=1$

32. $y=\sqrt{x}$; $x=0$ to $x=4$

33. $y=(x-3)^4$; $x=1$ to $x=4$

34. $y=e^{3x}$; $x=-\frac{1}{3}$ to $x=0$

35. Let $f(x)$ be the function pictured in Fig. 11. Determine whether $\int_0^7 f(x)\,dx$ is positive, negative, or zero.

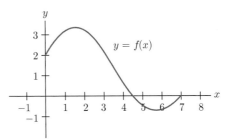

Figure 11

36. Let $g(x)$ be the function pictured in Fig. 12. Determine whether $\int_0^7 g(x)\,dx$ is positive, negative, or zero.

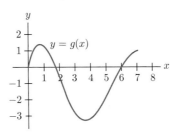

Figure 12

37. Cigarette Consumption The worldwide rate of cigarette consumption (in trillions of cigarettes per year) since 1960 is given approximately by the function $c(t) = .1t + 2.4$, where $t = 0$ corresponds to 1960. Determine the number of cigarettes sold from 1980 to 1998.

38. Suppose that $p(t)$ is the rate (in tons per year) at which pollutants are discharged into a lake, where t is the number of years since 1990. Interpret $\int_5^7 p(t)\,dt$.

39. A helicopter is rising straight up in the air. Its velocity at time t is $v(t) = 2t + 1$ feet per second.

    (a) How high does the helicopter rise during the first 5 seconds?

    (b) Represent the answer to part (a) as an area.

40. Production After t hours of operation, an assembly line is producing power lawn mowers at the rate of $r(t) = 21 - \frac{4}{5}t$ mowers per hour.

    (a) How many mowers are produced during the time from $t = 2$ to $t = 5$ hours?

    (b) Represent the answer to part (a) as an area.

41. Cost Suppose that the marginal cost function of a handbag manufacturer is $C'(x) = \frac{3}{32}x^2 - x + 200$ dollars per unit at production level x (where x is measured in units of 100 handbags).

    (a) Find the total cost of producing 6 additional units if 2 units are currently being produced.

    (b) Describe the answer to part (a) as an area. (Give a written description rather than a sketch.)

42. Profit Suppose that the marginal profit function for a company is $P'(x) = 100 + 50x - 3x^2$ at production level x.

(a) Find the extra profit earned from the sale of 3 additional units if 5 units are currently being produced.

(b) Describe the answer to part (a) as an area. (Do not make a sketch.)

43. Let $MP(x)$ be a company's marginal profit at production level x. Give an economic interpretation of the number $\int_{44}^{48} MP(x)\, dx$.

44. Let $MC(x)$ be a company's marginal cost at production level x. Give an economic interpretation of the number $\int_{0}^{100} MC(x)\, dx$. [*Note:* At any production level, the total cost equals the fixed cost plus the total variable cost.]

45. Heat Diffusion Some food is placed in a freezer. After t hours the temperature of the food is dropping at the rate of $r(t)$ degrees Fahrenheit per hour, where $r(t) = 12 + 4/(t+3)^2$.

(a) Compute the area under the graph of $y = r(t)$ over the interval $0 \le t \le 2$.

(b) What does the area in part (a) represent?

46. Velocity Suppose that the velocity of a car at time t is $v(t) = 40 + 8/(t+1)^2$ kilometers per hour.

(a) Compute the area under the velocity curve from $t = 1$ to $t = 9$.

(b) What does the area in part (a) represent?

47. Deforestation and Fuel Wood Deforestation is one of the major problems facing sub-Saharan Africa. Although the clearing of land for farming has been the major cause, the steadily increasing demand for fuel wood has become a significant factor. Figure 13 summarizes projections of the World Bank. The rate of fuel wood consumption (in millions of cubic meters per year) in the Sudan t years after 1980 is given approximately by

the function $c(t) = 76.2e^{.03t}$. Determine the amount of fuel wood that will be consumed from 1980 to 2000.

Figure 13. Data from "Sudan and Options in the Energy Sector," World Bank/UNDP, July 1983.

48. (a) Compute $\int_{1}^{b} \frac{1}{t}\, dt$, where $b > 1$.

(b) Explain how the logarithm of a number greater than 1 may be interpreted as the area of a region under a curve. (What is the curve?)

49. For each positive number x, let $A(x)$ be the area of the region under the curve $y = x^2 + 1$ from 0 to x. Find $A'(3)$.

50. For each number $x > 2$, let $A(x)$ be the area of the region under the curve $y = x^3$ from 2 to x. Find $A'(6)$.

Technology Exercises

In Exercises 51–54, first use the fundamental theorem of calculus to compute the definite integral and then verify your answer numerically with a graphing utility.

51. $\int_{2}^{3} \frac{1}{x}\, dx$

52. $\int_{1}^{3} 10x^4\, dx$

53. $\int_{0}^{4} 5\sqrt{x}\, dx$

54. $\int_{0}^{3} 6e^{2x}\, dx$

Solutions to Practice Problems 6.3

1. The desired area is
$$\int_{-3}^{2} e^{x/2}\, dx = 2e^{x/2}\Big|_{-3}^{2} = 2e - 2e^{-3/2} \approx 4.99.$$

2. The number $\int_{75}^{80} MR(x)\, dx$ is the net change in revenue received when the production level is raised from $x = 75$ to $x = 80$ units.

6.4 Areas in the *xy*-Plane

In this section we show how to use the definite integral to compute the area of a region that lies between the graphs of two or more functions. Three simple but important properties of the integral will be used repeatedly.

Let $f(x)$ and $g(x)$ be functions and a, b, and k be any constants. Then

$$\int_a^b f(x)\, dx + \int_a^b g(x)\, dx = \int_a^b [f(x) + g(x)]\, dx \tag{1}$$

$$\int_a^b f(x)\, dx - \int_a^b g(x)\, dx = \int_a^b [f(x) - g(x)]\, dx \tag{2}$$

$$\int_a^b k f(x)\, dx = k \int_a^b f(x)\, dx. \tag{3}$$

To verify property (1), let $F(x)$ and $G(x)$ be antiderivatives of $f(x)$ and $g(x)$, respectively. Then $F(x) + G(x)$ is an antiderivative of $f(x) + g(x)$. By the fundamental theorem of calculus,

$$\int_a^b [f(x) + g(x)]\, dx = [F(x) + G(x)]\Big|_a^b$$

$$= [F(b) + G(b)] - [F(a) + G(a)]$$

$$= [F(b) - F(a)] + [G(b) - G(a)]$$

$$= \int_a^b f(x)\, dx + \int_a^b g(x)\, dx.$$

The verifications of properties (2) and (3) are similar and use the facts that $F(x) - G(x)$ is an antiderivative of $f(x) - g(x)$ and $kF(x)$ is an antiderivative of $kf(x)$.

Let us now consider regions that are bounded both above and below by graphs of functions. Referring to Fig. 1, we would like to find a simple expression for the area of the shaded region under the graph of $y = f(x)$ and above the graph of $y = g(x)$ from $x = a$ to $x = b$. It is the region under the graph of $y = f(x)$ with the region under the graph of $y = g(x)$ taken away. Therefore,

$$[\text{area of shaded region}] = [\text{area under } f(x)] - [\text{area under } g(x)]$$

$$= \int_a^b f(x)\, dx - \int_a^b g(x)\, dx$$

$$= \int_a^b [f(x) - g(x)]\, dx \qquad [\text{by property (2)}].$$

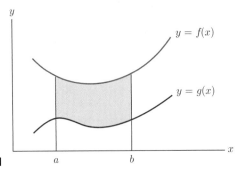

Figure 1

Area between Two Curves   If $y = f(x)$ lies above $y = g(x)$ from $x = a$ to $x = b$, the area of the region between $f(x)$ and $g(x)$ from $x = a$ to $x = b$ is

$$\int_a^b [f(x) - g(x)]\, dx.$$

EXAMPLE 1

Find the area of the region between $y = 2x^2 - 4x + 6$ and $y = -x^2 + 2x + 1$ from $x = 1$ to $x = 2$.

Solution   Upon sketching the two graphs (Fig. 2), we see that $f(x) = 2x^2 - 4x + 6$ lies above $g(x) = -x^2 + 2x + 1$ for $1 \leq x \leq 2$. Therefore, our formula gives the area of the shaded region as

$$\int_1^2 [(2x^2 - 4x + 6) - (-x^2 + 2x + 1)]\, dx = \int_1^2 (3x^2 - 6x + 5)\, dx$$

$$= (x^3 - 3x^2 + 5x) \Big|_1^2 = 6 - 3 = 3. \quad \blacksquare$$

Figure 2

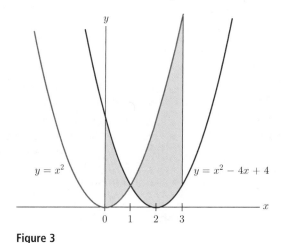

Figure 3

EXAMPLE 2

Find the area of the region between $y = x^2$ and $y = (x - 2)^2 = x^2 - 4x + 4$ from $x = 0$ to $x = 3$.

Solution   Upon sketching the graphs (Fig. 3), we see that the two graphs cross; by setting $x^2 = x^2 - 4x + 4$, we find that they cross when $x = 1$. Thus one graph does not always lie above the other from $x = 0$ to $x = 3$, so we cannot directly apply our rule for finding the area between two curves. However, the difficulty is easily surmounted if we break the region into two parts, the area from $x = 0$ to $x = 1$ and the area from $x = 1$ to $x = 3$. From $x = 0$ to $x = 1$, $y = x^2 - 4x + 4$ is on top; and from $x = 1$ to $x = 3$, $y = x^2$ is on top. Consequently,

$$[\text{area from } x = 0 \text{ to } x = 1] = \int_0^1 [(x^2 - 4x + 4) - (x^2)]\, dx$$

$$= \int_0^1 (-4x + 4)\, dx$$

$$= (-2x^2 + 4x) \Big|_0^1 = 2 - 0 = 2.$$

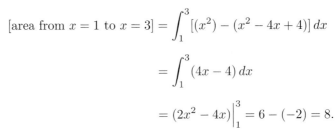

$$[\text{area from } x = 1 \text{ to } x = 3] = \int_1^3 [(x^2) - (x^2 - 4x + 4)]\, dx$$

$$= \int_1^3 (4x - 4)\, dx$$

$$= (2x^2 - 4x)\Big|_1^3 = 6 - (-2) = 8.$$

Thus the total area is $2 + 8 = 10$.   ◼

(a)

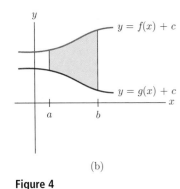

(b)

Figure 4

In our derivation of the formula for the area between two curves, we examined functions that are nonnegative. However, the statement of the rule does not contain this stipulation, and rightfully so. Consider the case where $f(x)$ and $g(x)$ are not always positive. Let us determine the area of the shaded region in Fig. 4(a). Select some constant c such that the graphs of the functions $f(x) + c$ and $g(x) + c$ lie completely above the x-axis [Fig. 4(b)]. The region between them will have the same area as the original region. Using the rule as applied to nonnegative functions, we have

$$[\text{area of the region}] = \int_a^b [(f(x) + c) - (g(x) + c)]\, dx$$

$$= \int_a^b [f(x) - g(x)]\, dx.$$

Therefore, we see that our rule is valid for any functions $f(x)$ and $g(x)$ as long as the graph of $f(x)$ lies above the graph of $g(x)$ for all x from $x = a$ to $x = b$.

EXAMPLE 3   Set up the integral that gives the area between the curves $y = x^2 - 2x$ and $y = -e^x$ from $x = -1$ to $x = 2$.

Solution   Since $y = x^2 - 2x$ lies above $y = -e^x$ (Fig. 5), the rule for finding the area between two curves can be applied directly. The area between the curves is

$$\int_{-1}^2 (x^2 - 2x + e^x)\, dx.$$ ◼

Sometimes we are asked to find the area between two curves without being given the values of a and b. In such cases there is a region that is completely enclosed by the two curves. As the next examples illustrate, we must first find the points of intersection of the two curves to obtain the values of a and b. In such problems, careful curve sketching is especially important.

EXAMPLE 4   Set up the integral that gives the area bounded by the curves $y = x^2 + 2x + 3$ and $y = 2x + 4$.

Solution   The two curves are sketched in Fig. 6, and the region bounded by them is shaded. To find the points of intersection, we set $x^2 + 2x + 3 = 2x + 4$ and solve for x. We obtain $x^2 = 1$, or $x = -1$ and $x = +1$. When $x = -1$, $2x + 4 = 2(-1) + 4 = 2$. When $x = 1$, $2x + 4 = 2(1) + 4 = 6$. Thus the curves intersect at the points $(1, 6)$ and $(-1, 2)$.

Since $y = 2x + 4$ lies above $y = x^2 + 2x + 3$ from $x = -1$ to $x = 1$, the area between the curves is given by

$$\int_{-1}^1 [(2x + 4) - (x^2 + 2x + 3)]\, dx = \int_{-1}^1 (1 - x^2)\, dx.$$ ◼

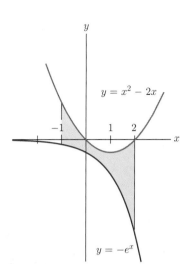

Figure 5

Figure 6

EXAMPLE 5   Set up the integral that gives the area bounded by the two curves $y = 2x^2$ and $y = x^3 - 3x$.

Solution   First we make a rough sketch of the two curves. (See Fig. 7.) The curves intersect where $x^3 - 3x = 2x^2$, or $x^3 - 2x^2 - 3x = 0$. Note that

$$x^3 - 2x^2 - 3x = x(x^2 - 2x - 3) = x(x - 3)(x + 1).$$

So the solutions to $x^3 - 2x^2 - 3x = 0$ are $x = 0, 3, -1$, and the curves intersect at $(-1, 2)$, $(0, 0)$, and $(3, 18)$. From $x = -1$ to $x = 0$, the curve $y = x^3 - 3x$ lies above $y = 2x^2$. But from $x = 0$ to $x = 3$, the reverse is true. Thus the area between the curves is given by

$$\int_{-1}^{0} (x^3 - 3x - 2x^2)\, dx + \int_{0}^{3} (2x^2 - x^3 + 3x)\, dx. \qquad \blacksquare$$

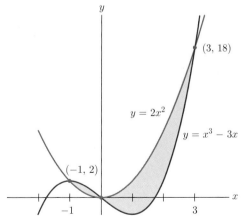

Figure 7

EXAMPLE 6   **Oil consumption**   Beginning in 1974, with the advent of dramatically higher oil prices, the exponential rate of growth of world oil consumption slowed down from a growth constant of 7% to a growth constant of 4% per year. A fairly good model for the annual rate of oil consumption since 1974 is given by

$$R_1(t) = 21.3e^{.04(t-4)}, \qquad t \geq 4,$$

where $t = 0$ corresponds to 1970. Determine the total amount of oil saved between 1976 and 1980 by not consuming oil at the rate predicted by the model of Example 8, Section 6.3:

$$R(t) = 16.1e^{.07t}, \qquad t \geq 0.$$

Solution If oil consumption had continued to grow as it did prior to 1974, the total oil consumption between 1976 and 1980 would have been

$$\int_6^{10} R(t)\, dt. \tag{4}$$

However, taking into account the slower increase in the rate of oil consumption since 1974, we find that the total oil consumed between 1976 and 1980 was approximately

$$\int_6^{10} R_1(t)\, dt. \tag{5}$$

The integrals in formulas (4) and (5) may be interpreted as the areas under the curves $y = R(t)$ and $y = R_1(t)$, respectively, from $t = 6$ to $t = 10$. (See Fig. 8.) By superimposing the two curves, we see that the area between them from $t = 6$ to $t = 10$ represents the total oil that was saved by consuming oil at the rate given by $R_1(t)$ instead of $R(t)$. (See Fig. 9.) The area between the two curves equals

$$\int_6^{10} [R(t) - R_1(t)]\, dt = \int_6^{10} \left[16.1e^{.07t} - 21.3e^{.04(t-4)}\right] dt$$

$$= \left.\left(\frac{16.1}{.07}e^{.07t} - \frac{21.3}{.04}e^{.04(t-4)}\right)\right|_6^{10}$$

$$\approx 13.02.$$

Thus about 13 billion barrels of oil were saved between 1976 and 1980. ■

(a) Consumption rate predicted by pre-1974 data

(b) Consumption rate changed by 1974 oil price rise

Figure 8

Figure 9

INCORPORATING TECHNOLOGY

Area between Two Curves We illustrate how to shade the area between two curves using the functions from Example 1. First, set $Y_1 = -X^2 + 2X + 1$ and $Y_2 = 2X^2 - 4X + 6$. Now press [2nd] [DRAW] [7] to display **Shade(**. Complete the expression to read **Shade$(Y_1, Y_2, 1, 2)$** and press [ENTER]. This will shade the area above Y_1 and below Y_2 with $1 \leq X \leq 2$. The result is shown in Fig. 10(a). Note that in the **Shade** routine it is necessary to list the lower function first and the upper function second.

The definite integral from Example 1 is evaluated to find the area between the curves. Select [MATH] [9] to display **fnInt(**. Now complete the expression to read **fnInt$(Y_2 - Y_1, 1, 2)$** and press [ENTER]. The result is shown in Fig. 10(b). ∎

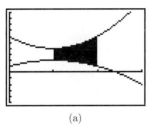

(a)   (b)

Figure 10

Practice Problems 6.4

1. Find the area between the curves $y = x + 3$ and $y = \frac{1}{2}x^2 + x - 7$ from $x = -2$ to $x = 1$.

2. A company plans to increase its production from 10 to 15 units per day. The present marginal cost function is $MC_1(x) = x^2 - 20x + 108$. By redesigning the production process and purchasing new equipment, the company can change the marginal cost function to $MC_2(x) = \frac{1}{2}x^2 - 12x + 75$. Determine the area between the graphs of the two marginal cost curves from $x = 10$ to $x = 15$. Interpret this area in economic terms.

EXERCISES 6.4

1. Write down a definite integral or sum of definite integrals that gives the area of the shaded portions of Fig. 11.

Fig. 12.

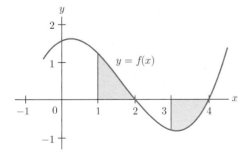

Figure 11

2. Write down a definite integral or sum of definite integrals that gives the area of the shaded portions of

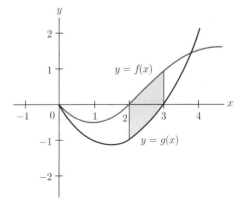

Figure 12

3. Shade the portion of Fig. 13 whose area is given by the integral

$$\int_0^2 [f(x) - g(x)]\,dx + \int_2^4 [h(x) - g(x)]\,dx.$$

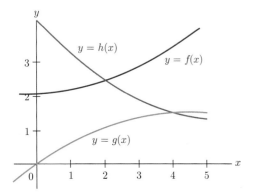

Figure 13

4. Shade the portion of Fig. 14 whose area is given by the integral

$$\int_0^1 [f(x) - g(x)] + \int_1^2 [g(x) - f(x)]\,dx.$$

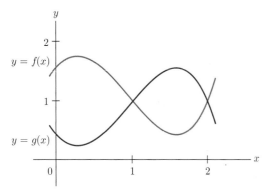

Figure 14

Find the area of the region between the curves.

5. $y = 2x^2$ and $y = 8$ (a horizontal line) from $x = -2$ to $x = 2$

6. $y = x^2 + 1$ and $y = -x^2 - 1$ from $x = -1$ to $x = 1$

7. $y = x^2 - 6x + 12$ and $y = 1$ from $x = 0$ to $x = 4$

8. $y = x(2 - x)$ and $y = 2$ from $x = 0$ to $x = 2$

9. $y = e^x$ and $y = \dfrac{1}{x^2}$ from $x = 1$ to $x = 2$

10. $y = e^{2x}$ and $y = 1 - x$ from $x = 0$ to $x = 1$

Find the area of the region bounded by the curves.

11. $y = x^2$ and $y = x$

12. $y = 4x(1 - x)$ and $y = \frac{3}{4}$

13. $y = -x^2 + 6x - 5$ and $y = 2x - 5$

14. $y = x^2 - 1$ and $y = 3$

15. $y = x(x^2 - 1)$ and the x-axis

16. $y = x^3$ and $y = 2x^2$

17. $y = 8x^2$ and $y = \sqrt{x}$

18. $y = \dfrac{4}{x}$ and $y = 5 - x$

19. Find the area of the region between $y = x^2 - 3x$ and the x-axis
   (a) from $x = 0$ to $x = 3$,
   (b) from $x = 0$ to $x = 4$,
   (c) from $x = -2$ to $x = 3$.

20. Find the area of the region between $y = x^2$ and $y = 1/x^2$
   (a) from $x = 1$ to $x = 4$,
   (b) from $x = \frac{1}{2}$ to $x = 4$.

21. Find the area of the region bounded by $y = 1/x^2$, $y = x$, and $y = 8x$, for $x \geq 0$. (See Fig. 15.)

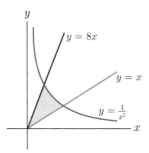

Figure 15

22. Find the area of the region bounded by $y = 1/x$, $y = 4x$, and $y = x/2$, for $x \geq 0$. (The region resembles the shaded region in Fig. 15.)

23. Find the area of the shaded region in Fig. 16(a) bounded by $y = 12/x$, $y = \frac{3}{2}\sqrt{x}$, and $y = x/3$.

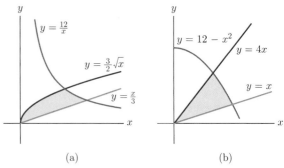

(a)   (b)

Figure 16

24. Find the area of the shaded region in Fig. 16(b) bounded by $y = 12 - x^2$, $y = 4x$, and $y = x$.

25. Refer to Exercise 47 of Section 6.3. The rate of new tree growth (in millions of cubic meters per year) in the Sudan t years after 1980 is given approximately by the function $g(t) = 50 - 6.03e^{.09t}$. Set up the definite integral giving the amount of depletion of the forests due to the excess of fuel wood consumption over new growth from 1980 to 2000.

26. Refer to the oil-consumption data in Example 6. Suppose that in 1970 the growth constant for the annual rate of oil consumption had been held to .04. What effect would this action have had on oil consumption from 1970 to 1974?

27. Profit The marginal profit for a certain company is $MP_1(x) = -x^2 + 14x - 24$. The company expects the daily production level to rise from $x = 6$ to $x = 8$ units. The management is considering a plan that would have the effect of changing the marginal profit to $M_2(x) = -x^2 + 12x - 20$. Should the company adopt the plan? Determine the area between the graphs of the two marginal profit functions from $x = 6$ to $x = 8$. Interpret this area in economic terms.

28. Two rockets are fired simultaneously straight up into the air. Their velocities (in meters per second) are $v_1(t)$ and $v_2(t)$, respectively, and $v_1(t) \geq v_2(t)$ for $t \geq 0$. Let A denote the area of the region between the graphs of $y = v_1(t)$ and $y = v_2(t)$ for $0 \leq t \leq 10$. What physical interpretation may be given to the value of A?

29. Distance Traveled Cars A and B start at the same place and travel in the same direction, with velocities after t

hours given by the functions $v_A(t)$ and $v_B(t)$ in Fig. 17.

(a) What does the area between the two curves from $t = 0$ to $t = 1$ represent?

(b) At what time will the distance between the cars be greatest?

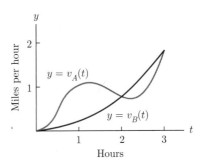

Figure 17

Technology Exercises

In Exercises 30–33, use a graphing utility to find the intersection points of the curves. Then use the utility to find the area of the region bounded by the curves.

30. $y = e^x$, $y = 4x + 1$

31. $y = 5 - (x - 2)^2$, $y = e^x$

32. $y = \sqrt{x + 1}$, $y = (x - 1)^2$

33. $y = 1/x$, $y = 3 - x$

Solutions to Practice Problems 6.4

1. First graph the two curves, as shown in Fig. 18. The curve $y = x + 3$ lies on top. So the area between the curves is

$$\int_{-2}^{1} \left[(x + 3) - \left(\tfrac{1}{2}x^2 + x - 7 \right) \right] dx$$

$$= \int_{-2}^{1} \left(-\tfrac{1}{2}x^2 + 10 \right) dx$$

$$= \left(-\tfrac{1}{6}x^3 + 10x \right) \Big|_{-2}^{1} = 28.5.$$

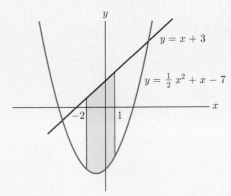

Figure 18

(Continued)

2. Graphing the two marginal cost functions yields the results shown in Fig. 19. So the area between the curves equals

$$\int_{10}^{15} [MC_1(x) - MC_2(x)]\, dx$$

$$= \int_{10}^{15} [(x^2 - 20x + 108) - (\tfrac{1}{2}x^2 - 12x + 75)]\, dx$$

$$= \int_{10}^{15} \left[\tfrac{1}{2}x^2 - 8x + 33\right]\, dx$$

$$= \left(\tfrac{1}{6}x^3 - 4x^2 + 33x\right)\Big|_{10}^{15} = 60\tfrac{5}{6}.$$

This amount represents the cost savings on the in-

creased production (from 10 to 15), provided the new production process is used.

Figure 19

6.5 Applications of the Definite Integral

The applications in this section have two features in common. First, each example contains a quantity that is computed by evaluating a definite integral. Second, the formula for the definite integral is derived by looking at Riemann sums.

In each application, we show that a certain quantity, call it Q, may be approximated by dividing an interval into equal subintervals and forming an appropriate sum. We then observe that this sum is a Riemann sum for some function $f(x)$. [Sometimes $f(x)$ is not given beforehand.] The Riemann sums approach Q as the number of subintervals becomes large. Since the Riemann sums also approach a definite integral, we conclude that the value of Q is given by the definite integral.

Once we have expressed Q as a definite integral, we may calculate its value using the fundamental theorem of calculus, which reduces the calculation to antidifferentiation. It is not necessary to calculate the value of any Riemann sum. Rather, we use the Riemann sums only as a device to express Q as a definite integral. The key step is to recognize a given sum as a Riemann sum and to determine the corresponding definite integral. Example 1 illustrates this step.

EXAMPLE 1   Suppose that the interval $1 \le x \le 2$ is divided into 50 subintervals, each of length Δx. Let $x_1, x_2, \ldots, x_{50}$ denote points selected from these subintervals. Find an approximate value for the sum

$$(8x_1^7 + 6x_1)\, \Delta x + (8x_2^7 + 6x_2)\, \Delta x + \cdots + (8x_{50}^7 + 6x_{50})\, \Delta x.$$

Solution   The sum is clearly a Riemann sum for the function $f(x) = 8x^7 + 6x$ on the interval $1 \le x \le 2$. Therefore, an approximation to the sum is given by the integral

$$\int_1^2 (8x^7 + 6x)\, dx.$$

We may evaluate this integral using the fundamental theorem of calculus:

$$\int_1^2 (8x^7 + 6x)\,dx = (x^8 + 3x^2)\Big|_1^2$$

$$= [2^8 + 3(2^2)] - [1^8 + 3(1^2)]$$

$$= 268 - 4 = 264.$$

Therefore, the sum is approximately 264. ∎

The Average Value of a Function Let $f(x)$ be a continuous function on the interval $a \le x \le b$. The definite integral may be used to define the *average value* of $f(x)$ on this interval. To calculate the average of a collection of numbers $y_1, y_2, \ldots, y_n$, we add the numbers and divide by n to obtain

$$\frac{y_1 + y_2 + \cdots + y_n}{n}.$$

To determine the average value of $f(x)$, we proceed similarly. Choose n values of x, say $x_1, x_2, \ldots, x_n$, and calculate the corresponding function values $f(x_1)$, $f(x_2)$, $\ldots$, $f(x_n)$. The average of these values is

$$\frac{f(x_1) + f(x_2) + \cdots + f(x_n)}{n}. \tag{1}$$

Our goal now is to obtain a reasonable definition of the average of all the values of $f(x)$ on the interval $a \le x \le b$. If the points $x_1, x_2, \ldots, x_n$ are spread "evenly" throughout the interval, the average (1) should be a good approximation to our intuitive concept of the average value of $f(x)$. In fact, as n becomes large, the average (1) should approximate the average value of $f(x)$ to any arbitrary degree of accuracy. To guarantee that the points $x_1, x_2, \ldots, x_n$ are "evenly" spread out from a to b, we divide the interval from $x = a$ to $x = b$ into n subintervals of equal length $\Delta x = (b-a)/n$. Then choose x_1 from the first subinterval, x_2 from the second, and so forth. The average (1) that corresponds to these points may be arranged in the form of a Riemann sum as follows:

$$\frac{f(x_1) + f(x_2) + \cdots + f(x_n)}{n}$$

$$= f(x_1) \cdot \frac{1}{n} + f(x_2) \cdot \frac{1}{n} + \cdots + f(x_n) \cdot \frac{1}{n}$$

$$= \frac{1}{b-a} \left[f(x_1) \cdot \frac{b-a}{n} + f(x_2) \cdot \frac{b-a}{n} + \cdots + f(x_n) \cdot \frac{b-a}{n} \right]$$

$$= \frac{1}{b-a} [f(x_1)\,\Delta x + f(x_2)\,\Delta x + \cdots + f(x_n)\,\Delta x].$$

The sum inside the brackets is a Riemann sum for the definite integral of $f(x)$. Thus we see that, for a large number of points x_i, the average in (1) approaches the quantity

$$\frac{1}{b-a} \int_a^b f(x)\,dx.$$

This argument motivates the following definition:

The *average value* of a continuous function $f(x)$ over the interval $a \le x \le b$ is defined as the quantity

$$\frac{1}{b-a} \int_a^b f(x)\,dx. \tag{2}$$

EXAMPLE 2   Compute the average value of $f(x) = \sqrt{x}$ over the interval $0 \le x \le 9$.

Solution   Using (2) with $a = 0$ and $b = 9$, the average value of $f(x) = \sqrt{x}$ over the interval $0 \le x \le 9$ is equal to

$$\frac{1}{9-0} \int_0^9 \sqrt{x}\, dx.$$

Since $\sqrt{x} = x^{1/2}$, an antiderivative of $f(x) = \sqrt{x}$ is $F(x) = \frac{2}{3}x^{3/2}$. Therefore,

$$\frac{1}{9}\int_0^9 \sqrt{x}\, dx = \frac{1}{9}\left(\frac{2}{3}x^{3/2}\right)\Big|_0^9 = \frac{1}{9}\left(\frac{2}{3}\cdot 9^{3/2} - 0\right) = \frac{1}{9}\left(\frac{2}{3}\cdot 27\right) = 2,$$

so that the average value of $f(x) = \sqrt{x}$ over the interval $0 \le x \le 9$ is 2. The area of the shaded region is the same as the area of the rectangle pictured in Fig. 1. ∎

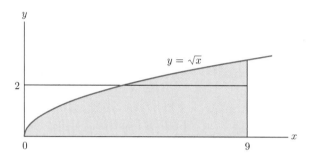

Figure 1.   Average value of a function.

EXAMPLE 3   On January 1, 1998, the world population was 5.9 billion and growing at the rate of 1.34% annually. Assuming that this growth rate continues, the population t years from then will be given by the exponential growth law

$$P(t) = 5.9e^{.0134t}.$$

Determine the average world population during the next 30 years. (This average is important in long-range planning for agricultural production and the allocation of goods and services.)

Solution   The average value of the population $P(t)$ from $t = 0$ to $t = 30$ is

$$\frac{1}{30-0}\int_0^{30} P(t)\, dt = \frac{1}{30}\int_0^{30} 5.9e^{.0134t}\, dt$$

$$= \frac{1}{30}\left(\frac{5.9}{.0134}e^{.0134t}\right)\Big|_0^{30} = \frac{5.9}{.402}(e^{.402} - 1)$$

$$\approx 7.26 \text{ billion.}$$ ∎

Consumers' Surplus   Using a demand curve from economics, we can derive a formula showing the amount that consumers benefit from an open system that has no price discrimination. Figure 2(a) is a *demand curve* for a commodity. It is determined by complex economic factors and gives a relationship between the quantity sold and the unit price of a commodity. Specifically, it says that to sell x units the price must be set at $f(x)$ dollars per unit. Since, for most commodities, selling larger quantities requires a lowering of the price, demand curves are usually decreasing. Interactions between supply and demand determine the amount of a quantity available. Let A designate the amount of the commodity currently available and $B = f(A)$ the current selling price.

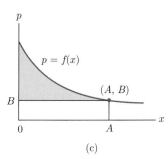

Figure 2. Consumers' surplus.

Divide the interval from 0 to A into n subintervals, each of length $\Delta x = (A - 0)/n$, and take x_i to be the right-hand endpoint of the ith interval. Consider the first subinterval, from 0 to x_1. [See Fig. 2(b).] Suppose that only x_1 units had been available. Then the price per unit could have been set at $f(x_1)$ dollars and these x_1 units sold. Of course, at this price we could not have sold any more units. However, those people who paid $f(x_1)$ dollars had a great demand for the commodity. It was extremely valuable to them, and there was no advantage in substituting another commodity at that price. They were actually paying what the commodity was worth to them. In theory, then, the first x_1 units of the commodity could be sold to these people at $f(x_1)$ dollars per unit, yielding

$$[\text{price per unit}] \cdot [\text{number of units}] = f(x_1) \cdot (x_1) = f(x_1) \cdot \Delta x \text{ dollars.}$$

After selling the first x_1 units, suppose that more units become available, so that now a total of x_2 units have been produced. Setting the price at $f(x_2)$, the remaining $x_2 - x_1 = \Delta x$ units can be sold, yielding $f(x_2) \cdot \Delta x$ dollars. Here, again, the second group of buyers would have paid as much for the commodity as it was worth to them. Continuing this process of price discrimination, the amount of money paid by consumers would be

$$f(x_1) \, \Delta x + f(x_2) \, \Delta x + \cdots + f(x_n) \, \Delta x.$$

Taking n large, we see that this Riemann sum approaches $\int_0^A f(x) \, dx$. Since $f(x)$ is positive, this integral equals the area under the graph of $f(x)$ from $x = 0$ to $x = A$.

Of course, in an open system, everyone pays the same price, B, so the total amount paid is $[\text{price per unit}] \cdot [\text{number of units}] = BA$. Since BA is the area of the rectangle under the graph of the line $p = B$ from $x = 0$ to $x = A$, the amount of money saved by the consumers is the area of the shaded region in Fig. 2(c). That is, the area between the curves $p = f(x)$ and $p = B$ gives a numerical value to one benefit of a modern efficient economy.

DEFINITION The *consumers' surplus* for a commodity having demand curve $p = f(x)$ is

$$\int_0^A [f(x) - B] \, dx,$$

where the quantity demanded is A and the price is $B = f(A)$.

EXAMPLE 4   Find the consumers' surplus for the demand curve $p = 50 - .06x^2$ at the sales level $x = 20$.

Solution   Since 20 units are sold, the price must be
$$B = 50 - .06(20)^2 = 50 - 24 = 26.$$
Therefore, the consumers' surplus is
$$\int_0^{20} [(50 - .06x^2) - 26] \, dx = \int_0^{20} (24 - .06x^2) \, dx$$
$$= (24x - .02x^3)\Big|_0^{20}$$
$$= 24(20) - .02(20)^3$$
$$= 480 - 160 = 320.$$
That is, the consumers' surplus is \$320. ∎

Future Value of an Income Stream The next example shows how the definite integral can be used to approximate the sum of a large number of terms.

EXAMPLE 5

Suppose that money is deposited daily into a savings account at an annual rate of $1000. The account pays 6% interest compounded continuously. Approximate the amount of money in the account at the end of 5 years.

Solution Divide the time interval from 0 to 5 years into daily subintervals. Each subinterval is then of duration $\Delta t = \frac{1}{365}$ years. Let $t_1, t_2, \ldots, t_n$ be points chosen from these subintervals. Since we deposit money at an annual rate of $1000, the amount deposited during one subinterval is $1000 \, \Delta t$ dollars. If this amount is deposited at time t_i, the $1000 \, \Delta t$ dollars will earn interest for the remaining $5 - t_i$ years. The total amount resulting from this one deposit at time t_i is then

$$1000 \, \Delta t \, e^{.06(5 - t_i)}.$$

Add the effects of the deposits at times $t_1, t_2, \ldots, t_n$ to arrive at the total balance in the account:

$$A = 1000 e^{.06(5 - t_1)} \Delta t + 1000 e^{.06(5 - t_2)} \Delta t + \cdots + 1000 e^{.06(5 - t_n)} \Delta t.$$

This is a Riemann sum for the function $f(t) = 1000 e^{.06(5 - t)}$ on the interval $0 \le t \le 5$. Since Δt is very small when compared with the interval, the total amount in the account, A, is approximately

$$\int_0^5 1000 e^{.06(5 - t)} \, dt = \frac{1000}{-.06} e^{.06(5 - t)} \Big|_0^5 = \frac{1000}{-.06}(1 - e^{.3}) \approx 5831.$$

That is, the approximate balance in the account at the end of 5 years is $5831. ■

NOTE The antiderivative was computed by observing that since the derivative of $e^{.06(5 - t)}$ is $e^{.06(5 - t)}(-.06)$, we must divide the integrand by $-.06$ to obtain an antiderivative.

■

In Example 5, money was deposited daily into the account. If the money had been deposited several times a day, the definite integral would have given an even better approximation to the balance. Actually, the more frequently the money is deposited, the better the approximation. Economists consider a hypothetical situation where money is deposited steadily throughout the year. This flow of money is called a *continuous income stream*, and the balance in the account is given exactly by the definite integral.

$y = g(x)$

a b

(a)

(b)

Figure 3

DEFINITION The *future value of a continuous income stream* of K dollars per year for N years at interest rate r compounded continuously is

$$\int_0^N K e^{r(N - t)} \, dt.$$

Solids of Revolution When the region of Fig. 3(a) is revolved about the x-axis, it sweeps out a solid. [See Fig. 3(b).] Riemann sums can be used to derive a formula for the volume of this *solid of revolution*. Let us break the x-axis between a and b into a large number n of equal subintervals, each of length $\Delta x = (b - a)/n$. Using each subinterval as a base, we can divide the region into strips. (See Fig. 4.)

Figure 4

Figure 5

Let x_i be a point in the ith subinterval. Then the volume swept out by revolving the ith strip is approximately the same as the volume of the cylinder swept out by revolving the rectangle of height $g(x_i)$ and base Δx around the x-axis. (See Fig. 5.) The volume of the cylinder is

$$[\text{area of circular side}] \cdot [\text{width}] = \pi[g(x_i)]^2 \cdot \Delta x.$$

The total volume swept out by all the strips is approximated by the total volume swept out by the rectangles, which is

$$[\text{volume}] \approx \pi[g(x_1)]^2 \Delta x + \pi[g(x_2)]^2 \Delta x + \cdots + \pi[g(x_n)]^2 \Delta x.$$

As n gets larger and larger, this approximation becomes arbitrarily close to the true volume. The expression on the right is a Riemann sum for the definite integral of $f(x) = \pi[g(x)]^2$. Therefore, the volume of the solid equals the value of the definite integral.

> The volume of the *solid of revolution* obtained from revolving the region below the graph of $y = g(x)$ from $x = a$ to $x = b$ about the x-axis is
>
> $$\int_a^b \pi[g(x)]^2 \, dx.$$

EXAMPLE 6   Find the volume of the solid of revolution obtained by revolving the region of Fig. 6 about the x-axis.

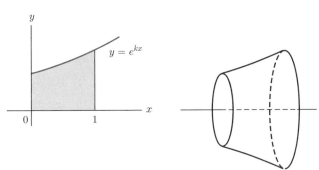

Figure 6

Solution Here $g(x) = e^{kx}$, and

$$[\text{volume}] = \int_0^1 \pi(e^{kx})^2 \, dx = \int_0^1 \pi e^{2kx} \, dx = \frac{\pi}{2k} e^{2kx} \Big|_0^1 = \frac{\pi}{2k}(e^{2k} - 1). \qquad \blacksquare$$

EXAMPLE 7   Find the volume of a right circular cone of radius r and height h.

Solution   The cone in Fig. 7(a) is the solid of revolution swept out when the shaded region in Fig. 7(b) is revolved about the x-axis. Using the formula developed previously, the volume of the cone is

$$\int_0^h \pi \left(\frac{r}{h}x\right)^2 dx = \frac{\pi r^2}{h^2}\int_0^h x^2 \, dx = \frac{\pi r^2}{h^2}\frac{x^3}{3}\Big|_0^h = \frac{\pi r^2 h}{3}. \qquad \blacksquare$$

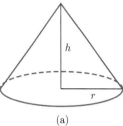

Figure 7   (a)   (b)

Practice Problems 6.5

1. A rock dropped from a bridge has a velocity of $-32t$ feet per second after t seconds. Find the average velocity of the rock during the first 3 seconds.

2. An investment yields \$300 per year compounded continuously for 10 years at 10% interest. What is the (future) value of this income stream at the end of 10 years?

EXERCISES 6.5

Determine the average value of $f(x)$ over the interval from $x = a$ to $x = b$, where:

1. $f(x) = x^2$; $a = 0$, $b = 3$
2. $f(x) = 1 - x$; $a = -1$, $b = 1$
3. $f(x) = 100e^{-.5x}$; $a = 0$, $b = 4$
4. $f(x) = 2$; $a = 0$, $b = 1$
5. $f(x) = 1/x$; $a = 1/3$, $b = 3$
6. $f(x) = \dfrac{1}{\sqrt{x}}$; $a = 1$, $b = 9$

7. **Average Temperature**   During a certain 12-hour period the temperature at time t (measured in hours from the start of the period) was $T(t) = 47 + 4t - \frac{1}{3}t^2$ degrees. What was the average temperature during that period?

8. **Average Population**   Assuming that a country's population is now 3 million and is growing exponentially with growth constant .02, what will be the average population during the next 50 years?

9. **Average Amount of Radium**   One hundred grams of radioactive radium having a half-life of 1690 years is placed in a concrete vault. What will be the average amount of radium in the vault during the next 1000 years?

10. **Average Amount of Money**   One hundred dollars are deposited in the bank at 5% interest compounded continuously. What will be the average value of the money in the account during the next 20 years?

Find the consumers' surplus for each of the following demand curves at the given sales level x.

11. $p = 3 - \dfrac{x}{10}$; $x = 20$

12. $p = \dfrac{x^2}{200} - x + 50$; $x = 20$

13. $p = \dfrac{500}{x + 10} - 3$; $x = 40$

14. $p = \sqrt{16 - .02x}$; $x = 350$

Figure 8 shows a supply curve for a commodity. It gives the relationship between the selling price of the commodity and the quantity that producers will manufacture. At a higher selling price, a greater quantity will be produced. Therefore, the curve is increasing. If (A, B) is a point on the curve, then, to stimulate the production of A units of the commodity, the price per unit must be B dollars. Of course, some producers will be willing to produce the commodity even with *a lower selling price. Since everyone receives the same price in an open efficient economy, most producers are receiving more than their minimal required price. The excess is called the* producers' surplus. *Using an argument analogous to that of the* consumers' surplus, *we can show that the total producers' surplus when the price is B is the area of the shaded region in Fig. 8. Find the producers' surplus for each of the following supply curves at the given sales level* x.

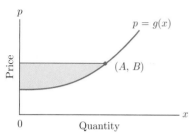

Figure 8. Producers' surplus.

15. $p = .01x + 3$; $x = 200$ **16.** $p = \dfrac{x^2}{9} + 1$; $x = 3$

17. $p = \dfrac{x}{2} + 7$; $x = 10$ **18.** $p = 1 + \frac{1}{2}\sqrt{x}$; $x = 36$

For a particular commodity, the quantity produced and the unit price are given by the coordinates of the point where the supply and demand curves intersect. For each pair of supply and demand curves, determine the point of intersection (A, B) and the consumers' and producers' surplus. (See Fig. 9.)

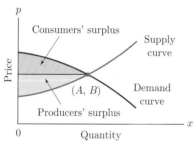

Figure 9

19. Demand curve: $p = 12 - (x/50)$; supply curve: $p = (x/20) + 5$.

20. Demand curve: $p = \sqrt{25 - .1x}$; supply curve: $p = \sqrt{.1x + 9} - 2$.

21. Future Value Suppose that money is deposited daily into a savings account at an annual rate of $1000. If the account pays 5% interest compounded continuously, estimate the balance in the account at the end of 3 years.

22. Future Value Suppose that money is deposited daily into a savings account at an annual rate of $2000. If the account pays 6% interest compounded continuously, approximately how much will be in the account at the end of 2 years?

23. Future Value Suppose that money is deposited steadily into a savings account at the rate of $16,000 per year. Determine the balance at the end of 4 years if the account pays 8% interest compounded continuously.

24. Future Value Suppose that money is deposited steadily into a savings account at the rate of $14,000 per year. Determine the balance at the end of 6 years if the account pays 4.5% interest compounded continuously.

25. An investment pays 10% interest compounded continuously. If money is invested steadily at the rate of $5000 per year, how much time is required until the value of the investment reaches $140,000?

26. A savings account pays 4.25% interest compounded continuously. At what rate per year must money be deposited steadily into the account to accumulate a balance of $100,000 after 10 years?

27. Suppose that money is to be deposited daily for 5 years into a savings account at an annual rate of $1000 and the account pays 4% interest compounded continuously. Let the interval from 0 to 5 be divided into daily subintervals, with each subinterval of duration $\Delta t = \frac{1}{365}$ years. Let $t_1, \ldots, t_n$ be points chosen from the subintervals.

(a) Show that the present value of a daily deposit at time t_i is $1000 \, \Delta t \, e^{-.04 t_i}$.

(b) Find the Riemann sum corresponding to the sum of the present values of all the deposits.

(c) What is the function and interval corresponding to the Riemann sum in part (b)?

(d) Give the definite integral that approximates the Riemann sum in part (b).

(e) Evaluate the definite integral in part (d). This number is the *present value of a continuous income stream.*

28. Use the result of Exercise 27 to calculate the present value of a continuous income stream of $5000 per year for 10 years at an interest rate of 5% compounded continuously.

Find the volume of the solid of revolution generated by revolving about the x-axis the region under each of the following curves.

29. $y = \sqrt{4 - x^2}$ from $x = -2$ to $x = 2$ (generates a sphere of radius 2)

30. $y = \sqrt{r^2 - x^2}$ from $x = -r$ to $x = r$ (generates a sphere of radius r)

31. $y = x^2$ from $x = 1$ to $x = 2$

32. $y = kx$ from $x = 0$ to $x = h$ (generates a cone)

33. $y = \sqrt{x}$ from $x = 0$ to $x = 4$ (The solid generated is called a *paraboloid.*)

34. $y = 2x - x^2$ from $x = 0$ to $x = 2$

35. $y = e^{-x}$ from $x = 0$ to $x = 1$

36. $y = 2x + 1$ from $x = 0$ to $x = 1$ (The solid generated is called a *truncated cone*.)

For the Riemann sums in Exercises 37–40, determine n, b, and $f(x)$.

37. $\left[(8.25)^3 + (8.75)^3 + (9.25)^3 + (9.75)^3\right](.5)$; $a = 8$

38. $\left[\dfrac{3}{1} + \dfrac{3}{1.5} + \dfrac{3}{2} + \dfrac{3}{2.5} + \dfrac{3}{3} + \dfrac{3}{3.5}\right](.5)$; $a = 1$

39. $\left[(5 + e^5) + (6 + e^6) + (7 + e^7)\right](1)$; $a = 4$

40. $\left[3(.3)^2 + 3(.9)^2 + 3(1.5)^2 + 3(2.1)^2 + 3(2.7)^2\right](.6)$; $a = 0$

41. Suppose that the interval $0 \le x \le 3$ is divided into 100 subintervals of width $\Delta x = .03$. Let $x_1, x_2, \ldots, x_{100}$ be points in these subintervals. Suppose that in a particular application we need to estimate the sum

$$(3 - x_1)^2 \Delta x + (3 - x_2)^2 \Delta x + \cdots + (3 - x_{100})^2 \Delta x.$$

Show that this sum is close to 9.

42. Suppose that the interval $0 \le x \le 1$ is divided into 100 subintervals of width $\Delta x = .01$. Show that the following sum is close to $5/4$.

$$\left[2(.01) + (.01)^3\right] \Delta x + \left[2(.02) + (.02)^3\right] \Delta x$$
$$+ \cdots + \left[2(1.0) + (1.0)^3\right] \Delta x$$

Technology Exercises

The following exercises ask for an unknown quantity x. After setting up the appropriate formula involving a definite integral, use the fundamental theorem to evaluate the definite integral as an expression in x. Because the resulting equation will be too complicated to solve algebraically, you must use a graphing utility to obtain the solution. (Note: If the quantity x is an interest rate paid by a savings account, it will most likely be between 0 and .10.)

43. A single deposit of $1000 is to be made into a savings account and the interest (compounded continuously) is allowed to accumulate for 3 years. Therefore, the amount at the end of t years is $1000e^{rt}$.

(a) Find an expression (involving r) that gives the average value of the money in the account during the 3-year time period $0 \le t \le 3$.

(b) Find the interest rate r at which the average amount in the account during the 3-year period is $1070.60.

44. A single deposit of $100 is made into a savings account paying 4% interest compounded continuously. How long must the money be held in the account so that the average amount of money during that time period will be $122.96?

45. Money is deposited steadily at the rate of $1000 per year into a savings account.

(a) Find the expression (involving r) that gives the (future) balance in the account at the end of 6 years.

(b) Find the interest rate that will result in a balance of $6997.18 after 6 years.

46. Money is deposited steadily at the rate of $3000 per year into a savings account. After 10 years the balance is $36,887. What interest rate, with interest compounded continuously, did the money earn?

Solutions to Practice Problems 6.5

1. By definition, the average value of the function $v(t) = -32t$ for $t = 0$ to $t = 3$ is

$$\frac{1}{3 - 0} \int_0^3 -32t\, dt = \frac{1}{3}(-16t^2)\Big|_0^3 = \frac{1}{3}(-16 \cdot 3^2)$$

$$= -48 \text{ feet per second.}$$

Note: There is another way to approach this problem:

$$[\text{average velocity}] = \frac{[\text{distance traveled}]}{[\text{time elapsed}]}.$$

As we discussed in Section 6.2, distance traveled equals the area under the velocity curve. Therefore,

$$[\text{average velocity}] = \frac{\int_0^3 -32t\, dt}{3}.$$

2. According to the formula developed in the text, the future value of the income stream after 10 years is equal to

$$\int_0^{10} 300e^{.1(10-t)}\, dt = -\frac{300}{.1}e^{.1(10-t)}\Big|_0^{10}$$

$$= -3000e^0 - (-3000e^1)$$

$$= 3000e - 3000$$

$$\approx \$5154.85.$$

REVIEW OF FUNDAMENTAL CONCEPTS

1. What does it mean to antidifferentiate a function?

2. State the formula for $\int h(x)\,dx$ for each of the following functions.

   (a) $h(x) = x^r$, $r \neq -1$ **(b)** $h(x) = e^{kx}$

   (c) $h(x) = \dfrac{1}{x}$ **(d)** $h(x) = f(x) + g(x)$

   (e) $h(x) = kf(x)$

3. In the formula $\Delta x = \dfrac{b-a}{n}$, what do a, b, n, and Δx denote?

4. What is a Riemann sum?

5. Give an interpretation of the area under a rate of change function. Give a concrete example.

6. What is a definite integral?

7. What is the difference between a definite integral and an indefinite integral?

8. State the fundamental theorem of calculus.

9. How is $F(x)\big|_a^b$ calculated and what is it called?

10. Outline a procedure for finding the area of a region bounded by two curves.

11. State the formula for each of the following quantities:

    (a) average value of a function

    (b) consumers' surplus

    (c) future value of an income stream

    (d) volume of a solid of revolution

SUPPLEMENTARY EXERCISES

Calculate the following integrals.

1. $\displaystyle\int 3^2\,dx$

2. $\displaystyle\int (x^2 - 3x + 2)\,dx$

3. $\displaystyle\int \sqrt{x+1}\,dx$

4. $\displaystyle\int \frac{2}{x+4}\,dx$

5. $2\displaystyle\int (x^3 + 3x^2 - 1)\,dx$

6. $\displaystyle\int \sqrt[5]{x+3}\,dx$

7. $\displaystyle\int e^{-x/2}\,dx$

8. $\displaystyle\int \frac{5}{\sqrt{x-7}}\,dx$

9. $\displaystyle\int (3x^4 - 4x^3)\,dx$

10. $\displaystyle\int (2x+3)^7\,dx$

11. $\displaystyle\int \sqrt{4-x}\,dx$

12. $\displaystyle\int \left(\frac{5}{x} - \frac{x}{5}\right)\,dx$

13. $\displaystyle\int_{-1}^{1} (x+1)^2\,dx$

14. $\displaystyle\int_{0}^{1/8} \sqrt[3]{x}\,dx$

15. $\displaystyle\int_{-1}^{2} \sqrt{2x+4}\,dx$

16. $2\displaystyle\int_{0}^{1} \left(\frac{2}{x+1} - \frac{1}{x+4}\right)\,dx$

17. $\displaystyle\int_{1}^{2} \frac{4}{x^5}\,dx$

18. $\dfrac{2}{3}\displaystyle\int_{0}^{8} \sqrt{x+1}\,dx$

19. $\displaystyle\int_{1}^{4} \frac{1}{x^2}\,dx$

20. $\displaystyle\int_{3}^{6} e^{2-(x/3)}\,dx$

21. $\displaystyle\int_{0}^{5} (5+3x)^{-1}\,dx$

22. $\displaystyle\int_{-2}^{2} \frac{3}{2e^{3x}}\,dx$

23. $\displaystyle\int_{0}^{\ln 2} (e^x - e^{-x})\,dx$

24. $\displaystyle\int_{\ln 2}^{\ln 3} (e^x + e^{-x})\,dx$

25. $\displaystyle\int_{0}^{\ln 3} \frac{e^x + e^{-x}}{e^{2x}}\,dx$

26. $\displaystyle\int_{0}^{1} \frac{3 + e^{2x}}{e^x}\,dx$

27. Find the area under the curve $y = (3x-2)^{-3}$ from $x = 1$ to $x = 2$.

28. Find the area under the curve $y = 1 + \sqrt{x}$ from $x = 1$ to $x = 9$.

In Exercises 29–36, find the area of the shaded region.

29.

30.

31.

32.

33.

34.

35.

36.

37. Find the area of the region bounded by the curves $y = x^3 - 3x + 1$ and $y = x + 1$.

38. Find the area of the region between the curves $y = 2x^2 + x$ and $y = x^2 + 2$ from $x = 0$ to $x = 2$.

39. Find the function $f(x)$ for which $f'(x) = (x - 5)^2$, $f(8) = 2$.

40. Find the function $f(x)$ for which $f'(x) = e^{-5x}$, $f(0) = 1$.

41. Describe all solutions of the following differential equations, where y represents a function of t.
   (a) $y' = 4t$ **(b)** $y' = 4y$ **(c)** $y' = e^{4t}$

42. Let k be a constant, and let $y = f(t)$ be a function such that $y' = kty$. Show that $y = Ce^{kt^2/2}$, for some constant C. [*Hint:* Use the product rule to evaluate $\frac{d}{dt}[f(t)e^{-kt^2/2}]$, and then apply Theorem II of Section 6.1.]

43. An airplane tire plant finds that its marginal cost of producing tires is $.04x + 150$ dollars at a production level of x tires per day. If fixed costs are \$500 per day, find the cost of producing x tires per day.

44. If the marginal revenue function for a company is $400 - 3x^2$, find the additional revenue received from doubling production if 10 units are currently being produced.

45. A drug is injected into a patient at the rate of $f(t)$ cubic centimeters per minute at time t. What does the area under the graph of $y = f(t)$ from $t = 0$ to $t = 4$ represent?

46. A rock thrown straight up into the air has a velocity of $v(t) = -9.8t + 20$ meters per second after t seconds.
   (a) Determine the distance the rock travels during the first 2 seconds.
   (b) Represent the answer to part (a) as an area.

47. Use a Riemann sum with $n = 4$ and left endpoints to estimate the area under the graph in Fig. 1 for $0 \le x \le 2$.

48. Redo Exercise 47 using right endpoints.

Figure 1

49. Use a Riemann sum with $n = 2$ and midpoints to estimate the area under the graph of

$$f(x) = \frac{1}{x+2}$$

on the interval $0 \leq x \leq 2$. Then use a definite integral to find the exact value of the area to five decimal places.

50. Use a Riemann sum with $n = 5$ and midpoints to estimate the area under the graph of $f(x) = e^{2x}$ on the interval $0 \leq x \leq 1$. Then use a definite integral to find the exact value of the area to five decimal places.

51. Find the consumers' surplus for the demand curve $p = \sqrt{25 - .04x}$ at the sales level $x = 400$.

52. Three thousand dollars is deposited in the bank at 4% interest compounded continuously. What will be the average value of the money in the account during the next 10 years?

53. Find the average value of $f(x) = 1/x^3$ from $x = \frac{1}{3}$ to $x = \frac{1}{2}$.

54. Suppose that the interval $0 \leq x \leq 1$ is divided into 100 subintervals with a width of $\Delta x = .01$. Show that the sum

$$\left[3e^{-.01}\right]\Delta x + \left[3e^{-.02}\right]\Delta x + \left[3e^{-.03}\right]\Delta x$$
$$+ \cdots + \left[3e^{-1}\right]\Delta x$$

is close to $3(1 - e^{-1})$.

55. In Fig. 2, three regions are labeled with their areas. Determine $\int_a^c f(x)\,dx$ and determine $\int_a^d f(x)\,dx$.

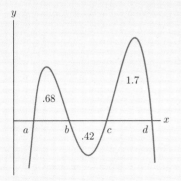

Figure 2

56. Find the volume of the solid of revolution generated by revolving about the x-axis the region under the curve $y = 1 - x^2$ from $x = 0$ to $x = 1$.

57. A store has an inventory of Q units of a certain product at time $t = 0$. The store sells the product at the steady rate of Q/A units per week, and exhausts the inventory in A weeks.

   (a) Find a formula $f(t)$ for the amount of product in inventory at time t.

   (b) Find the average inventory level during the period $0 \leq t \leq A$.

58. A retail store sells a certain product at the rate of $g(t)$ units per week at time t, where $g(t) = rt$. At time $t = 0$, the store has Q units of the product in inventory.

   (a) Find a formula $f(t)$ for the amount of product in inventory at time t.

   (b) Determine the value of r in part (a) such that the inventory is exhausted in A weeks.

   (c) Using $f(t)$, with r as in part (b), find the average inventory level during the period $0 \leq t \leq A$.

59. Let x be any positive number, and define $g(x)$ to be the number determined by the definite integral

$$g(x) = \int_0^x \frac{1}{1+t^2}\,dt.$$

   (a) Give a geometric interpretation of the number $g(3)$.

   (b) Find the derivative $g'(x)$.

60. For each number x satisfying $-1 \leq x \leq 1$, define $h(x)$ by

$$h(x) = \int_{-1}^x \sqrt{1 - t^2}\,dt.$$

   (a) Give a geometric interpretation of the values $h(0)$ and $h(1)$.

   (b) Find the derivative $h'(x)$.

61. Suppose that the interval $0 \leq t \leq 3$ is divided into 1000 subintervals of width Δt. Let $t_1, t_2, \ldots, t_{1000}$ denote the right endpoints of these subintervals. If we need to estimate the sum

$$5000e^{-.1t_1}\Delta t + 5000e^{-.1t_2}\Delta t$$
$$+ \cdots + 5000e^{-.1t_{1000}}\Delta t,$$

show that this sum is close to 13,000. [*Note:* A sum such as this would arise if we wanted to compute the present value of a continuous stream of income of $5000 per year for 3 years, with interest compounded continuously at 10%.]

62. What number does

$$\left[e^0 + e^{1/n} + e^{2/n} + e^{3/n} + \cdots + e^{(n-1)/n}\right] \cdot \frac{1}{n}$$

approach as n gets very large?

63. What number does the sum

$$\left[1^3 + \left(1 + \frac{1}{n}\right)^3 + \left(1 + \frac{2}{n}\right)^3 + \left(1 + \frac{3}{n}\right)^3\right.$$
$$\left. + \cdots + \left(1 + \frac{n-1}{n}\right)^3\right] \cdot \frac{1}{n}$$

approach as n gets very large?

64. In Fig. 3, the rectangle has the same area as the region under the graph of $f(x)$. What is the average value of $f(x)$ on the interval $2 \le x \le 6$?

Figure 3

65. *True or false*: If $3 \le f(x) \le 4$ whenever $0 \le x \le 5$, then $3 \le \frac{1}{5} \int_0^5 f(x)\, dx \le 4$.

66. Suppose that water is flowing into a tank at a rate of $r(t)$ gallons per hour, where the rate depends on the time t according to the formula

$$r(t) = 20 - 4t, \qquad 0 \le t \le 5.$$

(a) Consider a brief period of time, say from t_1 to t_2. The length of this time period is $\Delta t = t_2 - t_1$. During this period the rate of flow does not change much and is approximately $20 - 4t_1$ (the rate at the beginning of the brief time interval). Approximately how much water flows into the tank during the time from t_1 to t_2?

(b) Explain why the total amount of water added to the tank during the time interval from $t = 0$ to $t = 5$ is given by $\int_0^5 r(t)\, dt$.

67. The annual world rate of water use t years after 1960, for $t \le 35$, was approximately $860e^{.04t}$ cubic kilometers per year. How much water was used between 1960 and 1995?

68. If money is deposited steadily into a savings account at the rate of \$4500 per year, determine the balance at the end of 1 year if the account pays 9% interest compounded continuously.

69. Find a function $f(x)$ whose graph goes through the point $(1, 1)$ and whose slope at any point $(x, f(x))$ is $3x^2 - 2x + 1$.

70. For what value of a is the shaded area in Fig. 4 equal to 1?

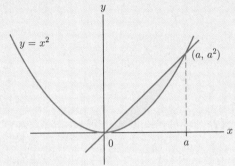

Figure 4

71. Show that for any positive number b we have

$$\int_0^{b^2} \sqrt{x}\, dx + \int_0^b x^2\, dx = b^3.$$

72. Generalize the result of Exercise 71 as follows: Let n be a positive integer. Show that for any positive number b we have

$$\int_0^{b^n} \sqrt[n]{x}\, dx + \int_0^b x^n\, dx = b^{n+1}.$$

73. Show that

$$\int_0^1 (\sqrt{x} - x^2)\, dx = \frac{1}{3}.$$

74. Generalize the result of Exercise 73 as follows: Let n be a positive integer. Show that

$$\int_0^1 (\sqrt[n]{x} - x^n)\, dx = \frac{n-1}{n+1}.$$

FUNCTIONS OF SEVERAL VARIABLES

Until now, most of our applications of calculus have involved functions of one variable. In real life, however, a quantity of interest often depends on more than one variable. For instance, the sales level of a product may depend not only on its price, but also on the prices of competing products, the amount spent on advertising, and perhaps the time of year. The total cost of manufacturing the product depends on the cost of raw materials, labor, plant maintenance, and so on.

This chapter introduces the basic ideas of calculus for functions of more than one variable. Section 7.1 presents two examples that will be used throughout the chapter. Derivatives are treated in Section 7.2 and then used in Sections 7.3 and 7.4 to solve optimization problems more general than those in Chapter 2. The final two sections are devoted to least-squares problems and a brief introduction to the integration of functions of two variables.

7.1 Examples of Functions of Several Variables

A function $f(x, y)$ of the two variables x and y is a rule that assigns a number to each pair of values for the variables; for instance,

$$f(x, y) = e^x(x^2 + 2y).$$

An example of a function of three variables is

$$f(x, y, z) = 5xy^2z.$$

EXAMPLE 1   A store sells butter at \$2.50 per pound and margarine at \$1.40 per pound. The revenue from the sale of x pounds of butter and y pounds of margarine is given by the function

$$f(x, y) = 2.50x + 1.40y.$$

Determine and interpret $f(200, 300)$.

Solution $f(200, 300) = 2.50(200) + 1.40(300) = 500 + 420 = 920$. The revenue from the sale of 200 pounds of butter and 300 pounds of margarine is \$920.   ■

A function $f(x, y)$ of two variables may be graphed in a manner analogous to that for functions of one variable. It is necessary to use a three-dimensional coordinate system, where each point is identified by three coordinates (x, y, z). For each choice of x, y, the graph of $f(x, y)$ includes the point $(x, y, f(x, y))$. This graph is usually a surface in three-dimensional space, with equation $z = f(x, y)$. (See Fig. 1.) Three graphs of specific functions are shown in Fig. 2.

Application to Architectural Design   When designing a building, we would like to know, at least approximately, how much heat the building loses per day. The heat loss affects many aspects of the design, such as the size of the heating plant, the size and location of duct work, and so on. A building loses heat through its sides, roof, and floor. How much heat is lost will generally differ for each face of the building and will depend on such factors as insulation, materials used in construction, exposure (north, south, east, or west), and climate. It is possible to estimate how much heat is lost per square foot of each face. Using these data, we can construct a heat-loss function as in the following example.

EXAMPLE 2   A rectangular industrial building of dimensions x, y, and z is shown in Fig. 3(a). In Fig. 3(b) we give the amount of heat lost per day by each side of the building, measured in suitable units of heat per square foot. Let $f(x, y, z)$ be the total daily heat loss for such a building.

(a) Find a formula for $f(x, y, z)$.

(b) Find the total daily heat loss if the building has length 100 feet, width 70 feet, and height 50 feet.

Solution   (a) The total heat loss is the sum of the amount of heat loss through each face of the building. The heat loss through the roof is

[heat loss per square foot of roof] · [area of roof in square feet] $= 10xy$.

Similarly, the heat loss through the east side is $8yz$. Continuing in this way, we see that the total daily heat loss is

$$f(x, y, z) = 10xy + 8yz + 6yz + 10xz + 5xz + 1 \cdot xy.$$

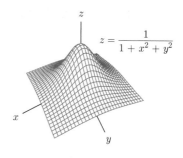

$$z = \frac{1}{1 + x^2 + y^2}$$

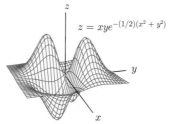

$$z = xye^{-(1/2)(x^2 + y^2)}$$

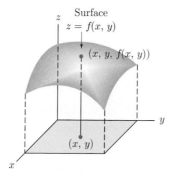

Surface
$z = f(x, y)$

$(x, y, f(x, y))$

(x, y)

Figure 1. Graph of $f(x, y)$.

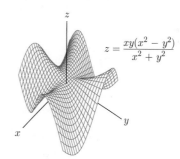

$$z = \frac{xy(x^2 - y^2)}{x^2 + y^2}$$

Figure 2

We collect terms to obtain

$$f(x, y, z) = 11xy + 14yz + 15xz.$$

(b) The amount of heat loss when $x = 100$, $y = 70$, and $z = 50$ is given by $f(100, 70, 50)$, which equals

$$f(100, 70, 50) = 11(100)(70) + 14(70)(50) + 15(100)(50)$$
$$= 77{,}000 + 49{,}000 + 75{,}000 = 201{,}000.$$ ∎

In Section 7.3 we will determine the dimensions x, y, z that minimize the heat loss for a building of specific volume.

Production Functions in Economics The costs of a manufacturing process can generally be classified as one of two types: cost of labor and cost of capital. The meaning of the cost of labor is clear. By the cost of capital, we mean the cost of buildings, tools, machines, and similar items used in the production process. A manufacturer usually has some control over the relative portions of labor and capital utilized in its production process. It can completely automate production so that labor is at a minimum or utilize mostly labor and little capital. Suppose that x units of labor and y units of capital are used.* Let $f(x, y)$ denote the number of units

---

*Economists normally use L and K, respectively, for labor and capital. However, for simplicity, we use x and y.

	Roof	East side	West side	North side	South side	Floor
Heat loss (per sq ft)	10	8	6	10	5	1
Area (sq ft)	xy	yz	yz	xz	xz	xy

(a)   (b)

Figure 3.   Heat loss from an industrial building.

of finished product that are manufactured. Economists have found that $f(x, y)$ is often a function of the form

$$f(x, y) = Cx^A y^{1-A},$$

where A and C are constants, $0 < A < 1$. Such a function is called a *Cobb–Douglas production function*.

EXAMPLE 3   **Production in a firm**   Suppose that during a certain time period the number of units of goods produced when utilizing x units of labor and y units of capital is $f(x, y) = 60x^{3/4}y^{1/4}$.

(a) How many units of goods will be produced by using 81 units of labor and 16 units of capital?

(b) Show that whenever the amounts of labor and capital being used are doubled so is the production. (Economists say that the production function has "constant returns to scale.")

Solution (a) $f(81, 16) = 60(81)^{3/4} \cdot (16)^{1/4} = 60 \cdot 27 \cdot 2 = 3240$. There will be 3240 units of goods produced.

(b) Utilization of a units of labor and b units of capital results in the production of $f(a, b) = 60a^{3/4}b^{1/4}$ units of goods. Utilizing $2a$ and $2b$ units of labor and capital, respectively, results in $f(2a, 2b)$ units produced. Set $x = 2a$ and $y = 2b$. Then we see that

$$f(2a, 2b) = 60(2a)^{3/4}(2b)^{1/4}$$

$$= 60 \cdot 2^{3/4} \cdot a^{3/4} \cdot 2^{1/4} \cdot b^{1/4}$$

$$= 60 \cdot 2^{(3/4 + 1/4)} \cdot a^{3/4}b^{1/4}$$

$$= 2^1 \cdot 60a^{3/4}b^{1/4}$$

$$= 2f(a, b). \qquad \blacksquare$$

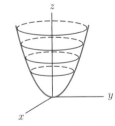

Graph of $f(x, y) = x^2 + y^2$

Level curves of $f(x, y) = x^2 + y^2$

Figure 4.   Level curves.

Level Curves   It is possible graphically to depict a function $f(x, y)$ of two variables using a family of curves called level curves. Let c be any number. Then the graph of the equation $f(x, y) = c$ is a curve in the xy-plane called the *level curve of height* c. This curve describes all points of height c on the graph of the function $f(x, y)$. As c varies, we have a family of level curves indicating the sets of points on which $f(x, y)$ assumes various values c. In Fig. 4 we have drawn the graph and various level curves for the function $f(x, y) = x^2 + y^2$.

Level curves often have interesting physical interpretations. For example, surveyors draw *topographic maps* that use level curves to represent points having equal altitude. Here $f(x, y) = $ the altitude at point (x, y). Figure 5(a) shows the graph of

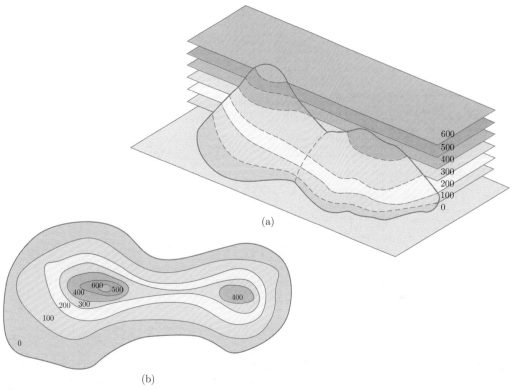

(a)

(b)

Figure 5.   Topographic level curves show altitudes.

$f(x, y)$ for a typical hilly region. Figure 5(b) shows the level curves corresponding to various altitudes. Note that when the level curves are closer together the surface is steeper.

EXAMPLE 4   Determine the level curve at height 600 for the production function $f(x, y) = 60x^{3/4}y^{1/4}$ of Example 3.

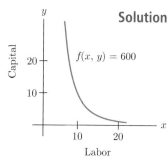

Figure 6.   A level curve of a production function.

Solution   The level curve is the graph of $f(x, y) = 600$, or

$$60x^{3/4}y^{1/4} = 600$$

$$y^{1/4} = \frac{10}{x^{3/4}}$$

$$y = \frac{10{,}000}{x^3}.$$

Of course, since x and y represent quantities of labor and capital, they must both be positive. We have sketched the graph of the level curve in Fig. 6. The points on the curve are precisely those combinations of capital and labor that yield 600 units of production. Economists call this curve an *isoquant* . ∎

Practice Problems 7.1

1. Let $f(x, y, z) = x^2 + y/(x - z) - 4$. Compute $f(3, 5, 2)$.

2. In a certain country the daily demand for coffee is given by $f(p_1, p_2) = 16p_1/p_2$ thousand pounds, where p_1 and p_2 are the respective prices of tea and coffee in dollars per pound. Compute and interpret $f(3, 4)$.

EXERCISES 7.1

1. Let $f(x, y) = x^2 - 3xy - y^2$. Compute $f(5, 0)$, $f(5, -2)$, and $f(a, b)$.

2. Let $g(x, y) = \sqrt{x^2 + 2y^2}$. Compute $g(1, 1)$, $g(0, -1)$, and $g(a, b)$.

3. Let $g(x, y, z) = x/(y - z)$. Compute $g(2, 3, 4)$ and $g(7, 46, 44)$.

4. Let $f(x, y, z) = x^2 e^{\sqrt{y^2 + z^2}}$. Compute $f(1, -1, 1)$ and $f(2, 3, -4)$.

5. Let $f(x, y) = xy$. Show that $f(2 + h, 3) - f(2, 3) = 3h$.

6. Let $f(x, y) = xy$. Show that $f(2, 3 + k) - f(2, 3) = 2k$.

7. Cost Find a formula $C(x, y, z)$ that gives the cost of materials for the closed rectangular box in Fig. 7(a), with dimensions in feet. Assume that the material for the top and bottom costs $3 per square foot and the material for the sides costs $5 per square foot.

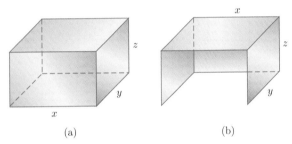

(a)   (b)

Figure 7

8. Cost Find a formula $C(x, y, z)$ that gives the cost of material for the rectangular enclosure in Fig. 7(b), with dimensions in feet. Assume that the material for the top costs $3 per square foot and the material for the back and two sides costs $5 per square foot.

9. Consider the Cobb–Douglas production function $f(x, y) = 20x^{1/3}y^{2/3}$. Compute $f(8, 1)$, $f(1, 27)$, and $f(8, 27)$. Show that, for any positive constant k, $f(8k, 27k) = kf(8, 27)$.

10. Let $f(x, y) = 10x^{2/5}y^{3/5}$. Show that $f(3a, 3b) = 3f(a, b)$.

11. The present value of A dollars to be paid t years in the future (assuming a 5% continuous interest rate) is $P(A, t) = Ae^{-.05t}$. Find and interpret $P(100, 13.8)$.

12. Refer to Example 3. If labor costs $100 per unit and capital costs $200 per unit, express as a function of two variables, $C(x, y)$, the cost of utilizing x units of labor and y units of capital.

13. Tax and Homeowner Exemption The value of residential property for tax purposes is usually much lower than its actual market value. If v is the market value, the *assessed value* for real estate taxes might be only 40%

of v. Suppose that the property tax, T, in a community is given by the function

$$T = f(r, v, x) = \frac{r}{100}(.40v - x),$$

where v is the estimated market value of a property (in dollars), x is a *homeowner's exemption* (a number of dollars depending on the type of property), and r is the tax rate (stated in dollars per hundred dollars) of net assessed value.

(a) Determine the real estate tax on a property valued at $200,000 with a homeowner's exemption of $5000, assuming a tax rate of $2.50 per hundred dollars of net assessed value.

(b) Determine the tax due if the tax rate increases by 20% to $3.00 per hundred dollars of net assessed value. Assume the same property value and homeowner's exemption. Does the tax due also increase by 20%?

14. Let $f(r, v, x)$ be the real estate tax function of Exercise 13.

(a) Determine the real estate tax on a property valued at $100,000 with a homeowner's exemption of $5000, assuming a tax rate of $2.20 per hundred dollars of net assessed value.

(b) Determine the real estate tax when the market value rises 20% to $120,000. Assume the same homeowner's exemption and a tax rate of $2.20 per hundred dollars of net assessed value. Does the tax due also increase by 20%?

Draw the level curves of heights 0, 1, and 2 for the functions in Exercises 15 and 16.

15. $f(x, y) = 2x - y$ **16.** $f(x, y) = -x^2 + 2y$

17. Draw the level curve of the function $f(x, y) = x - y$ containing the point $(0, 0)$.

18. Draw the level curve of the function $f(x, y) = xy$ containing the point $\left(\frac{1}{2}, 4\right)$.

19. Find a function $f(x, y)$ that has the line $y = 3x - 4$ as a level curve.

20. Find a function $f(x, y)$ that has the curve $y = 2/x^2$ as a level curve.

21. Suppose that a topographic map is viewed as the graph of a certain function $f(x, y)$. What are the level curves?

22. A certain production process uses labor and capital. If the quantities of these commodities are x and y, respectively, the total cost is $100x + 200y$ dollars. Draw the level curves of height 600, 800, and 1000 for this function. Explain the significance of these curves. (Economists frequently refer to these lines as *budget lines* or *isocost lines*.)

Match the graphs of the functions in Exercises 23–26 to the systems of level curves shown in Figs. 8(a)–(d).

23.

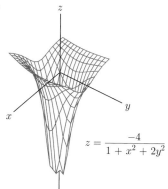

$$z = \frac{-4}{1 + x^2 + 2y^2}$$

24.

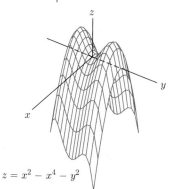

$$z = x^2 - x^4 - y^2$$

25.

$$z = \frac{15x^2 y^2 e^{-x^2 - y^2}}{x^2 + y^2}$$

26.

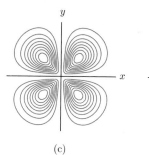

$$z = e^{-x^2} + e^{-4y^2}$$

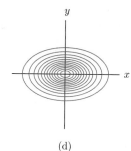

| (a) | (b) | (c) | (d) |

Figure 8

Solutions to Practice Problems 7.1

1. Substitute 3 for x, 5 for y, and 2 for z.

$$f(3, 5, 2) = 3^2 + \frac{5}{3 - 2} - 4 = 10.$$

2. To compute $f(3, 4)$, substitute 3 for p_1 and 4 for p_2 into $f(p_1, p_2) = 16 p_1 / p_2$. Thus

$$f(3, 4) = 16 \cdot \tfrac{3}{4} = 12.$$

Therefore, if the price of tea is $3 per pound and the price of coffee is $4 per pound, 12,000 pounds of coffee will be sold each day. (Notice that as the price of coffee increases the demand decreases.)

7.2 Partial Derivatives

In Chapter 1 we introduced the notion of a derivative to measure the rate at which a function $f(x)$ is changing with respect to changes in the variable x. Let us now study the analog of the derivative for functions of two (or more) variables.

Let $f(x, y)$ be a function of the two variables x and y. Since we want to know how $f(x, y)$ changes with respect to the changes in both the variable x and the variable y, we shall define two derivatives of $f(x, y)$ (to be called partial derivatives), one with respect to each variable.

> **DEFINITION**   The *partial derivative of $f(x, y)$ with respect to x*, written $\dfrac{\partial f}{\partial x}$, is the derivative of $f(x, y)$, where y is treated as a constant and $f(x, y)$ is considered as a function of x alone. The *partial derivative of $f(x, y)$ with respect to y*, written $\dfrac{\partial f}{\partial y}$, is the derivative of $f(x, y)$, where x is treated as a constant.

EXAMPLE 1   Let $f(x, y) = 5x^3 y^2$. Compute

$$\frac{\partial f}{\partial x} \quad \text{and} \quad \frac{\partial f}{\partial y}.$$

Solution To compute $\dfrac{\partial f}{\partial x}$, we think of $f(x, y)$ written as

$$f(x, y) = \left[5y^2\right] x^3,$$

where the brackets emphasize that $5y^2$ is to be treated as a constant. Therefore, when differentiating with respect to x, $f(x, y)$ is just a constant times x^3. Recall that if k is any constant then

$$\frac{d}{dx}(kx^3) = 3 \cdot k \cdot x^2.$$

Thus

$$\frac{\partial f}{\partial x} = 3 \cdot \left[5y^2\right] \cdot x^2 = 15x^2 y^2.$$

After some practice, it is unnecessary to place the y^2 in front of the x^3 before differentiating.

Now, to compute $\dfrac{\partial f}{\partial y}$, we think of

$$f(x, y) = \left[5x^3\right] y^2.$$

When differentiating with respect to y, $f(x, y)$ is simply a constant (that is, $5x^3$) times y^2. Hence

$$\frac{\partial f}{\partial y} = 2 \cdot \left[5x^3\right] \cdot y = 10x^3 y. \qquad \blacksquare$$

EXAMPLE 2   Let $f(x, y) = 3x^2 + 2xy + 5y$. Compute

$$\frac{\partial f}{\partial x} \quad \text{and} \quad \frac{\partial f}{\partial y}.$$

Solution To compute $\dfrac{\partial f}{\partial x}$, we think of

$$f(x, y) = 3x^2 + [2y]x + [5y].$$

Now we differentiate $f(x, y)$ as if it were a quadratic polynomial in x:

$$\frac{\partial f}{\partial x} = 6x + [2y] + 0 = 6x + 2y.$$

Note that $5y$ is treated as a constant when differentiating with respect to x, so the partial derivative of $5y$ with respect to x is zero.

To compute $\dfrac{\partial f}{\partial y}$, we think of

$$f(x, y) = \left[3x^2\right] + [2x]y + 5y.$$

Then

$$\frac{\partial f}{\partial y} = 0 + [2x] + 5 = 2x + 5.$$

Note that $3x^2$ is treated as a constant when differentiating with respect to y, so the partial derivative of $3x^2$ with respect to y is zero. ∎

EXAMPLE 3

Compute

$$\frac{\partial f}{\partial x} \quad \text{and} \quad \frac{\partial f}{\partial y}$$

for each of the following.

(a) $f(x, y) = (4x + 3y - 5)^8$ (b) $f(x, y) = e^{xy^2}$ (c) $f(x, y) = y/(x + 3y)$

Solution (a) To compute $\dfrac{\partial f}{\partial x}$, we think of

$$f(x, y) = (4x + [3y - 5])^8.$$

By the general power rule,

$$\frac{\partial f}{\partial x} = 8 \cdot (4x + [3y - 5])^7 \cdot 4 = 32(4x + 3y - 5)^7.$$

Here we used the fact that the derivative of $4x + 3y - 5$ with respect to x is just 4.

To compute $\dfrac{\partial f}{\partial y}$, we think of

$$f(x, y) = ([4x] + 3y - 5)^8.$$

Then

$$\frac{\partial f}{\partial y} = 8 \cdot ([4x] + 3y - 5)^7 \cdot 3 = 24(4x + 3y - 5)^7.$$

(b) To compute $\dfrac{\partial f}{\partial x}$, we observe that

$$f(x, y) = e^{x[y^2]},$$

so that

$$\frac{\partial f}{\partial x} = \left[y^2\right] e^{x[y^2]} = y^2 e^{xy^2}.$$

To compute $\dfrac{\partial f}{\partial y}$, we think of

$$f(x, y) = e^{[x]y^2}.$$

Thus

$$\frac{\partial f}{\partial y} = e^{[x]y^2} \cdot 2[x]y = 2xye^{xy^2}.$$

(c) To compute $\dfrac{\partial f}{\partial x}$, we use the general power rule to differentiate $[y](x + [3y])^{-1}$ with respect to x:

$$\frac{\partial f}{\partial x} = (-1) \cdot [y](x + [3y])^{-2} \cdot 1 = -\frac{y}{(x + 3y)^2}.$$

To compute $\dfrac{\partial f}{\partial y}$, we use the quotient rule to differentiate

$$f(x, y) = \frac{y}{[x] + 3y}$$

with respect to y. We find that

$$\frac{\partial f}{\partial y} = \frac{([x] + 3y) \cdot 1 - y \cdot 3}{([x] + 3y)^2} = \frac{x}{(x + 3y)^2}.$$

The use of brackets to highlight constants is helpful initially to compute partial derivatives. From now on we shall merely form a mental picture of those terms to be treated as constants and dispense with brackets. ∎

A partial derivative of a function of several variables is also a function of several variables and hence can be evaluated at specific values of the variables. We write

$$\frac{\partial f}{\partial x}(a, b)$$

for $\dfrac{\partial f}{\partial x}$ evaluated at $x = a$, $y = b$. Similarly,

$$\frac{\partial f}{\partial y}(a, b)$$

denotes the function $\dfrac{\partial f}{\partial y}$ evaluated at $x = a$, $y = b$.

EXAMPLE 4   Let $f(x, y) = 3x^2 + 2xy + 5y$.

(a) Calculate $\dfrac{\partial f}{\partial x}(1, 4)$.   **(b)** Evaluate $\dfrac{\partial f}{\partial y}$ at $(x, y) = (1, 4)$.

Solution **(a)** $\dfrac{\partial f}{\partial x} = 6x + 2y$, $\dfrac{\partial f}{\partial x}(1, 4) = 6 \cdot 1 + 2 \cdot 4 = 14$.

(b) $\dfrac{\partial f}{\partial y} = 2x + 5$, $\dfrac{\partial f}{\partial y}(1, 4) = 2 \cdot 1 + 5 = 7$. ∎

Geometric Interpretation of Partial Derivatives   Consider the three-dimensional surface $z = f(x, y)$ in Fig. 1. If y is held constant at b and x is allowed to vary, the equation

$$z = f(x, b)$$

$$\underset{\text{constant}}{\underline{\qquad}}$$

describes a curve on the surface. [The curve is formed by cutting the surface $z = f(x, y)$ with a vertical plane parallel to the xz-plane.] The value of $\dfrac{\partial f}{\partial x}(a, b)$ is the slope of the tangent line to the curve at the point where $x = a$ and $y = b$.

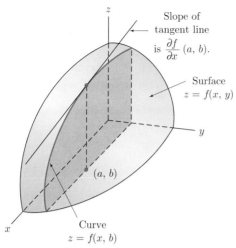

Figure 1. $\frac{\partial f}{\partial x}$ gives the slope of a curve formed by holding y constant.

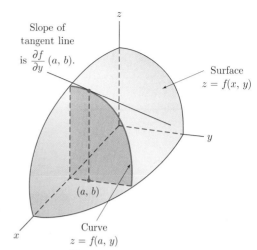

Figure 2. $\frac{\partial f}{\partial y}$ gives the slope of a curve formed by holding x constant.

Likewise, if x is held constant at a and y is allowed to vary, the equation

$$z = f(a, y)$$

$$\underset{\text{constant}}{\underline{\qquad}}$$

describes the curve on the surface $z = f(x, y)$ shown in Fig. 2. The value of the partial derivative $\frac{\partial f}{\partial y}(a, b)$ is the slope of this curve at the point where $x = a$ and $y = b$.

Partial Derivatives and Rates of Change Since $\frac{\partial f}{\partial x}$ is simply the ordinary derivative with y held constant, $\frac{\partial f}{\partial x}$ gives the rate of change of $f(x, y)$ with respect to x for y held constant. In other words, keeping y constant and increasing x by one (small) unit produces a change in $f(x, y)$ that is approximately given by $\frac{\partial f}{\partial x}$. An analogous interpretation holds for $\frac{\partial f}{\partial y}$.

EXAMPLE 5   Interpret the partial derivatives of $f(x, y) = 3x^2 + 2xy + 5y$ calculated in Example 4.

Solution We showed in Example 4 that

$$\frac{\partial f}{\partial x}(1, 4) = 14, \qquad \frac{\partial f}{\partial y}(1, 4) = 7.$$

The fact that

$$\frac{\partial f}{\partial x}(1, 4) = 14$$

means that if y is kept constant at 4 and x is allowed to vary near 1, then $f(x, y)$ changes at a rate 14 times the change in x. That is, if x increases by one small unit, $f(x, y)$ increases by approximately 14 units. If x increases by h units (where h is small), $f(x, y)$ increases by approximately $14 \cdot h$ units. That is,

$$f(1 + h, 4) - f(1, 4) \approx 14 \cdot h.$$

Similarly, the fact that

$$\frac{\partial f}{\partial y}(1,4) = 7$$

means that, if we keep x constant at 1 and let y vary near 4, then $f(x,y)$ changes at a rate equal to seven times the change in y. So, for a small value of k, we have

$$f(1, 4+k) - f(1,4) \approx 7 \cdot k. \qquad \blacksquare$$

We can generalize the interpretations of $\frac{\partial f}{\partial x}$ and $\frac{\partial f}{\partial y}$ given in Example 5 to yield the following general fact:

Let $f(x,y)$ be a function of two variables. Then, if h and k are small, we have

$$f(a+h, b) - f(a,b) \approx \frac{\partial f}{\partial x}(a,b) \cdot h,$$

$$f(a, b+k) - f(a,b) \approx \frac{\partial f}{\partial y}(a,b) \cdot k.$$

Partial derivatives can be computed for functions of any number of variables. When taking the partial derivative with respect to one variable, we treat the other variables as constant.

EXAMPLE 6   Let $f(x,y,z) = x^2yz - 3z$.

(a) Compute $\frac{\partial f}{\partial x}, \frac{\partial f}{\partial y}$, and $\frac{\partial f}{\partial z}$.   (b) Calculate $\frac{\partial f}{\partial z}(2,3,1)$.

Solution (a) $\frac{\partial f}{\partial x} = 2xyz, \frac{\partial f}{\partial y} = x^2z, \frac{\partial f}{\partial z} = x^2y - 3.$

(b) $\frac{\partial f}{\partial z}(2,3,1) = 2^2 \cdot 3 - 3 = 12 - 3 = 9.$ $\blacksquare$

EXAMPLE 7   Let $f(x,y,z)$ be the heat-loss function computed in Example 2 of Section 7.1. That is, $f(x,y,z) = 11xy + 14yz + 15xz$. Calculate and interpret $\frac{\partial f}{\partial x}(10,7,5)$.

Solution We have

$$\frac{\partial f}{\partial x} = 11y + 15z$$

$$\frac{\partial f}{\partial x}(10,7,5) = 11 \cdot 7 + 15 \cdot 5 = 77 + 75 = 152.$$

The quantity $\frac{\partial f}{\partial x}$ is commonly referred to as the *marginal heat loss with respect to change in x*. Specifically, if x is changed from 10 by h units (where h is small) and the values of y and z remain fixed at 7 and 5, the amount of heat loss will change by approximately $152 \cdot h$ units. $\blacksquare$

EXAMPLE 8   **Production**   Consider the production function $f(x,y) = 60x^{3/4}y^{1/4}$, which gives the number of units of goods produced when utilizing x units of labor and y units of capital.

(a) Find $\dfrac{\partial f}{\partial x}$ and $\dfrac{\partial f}{\partial y}$.

(b) Evaluate $\dfrac{\partial f}{\partial x}$ and $\dfrac{\partial f}{\partial y}$ at $x = 81$, $y = 16$.

(c) Interpret the numbers computed in part (b).

Solution (a) $\dfrac{\partial f}{\partial x} = 60 \cdot \dfrac{3}{4} x^{-1/4} y^{1/4} = 45 x^{-1/4} y^{1/4} = 45 \dfrac{y^{1/4}}{x^{1/4}},$

$\dfrac{\partial f}{\partial y} = 60 \cdot \dfrac{1}{4} x^{3/4} y^{-3/4} = 15 x^{3/4} y^{-3/4} = 15 \dfrac{x^{3/4}}{y^{3/4}}.$

(b) $\dfrac{\partial f}{\partial x}(81, 16) = 45 \cdot \dfrac{16^{1/4}}{81^{1/4}} = 45 \cdot \dfrac{2}{3} = 30,$

$\dfrac{\partial f}{\partial y}(81, 16) = 15 \cdot \dfrac{81^{3/4}}{16^{3/4}} = 15 \cdot \dfrac{27}{8} = \dfrac{405}{8} = 50\tfrac{5}{8}.$

(c) The quantities $\dfrac{\partial f}{\partial x}$ and $\dfrac{\partial f}{\partial y}$ are referred to as the *marginal productivity of labor* and the *marginal productivity of capital*. If the amount of capital is held fixed at $y = 16$ and the amount of labor increases by 1 unit, the quantity of goods produced will increase by approximately 30 units. Similarly, an increase in capital of 1 unit (with labor fixed at 81) results in an increase in production of approximately $50\tfrac{5}{8}$ units of goods. ∎

Just as we formed second derivatives in the case of one variable, we can form second partial derivatives of a function $f(x,y)$ of two variables. Since $\dfrac{\partial f}{\partial x}$ is a function of x and y, we can differentiate it with respect to x or y. The partial derivative of $\dfrac{\partial f}{\partial x}$ with respect to x is denoted by $\dfrac{\partial^2 f}{\partial x^2}$. The partial derivative of $\dfrac{\partial f}{\partial x}$ with respect to y is denoted by $\dfrac{\partial^2 f}{\partial y\, \partial x}$. Similarly, the partial derivative of the function $\dfrac{\partial f}{\partial y}$ with respect to x is denoted by $\dfrac{\partial^2 f}{\partial x\, \partial y}$, and the partial derivative of $\dfrac{\partial f}{\partial y}$ with respect to y is denoted by $\dfrac{\partial^2 f}{\partial y^2}$. Almost all functions $f(x,y)$ encountered in applications [and all functions $f(x,y)$ in this text] have the property that

$$\frac{\partial^2 f}{\partial y\, \partial x} = \frac{\partial^2 f}{\partial x\, \partial y}.$$

When computing $\dfrac{\partial^2 f}{\partial y\, \partial x}$ and $\dfrac{\partial^2 f}{\partial x\, \partial y}$, note that verifying the last equation is a check that you have done the differentiation correctly.

EXAMPLE 9

Let $f(x,y) = x^2 + 3xy + 2y^2$. Calculate

$$\frac{\partial^2 f}{\partial x^2}, \quad \frac{\partial^2 f}{\partial y^2}, \quad \frac{\partial^2 f}{\partial x\, \partial y}, \quad \text{and} \quad \frac{\partial^2 f}{\partial y\, \partial x}.$$

Solution First we compute $\dfrac{\partial f}{\partial x}$ and $\dfrac{\partial f}{\partial y}$.

$$\frac{\partial f}{\partial x} = 2x + 3y, \qquad \frac{\partial f}{\partial y} = 3x + 4y.$$

To compute $\dfrac{\partial^2 f}{\partial x^2}$, we differentiate $\dfrac{\partial f}{\partial x}$ with respect to x:

$$\frac{\partial^2 f}{\partial x^2} = 2.$$

Similarly, to compute $\dfrac{\partial^2 f}{\partial y^2}$, we differentiate $\dfrac{\partial f}{\partial y}$ with respect to y:

$$\frac{\partial^2 f}{\partial y^2} = 4.$$

To compute $\dfrac{\partial^2 f}{\partial x\,\partial y}$, we differentiate $\dfrac{\partial f}{\partial y}$ with respect to x:

$$\frac{\partial^2 f}{\partial x\,\partial y} = 3.$$

Finally, to compute $\dfrac{\partial^2 f}{\partial y\,\partial x}$, we differentiate $\dfrac{\partial f}{\partial x}$ with respect to y:

$$\frac{\partial^2 f}{\partial y\,\partial x} = 3. \qquad \blacksquare$$

INCORPORATING TECHNOLOGY

Evaluating Partial Derivatives The function from Example 4 and its first partial derivatives are specified in Fig. 3(a) and evaluated in Fig. 3(b). Recall that the expression $1 \to X$ is entered with $\boxed{1}$ $\boxed{\text{STO} \triangleright}$ $\boxed{\text{X,T,}\theta\text{,}n}$ and indicates that we are setting $X = 1$. The expression $4 \to Y$ has a similar meaning, but the variable Y is entered using $\boxed{\text{ALPHA}}$ [Y]. We can also evaluate other partial derivatives. For example, the partial derivative $\dfrac{\partial^2 f}{\partial x\,\partial y}$ in this case could be found by setting

$$Y_4 = \text{nDeriv}(Y_3, X, X).$$
$\blacksquare$

Figure 3 (a) (b)

Practice Problems 7.2

1. The number of TV sets an appliance store sells per week is given by a function of two variables, $f(x, y)$, where x is the price per TV set and y is the amount of money spent weekly on advertising. Suppose that the current price is \$400 per set and that currently \$2000 per week is being spent for advertising.

   (a) Would you expect $\dfrac{\partial f}{\partial x}(400, 2000)$ to be positive or negative?

   (b) Would you expect $\dfrac{\partial f}{\partial y}(400, 2000)$ to be positive or negative?

2. The monthly mortgage payment for a house is a function of two variables, $f(A, r)$, where A is the amount of the mortgage and the interest rate is $r\%$. For a 30-year mortgage, $f(92{,}000, 9) = 740.25$ and $\dfrac{\partial f}{\partial r}(92{,}000, 9) = 66.20$. What is the significance of the number 66.20?

EXERCISES 7.2

Find $\dfrac{\partial f}{\partial x}$ and $\dfrac{\partial f}{\partial y}$ for each of the following functions.

1. $f(x,y) = 5xy$

2. $f(x,y) = x^2 - y^2$

3. $f(x,y) = 2x^2 e^y$

4. $f(x,y) = xe^{xy}$

5. $f(x,y) = \dfrac{x}{y} + \dfrac{y}{x}$

6. $f(x,y) = \dfrac{1}{x+y}$

7. $f(x,y) = (2x - y + 5)^2$

8. $f(x,y) = \dfrac{e^x}{1 + e^y}$

9. $f(x,y) = x^2 e^{3x} \ln y$

10. $f(x,y) = \ln(xy)$

11. $f(x,y) = \dfrac{x - y}{x + y}$

12. $f(x,y) = \sqrt{x^2 + y^2}$

13. Let $f(L,K) = 3\sqrt{LK}$. Find $\dfrac{\partial f}{\partial L}$.

14. Let $f(p,q) = 1 - p(1+q)$. Find $\dfrac{\partial f}{\partial q}$ and $\dfrac{\partial f}{\partial p}$.

15. Let $f(x,y,z) = (1 + x^2 y)/z$. Find $\dfrac{\partial f}{\partial x}, \dfrac{\partial f}{\partial y}$, and $\dfrac{\partial f}{\partial z}$.

16. Let $f(x,y,z) = ze^{x/y}$. Find $\dfrac{\partial f}{\partial x}, \dfrac{\partial f}{\partial y}$, and $\dfrac{\partial f}{\partial z}$.

17. Let $f(x,y,z) = xze^{yz}$. Find $\dfrac{\partial f}{\partial x}, \dfrac{\partial f}{\partial y}$, and $\dfrac{\partial f}{\partial z}$.

18. Let $f(x,y,z) = \dfrac{xy}{z}$. Find $\dfrac{\partial f}{\partial x}, \dfrac{\partial f}{\partial y}$, and $\dfrac{\partial f}{\partial z}$.

19. Let $f(x,y) = x^2 + 2xy + y^2 + 3x + 5y$. Find $\dfrac{\partial f}{\partial x}(2, -3)$ and $\dfrac{\partial f}{\partial y}(2, -3)$.

20. Let $f(x,y) = (x + y^2)^3$. Evaluate $\dfrac{\partial f}{\partial x}$ and $\dfrac{\partial f}{\partial y}$ at $(x,y) = (1,2)$.

21. Let $f(x,y,z) = xy^2 z + 5$. Evaluate $\dfrac{\partial f}{\partial y}$ at $(x,y,z) = (2, -1, 3)$.

22. Let $f(x,y,z) = \dfrac{x}{y - z}$. Compute $\dfrac{\partial f}{\partial y}(2, -1, 3)$.

23. Let $f(x,y) = x^3 y + 2xy^2$. Find $\dfrac{\partial^2 f}{\partial x^2}, \dfrac{\partial^2 f}{\partial y^2}, \dfrac{\partial^2 f}{\partial x\,\partial y}$, and $\dfrac{\partial^2 f}{\partial y\,\partial x}$.

24. Let $f(x,y) = xe^y + x^4 y + y^3$. Find $\dfrac{\partial^2 f}{\partial x^2}, \dfrac{\partial^2 f}{\partial y^2}, \dfrac{\partial^2 f}{\partial x\,\partial y}$, and $\dfrac{\partial^2 f}{\partial y\,\partial x}$.

25. Production   A farmer can produce $f(x,y) = 200\sqrt{6x^2 + y^2}$ units of produce by utilizing x units of labor and y units of capital. (The capital is used to rent or purchase land, materials, and equipment.)

  (a) Calculate the marginal productivities of labor and capital when $x = 10$ and $y = 5$.

  (b) Let h be a small number. Use the result of part (a) to determine the approximate effect on production of changing labor from 10 to $10 + h$ units while keeping capital fixed at 5 units.

  (c) Use part (b) to estimate the change in production when labor decreases from 10 to 9.5 units and capital stays fixed at 5 units.

26. The productivity of a country is given by $f(x,y) = 300x^{2/3}y^{1/3}$, where x and y are the amount of labor and capital.

  (a) Compute the marginal productivities of labor and capital when $x = 125$ and $y = 64$.

  (b) Use part (a) to determine the approximate effect on productivity of increasing capital from 64 to 66 units, while keeping labor fixed at 125 units.

  (c) What would be the approximate effect of decreasing labor from 125 to 124 units while keeping capital fixed at 64 units?

27. Demand In a certain suburban community, commuters have the choice of getting into the city by bus or train. The demand for these modes of transportation varies with their cost. Let $f(p_1, p_2)$ be the number of people who will take the bus when p_1 is the price of the bus ride and p_2 is the price of the train ride. For example, if $f(4.50, 6) = 7000$, then 7000 commuters will take the bus when the price of a bus ticket is \$4.50 and the price of a train ticket is \$6.00. Explain why $\dfrac{\partial f}{\partial p_1} < 0$ and $\dfrac{\partial f}{\partial p_2} > 0$.

28. Refer to Exercise 27. Let $g(p_1, p_2)$ be the number of people who will take the train when p_1 is the price of the bus ride and p_2 is the price of the train ride. Would you expect $\dfrac{\partial g}{\partial p_1}$ to be positive or negative? How about $\dfrac{\partial g}{\partial p_2}$?

29. Let p_1 be the average price of DVD players, p_2 the average price of DVDs, $f(p_1, p_2)$ the demand for DVD players, and $g(p_1, p_2)$ the demand for DVDs. Explain why $\dfrac{\partial f}{\partial p_2} < 0$ and $\dfrac{\partial g}{\partial p_1} < 0$.

30. The demand for a certain gas-guzzling car is given by $f(p_1, p_2)$, where p_1 is the price of the car and p_2 is the price of gasoline. Explain why $\dfrac{\partial f}{\partial p_1} < 0$ and $\dfrac{\partial f}{\partial p_2} < 0$.

31. The volume (V) of a certain amount of a gas is determined by the temperature (T) and the pressure (P) by the formula $V = .08(T/P)$. Calculate and interpret $\dfrac{\partial V}{\partial P}$ and $\dfrac{\partial V}{\partial T}$ when $P = 20$, $T = 300$.

32. Beer Consumption Using data collected from 1929 to 1941, Richard Stone* determined that the yearly quantity Q of beer consumed in the United Kingdom was approximately given by the formula $Q = f(m, p, r, s)$, where

$$f(m, p, r, s) = (1.058)m^{.136}p^{-.727}r^{.914}s^{.816}$$

and m is the aggregate real income (personal income after direct taxes, adjusted for retail price changes), p is the average retail price of the commodity (in this case, beer), r is the average retail price level of all other consumer goods and services, and s is a measure of the strength of the beer. Determine which partial derivatives are positive and which are negative and give interpretations. (For example, since $\dfrac{\partial f}{\partial r} > 0$, people buy more beer when the prices of other goods increase and the other factors remain constant.)

33. Richard Stone (see Exercise 32) determined that the yearly consumption of food in the United States was given by

$$f(m, p, r) = (2.186)m^{.595}p^{-.543}r^{.922}.$$

Determine which partial derivatives are positive and which are negative and give interpretations of these facts.

34. Distribution of Revenue For the production function $f(x, y) = 60x^{3/4}y^{1/4}$ considered in Example 8, think of $f(x, y)$ as the revenue when utilizing x units of labor and y units of capital. Under actual operating conditions, say $x = a$ and $y = b$, $\dfrac{\partial f}{\partial x}(a, b)$ is referred to as the *wage per unit of labor* and $\dfrac{\partial f}{\partial y}(a, b)$ is referred to as the *wage per unit of capital*. Show that

$$f(a, b) = a \cdot \left[\frac{\partial f}{\partial x}(a, b)\right] + b \cdot \left[\frac{\partial f}{\partial y}(a, b)\right].$$

(This equation shows how the revenue is distributed between labor and capital.)

35. Compute $\dfrac{\partial^2 f}{\partial x^2}$, where $f(x, y) = 60x^{3/4}y^{1/4}$, a production function (where x is units of labor). Explain why $\dfrac{\partial^2 f}{\partial x^2}$ is always negative.

36. Compute $\dfrac{\partial^2 f}{\partial y^2}$, where $f(x, y) = 60x^{3/4}y^{1/4}$, a production function (where y is units of capital). Explain why $\dfrac{\partial^2 f}{\partial y^2}$ is always negative.

37. Let $f(x, y) = 3x^2 + 2xy + 5y$, as in Example 5. Show that

$$f(1 + h, 4) - f(1, 4) = 14h + 3h^2.$$

Thus the error in approximating $f(1+h, 4) - f(1, 4)$ by $14h$ is $3h^2$. (If $h = .01$, for instance, the error is only .0003.)

38. Body Surface Area Physicians, particularly pediatricians, sometimes need to know the body surface area of a patient. For instance, the surface area is used to adjust the results of certain tests of kidney performance. Tables are available that give the approximate body surface area A in square meters of a person who weighs W kilograms and is H centimeters tall. The following empirical formula† is also used:

$$A = .007W^{.425}H^{.725}.$$

Evaluate $\dfrac{\partial A}{\partial W}$ and $\dfrac{\partial A}{\partial H}$ when $W = 54$ and $H = 165$, and give a physical interpretation of your answers. You may use the approximations $(54)^{.425} \approx 5.4$, $(54)^{-.575} \approx .10$, $(165)^{.725} \approx 40.5$, and $(165)^{-.275} \approx .25$.

Solutions to Practice Problems 7.2

1. **(a)** Negative. $\dfrac{\partial f}{\partial x}(400, 2000)$ is approximately the change in sales due to a \$1 increase in x (price). Since raising prices lowers sales, we would expect $\dfrac{\partial f}{\partial x}(400, 2000)$ to be negative.

(b) Positive. $\dfrac{\partial f}{\partial y}(400, 2000)$ is approximately the change in sales due to a \$1 increase in advertising. Since spending more money on advertising brings in more customers, we would expect sales to increase; that is, $\dfrac{\partial f}{\partial y}(400, 2000)$ is most likely positive.

2. If the interest rate is raised from 9% to 10%, the monthly payment will increase by about \$66.20. [An increase to $9\frac{1}{2}\%$ causes an increase in the monthly payment of about $\frac{1}{2} \cdot (66.20)$ or \$33.10, and so on.]

7.3 Maxima and Minima of Functions of Several Variables

Previously, we studied how to determine the maxima and minima of functions of a single variable. Let us extend that discussion to functions of several variables.

If $f(x, y)$ is a function of two variables, we say that $f(x, y)$ has a *relative maximum* when $x = a$, $y = b$ if $f(x, y)$ is at most equal to $f(a, b)$ whenever x is near a and y is near b. Geometrically, the graph of $f(x, y)$ has a peak at $(x, y) = (a, b)$. [See Fig. 1(a).] Similarly, we say that $f(x, y)$ has a *relative minimum* when $x = a$, $y = b$ if $f(x, y)$ is at least equal to $f(a, b)$ whenever x is near a and y is near b. Geometrically, the graph of $f(x, y)$ has a pit whose bottom occurs at $(x, y) = (a, b)$. [See Fig. 1(b).]

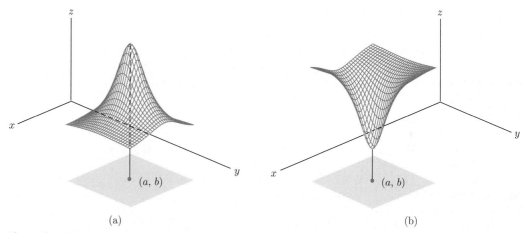

(a)   (b)

Figure 1.   Maximum and minimum points.

Suppose that the function $f(x, y)$ has a relative minimum at $(x, y) = (a, b)$, as in Fig. 2. When y is held constant at b, $f(x, y)$ is a function of x with a relative minimum at $x = a$. Therefore, the tangent line to the curve $z = f(x, b)$ is horizontal at $x = a$ and hence has slope 0. That is,

$$\frac{\partial f}{\partial x}(a, b) = 0.$$

Likewise, when x is held constant at a, then $f(x, y)$ is a function of y with a relative minimum at $y = b$. Therefore, its derivative with respect to y is zero at $y = b$. That

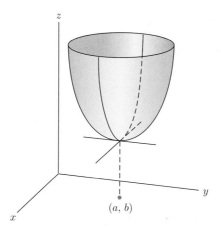

Figure 2.   Horizontal tangent lines at a relative minimum.

is,

$$\frac{\partial f}{\partial y}(a, b) = 0.$$

Similar considerations apply when $f(x, y)$ has a relative maximum at $(x, y) = (a, b)$.

> **First-Derivative Test for Functions of Two Variables** If $f(x, y)$ has either a relative maximum or minimum at $(x, y) = (a, b)$, then
>
> $$\frac{\partial f}{\partial x}(a, b) = 0$$
>
> and
>
> $$\frac{\partial f}{\partial y}(a, b) = 0.$$

A relative maximum or minimum may or may not be an absolute maximum or minimum. However, to simplify matters in this text, the examples and exercises have been chosen so that, if an absolute extremum of $f(x, y)$ exists, it will occur at a point where $f(x, y)$ has a relative extremum.

EXAMPLE 1   The function $f(x, y) = 3x^2 - 4xy + 3y^2 + 8x - 17y + 30$ has the graph pictured in Fig. 2. Find the point (a, b) at which $f(x, y)$ attains its minimum value.

Solution   We look for those values of x and y at which both partial derivatives are zero. The partial derivatives are

$$\frac{\partial f}{\partial x} = 6x - 4y + 8,$$

$$\frac{\partial f}{\partial y} = -4x + 6y - 17.$$

Setting $\dfrac{\partial f}{\partial x} = 0$ and $\dfrac{\partial f}{\partial y} = 0$, we obtain

$$6x - 4y + 8 = 0 \quad \text{or} \quad y = \frac{6x + 8}{4},$$

$$-4x + 6y - 17 = 0 \quad \text{or} \quad y = \frac{4x + 17}{6}.$$

By equating these two expressions for y, we have

$$\frac{6x + 8}{4} = \frac{4x + 17}{6}.$$

Cross-multiplying, we see that

$$36x + 48 = 16x + 68$$
$$20x = 20$$
$$x = 1.$$

When we substitute this value for x into our first equation for y in terms of x, we obtain

$$y = \frac{6x + 8}{4} = \frac{6 \cdot 1 + 8}{4} = \frac{7}{2}.$$

If $f(x, y)$ has a minimum, it must occur where $\dfrac{\partial f}{\partial x} = 0$ and $\dfrac{\partial f}{\partial y} = 0$. We have determined that the partial derivatives are zero only when $x = 1$, $y = \frac{7}{2}$. From Fig. 2 we know that $f(x, y)$ has a minimum, so it must be at $(x, y) = (1, \frac{7}{2})$. ∎

EXAMPLE 2

Price discrimination A monopolist markets a product in two countries and can charge different amounts in each country. Let x be the number of units to be sold in the first country and y the number of units to be sold in the second country. Due to the laws of demand, the monopolist must set the price at $97 - (x/10)$ dollars in the first country and $83 - (y/20)$ dollars in the second country to sell all the units. The cost of producing these units is $20{,}000 + 3(x + y)$. Find the values of x and y that maximize the profit.

Solution Let $f(x, y)$ be the profit derived from selling x units in the first country and y in the second. Then

$$f(x, y) = [\text{revenue from first country}] + [\text{revenue from second country}] - [\text{cost}]$$

$$= \left(97 - \frac{x}{10}\right)x + \left(83 - \frac{y}{20}\right)y - [20{,}000 + 3(x + y)]$$

$$= 97x - \frac{x^2}{10} + 83y - \frac{y^2}{20} - 20{,}000 - 3x - 3y$$

$$= 94x - \frac{x^2}{10} + 80y - \frac{y^2}{20} - 20{,}000.$$

To find where $f(x, y)$ has its maximum value, we look for those values of x and y at which both partial derivatives are zero.

$$\frac{\partial f}{\partial x} = 94 - \frac{x}{5},$$

$$\frac{\partial f}{\partial y} = 80 - \frac{y}{10}.$$

We set $\dfrac{\partial f}{\partial x} = 0$ and $\dfrac{\partial f}{\partial y} = 0$ to obtain

$$94 - \frac{x}{5} = 0 \quad \text{or} \quad x = 470,$$

$$80 - \frac{y}{10} = 0 \quad \text{or} \quad y = 800.$$

Therefore, the firm should adjust its prices to levels where it will sell 470 units in the first country and 800 units in the second country. ∎

EXAMPLE 3

Heat loss Suppose that we want to design a rectangular building having a volume of 147,840 cubic feet. Assuming that the daily loss of heat is given by

$$w = 11xy + 14yz + 15xz,$$

where x, y, and z are, respectively, the length, width, and height of the building, find the dimensions of the building for which the daily heat loss is minimal.

Solution We must minimize the function

$$w = 11xy + 14yz + 15xz, \tag{1}$$

where x, y, z satisfy the constraint equation (refer to Section 2.5)

$$xyz = 147{,}840.$$

For simplicity, let us denote 147,840 by V. Then $xyz = V$, so $z = V/xy$. We substitute this expression for z into the objective function (1) to obtain a heat-loss function $g(x, y)$ of two variables:

$$g(x, y) = 11xy + 14y\frac{V}{xy} + 15x\frac{V}{xy} = 11xy + \frac{14V}{x} + \frac{15V}{y}.$$

To minimize this function, we first compute the partial derivatives with respect to x and y; then we equate them to zero.

$$\frac{\partial g}{\partial x} = 11y - \frac{14V}{x^2} = 0,$$

$$\frac{\partial g}{\partial y} = 11x - \frac{15V}{y^2} = 0.$$

These two equations yield

$$y = \frac{14V}{11x^2}, \tag{2}$$

$$11xy^2 = 15V. \tag{3}$$

If we substitute the value of y from (2) into (3), we see that

$$11x\left(\frac{14V}{11x^2}\right)^2 = 15V$$

$$\frac{14^2V^2}{11x^3} = 15V$$

$$x^3 = \frac{14^2 \cdot V^2}{11 \cdot 15 \cdot V} = \frac{14^2 \cdot V}{11 \cdot 15}$$

$$= \frac{14^2 \cdot 147,840}{11 \cdot 15}$$

$$= 175,616.$$

Therefore, we see (using a calculator) that

$$x = 56.$$

From equation (2) we find that

$$y = \frac{14 \cdot V}{11x^2} = \frac{14 \cdot 147,840}{11 \cdot 56^2} = 60.$$

Finally,

$$z = \frac{V}{xy} = \frac{147,840}{56 \cdot 60} = 44.$$

Thus the building should be 56 feet long, 60 feet wide, and 44 feet high to minimize the heat loss.* ∎

---

*For further discussion of this heat-loss problem, as well as other examples of optimization in architectural design, see L. March, "Elementary Models of Built Forms," Chapter 3 in *Urban Space and Structures*, L. Martin and L. March, eds. (New York: Cambridge University Press, 1972).

When considering a function of two variables, we find points (x, y) at which $f(x, y)$ has a potential relative maximum or minimum by setting $\dfrac{\partial f}{\partial x}$ and $\dfrac{\partial f}{\partial y}$ equal to zero and solving for x and y. However, if we are given no additional information about $f(x, y)$, it may be difficult to determine whether we have found a maximum or a minimum (or neither). In the case of functions of one variable, we studied concavity and deduced the second-derivative test. There is an analog of the second derivative test for functions of two variables, but it is much more complicated than the one-variable test. We state it without proof.

Second-Derivative Test for Functions of Two Variables   Suppose that $f(x, y)$ is a function and (a, b) is a point at which

$$\frac{\partial f}{\partial x}(a, b) = 0 \quad \text{and} \quad \frac{\partial f}{\partial y}(a, b) = 0,$$

and let

$$D(x, y) = \frac{\partial^2 f}{\partial x^2} \cdot \frac{\partial^2 f}{\partial y^2} - \left(\frac{\partial^2 f}{\partial x \, \partial y}\right)^2.$$

1. If
$$D(a, b) > 0 \quad \text{and} \quad \frac{\partial^2 f}{\partial x^2}(a, b) > 0,$$
then $f(x, y)$ has a relative minimum at (a, b).

2. If
$$D(a, b) > 0 \quad \text{and} \quad \frac{\partial^2 f}{\partial x^2}(a, b) < 0,$$
then $f(x, y)$ has a relative maximum at (a, b).

3. If
$$D(a, b) < 0,$$
then $f(x, y)$ has neither a relative maximum nor a relative minimum at (a, b).

4. If $D(a, b) = 0$, no conclusion can be drawn from this test.

The saddle-shaped graph in Fig. 3 illustrates a function $f(x, y)$ for which $D(a, b) < 0$. Both partial derivatives are zero at $(x, y) = (a, b)$, and yet the function has neither a relative maximum nor a relative minimum there. (Observe that the function has a relative maximum with respect to x when y is held constant and a relative minimum with respect to y when x is held constant.)

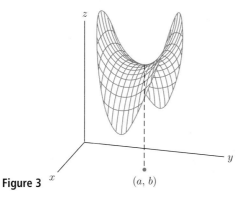

Figure 3

EXAMPLE 4   Let $f(x, y) = x^3 - y^2 - 12x + 6y + 5$. Find all possible relative maximum and minimum points of $f(x, y)$. Use the second-derivative test to determine the nature of each such point.

Solution Since

$$\frac{\partial f}{\partial x} = 3x^2 - 12, \qquad \frac{\partial f}{\partial y} = -2y + 6,$$

we find that $f(x, y)$ has a potential relative extreme point when

$$3x^2 - 12 = 0,$$
$$-2y + 6 = 0.$$

From the first equation, $3x^2 = 12$, $x^2 = 4$, and $x = \pm 2$. From the second equation, $y = 3$. Thus, $\dfrac{\partial f}{\partial x}$ and $\dfrac{\partial f}{\partial y}$ are both zero when $(x, y) = (2, 3)$ and when $(x, y) = (-2, 3)$. To apply the second-derivative test, compute

$$\frac{\partial^2 f}{\partial x^2} = 6x, \qquad \frac{\partial^2 f}{\partial y^2} = -2, \qquad \frac{\partial^2 f}{\partial x\, \partial y} = 0,$$

and

$$D(x, y) = \frac{\partial^2 f}{\partial x^2} \cdot \frac{\partial^2 f}{\partial y^2} - \left(\frac{\partial^2 f}{\partial x\, \partial y} \right)^2 = (6x)(-2) - 0^2 = -12x. \tag{4}$$

Since $D(2, 3) = -12(2) = -24$, which is negative, case 3 of the second-derivative test says that $f(x, y)$ has neither a relative maximum nor a relative minimum at $(2, 3)$. However, $D(-2, 3) = -12(-2) = 24$. Since $D(-2, 3)$ is positive, the function $f(x, y)$ has either a relative maximum or a relative minimum at $(-2, 3)$. To determine which, we compute

$$\frac{\partial^2 f}{\partial x^2}(-2, 3) = 6(-2) = -12 < 0.$$

By case 2 of the second-derivative test, the function $f(x, y)$ has a relative maximum at $(-2, 3)$. ∎

In this section we have restricted ourselves to functions of two variables, but the case of three or more variables is handled in a similar fashion. For instance, here is the first-derivative test for a function of three variables.

If $f(x, y, z)$ has a relative maximum or minimum at $(x, y, z) = (a, b, c)$, then

$$\frac{\partial f}{\partial x}(a, b, c) = 0,$$

$$\frac{\partial f}{\partial y}(a, b, c) = 0,$$

$$\frac{\partial f}{\partial z}(a, b, c) = 0.$$

Practice Problems 7.3

1. Find all points (x, y) where $f(x, y) = x^3 - 3xy + \frac{1}{2}y^2 + 8$ has a possible relative maximum or minimum.

2. Apply the second-derivative test to the function $g(x, y)$ of Example 3 to confirm that a relative minimum actually occurs when $x = 56$ and $y = 60$.

EXERCISES 7.3

Find all points (x, y) where $f(x, y)$ has a possible relative maximum or minimum.

1. $f(x, y) = x^2 - 3y^2 + 4x + 6y + 8$
2. $f(x, y) = \frac{1}{2}x^2 + y^2 - 3x + 2y - 5$
3. $f(x, y) = x^2 - 5xy + 6y^2 + 3x - 2y + 4$
4. $f(x, y) = -3x^2 + 7xy - 4y^2 + x + y$
5. $f(x, y) = x^3 + y^2 - 3x + 6y$
6. $f(x, y) = x^2 - y^3 + 5x + 12y + 1$
7. $f(x, y) = \frac{1}{3}x^3 - 2y^3 - 5x + 6y - 5$
8. $f(x, y) = x^4 - 8xy + 2y^2 - 3$
9. The function $f(x, y) = 2x + 3y + 9 - x^2 - xy - y^2$ has a maximum at some point (x, y). Find the values of x and y where this maximum occurs.
10. The function $f(x, y) = \frac{1}{2}x^2 + 2xy + 3y^2 - x + 2y$ has a minimum at some point (x, y). Find the values of x and y where this minimum occurs.

In Exercises 11–16, both first partial derivatives of the function $f(x, y)$ are zero at the given points. Use the second-derivative test to determine the nature of $f(x, y)$ at each of these points. If the second-derivative test is inconclusive, so state.

11. $f(x, y) = 3x^2 - 6xy + y^3 - 9y$; $(3, 3)$, $(-1, -1)$
12. $f(x, y) = 6xy^2 - 2x^3 - 3y^4$; $(0, 0)$, $(1, 1)$, $(1, -1)$
13. $f(x, y) = 2x^2 - x^4 - y^2$; $(-1, 0)$, $(0, 0)$, $(1, 0)$
14. $f(x, y) = x^4 - 4xy + y^4$; $(0, 0)$, $(1, 1)$, $(-1, -1)$
15. $f(x, y) = ye^x - 3x - y + 5$; $(0, 3)$
16. $f(x, y) = \dfrac{1}{x} + \dfrac{1}{y} + xy$; $(1, 1)$

Find all points (x, y) where $f(x, y)$ has a possible relative maximum or minimum. Then use the second-derivative test to determine, if possible, the nature of $f(x, y)$ at each of these points. If the second-derivative test is inconclusive, so state.

17. $f(x, y) = x^2 - 2xy + 4y^2$
18. $f(x, y) = 2x^2 + 3xy + 5y^2$
19. $f(x, y) = -2x^2 + 2xy - y^2 + 4x - 6y + 5$
20. $f(x, y) = -x^2 - 8xy - y^2$
21. $f(x, y) = x^2 + 2xy + 5y^2 + 2x + 10y - 3$
22. $f(x, y) = x^2 - 2xy + 3y^2 + 4x - 16y + 22$
23. $f(x, y) = x^3 - y^2 - 3x + 4y$
24. $f(x, y) = x^3 - 2xy + 4y$
25. $f(x, y) = 2x^2 + y^3 - x - 12y + 7$
26. $f(x, y) = x^2 + 4xy + 2y^4$
27. Find the possible values of x, y, z at which
$$f(x, y, z) = 2x^2 + 3y^2 + z^2 - 2x - y - z$$
assumes its minimum value.

28. Find the possible values of x, y, z at which
$$f(x, y, z) = 5 + 8x - 4y + x^2 + y^2 + z^2$$
assumes its minimum value.

29. United States postal rules require that the length plus the girth of a package cannot exceed 84 inches. Find the dimensions of the rectangular package of greatest volume that can be mailed. [*Note*: From Fig. 4 we see that $84 = (\text{length}) + (\text{girth}) = l + (2x + 2y)$.]

Figure 4

30. Find the dimensions of the rectangular box of least surface area that has a volume of 1000 cubic inches.

31. A company manufactures and sells two products, I and II, that sell for \$10 and \$9 per unit, respectively. The cost of producing x units of product I and y units of product II is
$$400 + 2x + 3y + .01(3x^2 + xy + 3y^2).$$
Find the values of x and y that maximize the company's profits. [*Note*: Profit = (revenue) − (cost).]

32. A monopolist manufactures and sells two competing products, I and II, that cost \$30 and \$20 per unit, respectively, to produce. The revenue from marketing x units of product I and y units of product II is $98x + 112y - .04xy - .1x^2 - .2y^2$. Find the values of x and y that maximize the monopolist's profits.

33. A company manufactures and sells two products, I and II, that sell for $\$p_{\mathrm{I}}$ and $\$p_{\mathrm{II}}$ per unit, respectively. Let $C(x, y)$ be the cost of producing x units of product I and y units of product II. Show that if the company's profit is maximized when $x = a$, $y = b$ then
$$\frac{\partial C}{\partial x}(a, b) = p_{\mathrm{I}} \quad \text{and} \quad \frac{\partial C}{\partial y}(a, b) = p_{\mathrm{II}}.$$

34. A monopolist manufactures and sells two competing products, I and II, that cost $\$p_{\mathrm{I}}$ and $\$p_{\mathrm{II}}$ per unit, respectively, to produce. Let $R(x, y)$ be the revenue from marketing x units of product I and y units of product II. Show that if the monopolist's profit is maximized when $x = a$, $y = b$ then
$$\frac{\partial R}{\partial x}(a, b) = p_{\mathrm{I}} \quad \text{and} \quad \frac{\partial R}{\partial y}(a, b) = p_{\mathrm{II}}.$$

Solutions to Practice Problems 7.3

1. Compute the first partial derivatives of $f(x, y)$ and solve the system of equations that results from setting the partials equal to zero.

$$\frac{\partial f}{\partial x} = 3x^2 - 3y = 0,$$

$$\frac{\partial f}{\partial y} = -3x + y = 0.$$

Solve each equation for y in terms of x.

$$\begin{cases} y = x^2 \\ y = 3x. \end{cases}$$

Equate expressions for y and solve for x.

$$x^2 = 3x$$

$$x^2 - 3x = 0$$

$$x(x - 3) = 0$$

$$x = 0 \quad \text{or} \quad x = 3.$$

When $x = 0$, $y = 0^2 = 0$. When $x = 3$, $y = 3^2 = 9$. Therefore, the possible relative maximum or minimum points are $(0, 0)$ and $(3, 9)$.

2. We have

$$g(x, y) = 11xy + \frac{14V}{x} + \frac{15V}{y},$$

$$\frac{\partial g}{\partial x} = 11y - \frac{14V}{x^2}, \quad \text{and} \quad \frac{\partial g}{\partial y} = 11x - \frac{15V}{y^2}.$$

Now,

$$\frac{\partial^2 g}{\partial x^2} = \frac{28V}{x^3}, \quad \frac{\partial^2 g}{\partial y^2} = \frac{30V}{y^3}, \quad \text{and} \quad \frac{\partial^2 g}{\partial x \, \partial y} = 11.$$

Therefore,

$$D(x, y) = \frac{28V}{x^3} \cdot \frac{30V}{y^3} - (11)^2$$

$$D(56, 60) = \frac{28(147{,}840)}{(56)^3} \cdot \frac{30(147{,}840)}{(60)^3} - 121$$

$$= 484 - 121 = 363 > 0,$$

and

$$\frac{\partial^2 g}{\partial x^2}(56, 60) = \frac{28(147{,}840)}{(56)^3} > 0.$$

It follows that $g(x, y)$ has a relative minimum at $x = 56$, $y = 60$.

7.4 Lagrange Multipliers and Constrained Optimization

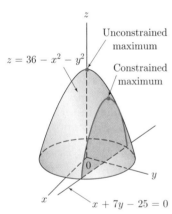

Figure 1. A constrained optimization problem.

We have seen a number of optimization problems in which we were required to minimize (or maximize) an objective function where the variables were subject to a constraint equation. For instance, in Example 4 of Section 2.5, we minimized the cost of a rectangular enclosure by minimizing the objective function $42x + 28y$, where x and y were subject to the constraint equation $600 - xy = 0$. In the preceding section (Example 3) we minimized the daily heat loss from a building by minimizing the objective function $11xy + 14yz + 15xz$, subject to the constraint equation $147{,}840 - xyz = 0$.

Figure 1 gives a graphical illustration of what happens when an objective function is maximized subject to a constraint. The graph of the objective function is the cone-shaped surface $z = 36 - x^2 - y^2$, and the colored curve on that surface consists of those points whose x- and y-coordinates satisfy the constraint equation $x + 7y - 25 = 0$. The constrained maximum is at the highest point on this curve. Of course, the surface itself has a higher "unconstrained maximum" at $(x, y, z) = (0, 0, 36)$, but these values of x and y do not satisfy the constraint equation.

In this section we introduce a powerful technique for solving problems of this type. Let us begin with the following general problem, which involves two variables.

Problem Let $f(x, y)$ and $g(x, y)$ be functions of two variables. Find values of x and y that maximize (or minimize) the objective function $f(x, y)$ and that also satisfy the constraint equation $g(x, y) = 0$.

Of course, if we can solve the equation $g(x, y) = 0$ for one variable in terms of the other and substitute the resulting expression into $f(x, y)$, we arrive at a function of a single variable that can be maximized (or minimized) by using the methods of Chapter 2. However, this technique can be unsatisfactory for two reasons. First, it may be difficult to solve the equation $g(x, y) = 0$ for x or for y. For example, if $g(x, y) = x^4 + 5x^3y + 7x^2y^3 + y^5 - 17 = 0$, it is difficult to write y as a function of x or x as a function of y. Second, even if $g(x, y) = 0$ can be solved for one variable in terms of the other, substitution of the result into $f(x, y)$ may yield a complicated function.

One clever idea for handling the preceding problem was discovered by the eighteenth-century mathematician Lagrange, and the technique that he pioneered today bears his name, the method of *Lagrange multipliers*. The basic idea of this method is to replace $f(x, y)$ by an auxiliary function of three variables $F(x, y, \lambda)$, defined as

$$F(x, y, \lambda) = f(x, y) + \lambda g(x, y).$$

The new variable λ (lambda) is called a *Lagrange multiplier* and always multiplies the constraint function $g(x, y)$. The following theorem is stated without proof.

> **Theorem**   Suppose that, subject to the constraint $g(x, y) = 0$, the function $f(x, y)$ has a relative maximum or minimum at $(x, y) = (a, b)$. Then there is a value of λ, say $\lambda = c$, such that the partial derivatives of $F(x, y, \lambda)$ all equal zero at $(x, y, \lambda) = (a, b, c)$.

The theorem implies that, if we locate all points (x, y, λ) where the partial derivatives of $F(x, y, \lambda)$ are all zero, among the corresponding points (x, y), we then will find all possible places where $f(x, y)$ may have a constrained relative maximum or minimum. Thus the first step in the method of Lagrange multipliers is to set the partial derivatives of $F(x, y, \lambda)$ equal to zero and solve for x, y, and λ:

$$\frac{\partial F}{\partial x} = 0 \tag{L-1}$$

$$\frac{\partial F}{\partial y} = 0 \tag{L-2}$$

$$\frac{\partial F}{\partial \lambda} = 0. \tag{L-3}$$

From the definition of $F(x, y, \lambda)$, we see that $\dfrac{\partial F}{\partial \lambda} = g(x, y)$. Thus the third equation (L-3) is just the original constraint equation $g(x, y) = 0$. So, when we find a point (x, y, λ) that satisfies (L-1), (L-2), and (L-3), the coordinates x and y will automatically satisfy the constraint equation.

The first example applies this method to the problem described in Fig. 1.

EXAMPLE 1   Maximize $36 - x^2 - y^2$ subject to the constraint $x + 7y - 25 = 0$.

Solution   Here $f(x, y) = 36 - x^2 - y^2$, $g(x, y) = x + 7y - 25$, and

$$F(x, y, \lambda) = 36 - x^2 - y^2 + \lambda(x + 7y - 25).$$

Equations (L-1) to (L-3) read

$$\frac{\partial F}{\partial x} = -2x + \lambda = 0, \tag{1}$$

$$\frac{\partial F}{\partial y} = -2y + 7\lambda = 0, \tag{2}$$

$$\frac{\partial F}{\partial \lambda} = x + 7y - 25 = 0. \tag{3}$$

We solve the first two equations for λ:

$$\lambda = 2x$$

$$\lambda = \tfrac{2}{7}y. \tag{4}$$

If we equate these two expressions for λ, we obtain

$$2x = \tfrac{2}{7}y$$

$$x = \tfrac{1}{7}y. \tag{5}$$

Substituting this expression for x into equation (3), we have

$$\tfrac{1}{7}y + 7y - 25 = 0$$

$$\tfrac{50}{7}y = 25$$

$$y = \tfrac{7}{2}.$$

With this value for y, equations (4) and (5) produce the values of x and λ:

$$x = \tfrac{1}{7}y = \tfrac{1}{7}\left(\tfrac{7}{2}\right) = \tfrac{1}{2},$$

$$\lambda = \tfrac{2}{7}y = \tfrac{2}{7}\left(\tfrac{7}{2}\right) = 1.$$

Therefore, the partial derivatives of $F(x, y, \lambda)$ are zero when $x = \tfrac{1}{2}$, $y = \tfrac{7}{2}$, and $\lambda = 1$. So the maximum value of $36 - x^2 - y^2$ subject to the constraint $x + 7y - 25 = 0$ is

$$36 - \left(\tfrac{1}{2}\right)^2 - \left(\tfrac{7}{2}\right)^2 = \tfrac{47}{2}. \qquad \blacksquare$$

The preceding technique for solving three equations in the three variables x, y, and λ can usually be applied to solve Lagrange multiplier problems. Here is the basic procedure.

1. Solve (L-1) and (L-2) for λ in terms of x and y; then equate the resulting expressions for λ.

2. Solve the resulting equation for one of the variables.

3. Substitute the expression so derived into the equation (L-3), and solve the resulting equation of one variable.

4. Use the one known variable and the equations of steps 1 and 2 to determine the other two variables.

In most applications we know that an absolute (constrained) maximum or minimum exists. In the event that the method of Lagrange multipliers produces exactly one possible relative extreme value, we will assume that it is indeed the sought after absolute extreme value. For instance, the statement of Example 1 is meant to imply that there is an absolute maximum value. Since we determined that there was just one possible relative extreme value, we concluded that it was the absolute maximum value.

EXAMPLE 2

Using Lagrange multipliers, minimize $42x + 28y$, subject to the constraint $600 - xy = 0$, where x and y are restricted to positive values. (This problem arose in Example 4 of Section 2.5, where $42x + 28y$ was the cost of building a 600-square-foot enclosure having dimensions x and y.)

Solution We have $f(x, y) = 42x + 28y$, $g(x, y) = 600 - xy$, and

$$F(x, y, \lambda) = 42x + 28y + \lambda(600 - xy).$$

The equations (L-1) to (L-3), in this case, are

$$\frac{\partial F}{\partial x} = 42 - \lambda y = 0,$$

$$\frac{\partial F}{\partial y} = 28 - \lambda x = 0,$$

$$\frac{\partial F}{\partial \lambda} = 600 - xy = 0.$$

From the first two equations we see that

$$\lambda = \frac{42}{y} = \frac{28}{x}. \qquad \text{(step 1)}$$

Therefore,

$$42x = 28y$$

and

$$x = \frac{2}{3}y. \qquad \text{(step 2)}$$

Substituting this expression for x into the third equation, we derive

$$600 - \left(\frac{2}{3}y\right)y = 0$$

$$y^2 = \frac{3}{2} \cdot 600 = 900$$

$$y = \pm 30. \qquad \text{(step 3)}$$

We discard the case $y = -30$ because we are interested only in positive values of x and y. Using $y = 30$, we find that

$$\left.\begin{array}{l} x = \dfrac{2}{3}(30) = 20 \\[2mm] \lambda = \dfrac{28}{20} = \dfrac{7}{5}. \end{array}\right\} \qquad \text{(step 4)}$$

So the minimum value of $42x + 28y$ with x and y subject to the constraint occurs when $x = 20$, $y = 30$, and $\lambda = \frac{7}{5}$. That minimum value is

$$42 \cdot (20) + 28 \cdot (30) = 1680. \qquad \blacksquare$$

EXAMPLE 3

Production Suppose that x units of labor and y units of capital can produce $f(x, y) = 60x^{3/4}y^{1/4}$ units of a certain product. Also suppose that each unit of labor costs \$100, whereas each unit of capital costs \$200. Assume that \$30,000 is available to spend on production. How many units of labor and how many units of capital should be utilized to maximize production?

Solution The cost of x units of labor and y units of capital equals $100x + 200y$. Therefore, since we want to use all the available money ($30,000), we must satisfy the constraint equation

$$100x + 200y = 30{,}000$$

or

$$g(x, y) = 30{,}000 - 100x - 200y = 0.$$

The objective function is $f(x, y) = 60x^{3/4}y^{1/4}$. In this case, we have

$$F(x, y, \lambda) = 60x^{3/4}y^{1/4} + \lambda(30{,}000 - 100x - 200y).$$

The equations (L-1) to (L-3) read

$$\frac{\partial F}{\partial x} = 45x^{-1/4}y^{1/4} - 100\lambda = 0, \tag{L-1}$$

$$\frac{\partial F}{\partial y} = 15x^{3/4}y^{-3/4} - 200\lambda = 0, \tag{L-2}$$

$$\frac{\partial F}{\partial \lambda} = 30{,}000 - 100x - 200y = 0. \tag{L-3}$$

By solving the first two equations for λ, we see that

$$\lambda = \frac{45}{100}x^{-1/4}y^{1/4} = \frac{9}{20}x^{-1/4}y^{1/4},$$

$$\lambda = \frac{15}{200}x^{3/4}y^{-3/4} = \frac{3}{40}x^{3/4}y^{-3/4}.$$

Therefore, we must have

$$\frac{9}{20}x^{-1/4}y^{1/4} = \frac{3}{40}x^{3/4}y^{-3/4}.$$

To solve for y in terms of x, let us multiply both sides of this equation by $x^{1/4}y^{3/4}$:

$$\frac{9}{20}y = \frac{3}{40}x$$

or

$$y = \frac{1}{6}x.$$

Inserting this result in (L-3), we find that

$$100x + 200\left(\frac{1}{6}x\right) = 30{,}000$$

$$\frac{400x}{3} = 30{,}000$$

$$x = 225.$$

Hence

$$y = \frac{225}{6} = 37.5.$$

So maximum production is achieved by using 225 units of labor and 37.5 units of capital.

In Example 3 it turns out that, at the optimum values of x and y,

$$\lambda = \frac{9}{20}x^{-1/4}y^{1/4} = \frac{9}{20}(225)^{-1/4}(37.5)^{1/4} \approx .2875,$$

$$\frac{\partial f}{\partial x} = 45x^{-1/4}y^{1/4} = 45(225)^{-1/4}(37.5)^{1/4}, \tag{6}$$

$$\frac{\partial f}{\partial y} = 15x^{3/4}y^{-3/4} = 15(225)^{3/4}(37.5)^{-3/4}. \tag{7}$$

It can be shown that the Lagrange multiplier λ can be interpreted as the *marginal productivity of money*. That is, if 1 additional dollar is available, approximately .2875 additional units of the product can be produced.

Recall that the partial derivatives $\dfrac{\partial f}{\partial x}$ and $\dfrac{\partial f}{\partial y}$ are called the marginal productivity of labor and capital, respectively. From equations (6) and (7) we have

$$\frac{[\text{marginal productivity of labor}]}{[\text{marginal productivity of capital}]} = \frac{45(225)^{-1/4}(37.5)^{1/4}}{15(225)^{3/4}(37.5)^{-3/4}}$$

$$= \frac{45}{15}(225)^{-1}(37.5)^{1}$$

$$= \frac{3(37.5)}{225} = \frac{37.5}{75} = \frac{1}{2}.$$

On the other hand,

$$\frac{[\text{cost per unit of labor}]}{[\text{cost per unit of capital}]} = \frac{100}{200} = \frac{1}{2}.$$

This result illustrates the following law of economics. *If labor and capital are at their optimal levels, the ratio of their marginal productivities equals the ratio of their unit costs.*

The method of Lagrange multipliers generalizes to functions of any number of variables. For instance, we can maximize $f(x, y, z)$, subject to the constraint equation $g(x, y, z) = 0$, by considering the Lagrange function

$$F(x, y, z, \lambda) = f(x, y, z) + \lambda g(x, y, z).$$

The analogs of equations (L-1) to (L-3) are

$$\frac{\partial F}{\partial x} = 0, \quad \frac{\partial F}{\partial y} = 0, \quad \frac{\partial F}{\partial z} = 0, \quad \frac{\partial F}{\partial \lambda} = 0.$$

Let us now show how we can solve the heat-loss problem of Section 7.3 by using this method.

EXAMPLE 4

Use Lagrange multipliers to find the values of x, y, z that minimize the objective function

$$f(x, y, z) = 11xy + 14yz + 15xz,$$

subject to the constraint

$$xyz = 147{,}840.$$

Solution The Lagrange function is

$$F(x, y, z, \lambda) = 11xy + 14yz + 15xz + \lambda(147{,}840 - xyz).$$

The conditions for a relative minimum are

$$\frac{\partial F}{\partial x} = 11y + 15z - \lambda yz = 0,$$

$$\frac{\partial F}{\partial y} = 11x + 14z - \lambda xz = 0,$$

$$\frac{\partial F}{\partial z} = 14y + 15x - \lambda xy = 0,$$

$$\frac{\partial F}{\partial \lambda} = 147{,}840 - xyz = 0. \tag{8}$$

From the first three equations we have

$$\left.\begin{array}{l} \lambda = \dfrac{11y + 15z}{yz} = \dfrac{11}{z} + \dfrac{15}{y} \\[2mm] \lambda = \dfrac{11x + 14z}{xz} = \dfrac{11}{z} + \dfrac{14}{x} \\[2mm] \lambda = \dfrac{14y + 15x}{xy} = \dfrac{14}{x} + \dfrac{15}{y} \end{array}\right\}. \tag{9}$$

Let us equate the first two expression for λ:

$$\frac{11}{z} + \frac{15}{y} = \frac{11}{z} + \frac{14}{x}$$

$$\frac{15}{y} = \frac{14}{x}$$

$$x = \frac{14}{15}y.$$

Next, we equate the second and third expressions for λ in (9):

$$\frac{11}{z} + \frac{14}{x} = \frac{14}{x} + \frac{15}{y}$$

$$\frac{11}{z} = \frac{15}{y}$$

$$z = \frac{11}{15}y.$$

We now substitute the expressions for x and z into the constraint equation (8) and obtain

$$\frac{14}{15}y \cdot y \cdot \frac{11}{15}y = 147{,}840$$

$$y^3 = \frac{(147{,}840)(15)^2}{(14)(11)} = 216{,}000$$

$$y = 60.$$

From this, we find that

$$x = \frac{14}{15}(60) = 56 \quad \text{and} \quad z = \frac{11}{15}(60) = 44.$$

We conclude that the heat loss is minimized when $x = 56$, $y = 60$, and $z = 44$. ∎

In the solution of Example 4, we found that, at the optimal values of x, y, and z,

$$\frac{14}{x} = \frac{15}{y} = \frac{11}{z}.$$

Referring to Example 2 of Section 7.1, we see that 14 is the combined heat loss through the east and west sides of the building, 15 is the heat loss through the north and south sides of the building, and 11 is the heat loss through the floor and roof. Thus we have that under optimal conditions

$$\frac{[\text{heat loss through east and west sides}]}{[\text{distance between east and west sides}]}$$

$$= \frac{[\text{heat loss through north and south sides}]}{[\text{distance between north and south sides}]}$$

$$= \frac{[\text{heat loss through floor and roof}]}{[\text{distance between floor and roof}]}.$$

This is a principle of optimal design: Minimal heat loss occurs when the distance between each pair of opposite sides is some fixed constant times the heat loss from the pair of sides.

The value of λ in Example 4 corresponding to the optimal values of x, y, and z is

$$\lambda = \frac{11}{z} + \frac{15}{y} = \frac{11}{44} + \frac{15}{60} = \frac{1}{2}.$$

We can show that the Lagrange multiplier λ is the marginal heat loss with respect to volume. That is, if a building of volume slightly more than 147,840 cubic feet is optimally designed, $\frac{1}{2}$ unit of additional heat will be lost for each additional cubic foot of volume.

Practice Problems 7.4

1. Let $F(x, y, \lambda) = 2x + 3y + \lambda(90 - 6x^{1/3}y^{2/3})$. Find $\frac{\partial F}{\partial x}$.

2. Refer to Exercise 29 of Section 7.3. What is the function $F(x, y, l, \lambda)$ when the exercise is solved using the method of Lagrange multipliers?

EXERCISES 7.4

Solve the following exercises by the method of Lagrange multipliers.

1. Minimize $x^2 + 3y^2 + 10$, subject to the constraint $8 - x - y = 0$.

2. Maximize $x^2 - y^2$, subject to the constraint $2x + y - 3 = 0$.

3. Maximize $x^2 + xy - 3y^2$, subject to the constraint $2 - x - 2y = 0$.

4. Minimize $\frac{1}{2}x^2 - 3xy + y^2 + \frac{1}{2}$, subject to the constraint $3x - y - 1 = 0$.

5. Find the values of x, y that maximize

$$-2x^2 - 2xy - \tfrac{3}{2}y^2 + x + 2y,$$

subject to the constraint $x + y - \frac{5}{2} = 0$.

6. Find the values of x, y that minimize

$$x^2 + xy + y^2 - 2x - 5y,$$

subject to the constraint $1 - x + y = 0$.

7. Find the two positive numbers whose product is 25 and whose sum is as small as possible.

8. Four hundred eighty dollars are available to fence in a rectangular garden. The fencing for the north and south sides of the garden costs $10 per foot and the fencing for the east and west sides costs $15 per foot. Find the dimensions of the largest possible garden.

9. Three hundred square inches of material are available to construct an open rectangular box with a square base. Find the dimensions of the box that maximize the volume.

10. The amount of space required by a particular firm is $f(x, y) = 1000\sqrt{6x^2 + y^2}$, where x and y are, respectively, the number of units of labor and capital utilized. Suppose that labor costs \$480 per unit and capital costs \$40 per unit and that the firm has \$5000 to spend. Determine the amounts of labor and capital that should be utilized in order to minimize the amount of space required.

11. Find the dimensions of the rectangle of maximum area that can be inscribed in the unit circle. [See Fig. 2(a).]

(a)

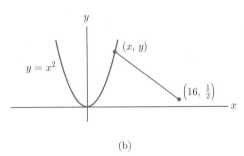

(b)

Figure 2

12. Find the point on the parabola $y = x^2$ that has minimal distance from the point $\left(16, \frac{1}{2}\right)$. [See Fig. 2(b).] [*Suggestion*: If d denotes the distance from (x, y) to $\left(16, \frac{1}{2}\right)$, then $d^2 = (x - 16)^2 + \left(y - \frac{1}{2}\right)^2$. If d^2 is minimized, then d will be minimized.]

13. Suppose that a firm makes two products A and B that use the same raw materials. Given a fixed amount of raw materials and a fixed amount of manpower, the firm must decide how much of its resources should be allocated to the production of A and how much to B. If x units of A and y units of B are produced, suppose that x and y must satisfy

$$9x^2 + 4y^2 = 18,000.$$

The graph of this equation (for $x \geq 0$, $y \geq 0$) is called a *production possibilities curve* (Fig. 3). A point (x, y) on this curve represents a *production schedule* for the firm, committing it to produce x units of A and y units of B. The reason for the relationship between x and y

involves the limitations on personnel and raw materials available to the firm. Suppose that each unit of A yields a \$3 profit, whereas each unit of B yields a \$4 profit. Then the profit of the firm is

$$P(x, y) = 3x + 4y.$$

Find the production schedule that maximizes the profit function $P(x, y)$.

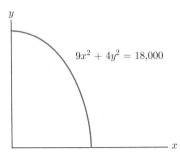

Figure 3. A production possibilities curve.

14. A firm makes x units of product A and y units of product B and has a production possibilities curve given by the equation $4x^2 + 25y^2 = 50,000$ for $x \geq 0$, $y \geq 0$. (See Exercise 13.) Suppose profits are \$2 per unit for product A and \$10 per unit for product B. Find the production schedule that maximizes the total profit.

15. The production function for a firm is $f(x, y) = 64x^{3/4}y^{1/4}$, where x and y are the number of units of labor and capital utilized. Suppose that labor costs \$96 per unit and capital costs \$162 per unit and that the firm decides to produce 3456 units of goods.

   (a) Determine the amounts of labor and capital that should be utilized in order to minimize the cost. That is, find the values of x, y that minimize $96x + 162y$, subject to the constraint $3456 - 64x^{3/4}y^{1/4} = 0$.

   (b) Find the value of λ at the optimal level of production.

   (c) Show that, at the optimal level of production, we have

$$\frac{[\text{marginal productivity of labor}]}{[\text{marginal productivity of capital}]}$$
$$= \frac{[\text{unit price of labor}]}{[\text{unit price of capital}]}.$$

16. Consider the monopolist of Example 2, Section 7.3, who sells his goods in two countries. Suppose that he must set the same price in each country. That is, $97 - (x/10) = 83 - (y/20)$. Find the values of x and y that maximize profits under this new restriction.

17. Find the values of x, y, and z that maximize xyz subject to the constraint $36 - x - 6y - 3z = 0$.

18. Find the values of x, y, and z that maximize $xy+3xz+3yz$ subject to the constraint $9 - xyz = 0$.

19. Find the values of x, y, z that maximize

$$3x + 5y + z - x^2 - y^2 - z^2,$$

subject to the constraint $6 - x - y - z = 0$.

20. Find the values of x, y, z that minimize

$$x^2 + y^2 + z^2 - 3x - 5y - z,$$

subject to the constraint $20 - 2x - y - z = 0$.

21. The material for a closed rectangular box costs $2 per square foot for the top and $1 per square foot for the sides and bottom. Using Lagrange multipliers, find the dimensions for which the volume of the box is 12 cubic feet and the cost of the materials is minimized. [Referring to Fig. 4(a), the cost will be $3xy + 2xz + 2yz$.]

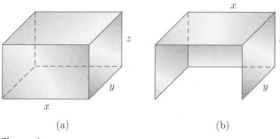

(a)   (b)

Figure 4

22. Use Lagrange multipliers to find the three positive numbers whose sum is 15 and whose product is as large as possible.

23. Find the dimensions of an open rectangular glass tank of volume 32 cubic feet for which the amount of material needed to construct the tank is minimized. [See Fig. 4(a).]

24. A shelter for use at the beach has a back, two sides, and a top made of canvas. [See Fig. 4(b).] Find the dimensions that maximize the volume and require 96 square feet of canvas.

25. Let $f(x, y)$ be any production function where x represents labor (costing $a per unit) and y represents capital (costing $b per unit). Assuming that $c is available, show that, at the values of x, y that maximize production,

$$\frac{\frac{\partial f}{\partial x}}{\frac{\partial f}{\partial y}} = \frac{a}{b}.$$

Note: Let $F(x, y, \lambda) = f(x, y) + \lambda(c - ax - by)$. The result follows from (L-1) and (L-2).

26. By applying the result in Exercise 25 to the production function $f(x, y) = kx^\alpha y^\beta$, show that, for the values of x, y that maximize production, we have

$$\frac{y}{x} = \frac{a\beta}{b\alpha}.$$

(This tells us that the ratio of capital to labor does not depend on the amount of money available nor on the level of production but only on the numbers a, b, α, and β.)

Solutions to Practice Problems 7.4

1. The function can be written as

$$F(x, y, \lambda) = 2x + 3y + \lambda \cdot 90 - \lambda \cdot 6x^{1/3}y^{2/3}.$$

When differentiating with respect to x, both y and λ should be treated as constants (so $\lambda \cdot 90$ and $\lambda \cdot 6$ are also regarded as constants).

$$\frac{\partial F}{\partial x} = 2 - \lambda \cdot 6 \cdot \frac{1}{3}x^{-2/3} \cdot y^{2/3}$$

$$= 2 - 2\lambda x^{-2/3}y^{2/3}.$$

(*Note:* It is not necessary to write out the multiplication by λ as we did. Most people just do this mentally and then differentiate.)

2. The quantity to be maximized is the volume xyl. The constraint is that length plus girth is 84. This translates to $84 = l + 2x + 2y$ or $84 - l - 2x - 2y = 0$. Therefore,

$$F(x, y, l, \lambda) = xyl + \lambda(84 - l - 2x - 2y).$$

7.5 The Method of Least Squares

Today, people can compile graphs of literally thousands of different quantities: the purchasing value of the dollar as a function of time, the pressure of a fixed volume of air as a function of temperature, the average income of people as a function of their years of formal education, or the incidence of strokes as a function of blood pressure. The observed points on such graphs tend to be irregularly distributed due to the complicated nature of the phenomena underlying them, as well as to errors made in observation. (For example, a given procedure for measuring average income may not count certain groups.)

In spite of the imperfect nature of the data, we are often faced with the problem of making assessments and predictions based on them. Roughly speaking, this problem amounts to filtering the sources of errors in the data and isolating the basic underlying trend. Frequently, on the basis of a suspicion or a working hypothesis, we may suspect that the underlying trend is linear; that is, the data should lie on a straight line. But which straight line? This is the problem that the *method of least squares* attempts to answer. To be more specific, let us consider the following problem:

Problem of Fitting a Straight Line to Data   Given observed data points (x_1, y_1), (x_2, y_2), ..., (x_N, y_N) on a graph, find the straight line that best fits these points.

To completely understand the statement of the problem being considered, we must define what it means for a line to "best" fit a set of points. If (x_i, y_i) is one of our observed points, we will measure how far it is from a given line $y = Ax + B$ by the vertical distance from the point to the line. Since the point on the line with x-coordinate x_i is $(x_i, Ax_i + B)$, this vertical distance is the distance between the y-coordinates $Ax_i + B$ and y_i. (See Fig. 1.) If $E_i = (Ax_i + B) - y_i$, either E_i or $-E_i$ is the vertical distance from (x_i, y_i) to the line. To avoid this ambiguity, we work with the square of this vertical distance:

$$E_i^2 = (Ax_i + B - y_i)^2.$$

The total error in approximating the data points $(x_1, y_1), \ldots, (x_N, y_N)$ by the line $y = Ax + B$ is usually measured by the sum E of the squares of the vertical distances from the points to the line,

$$E = E_1^2 + E_2^2 + \cdots + E_N^2.$$

E is called the *least-squares error* of the observed points with respect to the line. If all the observed points lie on the line $y = Ax + B$, all E_i are zero and the error E is zero. If a given observed point is far away from the line, the corresponding E_i^2 is large and hence makes a large contribution to the error E.

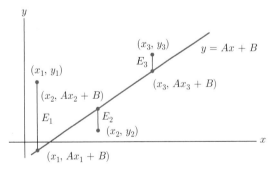

Figure 1.   Fitting a line to data points.

In general, we cannot expect to find a line $y = Ax + B$ that fits the observed points so well that the error E is zero. Actually, this situation will occur only if the observed points lie on a straight line. However, we can rephrase our original problem as follows:

Problem   Given observed data points $(x_1, y_1), (x_2, y_2), \ldots, (x_N, y_N)$, find a straight line $y = Ax + B$ for which the error E is as small as possible. This line is called the *least-squares line* or *regression line*.

It turns out that this problem is a minimization problem in the two variables A and B and can be solved by using the methods of Section 7.3. Let us consider an example.

EXAMPLE 1

Find the straight line that minimizes the least-squares error for the points $(1, 4)$, $(2, 5)$, $(3, 8)$.

Solution   Let the straight line be $y = Ax + B$. When $x = 1, 2, 3$, the y-coordinate of the corresponding point of the line is $A+B$, $2A+B$, $3A+B$, respectively. Therefore, the squares of the vertical distances from the points $(1, 4)$, $(2, 5)$, $(3, 8)$ are, respectively,

$$E_1^2 = (A + B - 4)^2,$$
$$E_2^2 = (2A + B - 5)^2,$$
$$E_3^2 = (3A + B - 8)^2.$$

(See Fig. 2.) Thus the least-squares error is

$$E = E_1^2 + E_2^2 + E_3^2 = (A + B - 4)^2 + (2A + B - 5)^2 + (3A + B - 8)^2.$$

This error obviously depends on the choice of A and B. Let $f(A, B)$ denote this least-squares error. We want to find values of A and B that minimize $f(A, B)$. To do so, we take partial derivatives with respect to A and B and set the partial derivatives equal to zero:

$$\frac{\partial f}{\partial A} = 2(A + B - 4) + 2(2A + B - 5) \cdot 2 + 2(3A + B - 8) \cdot 3$$
$$= 28A + 12B - 76 = 0,$$

$$\frac{\partial f}{\partial B} = 2(A + B - 4) + 2(2A + B - 5) + 2(3A + B - 8)$$
$$= 12A + 6B - 34 = 0.$$

To find A and B, we must solve the system of simultaneous linear equations

$$28A + 12B = 76$$
$$12A + 6B = 34.$$

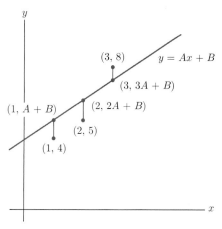

Figure 2

Multiplying the second equation by 2 and subtracting from the first equation, we have $4A = 8$, or $A = 2$. Therefore, $B = \frac{5}{3}$, and the straight line that minimizes the least-squares error is $y = 2x + \frac{5}{3}$. ◼

The minimization process used in Example 1 can be applied to a general set of data points $(x_1, y_1), \ldots, (x_N, y_N)$ to obtain the following algebraic formula for A and B:

$$A = \frac{N \cdot \Sigma\, xy - \Sigma\, x \cdot \Sigma\, y}{N \cdot \Sigma\, x^2 - (\Sigma\, x)^2},$$

$$B = \frac{\Sigma\, y - A \cdot \Sigma\, x}{N},$$

where

$$\Sigma\, x = \text{sum of the } x\text{-coordinates of the data points}$$

$$\Sigma\, y = \text{sum of the } y\text{-coordinates of the data points}$$

$$\Sigma\, xy = \text{sum of the products of the coordinates of the data points}$$

$$\Sigma\, x^2 = \text{sum of the squares of the } x\text{-coordinates of the data points}$$

$$N = \text{number of data points.}$$

That is,

$$\Sigma\, x = x_1 + x_2 + \cdots + x_N$$

$$\Sigma\, y = y_1 + y_2 + \cdots + y_N$$

$$\Sigma\, xy = x_1 \cdot y_1 + x_2 \cdot y_2 + \cdots + x_N \cdot y_N$$

$$\Sigma\, x^2 = x_1^2 + x_2^2 + \cdots + x_N^2.$$

EXAMPLE 2

The following table* gives the crude male death rate for lung cancer in 1950 and the per capita consumption of cigarettes in 1930 in various countries.

Country	Cigarette Consumption (Per Capita)	Lung Cancer Deaths (Per Million Males)
Norway	250	95
Sweden	300	120
Denmark	350	165
Australia	470	170

(a) Use the preceding formulas to obtain the straight line that best fits these data.

(b) In 1930 the per capita cigarette consumption in Finland was 1100. Use the straight line found in part (a) to estimate the male lung cancer death rate in Finland in 1950.

Solution (a) The points are plotted in Fig. 3. The sums are calculated in Table 1 and then used to determine the values of A and B.

---

*These data were obtained from *Smoking and Health*, Report of the Advisory Committee to the Surgeon General of the Public Health Service, U.S. Department of Health, Education, and Welfare, Washington, D.C., Public Health Service Publication No. 1103, p. 176.

Figure 3. Lung cancer data for least-squares analysis.

TABLE 1	Least-Squares Calculation for Smoking Data		
x	y	xy	x^2
250	95	23,750	62,500
300	120	36,000	90,000
350	165	57,750	122,500
470	170	79,900	220,900
$\Sigma x = 1370$	$\Sigma y = 550$	$\Sigma xy = 197,400$	$\Sigma x^2 = 495,900$

$$A = \frac{4 \cdot 197,400 - 1370 \cdot 550}{4 \cdot 495,900 - 1370^2} = \frac{36,100}{106,700} = \frac{361}{1067} \approx .338,$$

$$B = \frac{550 - \frac{361}{1067} \cdot 1370}{4} = \frac{1067 \cdot 550 - 361 \cdot 1370}{1067 \cdot 4} = \frac{92,280}{4268} \approx 21.621.$$

Therefore, the equation of the least-squares line is $y = .338x + 21.621$.

(b) We use the straight line to estimate the lung cancer death rate in Finland by setting $x = 1100$. Then we get

$$y = .338(1100) + 21.621 = 393.421 \approx 393.$$

Therefore, we estimate the lung cancer death rate in Finland to be 393 deaths per million males. (*Note*: The actual rate was 350 deaths per million males.) ■

INCORPORATING TECHNOLOGY

Least Squares Method   To implement the least-squares method on your TI-83/84, select $\boxed{\text{SELECT}}$ $\boxed{1}$ for the EDIT screen to obtain a table used for entering the data. If necessary, clear data from columns L_1 and/or L_2 by moving the cursor to the top of the column and pressing $\boxed{\text{CLEAR}}$ $\boxed{\text{ENTER}}$. [See Fig. 4(a).]

After the x- and y-values are placed into lists on a graphing calculator, we use the statistical routine LinReg to calculate the coefficients of the least-squares line. Now press $\boxed{\text{STAT}}$ $\boxed{\triangleright}$ for the CALC menu, and press $\boxed{4}$ to place **LinReg(ax+b)** on the home screen. Press $\boxed{\text{ENTER}}$ to obtain the slope and y-intercept of the least-squares line. [See Fig. 4(b).]

If desired, the equation for the line can automatically be assigned to a function and graphed along with the original points. First, we assign the equation for the least-squares line to a function. Select $\boxed{\text{Y=}}$, move to the function, and press $\boxed{\text{CLEAR}}$ to erase any current expression. Now press $\boxed{\text{VARS}}$ $\boxed{5}$ to select the **Statistics** variables. Move your cursor over to he EQ menu, and press $\boxed{1}$ for **RegEQ** (Regression Equation).

To graph this line, press GRAPH. To graph this line along with the original data points, we proceed as follows. From the Y=, and with only the least-squares line selected, press 2nd [STAT PLOT] ENTER to select **Plot1**, and press ENTER to turn **Plot1 ON**. Now select the first plot from the six icons for the plot **Type**. This corresponds to a scatter plot. Finally, press GRAPH. [See Fig. 4(c).] ■

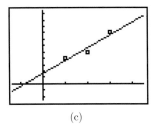

Figure 4   (a)   (b)   (c)

Practice Problems 7.5

1. Let $E = (A + B + 2)^2 + (3A + B)^2 + (6A + B - 8)^2$. What is $\dfrac{\partial E}{\partial A}$?

2. Find the formula (of the type in Problem 1) that gives the least-squares error E for the points $(1, 10)$, $(5, 8)$, and $(7, 0)$.

EXERCISES 7.5

1. Find the least-squares error E for the least-squares line fit to the four points in Fig. 5.

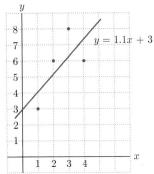

Figure 5

2. Find the least-squares error E for the least-squares line fit to the five points in Fig. 6.

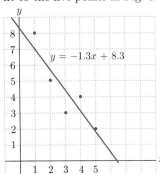

Figure 6

3. Find the formula (of the type in Practice Problem 1) that gives the least-squares error for the points $(2, 6)$, $(5, 10)$, and $(9, 15)$.

4. Find the formula (of the type in Practice Problem 1) that gives the least-squares error for the points $(8, 4)$, $(9, 2)$, and $(10, 3)$.

In Exercises 5–8, use partial derivatives to obtain the formula for the best least-squares fit to the data points.

5. $(1, 2)$, $(2, 5)$, $(3, 11)$

6. $(1, 8)$, $(2, 4)$, $(4, 3)$

7. $(1, 9)$, $(2, 8)$, $(3, 6)$, $(4, 3)$

8. $(1, 5)$, $(2, 7)$, $(3, 6)$, $(4, 10)$

9. Complete Table 2 and find the values of A and B for the straight line that provides the best least-squares fit to the data.

TABLE 2			
x	y	xy	x^2
1	7		
2	6		
3	4		
4	3		
$\Sigma x =$	$\Sigma y =$	$\Sigma xy =$	$\Sigma x^2 =$

10. Complete Table 3 and find the values of A and B for the straight line that provides the best least-squares fit to the data.

TABLE 3

x	y	xy	x^2
1	2		
2	4		
3	7		
4	9		
5	12		
$\Sigma x =$	$\Sigma y =$	$\Sigma xy =$	$\Sigma x^2 =$

In the remaining exercises, use one or more of the three methods discussed in this section (partial derivatives, formulas, or graphing utilities) to obtain the formula for the least-squares line.

11. Table 4* gives the U.S. per capita health care expenditures for the years 1990–1994.

TABLE 4 U.S. Per Capita Health Care Expenditures

Years (after 1990)	Dollars (in thousands)
0	2.688
1	2.902
2	3.144
3	3.331
4	3.510

(a) Find the least-squares line for these data.

(b) Use the least-squares line to predict the per capita health care expenditures for the year 2000.

(c) Use the least-squares line to predict when per capita health care expenditures will reach $6000.

12. Table 5 shows the 1994 price of a gallon of fuel (in U.S. dollars) and the average miles driven per automobile for several countries.[†]

(a) Find the straight line that provides the best least-squares fit to these data.

(b) In 1994, the price of gas in Japan was $4.14 per gallon. Use the straight line of part (a) to estimate the average number of miles automobiles were driven in Japan.

TABLE 5 Effect of Gas Prices on Miles Driven

Country	Price per Gallon	Average Miles per Auto
Canada	$1.57	10,371
England	$2.86	10,186
France	$3.31	8,740
Germany	$3.34	7,674
Sweden	$3.44	7,456
United States	$1.24	11,099

13. Table 6 gives the percent of persons 25 years and over who have completed four or more years of college.[‡]

TABLE 6 College Completion Rates

Year	1970	1975	1980	1985	1990	1995
Percent	10.7	13.9	16.2	19.4	21.3	23.0

(a) Use the method of least squares to obtain the straight line that best fits these data. [*Hint:* First convert *Year* to *Years after 1970.*]

(b) Estimate the percent for the year 1993.

(c) If the trend determined by the straight line in part (a) continues, when will the percent reach 27.1?

14. Table 7 gives the number of cars (in millions) in use in the United States[§] for certain years.

TABLE 7 Automobile Population

Year	Cars	Year	Cars
1980	104.6	1991	123.3
1985	114.7	1992	120.3
1989	122.8	1993	121.1
1990	123.3	1994	122

(a) Use the method of least squares to obtain the straight line that best fits these data. [*Hint:* First convert *Years* to *Years after 1980.*]

(b) Estimate the number of cars in use in 1983.

(c) If the trend determined by the straight line in part (a) continues, when will the number of cars in use reach 130 million?

---

*U.S. Health Care Financing Administration, *Health Care Financing Review*, Spring 1996.

†Source: Energy Information Administration, *International Energy Annual*. U.S. Highway Administration, *Highway Statistics*, 1994.

‡U.S. Bureau of the Census.

§American Automobile Manufacturers Association, Inc., Detroit, Michigan, *Motor Vehicle Facts and Figures*.

15. An ecologist wished to know whether certain species of aquatic insects have their ecological range limited by temperature. He collected the data in Table 8, relating the average daily temperature at different portions of a creek with the elevation (above sea level) of that portion of the creek.[*]

(a) Find the straight line that provides the best least-squares fit to these data.

(b) Use the linear function to estimate the average daily temperature for this creek at altitude 3.2 kilometers.

TABLE 8 Relationship between Elevation and Temperature in a Creek	
Elevation (kilometers)	Average Temperature (degrees Celsius)
2.7	11.2
2.8	10
3.0	8.5
3.5	7.5

Solutions to Practice Problems 7.5

1. $\dfrac{\partial E}{\partial A} = 2(A + B + 2) \cdot 1 + 2(3A + B) \cdot 3$
$\qquad + 2(6A + B - 8) \cdot 6$

$\qquad = (2A + 2B + 4) + (18A + 6B)$
$\qquad\quad + (72A + 12B - 96)$

$\qquad = 92A + 20B - 92.$

(Notice that we used the general power rule when differentiating and so had to always multiply by the derivative of the quantity inside the parentheses. Also, you might be tempted to first square the terms in the expression for E and then differentiate. We recommend that you resist this temptation.)

2. $E = (A + B - 10)^2 + (5A + B - 8)^2 + (7A + B)^2$. In general, E is a sum of squares, one for each point being fitted. The point (a, b) gives rise to the term $(aA + B - b)^2$.

7.6 Double Integrals

Up to this point, our discussion of the calculus of several variables has been confined to the study of differentiation. Let us now take up the topic of the integration of functions of several variables. As has been the case throughout most of this chapter, we restrict our discussion to functions $f(x, y)$ of two variables.

We begin with some motivation. Before we define the concept of an integral for functions of several variables, we review the essential features of the integral in one variable.

Consider the definite integral $\int_a^b f(x)\, dx$. To write down this integral takes two pieces of information. The first is the function $f(x)$. The second is the interval over which the integration is to be performed. In this case, the interval is the portion of the x-axis from $x = a$ to $x = b$. The value of the definite integral is a number. In case the function $f(x)$ is nonnegative throughout the interval from $x = a$ to $x = b$, this number equals the area under the graph of $f(x)$ from $x = a$ to $x = b$. (See Fig. 1.) If $f(x)$ is negative for some values of x in the interval, the integral still equals the area bounded by the graph, but areas below the x-axis are counted as negative.

Let us generalize the foregoing ingredients to a function $f(x, y)$ of two variables. First, we must provide a two-dimensional analog of the interval from $x = a$ to $x = b$. This is easy. We take a two-dimensional region R of the plane, such as the region shown in Fig. 2. As our generalization of $f(x)$, we take a function $f(x, y)$ of two

---

[*]The authors express their thanks to Dr. J. David Allen, formerly of the Department of Zoology at the University of Maryland, for providing the data for this exercise.

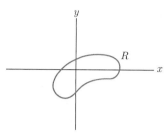

Figure 1

Figure 2.   A region in the xy-plane.

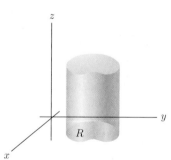

Figure 3.   Graph of $f(x, y)$ above the region R.

variables. Our generalization of the definite integral is denoted

$$\iint\limits_{R} f(x, y)\, dx\, dy$$

and is called the *double integral of $f(x, y)$ over the region R.* The value of the double integral is a number defined as follows. For the sake of simplicity, let us begin by assuming that $f(x, y) \geq 0$ for all points (x, y) in the region R. [This is the analog of the assumption that $f(x) \geq 0$ for all x in the interval from $x = a$ to $x = b$.] This means that the graph of f lies above the region R in three-dimensional space. (See Fig. 3.) The portion of the graph over R determines a solid figure. (See Fig. 4.) This figure is called the *solid bounded by $f(x, y)$ over the region R.* We define the double integral $\iint\limits_{R} f(x, y)\, dx\, dy$ to be the volume of this solid. In case the graph of $f(x, y)$ lies partially above the region R and partially below, we define the double integral to be the volume of the solid above the region minus the volume of the solid below the region. That is, we count volumes below the xy-plane as negative.

Now that we have defined the notion of a double integral, we must learn how to calculate its value. To do so, let us introduce the notion of an iterated integral. Let $f(x, y)$ be a function of two variables, let $g(x)$ and $h(x)$ be two functions of x alone, and let a and b be numbers. Then an *iterated integral* is an expression of the form

$$\int_{a}^{b} \left(\int_{g(x)}^{h(x)} f(x, y)\, dy \right) dx.$$

Figure 4.   Solid bounded by $f(x, y)$ over R.

To explain the meaning of this collection of symbols, we proceed from the inside out. We evaluate the integral

$$\int_{g(x)}^{h(x)} f(x, y)\, dy$$

by considering $f(x, y)$ as a function of y alone. This is indicated by the dy in the inner integral. We treat x as a constant in this integration. So we evaluate the integral by first finding an antiderivative $F(x, y)$ with respect to y. The integral above is then evaluated as

$$F(x, h(x)) - F(x, g(x)).$$

That is, we evaluate the antiderivative between the limits $y = g(x)$ and $y = h(x)$. This gives us a function of x alone. To complete the evaluation of the integral, we integrate this function from $x = a$ to $x = b$. The next two examples illustrate the procedure for evaluating iterated integrals.

EXAMPLE 1   Evaluate the iterated integral

$$\int_1^2 \left(\int_3^4 (y - x)\, dy \right) dx.$$

Solution Here $g(x)$ and $h(x)$ are constant functions: $g(x) = 3$ and $h(x) = 4$. We evaluate the inner integral first. The variable in this integral is y, so we treat x as a constant.

$$\int_3^4 (y - x)\, dy = \left(\frac{1}{2} y^2 - xy \right) \Big|_3^4$$

$$= \left(\frac{1}{2} \cdot 16 - x \cdot 4 \right) - \left(\frac{1}{2} \cdot 9 - x \cdot 3 \right)$$

$$= 8 - 4x - \frac{9}{2} + 3x$$

$$= \frac{7}{2} - x.$$

Now we carry out the integration with respect to x:

$$\int_1^2 \left(\frac{7}{2} - x \right) dx = \frac{7}{2} x - \frac{1}{2} x^2 \Big|_1^2$$

$$= \left(\frac{7}{2} \cdot 2 - \frac{1}{2} \cdot 4 \right) - \left(\frac{7}{2} - \frac{1}{2} \cdot 1 \right)$$

$$= (7 - 2) - (3) = 2.$$

So the value of the iterated integral is 2. ∎

EXAMPLE 2   Evaluate the iterated integral

$$\int_0^1 \left(\int_{\sqrt{x}}^{x+1} 2xy\, dy \right) dx.$$

Solution We evaluate the inner integral first.

$$\int_{\sqrt{x}}^{x+1} 2xy\, dy = xy^2 \Big|_{\sqrt{x}}^{x+1} = x(x+1)^2 - x(\sqrt{x})^2$$

$$= x(x^2 + 2x + 1) - x \cdot x$$

$$= x^3 + 2x^2 + x - x^2$$

$$= x^3 + x^2 + x.$$

Now we evaluate the outer integral.

$$\int_0^1 (x^3 + x^2 + x)\, dx = \frac{1}{4} x^4 + \frac{1}{3} x^3 + \frac{1}{2} x^2 \Big|_0^1 = \frac{1}{4} + \frac{1}{3} + \frac{1}{2} = \frac{13}{12}$$

So the value of the iterated integral is $\frac{13}{12}$. ∎

Let us now return to the discussion of the double integral $\iint\limits_R f(x, y)\, dx\, dy$.

When the region R has a special form, the double integral may be expressed as an iterated integral, as follows: Suppose that R is bounded by the graphs of $y = g(x)$, $y = h(x)$ and by the vertical lines $x = a$ and $x = b$. (See Fig. 5.) In this case, we have the following fundamental result, which we cite without proof.

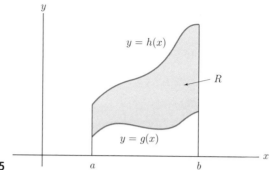

Figure 5

Let R be the region in the xy-plane bounded by the graphs of $y = g(x)$, $y = h(x)$, and the vertical lines $x = a$, $x = b$. Then

$$\iint\limits_{R} f(x,y)\,dx\,dy = \int_{a}^{b} \left(\int_{g(x)}^{h(x)} f(x,y)\,dy \right) dx.$$

Since the value of the double integral gives the volume of the solid bounded by the graph of $f(x,y)$ over the region R, the preceding result may be used to calculate volumes, as the next two examples show.

EXAMPLE 3   Calculate the volume of the solid bounded above by the function $f(x,y) = y - x$ and lying over the rectangular region R: $1 \le x \le 2$, $3 \le y \le 4$. (See Fig. 6.)

Solution   The desired volume is given by the double integral $\iint\limits_{R}(y-x)\,dx\,dy$. By the result just cited, this double integral is equal to the iterated integral

$$\int_{1}^{2} \left(\int_{3}^{4} (y-x)\,dy \right) dx.$$

The value of this iterated integral was shown in Example 1 to be 2, so the volume of the solid shown in Fig. 6 is 2. ∎

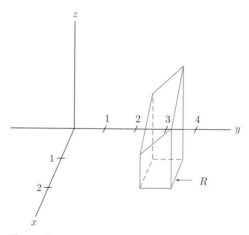

Figure 6

EXAMPLE 4

Calculate $\iint_R 2xy \, dx \, dy$, where R is the region shown in Fig. 7.

$y = x + 1$
$(1, 2)$

R

$(1, 1)$

$y = \sqrt{x}$

Figure 7

Solution The region R is bounded below by $y = \sqrt{x}$, above by $y = x + 1$, on the left by $x = 0$, and on the right by $x = 1$. Therefore,

$$\iint_R 2xy \, dx \, dy = \int_0^1 \left(\int_{\sqrt{x}}^{x+1} 2xy \, dy \right) dx = \frac{13}{12} \quad \text{(by Example 2).} \qquad \blacksquare$$

In our discussion, we have confined ourselves to iterated integrals in which the inner integral was with respect to y. In a completely analogous manner, we may treat iterated integrals in which the inner integral is with respect to x. Such iterated integrals may be used to evaluate double integrals over regions R bounded by curves of the form $x = g(y)$, $x = h(y)$, and horizontal lines $y = a$, $y = b$. The computations are analogous to those given in this section.

Practice Problems 7.6

1. Calculate the iterated integral

$$\int_0^2 \left(\int_0^{x/2} e^{2y-x} \, dy \right) dx.$$

2. Calculate

$$\iint_R e^{2y-x} \, dx \, dy$$

where R is the region in Fig. 8.

$(2, 1)$

$(0, 0)$ R

Figure 8

EXERCISES 7.6

Calculate the following iterated integrals.

1. $\int_0^1 \left(\int_0^1 e^{x+y} \, dy \right) dx$

2. $\int_{-1}^1 \left(\int_{-1}^1 xy \, dx \right) dy$

3. $\int_{-2}^0 \left(\int_{-1}^1 xe^{xy} \, dy \right) dx$

4. $\int_0^1 \left(\int_{-1}^1 \frac{1}{3} y^3 x \, dy \right) dx$

5. $\int_1^4 \left(\int_x^{x^2} xy \, dy \right) dx$

6. $\int_0^3 \left(\int_x^{2x} y \, dy \right) dx$

7. $\int_{-1}^1 \left(\int_x^{2x} (x + y) \, dy \right) dx$

8. $\int_0^1 \left(\int_0^x e^{x+y} \, dy \right) dx$

Let R be the rectangle consisting of all points (x, y) such that $0 \le x \le 2$, $2 \le y \le 3$. Calculate the following double integrals. Interpret each as a volume.

9. $\iint_R xy^2 \, dx \, dy$

10. $\iint_R (xy + y^2) \, dx \, dy$

11. $\iint_R e^{-x-y} \, dx \, dy$

12. $\iint_R e^{y-x} \, dx \, dy$

Calculate the volumes over the following regions R bounded above by the graph of $f(x, y) = x^2 + y^2$.

13. R is the rectangle bounded by the lines $x = 1$, $x = 3$, $y = 0$, and $y = 1$.

14. R is the region bounded by the lines $x = 0$, $x = 1$ and the curves $y = 0$ and $y = \sqrt[3]{x}$.

Solutions to Practice Problems 7.6

1. $\displaystyle\int_0^2 \left(\int_0^{x/2} e^{2y-x}\,dy\right) dx = \int_0^2 \left(\frac{1}{2}e^{2y-x}\Big|_0^{x/2}\right) dx$

$\displaystyle = \int_0^2 \left(\frac{1}{2}e^{2(x/2)-x} - \frac{1}{2}e^{2(0)-x}\right) dx$

$\displaystyle = \int_0^2 \left(\frac{1}{2} - \frac{1}{2}e^{-x}\right) dx$

$\displaystyle = \frac{1}{2}x + \frac{1}{2}e^{-x}\Big|_0^2$

$\displaystyle = \frac{1}{2}\cdot 2 + \frac{1}{2}e^{-2} - \left(\frac{1}{2}\cdot 0 + \frac{1}{2}e^{-0}\right)$

$\displaystyle = 1 + \frac{1}{2}e^{-2} - 0 - \frac{1}{2}$

$\displaystyle = \frac{1}{2} + \frac{1}{2}e^{-2}.$

2. The line passing through the points $(0,0)$ and $(2,1)$ has equation $y = x/2$. Hence the region R is bounded below by $y = 0$, above by $y = x/2$, on the left by $x = 0$, and on the right by $x = 2$. Therefore,

$$\iint_R e^{2y-x}\,dx\,dy = \int_0^2 \left(\int_0^{x/2} e^{2y-x}\,dy\right) dx$$

$$= \frac{1}{2} + \frac{1}{2}e^{-2}$$

by Problem 1.

REVIEW OF FUNDAMENTAL CONCEPTS

1. Give an example of a level curve of a function of two variables.

2. Explain how to find a first partial derivative of a function of two variables.

3. Explain how to find a second partial derivative of a function of two variables.

4. What expression involving a partial derivative gives an approximation to $f(a+h,b) - f(a,b)$?

5. Interpret $\dfrac{\partial f}{\partial y}(2,3)$ as a rate of change.

6. Give an example of a Cobb–Douglas production function. What is the marginal productivity of labor? Of capital?

7. Explain how to find possible relative extreme points for a function of several variables.

8. State the second-derivative test for functions of two variables.

9. Outline how the method of Lagrange multipliers is used to solve an optimization problem.

10. What is the least-squares line approximation to a set of data points? How is the line determined?

11. Give a geometric interpretation for $\iint_R f(x,y)\,dx\,dy$, where $f(x,y) \geq 0$.

12. Give a formula for evaluating a double integral in terms of an iterated integral.

SUPPLEMENTARY EXERCISES

1. Let $f(x,y) = x\sqrt{y}/(1+x)$. Compute $f(2,9)$, $f(5,1)$, and $f(0,0)$.

2. Let $f(x,y,z) = x^2 e^{y/z}$. Compute $f(-1,0,1)$, $f(1,3,3)$, and $f(5,-2,2)$.

3. If A dollars are deposited in a bank at a 6% continuous interest rate, the amount in the account after t years is $f(A,t) = Ae^{.06t}$. Find and interpret $f(10,11.5)$.

4. Let $f(x,y,\lambda) = xy + \lambda(5 - x - y)$. Find $f(1,2,3)$.

5. Let $f(x,y) = 3x^2 + xy + 5y^2$. Find $\dfrac{\partial f}{\partial x}$ and $\dfrac{\partial f}{\partial y}$.

6. Let $f(x,y) = 3x - \frac{1}{2}y^4 + 1$. Find $\dfrac{\partial f}{\partial x}$ and $\dfrac{\partial f}{\partial y}$.

7. Let $f(x,y) = e^{x/y}$. Find $\dfrac{\partial f}{\partial x}$ and $\dfrac{\partial f}{\partial y}$.

8. Let $f(x,y) = x/(x - 2y)$. Find $\dfrac{\partial f}{\partial x}$ and $\dfrac{\partial f}{\partial y}$.

9. Let $f(x,y,z) = x^3 - yz^2$. Find $\dfrac{\partial f}{\partial x}$, $\dfrac{\partial f}{\partial y}$, and $\dfrac{\partial f}{\partial z}$.

10. Let $f(x,y,\lambda) = xy + \lambda(5 - x - y)$. Find $\dfrac{\partial f}{\partial x}$, $\dfrac{\partial f}{\partial y}$, and $\dfrac{\partial f}{\partial \lambda}$.

11. Let $f(x,y) = x^3 y + 8$. Compute $\dfrac{\partial f}{\partial x}(1,2)$ and $\dfrac{\partial f}{\partial y}(1,2)$.

12. Let $f(x, y, z) = (x + y)z$. Evaluate $\dfrac{\partial f}{\partial y}$ at $(x, y, z) = (2, 3, 4)$.

13. Let $f(x, y) = x^5 - 2x^3y + \frac{1}{2}y^4$. Find $\dfrac{\partial^2 f}{\partial x^2}$, $\dfrac{\partial^2 f}{\partial y^2}$, $\dfrac{\partial^2 f}{\partial x\, \partial y}$, and $\dfrac{\partial^2 f}{\partial y\, \partial x}$.

14. Let $f(x, y) = 2x^3 + x^2y - y^2$. Compute $\dfrac{\partial^2 f}{\partial x^2}$, $\dfrac{\partial^2 f}{\partial y^2}$, and $\dfrac{\partial^2 f}{\partial x\, \partial y}$ at $(x, y) = (1, 2)$.

15. A dealer in a certain brand of electronic calculator finds that (within certain limits) the number of calculators she can sell per week is given by $f(p, t) = -p + 6t - .02pt$, where p is the price of the calculator and t is the number of dollars spent on advertising. Compute $\dfrac{\partial f}{\partial p}(25, 10{,}000)$ and $\dfrac{\partial f}{\partial t}(25, 10{,}000)$ and interpret these numbers.

16. The crime rate in a certain city can be approximated by a function $f(x, y, z)$, where x is the unemployment rate, y is the amount of social services available, and z is the size of the police force. Explain why $\dfrac{\partial f}{\partial x} > 0$, $\dfrac{\partial f}{\partial y} < 0$, and $\dfrac{\partial f}{\partial z} < 0$.

In Exercises 17–20, find all points (x, y) where $f(x, y)$ has a possible relative maximum or minimum.

17. $f(x, y) = -x^2 + 2y^2 + 6x - 8y + 5$
18. $f(x, y) = x^2 + 3xy - y^2 - x - 8y + 4$
19. $f(x, y) = x^3 + 3x^2 + 3y^2 - 6y + 7$
20. $f(x, y) = \frac{1}{2}x^2 + 4xy + y^3 + 8y^2 + 3x + 2$

In Exercises 21–23, find all points (x, y) where $f(x, y)$ has a possible relative maximum or minimum. Then use the second-derivative test to determine, if possible, the nature of $f(x, y)$ at each of these points. If the second-derivative test is inconclusive, so state.

21. $f(x, y) = x^2 + 3xy + 4y^2 - 13x - 30y + 12$
22. $f(x, y) = 7x^2 - 5xy + y^2 + x - y + 6$
23. $f(x, y) = x^3 + y^2 - 3x - 8y + 12$

24. Find the values of x, y, z at which

$$f(x, y, z) = x^2 + 4y^2 + 5z^2 - 6x + 8y + 3$$

assumes its minimum value.

Use the method of Lagrange multipliers to:

25. Maximize $3x^2 + 2xy - y^2$, subject to the constraint $5 - 2x - y = 0$.

26. Find the values of x, y that minimize $-x^2 - 3xy - \frac{1}{2}y^2 + y + 10$, subject to the constraint $10 - x - y = 0$.

27. Find the values of x, y, z that minimize $3x^2 + 2y^2 + z^2 + 4x + y + 3z$, subject to the constraint $4 - x - y - z = 0$.

28. Find the dimensions of a rectangular box of volume 1000 cubic inches for which the sum of the dimensions is minimized.

29. A person wants to plant a rectangular garden along one side of a house and put a fence on the other three sides. (See Fig. 1.) Using the method of Lagrange multipliers, find the dimensions of the garden of greatest area that can be enclosed by using 40 feet of fencing.

Figure 9. A garden.

30. The solution to Exercise 29 is $x = 10$, $y = 20$, $\lambda = 10$. If 1 additional foot of fencing becomes available, compute the new optimal dimensions and the new area. Show that the increase in area (compared with the area in Exercise 29) is approximately equal to 10 (the value of λ).

In Exercises 31–33, find the straight line that best fits the following data points, where "best" is meant in the sense of least squares.

31. $(1, 1)$, $(2, 3)$, $(3, 6)$

32. $(1, 1)$, $(3, 4)$, $(5, 7)$

33. $(0, 1)$, $(1, -1)$, $(2, -3)$, $(3, -5)$

In Exercises 34 and 35, calculate the iterated integral.

34. $\displaystyle \int_0^1 \left(\int_0^4 (x\sqrt{y} + y)\, dy \right) dx$

35. $\displaystyle \int_0^5 \left(\int_1^4 (2xy^4 + 3)\, dy \right) dx$

In Exercises 36 and 37, let R be the rectangle consisting of all points (x, y) such that $0 \le x \le 4$, $1 \le y \le 3$, and calculate the double integral.

36. $\displaystyle \iint_R (2x + 3y)\, dx\, dy$ **37.** $\displaystyle \iint_R 5\, dx\, dy$

38. The present value of y dollars after x years at 15% continuous interest is $f(x, y) = ye^{-.15x}$. Sketch some sample level curves. (Economists call this collection of level curves a *discount system*.)

THE TRIGONOMETRIC FUNCTIONS

In this chapter we expand the collection of functions to which we can apply calculus by introducing the trigonometric functions. As we shall see, these functions are *periodic*. That is, after a certain point their graphs repeat themselves. This repetitive phenomenon is not displayed by any of the functions that we have considered until now. Yet many natural phenomena are repetitive or cyclical, for example, the motion of the planets in our solar system, earthquake vibrations, and the natural rhythm of the heart. Thus the functions introduced in this chapter add considerably to our capacity to describe physical processes.

8.1 Radian Measure of Angles

The ancient Babylonians introduced angle measurement in terms of degrees, minutes, and seconds, and these units are still generally used today for navigation and practical measurements. In calculus, however, it is more convenient to measure angles in terms of *radians*, for in this case the differentiation formulas for the trigonometric functions are easier to remember and use. Also, the radian is becoming more widely used today in scientific work because it is the unit of angle measurement in the international metric system (Système International d'Unités).

To define a radian, we consider a circle of radius 1 and measure angles in terms of distances around the circumference. The central angle determined by an arc of length 1 along the circumference is said to have a measure of 1 *radian*. (See Fig. 1.) Since the circumference of the circle of radius 1 has length 2π, there are 2π radians in one full revolution of the circle. Equivalently,

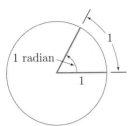

Figure 1

$$360° = 2\pi \text{ radians.} \tag{1}$$

The following important relations should be memorized (see Fig. 2):

$$90° = \frac{\pi}{2} \text{ radians} \qquad \text{(one quarter-revolution)}$$

$$180° = \pi \text{ radians} \qquad \text{(one half-revolution)}$$

$$270° = \frac{3\pi}{2} \text{ radians} \qquad \text{(three quarter-revolutions)}$$

$$360° = 2\pi \text{ radians} \qquad \text{(one full revolution).}$$

Figure 2

From formula (1) we see that

$$1° = \frac{2\pi}{360} \text{ radians} = \frac{\pi}{180} \text{ radians.}$$

If d is any number, then

$$d° = d \times \frac{\pi}{180} \text{ radians.} \tag{2}$$

That is, to convert degrees to radians, multiply the number of degrees by $\pi/180$.

EXAMPLE 1   Convert $45°$, $60°$, and $135°$ to radians.

Solution

$$45° = \overset{1}{\cancel{45}} \times \frac{\pi}{\underset{4}{\cancel{180}}} \text{ radians} = \frac{\pi}{4} \text{ radians,}$$

$$60° = \overset{1}{\cancel{60}} \times \frac{\pi}{\underset{3}{\cancel{180}}} \text{ radians} = \frac{\pi}{3} \text{ radians,}$$

$$135° = \overset{3}{\cancel{135}} \times \frac{\pi}{\underset{4}{\cancel{180}}} \text{ radians} = \frac{3\pi}{4} \text{ radians.}$$

These three angles are shown in Fig. 3.

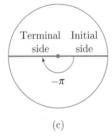

Figure 3

We usually omit the word "radian" when measuring angles because all our angle measurements will be in radians unless degrees are specifically indicated.

For our purposes, it is important to be able to speak of both negative and positive angles, so let us define what we mean by a negative angle. We shall usually consider angles that are in *standard position* on a coordinate system, with the vertex of the angle at $(0,0)$ and one side, called the "initial side," along the positive x-axis. We measure such an angle from the initial side to the "terminal side," where a *counterclockwise angle is positive* and a *clockwise angle is negative*. Some examples are given in Fig. 4.

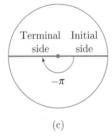

Figure 4　　　(a)　　　　　　　(b)　　　　　　　(c)

Notice in Figs. 4(a) and (b) how essentially the same picture can describe more than one angle.

By considering angles formed from more than one revolution (in the positive or negative direction), we can construct angles whose measure is of arbitrary size (that is, not necessarily between -2π and 2π). Three examples are illustrated in Fig. 5.

Figure 5

EXAMPLE 2　　　(a) What is the radian measure of the angle in Fig. 6?

(b) Construct an angle of $5\pi/2$ radians.

Solution　(a) The angle described in Fig. 6 consists of one full revolution (2π radians) plus three quarter-revolutions [$3 \times (\pi/2)$ radians]. That is,

$$t = 2\pi + 3 \times \frac{\pi}{2} = 4 \times \frac{\pi}{2} + 3 \times \frac{\pi}{2} = \frac{7\pi}{2}.$$

(b) Think of $5\pi/2$ radians as $5 \times (\pi/2)$ radians, that is, five quarter-revolutions of the circle. This is one full revolution plus one quarter-revolution. An angle of $5\pi/2$ radians is shown in Fig. 7. ∎

Figure 6

Figure 7

Practice Problems 8.1

1. A right triangle has one angle of $\pi/3$ radians. What are the other angles?

2. How many radians are there in an angle of $-780°$? Draw the angle.

EXERCISES 8.1

Convert the following to radian measure.

1. $30°, 120°, 315°$

2. $18°, 72°, 150°$

3. $450°, -210°, -90°$

4. $990°, -270°, -540°$

Give the radian measure of each angle described.

5.

6.

7.

8.

9.

10.

11.

12.

Construct angles with the following radian measure.

13. $3\pi/2, 3\pi/4, 5\pi$

14. $\pi/3, 5\pi/2, 6\pi$

15. $-\pi/3, -3\pi/4, -7\pi/2$

16. $-\pi/4, -3\pi/2, -3\pi$

17. $\pi/6, -2\pi/3, -\pi$

18. $2\pi/3, -\pi/6, 7\pi/2$

Solutions to Practice Problems 8.1

1. The sum of the angles of a triangle is $180°$ or π radians. Since a right angle is $\pi/2$ radians and one angle is $\pi/3$ radians, the remaining angle is $\pi - (\pi/2 + \pi/3) = \pi/6$ radians.

2. $-780° = -780 \times (\pi/180)$ radians
$$= -13\pi/3 \text{ radians}.$$

Since $-13\pi/3 = -4\pi - \pi/3$, we draw the angle by first making two revolutions in the negative direction and then a rotation of $\pi/3$ in the negative direction. (See Fig. 8.)

Figure 8

8.2 The Sine and the Cosine

Given a number t, we consider an angle of t radians placed in standard position, as in Fig. 1, and we let P be a point on the terminal side of this angle. Denote the coordinates of P by (x, y) and let r be the length of the segment OP; that is, $r = \sqrt{x^2 + y^2}$. The *sine* and *cosine* of t, denoted by $\sin t$ and $\cos t$, respectively, are defined by the ratios

$$\sin t = \frac{y}{r}$$

$$\cos t = \frac{x}{r}. \tag{1}$$

It does not matter which point on the ray through P we use to define $\sin t$ and $\cos t$. If $P' = (x', y')$ is another point on the same ray and if r' is the length of OP' (Fig. 2), then, by properties of similar triangles, we have

$$\frac{y'}{r'} = \frac{y}{r} = \sin t,$$

$$\frac{x'}{r'} = \frac{x}{r} = \cos t.$$

Three examples that illustrate the definition of $\sin t$ and $\cos t$ are shown in Fig. 3.

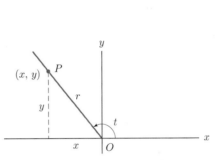

Figure 1. Diagram for the definitions of sine and cosine.

Figure 2

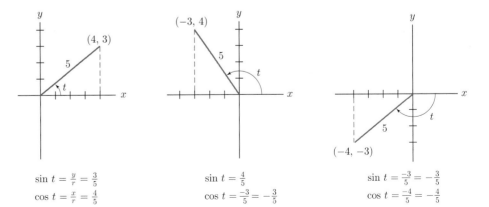

$$\sin t = \frac{y}{r} = \frac{3}{5}$$
$$\cos t = \frac{x}{r} = \frac{4}{5}$$

$$\sin t = \frac{4}{5}$$
$$\cos t = \frac{-3}{5} = -\frac{3}{5}$$

$$\sin t = \frac{-3}{5} = -\frac{3}{5}$$
$$\cos t = \frac{-4}{5} = -\frac{4}{5}$$

Figure 3. Calculation of $\sin t$ and $\cos t$.

When $0 < t < \pi/2$, the values of $\sin t$ and $\cos t$ may be expressed as ratios of the lengths of the sides of a right triangle. Indeed, if we are given the right triangle in Fig. 4, we have

$$\sin t = \frac{\text{opposite}}{\text{hypotenuse}}, \qquad \cos t = \frac{\text{adjacent}}{\text{hypotenuse}}. \qquad (2)$$

A typical application of (2) appears in Example 1.

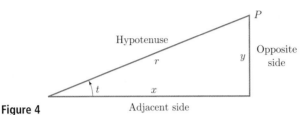

Figure 4

EXAMPLE 1   The hypotenuse of a right triangle is 4 units and one angle is .7 radian. Determine the length of the side opposite this angle.

Solution   See Fig. 5. Since $y/4 = \sin .7$, we have

$$y = 4 \sin .7$$
$$\approx 4(.64422) = 2.57688. \qquad \blacksquare$$

Figure 5

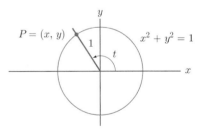

Figure 6

Another way to describe the sine and cosine functions is to choose the point P in Fig. 1 so that $r = 1$. That is, choose P on the unit circle. (See Fig. 6.) In this case

$$\sin t = \frac{y}{1} = y,$$

$$\cos t = \frac{x}{1} = x.$$

So the y-coordinate of P is $\sin t$, and the x-coordinate of P is $\cos t$. Thus we have the following result:

Figure 7

> **Alternative Definition of Sine and Cosine Functions**   We can think of $\cos t$ and $\sin t$ as the x- and y-coordinates of the point P on the unit circle that is determined by an angle of t radians. (See Fig. 7.)

EXAMPLE 2   Find a value of t such that $0 < t < \pi/2$ and $\cos t = \cos(-\pi/3)$.

Solution   On the unit circle we locate the point P that is determined by an angle of $-\pi/3$ radians. The x-coordinate of P is $\cos(-\pi/3)$. There is another point Q on the unit circle with the same x-coordinate. (See Fig. 8.) Let t be the radian measure of the angle determined by Q. Then

$$\cos t = \cos\left(-\frac{\pi}{3}\right)$$

because Q and P have the same x-coordinate. Also, $0 < t < \pi/2$. From the symmetry of the diagram, it is clear that $t = \pi/3$.   ■

Figure 8

Properties of the Sine and Cosine Functions   Each number t determines a point $(\cos t, \sin t)$ on the unit circle $x^2 + y^2 = 1$ as in Fig. 7. Therefore, $(\cos t)^2 + (\sin t)^2 = 1$. It is convenient (and traditional) to write $\sin^2 t$ instead of $(\sin t)^2$ and $\cos^2 t$ instead of $(\cos t)^2$. Thus we can write the last formula as follows:

$$\cos^2 t + \sin^2 t = 1. \tag{3}$$

The numbers t and $t \pm 2\pi$ determine the same point on the unit circle (because 2π represents a full revolution of the circle). But $t + 2\pi$ and $t - 2\pi$ correspond to the points $(\cos(t + 2\pi), \sin(t + 2\pi))$ and $(\cos(t - 2\pi), \sin(t - 2\pi))$, respectively. Hence

$$\cos(t \pm 2\pi) = \cos t, \qquad \sin(t \pm 2\pi) = \sin t. \tag{4}$$

Figure 9(a) illustrates another property of the sine and cosine:

$$\cos(-t) = \cos t, \qquad \sin(-t) = -\sin t. \tag{5}$$

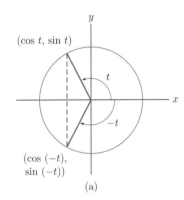

Figure 9(b) shows that the points P and Q corresponding to t and to $\pi/2 - t$ are reflections of each other through the line $y = x$. Consequently, the coordinates of Q are obtained by interchanging the coordinates of P. This means that

$$\cos\left(\frac{\pi}{2} - t\right) = \sin t, \qquad \sin\left(\frac{\pi}{2} - t\right) = \cos t. \tag{6}$$

The equations in (3) to (6) are called *identities* because they hold for all values of t. Another identity that holds for all numbers s and t is

$$\sin(s + t) = \sin s \cos t + \cos s \sin t. \tag{7}$$

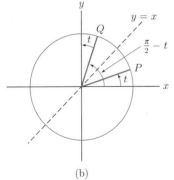

Figure 9. Diagrams for two identities.

A proof of (7) may be found in any introductory book on trigonometry. There are a number of other identities concerning trigonometric functions, but we shall not need them here.

The Graph of sin t   Let us analyze what happens to $\sin t$ as t increases from 0 to π. When $t = 0$, the point $P = (\cos t, \sin t)$ is at $(1, 0)$, as in Fig. 10(a). As t increases, P moves counterclockwise around the unit circle. [See Fig. 10(b).] The y-coordinate of P, that is, $\sin t$, increases until $t = \pi/2$, where $P = (0, 1)$. [See Fig. 10(c).] As t

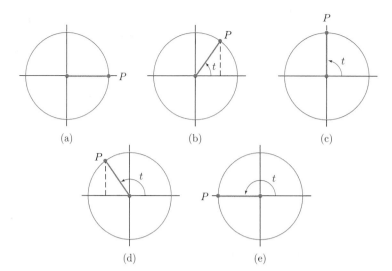

Figure 10. Movement along the unit circle.

increases from $\pi/2$ to π, the y-coordinate of P, that is, $\sin t$, decreases from 1 to 0. [See Figs. 10(d) and (e).]

Part of the graph of $\sin t$ is sketched in Fig. 11. For t between 0 and π, notice that the values of $\sin t$ increase from 0 to 1 and then decrease back to 0, just as we predicted from Fig. 10. For t between π and 2π, the values of $\sin t$ are negative. The graph of $y = \sin t$ for t between 2π and 4π is exactly the same as the graph for t between 0 and 2π. This result follows from formula (4). We say that the sine function is *periodic with period* 2π because the graph repeats itself every 2π units. We can use this fact to make a quick sketch of part of the graph for negative values of t. (See Fig. 12.)

Figure 11

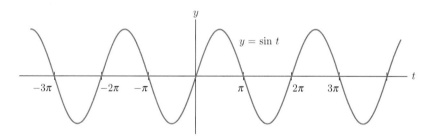

Figure 12. Graph of the sine function.

The Graph of cos *t*   By analyzing what happens to the first coordinate of the point $(\cos t, \sin t)$ as t varies, we obtain the graph of $\cos t$. Note from Fig. 13 that the graph of the cosine function is also periodic with period 2π.

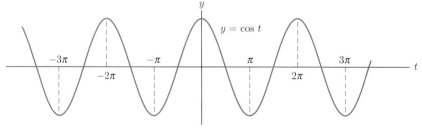

Figure 13. Graph of the cosine function.

The sine and cosine functions assign to each number t the values $\sin t$ and $\cos t$, respectively. There is nothing special, however, about the letter t. Although we chose to use the letters t, x, y, and r in the *definition* of the sine and cosine, other letters could have been used as well. Now that the sine and cosine of every number are defined, we are free to use *any* letter to represent the independent variable. ■

INCORPORATING TECHNOLOGY   **Graphing Trigonometric Functions**   The ZTrig window setting is optimized for displaying graphs of trigonometric functions. This can be accessed via $\boxed{\text{zoom}}$ $\boxed{7}$, and it sets the window dimensions to $[-2\pi, 2\pi]$ by $[-4, 4]$ with an x-scale of $\pi/2$. Figure 14 is the graph of $y = \sin x$ in this setting. ■

Figure 14

Practice Problems 8.2

1. Find $\cos t$, where t is the radian measure of the angle shown in Fig. 15.

Figure 15

2. Assume that $\cos(1.17) = .390$. Use properties of the cosine and sine to determine $\sin(1.17)$, $\cos(1.17+4\pi)$, $\cos(-1.17)$, and $\sin(-1.17)$.

EXERCISES 8.2

In Exercises 1–12, give the values of $\sin t$ *and* $\cos t$*, where t is the radian measure of the angle shown.*

1.

2.

3.

4.

5.

6.

7.

8.

9.

10.

11.

12.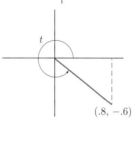

Exercises 13–20 refer to various right triangles whose sides and angles are labeled as in Fig. 16. Round off all lengths of sides to one decimal place.

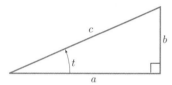

Figure 16

13. Estimate t if $a = 12$, $b = 5$, and $c = 13$.

14. If $t = 1.1$ and $c = 10.0$, find b.

15. If $t = 1.1$ and $b = 3.2$, find c.

16. If $t = .4$ and $c = 5.0$, find a.

17. If $t = .4$ and $a = 10.0$, find c.

18. If $t = .9$ and $c = 20.0$, find a and b.

19. If $t = .5$ and $a = 2.4$, find b and c.

20. If $t = 1.1$ and $b = 3.5$, find a and c.

Find t such that $0 \leq t \leq \pi$ and t satisfy the stated condition.

21. $\cos t = \cos(-\pi/6)$ **22.** $\cos t = \cos(3\pi/2)$

23. $\cos t = \cos(5\pi/4)$ **24.** $\cos t = \cos(-4\pi/6)$

25. $\cos t = \cos(-5\pi/8)$ **26.** $\cos t = \cos(-3\pi/4)$

Find t such that $-\pi/2 \leq t \leq \pi/2$ and t satisfy the stated condition.

27. $\sin t = \sin(3\pi/4)$ **28.** $\sin t = \sin(7\pi/6)$

29. $\sin t = \sin(-4\pi/3)$ **30.** $\sin t = -\sin(3\pi/8)$

31. $\sin t = -\sin(\pi/6)$ **32.** $\sin t = -\sin(-\pi/3)$

33. $\sin t = \cos t$ **34.** $\sin t = -\cos t$

35. Refer to Fig. 10. Describe what happens to $\cos t$ as t increases from 0 to π.

36. Use the unit circle to describe what happens to $\sin t$ as t increases from π to 2π.

37. Determine the value of $\sin t$ when $t = 5\pi$, -2π, $17\pi/2$, $-13\pi/2$.

38. Determine the value of $\cos t$ when $t = 5\pi$, -2π, $17\pi/2$, $-13\pi/2$.

39. Assume that $\cos(.19) = .98$. Use properties of the cosine and sine to determine $\sin(.19)$, $\cos(.19 - 4\pi)$, $\cos(-.19)$, and $\sin(-.19)$.

40. Assume that $\sin(.42) = .41$. Use properties of the cosine and sine to determine $\sin(-.42)$, $\sin(6\pi - .42)$, and $\cos(.42)$.

Technology Exercises

41. In any given locality, tap water temperature varies during the year. In Dallas, Texas, the tap water temperature (in degrees Fahrenheit) t days after the beginning of a year is given approximately by the formula*

$$T = 59 + 14 \cos \left[\frac{2\pi}{365}(t - 208) \right], \qquad 0 \leq t \leq 365.$$

(a) Graph the function in the window $[0, 365]$ by $[-10, 73]$.

(b) What is the temperature on February 14, that is, when $t = 45$?

(c) Use the fact that the value of the cosine function ranges from -1 to 1 to find the coldest and warmest tap water temperatures during the year.

(d) Use the **TRACE** feature or the **MINIMUM** command to estimate the day during which the tap water temperature is coldest. Find the exact day algebraically by using the fact that $\cos(-\pi) = -1$.

(e) Use the **TRACE** feature or the **MAXIMUM** command to estimate the day during which the tap water temperature is warmest. Find the exact day algebraically by using the fact that $\cos(0) = 1$.

(f) The average tap water temperature during the year is $59°$. Find the two days during which the average temperature is achieved. [*Note:* Answer this question both graphically and algebraically.]

42. In any given locality, the length of daylight varies during the year. In Des Moines, Iowa, the number of minutes of daylight in a day t days after the beginning of a year is given approximately by the formula†

$$D = 720 + 200 \sin \left[\frac{2\pi}{365}(t - 79.5) \right], \qquad 0 \leq t \leq 365.$$

*See D. Rapp, *Solar Energy* (Upper Saddle River, NJ: Prentice-Hall, Inc., 1981), p. 171.

†See D. R. Duncan et al., "Climate Curves," *School Science and Mathematics*, Vol. 76 (January 1976), pp. 41–49.

(a) Graph the function in the window $[0, 365]$ by $[-100, 940]$.

(b) How many minutes of daylight are there on February 14, that is, when $t = 45$?

(c) Use the fact that the value of the sine function ranges from -1 to 1 to find the shortest and longest amounts of daylight during the year.

(d) Use the **TRACE** feature or the **MINIMUM** command to estimate the day with the shortest amount of daylight. Find the exact day alge-

braically by using the fact that $\sin(3\pi/2) = -1$.

(e) Use the **TRACE** feature or the **MAXIMUM** command to estimate the day with the longest amount of daylight. Find the exact day algebraically by using the fact that $\sin(\pi/2) = 1$.

(f) Find the two days during which the amount of daylight equals the amount of darkness. (These days are called *equinoxes*.) [*Note*: Answer this question both graphically and algebraically.]

Solutions to Practice Problems 8.2

1. Here $P = (x, y) = (-3, -1)$. The length of the line segment OP is

$$r = \sqrt{x^2 + y^2} = \sqrt{(-3)^2 + (-1)^2} = \sqrt{10}.$$

Then

$$\cos t = \frac{x}{r} = \frac{-3}{\sqrt{10}} \approx -.94868.$$

2. Given $\cos(1.17) = .390$, use the relation $\cos^2 t + \sin^2 t = 1$ with $t = 1.17$ to solve for $\sin(1.17)$:

$$\cos^2(1.17) + \sin^2(1.17) = 1$$

$$\sin^2(1.17) = 1 - \cos^2(1.17)$$

$$= 1 - (.390)^2 = .8479.$$

So

$$\sin(1.17) = \sqrt{.8479} \approx .921.$$

Also, from properties (4) and (5),

$$\cos(1.17 + 4\pi) = \cos(1.17) = .390$$

$$\cos(-1.17) = \cos(1.17) = .390$$

$$\sin(-1.17) = -\sin(1.17) = -.921.$$

8.3 Differentiation and Integration of sin *t* and cos *t*

In this section we study the two differentiation rules

$$\frac{d}{dt} \sin t = \cos t, \tag{1}$$

$$\frac{d}{dt} \cos t = -\sin t. \tag{2}$$

It is not difficult to see why these rules might be true. Formula (1) says that the slope of the curve $y = \sin t$ at a particular value of t is given by the corresponding value of $\cos t$. To check it, we draw a careful graph of $y = \sin t$ and estimate the slope at various points. (See Fig. 1.) Let us plot the slope as a function of t. (See Fig. 2.) As can be seen, the "slope function" (the derivative) of $\sin t$ has a graph similar to the curve $y = \cos t$. Thus formula (1) seems to be reasonable. A similar analysis of the graph of $y = \cos t$ would show why (2) might be true. Proofs of these differentiation rules are outlined in an appendix at the end of this section.

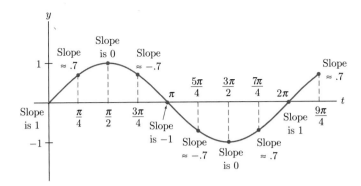

Figure 1. Slope estimates along the graph of $y = \sin t$.

Combining (1), (2), and the chain rule, we obtain the following general rules.

$$\frac{d}{dt}(\sin g(t)) = [\cos g(t)]\,g'(t), \tag{3}$$

$$\frac{d}{dt}(\cos g(t)) = [-\sin g(t)]\,g'(t). \tag{4}$$

EXAMPLE 1   Differentiate
(a) $\sin 3t$ (b) $(t^2 + 3\sin t)^5$.

Solution (a) $\dfrac{d}{dt}(\sin 3t) = (\cos 3t)\dfrac{d}{dt}(3t) = (\cos 3t) \cdot 3 = 3\cos 3t$

(b) $\dfrac{d}{dt}(t^2 + 3\sin t)^5 = 5(t^2 + 3\sin t)^4 \cdot \dfrac{d}{dt}(t^2 + 3\sin t)$

$$= 5(t^2 + 3\sin t)^4(2t + 3\cos t). \qquad \blacksquare$$

EXAMPLE 2   Differentiate
(a) $\cos(t^2 + 1)$ (b) $\cos^2 t$.

Solution (a) $\dfrac{d}{dt}\cos(t^2 + 1) = -\sin(t^2 + 1)\dfrac{d}{dt}(t^2 + 1) = -\sin(t^2 + 1) \cdot (2t)$

$$= -2t\sin(t^2 + 1).$$

(b) Recall that the notation $\cos^2 t$ means $(\cos t)^2$.

$$\frac{d}{dt}\cos^2 t = \frac{d}{dt}(\cos t)^2 = 2(\cos t)\frac{d}{dt}\cos t = -2\cos t \sin t. \qquad \blacksquare$$

EXAMPLE 3   Differentiate
(a) $t^2 \cos 3t$ (b) $(\sin 2t)/t$.

Solution (a) From the product rule we have

$$\frac{d}{dt}(t^2 \cos 3t) = t^2\frac{d}{dt}\cos 3t + (\cos 3t)\frac{d}{dt}t^2$$

$$= t^2(-3\sin 3t) + (\cos 3t)(2t)$$

$$= -3t^2\sin 3t + 2t\cos 3t.$$

$$du = \frac{d}{dx}(x^3 + 1)\, dx = 3x^2\, dx, \text{ so}$$

$$\int 3x^2 \sqrt{x^3 + 1}\, dx = \int \sqrt{u}\, du$$

$$= \frac{2}{3} u^{3/2} + C$$

$$= \frac{2}{3}(x^3 + 1)^{3/2} + C. \qquad \blacksquare$$

EXAMPLE 4   Find

$$\int \frac{(\ln x)^2}{x}\, dx.$$

Solution Let $u = \ln x$. Then $du = (1/x)\, dx$ and

$$\int \frac{(\ln x)^2}{x}\, dx = \int (\ln x)^2 \cdot \frac{1}{x}\, dx$$

$$= \int u^2\, du$$

$$= \frac{u^3}{3} + C$$

$$= \frac{(\ln x)^3}{3} + C \quad (\text{since } u = \ln x). \qquad \blacksquare$$

Knowing the correct substitution to make is a skill that develops through practice. Basically, we look for an occurrence of function composition, $f(g(x))$, where $f(x)$ is a function that we know how to integrate and where $g'(x)$ also appears in the integrand. Sometimes $g'(x)$ does not appear exactly, but can be obtained by multiplying by a constant. Such a shortcoming is easily remedied, as is illustrated in Examples 5 and 6.

EXAMPLE 5   Find

$$\int x^2 e^{x^3}\, dx.$$

Solution Let $u = x^3$; then $du = 3x^2\, dx$. The integrand involves $x^2\, dx$, not $3x^2\, dx$. To introduce the missing factor 3, we write

$$\int x^2 e^{x^3}\, dx = \int \frac{1}{3} \cdot 3x^2 e^{x^3}\, dx = \frac{1}{3} \int e^{x^3} 3x^2\, dx.$$

(Recall from Section 6.1 that constant multiples may be moved through the integral sign.) Substituting, we obtain

$$\int x^2 e^{x^3}\, dx = \frac{1}{3} \int e^u\, du = \frac{1}{3} e^u + C$$

$$= \frac{1}{3} e^{x^3} + C \quad (\text{since } u = x^3).$$

Another way to handle the missing factor 3 is to write

$$u = x^3, \quad du = 3x^2\, dx, \quad \text{and} \quad \frac{1}{3} du = x^2\, dx.$$

Then substitution yields

$$\int x^2 e^{x^3}\, dx = \int e^{x^3} \cdot x^2\, dx = \int e^u \cdot \frac{1}{3}\, du = \frac{1}{3} \int e^u\, du$$

$$= \frac{1}{3} e^u + C = \frac{1}{3} e^{x^3} + C.$$ ■

EXAMPLE 6   Find

$$\int \frac{2 - x}{\sqrt{2x^2 - 8x + 1}}\, dx.$$

Solution Let $u = 2x^2 - 8x + 1$; then $du = (4x - 8)\, dx$. Observe that $4x - 8 = -4(2 - x)$. So we multiply the integrand by -4 and compensate by placing a factor of $-\frac{1}{4}$ in front of the integral.

$$\int \frac{1}{\sqrt{2x^2 - 8x + 1}} \cdot (2 - x)\, dx = -\frac{1}{4} \int \frac{1}{\sqrt{2x^2 - 8x + 1}} \cdot (-4)(2 - x)\, dx$$

$$= -\frac{1}{4} \int \frac{1}{\sqrt{u}}\, du = -\frac{1}{4} \int u^{-1/2}\, du$$

$$= -\frac{1}{4} \cdot 2u^{1/2} + C = -\frac{1}{2} u^{1/2} + C$$

$$= -\frac{1}{2}(2x^2 - 8x + 1)^{1/2} + C.$$ ■

EXAMPLE 7   Find

$$\int \frac{2x}{x^2 + 1}\, dx.$$

Solution We note that the derivative of $x^2 + 1$ is $2x$. Thus we make the substitution $u = x^2 + 1$, $du = 2x\, dx$ to derive

$$\int \frac{2x}{x^2 + 1}\, dx = \int \frac{1}{u}\, du = \ln|u| + C = \ln(x^2 + 1) + C.$$ ■

EXAMPLE 8   Evaluate

$$\int \tan x\, dx.$$

Solution Since $\tan x = \dfrac{\sin x}{\cos x}$, we have

$$\int \tan x\, dx = \int \frac{\sin x}{\cos x}\, dx.$$

Let $u = \cos x$, so $du = -\sin x\, dx$. Then

$$\int \frac{\sin x}{\cos x}\, dx = -\int \frac{-\sin x}{\cos x}\, dx$$

$$= -\int \frac{1}{u}\, du$$

$$= -\ln|u| + C$$

$$= -\ln|\cos x| + C.$$

Note that

$$-\ln|\cos x| = \ln\left|\frac{1}{\cos x}\right| = \ln|\sec x|.$$

So the preceding formula can be written

$$\int \tan x \, dx = \ln|\sec x| + C.$$

∎

Practice Problems 9.1

1. (*Review*) Differentiate the following functions:

   (a) $e^{(2x^3+3x)}$ (b) $\ln x^5$

   (c) $\ln\sqrt{x}$ (d) $\ln 5|x|$

   (e) $x\ln x$ (f) $\ln(x^4+x^2+1)$

   (g) $\sin x^3$ (h) $\tan x$

2. Use the substitution $u = \dfrac{3}{x}$ to determine $\displaystyle\int \frac{e^{3/x}}{x^2}\,dx$.

EXERCISES 9.1

Determine the integrals in Exercises 1–36 by making appropriate substitutions.

1. $\displaystyle\int 2x(x^2+4)^5\,dx$

2. $\displaystyle\int 2(2x-1)^7\,dx$

3. $\displaystyle\int \frac{2x+1}{\sqrt{x^2+x+3}}\,dx$

4. $\displaystyle\int (x^2+2x+3)^6(x+1)\,dx$

5. $\displaystyle\int 3x^2 e^{(x^3-1)}\,dx$

6. $\displaystyle\int 2xe^{-x^2}\,dx$

7. $\displaystyle\int x\sqrt{4-x^2}\,dx$

8. $\displaystyle\int \frac{(1+\ln x)^3}{x}\,dx$

9. $\displaystyle\int \frac{1}{\sqrt{2x+1}}\,dx$

10. $\displaystyle\int (x^3-6x)^7(x^2-2)\,dx$

11. $\displaystyle\int xe^{x^2}\,dx$

12. $\displaystyle\int \frac{e^{\sqrt{x}}}{\sqrt{x}}\,dx$

13. $\displaystyle\int \frac{\ln(2x)}{x}\,dx$

14. $\displaystyle\int \frac{\sqrt{\ln x}}{x}\,dx$

15. $\displaystyle\int \frac{x^4}{x^5+1}\,dx$

16. $\displaystyle\int \frac{x}{\sqrt{x^2+1}}\,dx$

17. $\displaystyle\int \frac{x-3}{(1-6x+x^2)^2}\,dx$

18. $\displaystyle\int x^{-2}\left(\frac{1}{x}+2\right)^5\,dx$

19. $\displaystyle\int \frac{\ln\sqrt{x}}{x}\,dx$

20. $\displaystyle\int \frac{x^2}{3-x^3}\,dx$

21. $\displaystyle\int \frac{x^2-2x}{x^3-3x^2+1}\,dx$

22. $\displaystyle\int \frac{\ln(3x)}{3x}\,dx$

23. $\displaystyle\int \frac{8x}{e^{x^2}}\,dx$

24. $\displaystyle\int \frac{3}{(2x+1)^3}\,dx$

25. $\displaystyle\int \frac{1}{x\ln x^2}\,dx$

26. $\displaystyle\int \frac{2}{x(\ln x)^4}\,dx$

27. $\displaystyle\int (3-x)(x^2-6x)^4\,dx$

28. $\displaystyle\int \frac{dx}{3-5x}$

29. $\displaystyle\int e^x(1+e^x)^5\,dx$

30. $\displaystyle\int e^x\sqrt{1+e^x}\,dx$

31. $\displaystyle\int \frac{e^x}{1+2e^x}\,dx$

32. $\displaystyle\int \frac{e^x+e^{-x}}{e^x-e^{-x}}\,dx$

33. $\displaystyle\int \frac{e^{-x}}{1-e^{-x}}\,dx$

34. $\displaystyle\int \frac{(1+e^{-x})^3}{e^x}\,dx$

[*Hint:* In Exercises 35 and 36, multiply the numerator and denominator by e^{-x}.]

35. $\displaystyle\int \frac{1}{1+e^x}\,dx$

36. $\displaystyle\int \frac{e^{2x}-1}{e^{2x}+1}\,dx$

37. Figure 1 shows graphs of several functions $f(x)$ whose slope at each x is $x/\sqrt{x^2+9}$. Find the expression for the function $f(x)$ whose graph passes through $(4,8)$.

38. Figure 2 shows graphs of several functions $f(x)$ whose slope at each x is $(2\sqrt{x}+1)/\sqrt{x}$. Find the expression for the function $f(x)$ whose graph passes through $(4,15)$.

Determine the following integrals using the indicated substitution.

39. $\displaystyle\int (x+5)^{-1/2}e^{\sqrt{x+5}}\,dx;\ u=\sqrt{x+5}$

40. $\displaystyle\int \frac{x^4}{x^5-7}\ln(x^5-7)\,dx;\ u=\ln(x^5-7)$

41. $\displaystyle\int x\sec^2 x^2\,dx;\ u=x^2$

Figure 1

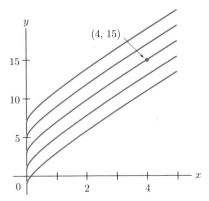

Figure 2

42. $\int (1 + \ln x) \sin(x \ln x)\, dx$; $u = x \ln x$

Determine the following integrals by making an appropriate substitution.

43. $\int \sin x \cos x\, dx$

44. $\int 2x \cos x^2\, dx$

45. $\int \dfrac{\cos \sqrt{x}}{\sqrt{x}}\, dx$

46. $\int \dfrac{\cos x}{(2 + \sin x)^3}\, dx$

47. $\int \cos^3 x \sin x\, dx$

48. $\int (\sin 2x) e^{\cos 2x}\, dx$

49. $\int \dfrac{\cos 3x}{\sqrt{2 - \sin 3x}}\, dx$

50. $\int \cot x\, dx$

51. $\int \dfrac{\sin x + \cos x}{\sin x - \cos x}\, dx$

52. $\int \tan x \sec^2 x\, dx$

53. Determine $\int 2x(x^2 + 5)\, dx$ by making a substitution. Then determine the integral by multiplying out the integrand and antidifferentiating. Account for the difference in the two results.

Solutions to Practice Problems 9.1

1. (a) $\dfrac{d}{dx} e^{(2x^3 + 3x)} = e^{(2x^3 + 3x)} \cdot (6x^2 + 3)$ (chain rule)

(b) $\dfrac{d}{dx} \ln x^5 = \dfrac{d}{dx} 5 \ln x = 5 \cdot \dfrac{1}{x}$ (Logarithm Property LIV)

(c) $\dfrac{d}{dx} \ln \sqrt{x} = \dfrac{d}{dx} \dfrac{1}{2} \ln x = \dfrac{1}{2} \cdot \dfrac{1}{x} = \dfrac{1}{2x}$ (Logarithm Property LIV)

(d) $\dfrac{d}{dx} \ln 5|x| = \dfrac{d}{dx} [\ln 5 + \ln |x|] = 0 + \dfrac{1}{x} = \dfrac{1}{x}$ (Logarithm Property LI)

(e) $\dfrac{d}{dx} x \ln x = x \cdot \dfrac{1}{x} + (\ln x) \cdot 1 = 1 + \ln x$ (product rule)

(f) $\dfrac{d}{dx} \ln(x^4 + x^2 + 1) = \dfrac{4x^3 + 2x}{x^4 + x^2 + 1}$ (chain rule)

(g) $\dfrac{d}{dx} \sin x^3 = (\cos x^3) \cdot (3x^2)$ (chain rule)

(h) $\dfrac{d}{dx} \tan x = \sec^2 x$ (formula for the derivative of $\tan x$)

2. Let $u = 3/x$, $du = (-3/x^2)\, dx$. Then

$$\int \dfrac{e^{3/x}}{x^2}\, dx = -\dfrac{1}{3} \int e^{3/x} \cdot \left(-\dfrac{3}{x^2}\right) dx$$

$$= -\dfrac{1}{3} \int e^u\, du = -\dfrac{1}{3} e^u + C = -\dfrac{1}{3} e^{3/x} + C.$$

9.2 Integration by Parts

In the preceding section we developed the method of integration by substitution by turning the chain rule into an integration formula. Let us do the same for the product rule. Let $f(x)$ and $g(x)$ be any two functions, and let $G(x)$ be an antiderivative of $g(x)$. The product rule asserts that

$$\frac{d}{dx}[f(x)G(x)] = f(x)G'(x) + f'(x)G(x)$$
$$= f(x)g(x) + f'(x)G(x) \quad [\text{since } G'(x) = g(x)].$$

Therefore,

$$f(x)G(x) = \int f(x)g(x)\,dx + \int f'(x)G(x)\,dx.$$

This last formula can be rewritten in the following more useful form.

$$\int f(x)g(x)\,dx = f(x)G(x) - \int f'(x)G(x)\,dx. \qquad (1)$$

Equation (1) is the principle of *integration by parts* and is one of the most important techniques of integration.

EXAMPLE 1 Evaluate

$$\int xe^x\,dx.$$

Solution Set $f(x) = x$, $g(x) = e^x$. Then $f'(x) = 1$, $G(x) = e^x$, and equation (1) yields

$$\int xe^x\,dx = xe^x - \int 1 \cdot e^x\,dx = xe^x - e^x + C. \qquad \blacksquare$$

The following principles underlie Example 1 and also illustrate general features of situations to which integration by parts may be applied:

1. The integrand is the product of two functions $f(x) = x$ and $g(x) = e^x$.

2. It is easy to compute $f'(x)$ and $G(x)$. That is, we can differentiate $f(x)$ and integrate $g(x)$.

3. The integral $\int f'(x)G(x)\,dx$ can be calculated.

Let us consider another example to see how these three principles work.

EXAMPLE 2 Evaluate

$$\int x(x+5)^8\,dx.$$

Solution Our calculations can be set up as follows:

$$f(x) = x, \qquad g(x) = (x+5)^8,$$

$$f'(x) = 1, \qquad G(x) = \frac{1}{9}(x+5)^9.$$

Then

$$\int x(x+5)^8 \, dx = x \cdot \frac{1}{9}(x+5)^9 - \int 1 \cdot \frac{1}{9}(x+5)^9 \, dx$$

$$= \frac{1}{9}x(x+5)^9 - \frac{1}{9}\int (x+5)^9 \, dx$$

$$= \frac{1}{9}x(x+5)^9 - \frac{1}{9} \cdot \frac{1}{10}(x+5)^{10} + C$$

$$= \frac{1}{9}x(x+5)^9 - \frac{1}{90}(x+5)^{10} + C.$$ ■

We were led to try integration by parts because our integrand is the product of two functions. We choose $f(x) = x$ [and not $(x+5)^8$] because $f'(x) = 1$, so the new integrand has only the factor $x+5$, thereby simplifying the integral.

EXAMPLE 3   Evaluate

$$\int x \sin x \, dx.$$

Solution Let us set

$$f(x) = x, \qquad g(x) = \sin x,$$
$$f'(x) = 1, \qquad G(x) = -\cos x.$$

Then

$$\int x \sin x \, dx = -x \cos x - \int 1 \cdot (-\cos x) \, dx$$

$$= -x \cos x + \int \cos x \, dx$$

$$= -x \cos x + \sin x + C.$$ ■

EXAMPLE 4   Evaluate

$$\int x^2 \ln x \, dx.$$

Solution Set

$$f(x) = \ln x, \qquad g(x) = x^2$$

$$f'(x) = \frac{1}{x}, \qquad G(x) = \frac{x^3}{3}.$$

Then

$$\int x^2 \ln x \, dx = \frac{x^3}{3}\ln x - \int \frac{1}{x} \cdot \frac{x^3}{3} \, dx$$

$$= \frac{x^3}{3}\ln x - \frac{1}{3}\int x^2 \, dx$$

$$= \frac{x^3}{3}\ln x - \frac{1}{9}x^3 + C.$$ ■

The next example shows how integration by parts can be used to compute a reasonably complicated integral.

EXAMPLE 5   Find

$$\int \frac{xe^x}{(x+1)^2}\, dx.$$

Solution Let $f(x) = xe^x$, $g(x) = \dfrac{1}{(x+1)^2}$. Then

$$f'(x) = xe^x + e^x \cdot 1 = (x+1)e^x, \qquad G(x) = \frac{-1}{x+1}.$$

As a result, we have

$$\int \frac{xe^x}{(x+1)^2}\, dx = xe^x \cdot \frac{-1}{x+1} - \int (x+1)e^x \cdot \frac{-1}{x+1}\, dx$$

$$= -\frac{xe^x}{x+1} + \int e^x\, dx$$

$$= -\frac{xe^x}{x+1} + e^x + C = \frac{e^x}{x+1} + C. \qquad \blacksquare$$

Sometimes we must use integration by parts more than once.

EXAMPLE 6   Find

$$\int x^2 \sin x\, dx.$$

Solution Let $f(x) = x^2$, $g(x) = \sin x$. Then $f'(x) = 2x$ and $G(x) = -\cos x$. Applying our formula for integration by parts, we have

$$\int x^2 \sin x\, dx = -x^2 \cos x - \int 2x \cdot (-\cos x)\, dx$$

$$= -x^2 \cos x + 2\int x \cos x\, dx. \qquad (2)$$

The integral $\int x \cos x\, dx$ can itself be handled by integration by parts. Let $f(x) = x$, $g(x) = \cos x$. Then $f'(x) = 1$ and $G(x) = \sin x$, so

$$\int x \cos x\, dx = x \sin x - \int 1 \cdot \sin x\, dx$$

$$= x \sin x + \cos x + C. \qquad (3)$$

Combining (2) and (3), we see that

$$\int x^2 \sin x\, dx = -x^2 \cos x + 2(x \sin x + \cos x) + C$$

$$= -x^2 \cos x + 2x \sin x + 2 \cos x + C. \qquad \blacksquare$$

EXAMPLE 7   Evaluate

$$\int \ln x\, dx.$$

Solution Since $\ln x = 1 \cdot \ln x$, we may view $\ln x$ as a product $f(x)g(x)$, where $f(x) = \ln x$, $g(x) = 1$. Then

$$f'(x) = \frac{1}{x}, \qquad G(x) = x.$$

EXAMPLE 3

y
b
$-a$

Figure 1

Integration by Parts in Definite Integrals

EXAMPLE 4   Evaluate

$$\int_0^{\pi/2} x \cos x \, dx.$$

Solution We use integration by parts to find an antiderivative of $x \cos x$. Let $f(x) = x$, $g(x) = \cos x$, $f'(x) = 1$, and $G(x) = \sin x$. Then

$$\int x \cos x \, dx = x \sin x - \int 1 \cdot \sin x \, dx = x \sin x + \cos x + C.$$

Solution

Hence

$$\int_0^{\pi/2} x \cos x \, dx = (x \sin x + \cos x) \Big|_0^{\pi/2}$$

$$= \left(\frac{\pi}{2} \sin \frac{\pi}{2} + \cos \frac{\pi}{2} \right) - (0 + \cos 0)$$

$$= \frac{\pi}{2} - 1. \quad \blacksquare$$

EXAMPLE 5   Evaluate

$$\int_0^5 \frac{x}{\sqrt{x+4}} \, dx.$$

Solution Let $f(x) = x$, $g(x) = (x+4)^{-1/2}$, $f'(x) = 1$, and $G(x) = 2(x+4)^{1/2}$. Then

$$\int_0^5 \frac{x}{\sqrt{x+4}} \, dx = 2x(x+4)^{1/2} \Big|_0^5 - \int_0^5 1 \cdot 2(x+4)^{1/2} \, dx$$

$$= 2x(x+4)^{1/2} \Big|_0^5 - \frac{4}{3}(x+4)^{3/2} \Big|_0^5$$

$$= \left[10(9)^{1/2} - 0 \right] - \left[\frac{4}{3}(9)^{3/2} - \frac{4}{3}(4)^{3/2} \right]$$

$$= [30] - \left[36 - \frac{32}{3} \right] = 4\frac{2}{3}. \quad \blacksquare$$

Practice Problems 9.3

Evaluate the following definite integrals:

1. $\int_0^1 (2x+3)e^{x^2+3x+6} \, dx$

2. $\int_e^{e^{\pi/2}} \frac{\sin(\ln x)}{x} \, dx$

EXERCISES 9.3

EXAMPLE 2

Evaluate the following definite integrals:

1. $\int_{5/2}^3 2(2x-5)^{14} \, dx$ **2.** $\int_2^6 \frac{1}{\sqrt{4x+1}} \, dx$

3. $\int_0^2 4x(1+x^2)^3 \, dx$ **4.** $\int_0^1 \frac{2x}{\sqrt{x^2+1}} \, dx$

5. $\int_0^3 \frac{x}{\sqrt{x+1}} \, dx$ **6.** $\int_0^1 x(3+x)^5 \, dx$

7. $\displaystyle\int_3^5 x\sqrt{x^2-9}\,dx$

8. $\displaystyle\int_0^1 \frac{1}{(1+2x)^4}\,dx$

9. $\displaystyle\int_{-1}^2 (x^2-1)(x^3-3x)^4\,dx$

10. $\displaystyle\int_0^1 (2x-1)(x^2-x)^{10}\,dx$

11. $\displaystyle\int_0^1 \frac{x}{x^2+3}\,dx$

12. $\displaystyle\int_0^4 8x(x+4)^{-3}\,dx$

13. $\displaystyle\int_1^3 x^2 e^{x^3}\,dx$

14. $\displaystyle\int_{-1}^1 2xe^x\,dx$

15. $\displaystyle\int_1^e \frac{\ln x}{x}\,dx$

16. $\displaystyle\int_1^e \ln x\,dx$

17. $\displaystyle\int_0^\pi e^{\sin x}\cos x\,dx$

18. $\displaystyle\int_0^{\pi/4} \tan x\,dx$

19. $\displaystyle\int_0^1 x\sin\pi x\,dx$

20. $\displaystyle\int_0^{\pi/2} \sin\left(2x-\frac{\pi}{2}\right)\,dx$

Use substitutions and the fact that a circle of radius r has area πr^2 to evaluate the following integrals:

21. $\displaystyle\int_{-\pi/2}^{\pi/2} \sqrt{1-\sin^2 x}\,\cos x\,dx$

22. $\displaystyle\int_0^{\sqrt{2}} \sqrt{4-x^4}\cdot 2x\,dx$

23. $\displaystyle\int_{-6}^0 \sqrt{-x^2-6x}\,dx$ [Complete the square: $-x^2-6x = 9-(x+3)^2$]

In Exercises 24 and 25, find the area of the shaded regions.

24.

$y = -x\sqrt{9-x^2}$

25.

$y = x\sqrt{4-x^2}$

9.4 Approximation of Definite Integrals

The definite integrals that arise in practical problems cannot always be evaluated by computing the net change in an antiderivative, as we did in the preceding section. Mathematicians have compiled extensive tables of antiderivatives. Moreover, many excellent software programs can be used to determine antiderivatives. However, the form of an antiderivative may be quite complex, and in some cases there may actually be no way to express an antiderivative in terms of elementary functions. In this section we discuss three methods for approximating the numerical value of the definite integral,

$$\int_a^b f(x)\,dx$$

without computing an antiderivative.

Given a positive integer n, divide the interval $a \leq x \leq b$ into n equal subintervals, each of length $\Delta x = (b - a)/n$. Denote the endpoints of the subintervals by $a_0, a_1, \ldots, a_n$, and denote the midpoints of the subintervals by $x_1, x_2, \ldots, x_n$. (See Fig. 1.) Recall from Chapter 6 that the definite integral is the limit of Riemann sums. When the midpoints of the subintervals in Fig. 1 are used to construct a Riemann sum, the resulting approximation to $\int_a^b f(x)\,dx$ is called the *midpoint rule*.

Figure 1

Midpoint Rule

$$\int_a^b f(x)\,dx \approx f(x_1)\Delta x + f(x_2)\Delta x + \cdots + f(x_n)\Delta x$$

$$= [f(x_1) + f(x_2) + \cdots + f(x_n)]\Delta x. \tag{1}$$

EXAMPLE 1   Use the midpoint rule with $n = 4$ to approximate

$$\int_0^2 \frac{1}{1 + e^x}\,dx.$$

Solution We have $\Delta x = (b - a)/n = (2 - 0)/4 = .5$. The endpoints of the four subintervals begin at $a = 0$ and are spaced .5 unit apart. The first midpoint is at $a + \Delta x/2 = .25$.

The midpoints are also spaced .5 unit apart. According to the midpoint rule, the integral is approximately equal to

$$\left[\frac{1}{1 + e^{.25}} + \frac{1}{1 + e^{.75}} + \frac{1}{1 + e^{1.25}} + \frac{1}{1 + e^{1.75}}\right](.5)$$

$$\approx .5646961 \quad \text{(to seven decimal places)}. \quad \blacksquare$$

A second method of approximation, the *trapezoidal rule*, uses the values of $f(x)$ at the endpoints of the subintervals of the interval $a \leq x \leq b$.

Trapezoidal Rule

$$\int_a^b f(x)\,dx \approx [f(a_0) + 2f(a_1) + \cdots + 2f(a_{n-1}) + f(a_n)]\frac{\Delta x}{2}. \tag{2}$$

We will discuss the origin of the trapezoidal rule and why we call it by that name later in this section.

EXAMPLE 2   Use the trapezoidal rule with $n = 4$ to approximate

$$\int_0^2 \frac{1}{1 + e^x}\,dx.$$

Solution As in Example 1, $\Delta x = .5$ and the endpoints of the subintervals are $a_0 = 0$, $a_1 = .5$, $a_2 = 1$, $a_3 = 1.5$, and $a_4 = 2$. The trapezoidal rule gives

$$\left[\frac{1}{1+e^0} + 2 \cdot \frac{1}{1+e^{.5}} + 2 \cdot \frac{1}{1+e^1} + 2 \cdot \frac{1}{1+e^{1.5}} + \frac{1}{1+e^2}\right] \frac{.5}{2}$$

$$\approx .5692545 \quad \text{(to seven decimal places).} \quad \blacksquare$$

When the function $f(x)$ is given explicitly, either the midpoint rule or the trapezoidal rule may be used to approximate the definite integral. However, occasionally the values of $f(x)$ may be known only at the endpoints of the subintervals. This may happen, for instance, when the values of $f(x)$ are obtained from experimental data. In this case, the midpoint rule cannot be used.

EXAMPLE 3

Measuring cardiac output[*]   Five milligrams of dye is injected into a vein leading to the heart. The concentration of the dye in the aorta, an artery leading from the heart, is determined every 2 seconds for 22 seconds. (See Table 1.) Let $c(t)$ be the concentration in the aorta after t seconds. Use the trapezoidal rule to estimate $\int_0^{22} c(t)\, dt$.

TABLE 1   Concentration of Dye in the Aorta

Seconds after injection	0	2	4	6	8	10	12	14	16	18	20	22
Concentration (mg/liter)	0	0	.6	1.4	2.7	3.7	4.1	3.8	2.9	1.5	.9	.5

Solution Let $n = 11$. Then $a = 0$, $b = 22$, and $\Delta t = (22 - 0)/11 = 2$. The endpoints of the subintervals are $a_0 = 0, a_1 = 2, a_2 = 4, \ldots, a_{10} = 20$, and $a_{11} = 22$. By the trapezoidal rule,

$$\int_0^{22} c(t)\, dt \approx [c(0) + 2c(2) + 2c(4) + 2c(6) + \cdots + 2c(20) + c(22)]\left(\frac{2}{2}\right)$$

$$= [0 + 2(0) + 2(.6) + 2(1.4) + \cdots + 2(.9) + .5](1)$$

$$\approx 43.7 \text{ liters.}$$

Note that cardiac output is the rate (usually measured in liters per minute) at which the heart pumps blood, and it may be computed by the formula

$$R = \frac{60D}{\int_0^{22} c(t)\, dt},$$

where D is the quantity of dye injected. For the preceding data, $R = 60(5)/43.7 \approx 6.9$ liters per minute. $\quad \blacksquare$

Let us return to the approximations to $\int_0^2 \frac{1}{1+e^x}\, dx$ found in Examples 1 and 2. These numbers are shown in Fig. 2 along with the exact value of the definite

---

[*]Data from B. Horelick and S. Koont, Project UMAP, *Measuring Cardiac Output* (Newton, MA: Educational Development Center, Inc., 1978).

Figure 2

integral to seven decimal places and the error of the two approximations. [The scale is greatly enlarged. The exact value of the integral is $\ln 2 - \ln(1 + e^{-2}) \approx .5662192$. To find it, see Exercise 35, Section 9.1.] It can be shown that, in general, the error from the midpoint rule is about one-half the error from the trapezoidal rule, and the estimates from these two rules are usually on opposite sides of the actual value of the definite integral. These observations suggest that we might improve our estimate of the value of a definite integral by using a "weighted average" of these two estimates. Let M and T denote the estimates from the midpoint and trapezoidal rules, respectively, and define

$$S = \frac{2}{3}M + \frac{1}{3}T = \frac{2M + T}{3}. \tag{3}$$

The use of S as an estimate of the value of a definite integral is called *Simpson's rule*. If we use Simpson's rule to estimate the definite integral in Example 1, we find that

$$S = \frac{2(.5646961) + .5692545}{3} \approx .5662156.$$

The error here is only .0000036. The error from the trapezoidal rule, in this example, is over 800 times as large!

As the number n of subintervals increases, Simpson's rule becomes more accurate than both the midpoint rule and the trapezoidal rule. For a given definite integral, the error in the midpoint and trapezoidal rules is proportional to $1/n^2$, so doubling n will divide the error by 4. However, the error in Simpson's rule is proportional to $1/n^4$, so doubling n will divide the error by 16, and multiplying n by a factor of 10 will divide the error by 10,000.

It is possible to combine the formulas for the midpoint and trapezoidal rules into a single formula for Simpson's rule by using the fact that $S = (4M + 2T)/6$.

Simpson's Rule

$$\int_a^b f(x)\,dx \approx [f(a_0) + 4f(x_1) + 2f(a_1) + 4f(x_2) + 2f(a_2)$$
$$+ \cdots + 2f(a_{n-1}) + 4f(x_n) + f(a_n)]\frac{\Delta x}{6}. \tag{4}$$

Distribution of IQs Psychologists use various standardized tests to measure intelligence. The method most commonly used to describe the results of such tests is an intelligence quotient (or IQ). An IQ is a positive number that, in theory, indicates how a person's mental age compares with the person's chronological age. The median IQ is arbitrarily set at 100, so half the population has an IQ less than 100 and half greater. IQs are distributed according to a bell-shaped curve called a *normal curve*, pictured in Fig. 3. The proportion of all people having IQs between A and

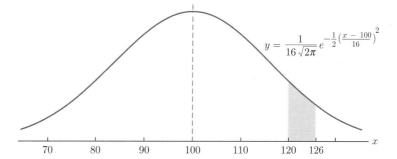

Figure 3. Proportion of IQs between 120 and 126.

B is given by the area under the curve from A to B, that is, by the integral

$$\frac{1}{16\sqrt{2\pi}} \int_A^B e^{-(1/2)[(x-100)/16]^2}\, dx.$$

EXAMPLE 4　　Estimate the proportion of all people having IQs between 120 and 126.

Solution　We have seen that this proportion is given by

$$\frac{1}{16\sqrt{2\pi}} \int_{120}^{126} f(x)\, dx, \quad \text{where } f(x) = e^{-(1/2)[(x-100)/16]^2}.$$

Let us approximate the definite integral by Simpson's rule with $n = 3$. Then $\Delta x = (126 - 120)/3 = 2$. The endpoints of the subintervals are 120, 122, 124, and 126; the midpoints of these subintervals are 121, 123, and 125. Simpson's rule gives

$$[f(120) + 4f(121) + 2f(122) + 4f(123) + 2f(124) + 4f(125) + f(126)]\frac{\Delta x}{6}$$

$$\approx [.4578 + 1.6904 + .7771 + 1.4235 + .6493 + 1.1801 + .2671]\left(\frac{2}{6}\right)$$

$$\approx 2.1484.$$

Multiplying this estimate by $1/(16\sqrt{2\pi})$, the constant in front of the integral, we get .0536. Thus approximately 5.36% of the population have IQs between 120 and 126. ∎

Geometric Interpretation of the Approximation Rules　Let $f(x)$ be a continuous nonnegative function on $a \le x \le b$. The approximation rules discussed previously may be interpreted as methods for estimating the area under the graph of $f(x)$. The midpoint rule arises from replacing this area by a collection of n rectangles, one lying over each subinterval, the first of height $f(x_1)$, the second of height $f(x_2)$, and so on. (See Fig. 4.)

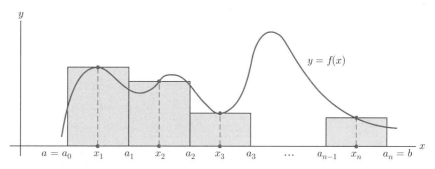

Figure 4. Approximation by rectangles.

If we approximate the area under the graph of $f(x)$ by trapezoids, as in Fig. 5, the total area of these trapezoids turns out to be the number given by the trapezoidal rule (hence the name of the rule).

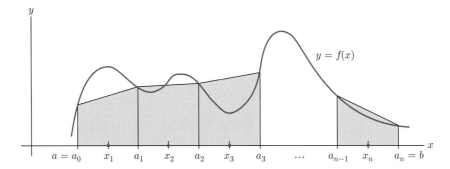

Figure 5. Approximation by trapezoids.

Simpson's rule corresponds to approximating the graph of $f(x)$ on each subinterval by a parabola instead of a straight line, as in the midpoint and trapezoidal rules. On each subinterval, the parabola is chosen so that it intersects the graph of $f(x)$ at the midpoint and both endpoints of the subinterval. (See Fig. 6.) It can be shown that the sum of the areas under these parabolas is the number given by Simpson's rule. Even more powerful approximation rules may be obtained by approximating the graph of $f(x)$ on each subinterval by cubic curves, or graphs of higher-order polynomials.

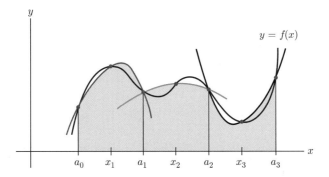

Figure 6. Approximation by parabolas.

Error Analysis A simple measure of the error of an approximation to a definite integral is the quantity

$$\big|[\text{approximate value}] - [\text{actual value}]\big|.$$

The following theorem gives an idea of how small this error must be for the various approximation rules. In a concrete example, the actual error of an approximation may be even substantially less than the "error bound" given in the theorem.

Error of Approximation Theorem Let n be the number of subintervals used in an approximation of the definite integral

$$\int_a^b f(x)\, dx.$$

1. The error for the midpoint rule is at most $\dfrac{A(b-a)^3}{24n^2}$, where A is a number such that $|f''(x)| \le A$ for all x satisfying $a \le x \le b$.

2. The error for the trapezoidal rule is at most $\dfrac{A(b-a)^3}{12n^2}$, where A is a number such that $|f''(x)| \leq A$ for all x satisfying $a \leq x \leq b$.

3. The error for Simpson's rule is at most $\dfrac{A(b-a)^5}{2880n^4}$, where A is a number such that $|f''''(x)| \leq A$ for all x satisfying $a \leq x \leq b$.

EXAMPLE 5

Obtain a bound on the error of using the trapezoidal rule with $n = 20$ to approximate

$$\int_0^1 e^{x^2}\, dx.$$

Solution Here $a = 0$, $b = 1$, and $f(x) = e^{x^2}$. Differentiating twice, we find that

$$f''(x) = (4x^2 + 2)e^{x^2}.$$

How large could $|f''(x)|$ be if x satisfies $0 \leq x \leq 1$? Since the function $(4x^2 + 2)e^{x^2}$ is clearly increasing on the interval from 0 to 1, its greatest value occurs at $x = 1$. (See Fig. 7.) Therefore, its greatest value is

$$(4 \cdot 1^2 + 2)e^{1^2} = 6e,$$

so we may take $A = 6e$ in the preceding theorem. The error of approximation using the trapezoidal rule is at most

$$\frac{6e(1 - 0)^3}{12(20)^2} = \frac{e}{800} \approx \frac{2.71828}{800} \approx .003398. \blacksquare$$

Figure 7

INCORPORATING TECHNOLOGY

Approximating Integrals In the Incorporating Technology part of Section 6.2 we showed how to use the TI-83/84's **sum** and **seq** functions to evaluate Riemann sums. Here we use that technique to demonstrate how to implement the approximations discussed in this section. In Fig. 8 we use the Midpoint Rule as in Example 1 and store the result in a variable M.

In Fig. 9 we implement the Trapezoidal Rule and store the result in a variable T. Here we calculate the first term separately, use **sum(seq** to total the middle three terms, and then we calculate the last term separately.

Figure 8. Calculation for Example 1.

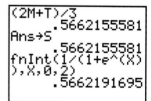

Figure 9. Calculation for Example 2.

Figure 10. Simpson's rule.

Figure 10 applies Simpson's rule to the same function and shows that it is very close to the highly accurate value given by **fnInt**.

We can also use our TI-83/84 to determine a suitable value for A in the error approximation theorem by finding the greatest value of either $f''(x)$ or $f''''(x)$. Graphing derivatives is explained in the Incorporating Technology section of Section 2.2. The method demonstrated there can be easily extended to graph the second and fourth derivatives of a function. $\blacksquare$

Practice Problems 9.4

Consider $\displaystyle\int_{1}^{3.4} (5x - 9)^2 \, dx.$

1. Divide the interval $1 \le x \le 3.4$ into three subintervals. List Δx and the endpoints and midpoints of the subintervals.

2. Approximate the integral by the midpoint rule with $n = 3$.

3. Approximate the integral by the trapezoidal rule with $n = 3$.

4. Approximate the integral by Simpson's rule with $n = 3$.

5. Find the exact value of the integral by integration.

EXERCISES 9.4

In Exercises 1 and 2, divide the given interval into n subintervals and list the value of Δx and the endpoints $a_0, a_1, \ldots, a_n$ of the subintervals.

1. $3 \le x \le 5$; $n = 5$ **2.** $-1 \le x \le 2$; $n = 5$

In Exercises 3 and 4, divide the interval into n subintervals and list the value of Δx and the midpoints $x_1, \ldots, x_n$ of the subintervals.

3. $-1 \le x \le 1$; $n = 4$ **4.** $0 \le x \le 3$; $n = 6$

5. Refer to the graph in Fig. 11. Draw the rectangles that approximate the area under the curve from 0 to 8 when using the midpoint rule with $n = 4$.

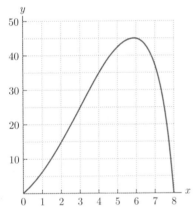

Figure 11

6. Refer to the graph in Fig. 11. Apply the trapezoidal rule with $n = 4$ to estimate the area under the curve.

Approximate the following integrals by the midpoint rule; then find the exact value by integration. Express your answers to five decimal places.

7. $\displaystyle\int_{0}^{4} (x^2 + 5) \, dx$; $n = 2, 4$

8. $\displaystyle\int_{1}^{5} (x - 1)^2 \, dx$; $n = 2, 4$

9. $\displaystyle\int_{0}^{1} e^{-x} \, dx$; $n = 5$ **10.** $\displaystyle\int_{1}^{2} \frac{1}{x + 1} \, dx$; $n = 5$

Approximate the following integrals by the trapezoidal rule; then find the exact value by integration. Express your answers to five decimal places.

11. $\displaystyle\int_{0}^{1} \left(x - \tfrac{1}{2}\right)^2 \, dx$; $n = 4$ **12.** $\displaystyle\int_{4}^{9} \frac{1}{x - 3} \, dx$; $n = 5$

13. $\displaystyle\int_{1}^{5} \frac{1}{x^2} \, dx$; $n = 3$ **14.** $\displaystyle\int_{-1}^{1} e^{2x} \, dx$; $n = 2, 4$

Approximate the following integrals by the midpoint rule, the trapezoidal rule, and Simpson's rule. Then find the exact value by integration. Express your answers to five decimal places.

15. $\displaystyle\int_{1}^{4} (2x - 3)^3 \, dx$; $n = 3$ **16.** $\displaystyle\int_{10}^{20} \frac{\ln x}{x} \, dx$; $n = 5$

17. $\displaystyle\int_{0}^{2} 2x e^{x^2} \, dx$; $n = 4$ **18.** $\displaystyle\int_{0}^{3} x\sqrt{4 - x} \, dx$; $n = 5$

19. $\displaystyle\int_{2}^{5} x e^x \, dx$; $n = 5$ **20.** $\displaystyle\int_{1}^{5} (4x^3 - 3x^2) \, dx$; $n = 2$

The following integrals cannot be evaluated in terms of elementary antiderivatives. Find an approximate value by Simpson's rule. Express your answers to five decimal places.

21. $\displaystyle\int_{0}^{2} \sqrt{1 + x^3} \, dx$; $n = 4$ **22.** $\displaystyle\int_{0}^{1} \frac{1}{x^3 + 1} \, dx$; $n = 2$

23. $\displaystyle\int_{0}^{2} \sqrt{\sin x} \, dx$; $n = 5$ **24.** $\displaystyle\int_{-1}^{1} \sqrt{1 + x^4} \, dx$; $n = 4$

25. Area In a survey of a piece of oceanfront property, measurements of the distance to the water were made every 50 feet along a 200-foot side. (See Fig. 12.) Use the trapezoidal rule to estimate the area of the property.

Figure 12. Survey of an oceanfront property.

26. Area To determine the amount of water flowing down a certain 100-yard-wide river, engineers need to know the area of a vertical cross section of the river. Measurements of the depth of the river were made every 20 yards from one bank to the other. The readings in fathoms were 0, 1, 2, 3, 1, 0. (One fathom equals 2 yards.) Use the trapezoidal rule to estimate the area of the cross section.

27. Distance Traveled Upon takeoff, the velocity readings of a rocket noted every second for 10 seconds were 0, 30, 75, 115, 155, 200, 250, 300, 360, 420, and 490 feet per second. Use the trapezoidal rule to estimate the distance the rocket traveled during the first 10 seconds. [*Hint:* If $s(t)$ is the distance traveled by time t and $v(t)$ is the velocity at time t, then $s(10) = \int_0^{10} v(t)\,dt$.]

28. Distance Traveled In a drive along a country road, the speedometer readings are recorded each minute during a 5-minute interval.

Time (minutes)	0	1	2	3	4	5
Velocity (mph)	33	32	28	30	32	35

Use the trapezoidal rule to estimate the distance traveled during the 5 minutes. [*Hint:* If time is measured in minutes, velocity should be expressed in distance per minute. For example, 35 mph is $\frac{35}{60}$ miles per minute. Also, see the hint for Exercise 27.]

29. Consider $\int_0^2 f(x)\,dx$, where $f(x) = \frac{1}{12}x^4 + 3x^2$.

  (a) Make a rough sketch of the graph of $f''(x)$ for $0 \le x \le 2$.

  (b) Find a number A such that $|f''(x)| \le A$ for all x satisfying $0 \le x \le 2$.

  (c) Obtain a bound on the error of using the midpoint rule with $n = 10$ to approximate the definite integral.

  (d) The exact value of the definite integral (to four decimal places) is 8.5333, and the midpoint rule with $n = 10$ gives 8.5089. What is the error for the midpoint approximation? Does this error satisfy the bound obtained in part (c)?

  (e) Redo part (c) with the number of intervals doubled to $n = 20$. Is the bound on the error halved? Quartered?

30. Consider $\int_1^2 f(x)\,dx$, where $f(x) = 3\ln x$.

  (a) Make a rough sketch of the graph of the fourth derivative of $f(x)$ for $1 \le x \le 2$.

  (b) Find a number A such that $|f''''(x)| \le A$ for all x satisfying $1 \le x \le 2$.

  (c) Obtain a bound on the error of using Simpson's rule with $n = 2$ to approximate the definite integral.

  (d) The exact value of the definite integral (to four decimal places) is 1.1589, and Simpson's rule with $n = 2$ gives 1.1588. What is the error for the approximation by Simpson's rule? Does this error satisfy the bound obtained in part (c)?

  (e) Redo part (c) with the number of intervals tripled to $n = 6$. Is the bound on the error divided by three?

31. (a) Show that the area of the trapezoid in Fig. 13(a) is $\frac{1}{2}(h + k)l$. [*Hint:* Divide the trapezoid into a rectangle and a triangle.]

  (b) Show that the area of the first trapezoid on the left in Fig. 13(b) is $\frac{1}{2}[f(a_0) + f(a_1)]\Delta x$.

  (c) Derive the trapezoidal rule for the case $n = 4$.

(a)   (b)

Figure 13. Derivation of the trapezoidal rule.

32. Approximate the value of $\int_a^b f(x)\,dx$, where $f(x) \ge 0$, by dividing the interval $a \le x \le b$ into four subintervals and constructing five rectangles. (See Fig. 14.) Note that the width of the three inside rectangles is Δx, while the width of the two outside rectangles is $\Delta x/2$. Compute the sum of the areas of these five rectangles and compare this sum with the trapezoidal rule for $n = 4$.

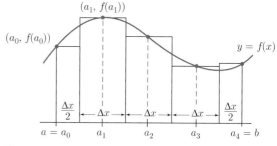

Figure 14. Another view of the trapezoidal rule.

33. (a) Suppose that the graph of $f(x)$ is above the x-axis and concave down on the interval $a_0 \le x \le a_1$. Let x_1 be the midpoint of this interval, let $\Delta x = a_1 - a_0$, and construct the line tangent to the graph of $f(x)$ at $(x_1, f(x_1))$, as in Fig. 15(a). Show that the area of the shaded trapezoid in Fig. 15(a) is the same as the area of the shaded rectangle in Fig. 15(c), that is, $f(x_1)\Delta x$. [*Hint:* Look at Fig. 15(b).] This shows that the area of the rectangle in Fig. 15(c) exceeds the area under the graph of $f(x)$ on the interval $a_0 \le x \le a_1$.

(b) Suppose that the graph of $f(x)$ is above the x-axis and concave down for all x in the interval $a \le x \le b$. Explain why $T \le \int_a^b f(x)\, dx \le M$, where T and M are the approximations given by the trapezoidal and midpoint rules, respectively.

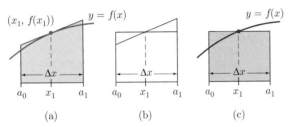

(a) (b) (c)

Figure 15

34. Riemann Sum Derivation of Formula for Cardiac Output; see Example 3. Subdivide the interval $0 \le t \le 22$ into n subintervals of length $\Delta t = 22/n$ seconds. Let t_i be a point in the ith subinterval.

(a) Show that $(R/60)\Delta t \approx$ [number of liters of blood flowing past the monitoring point during the ith time interval].

(b) Show that $c(t_i)(R/60)\Delta t \approx$ [quantity of dye flowing past the monitoring point during the ith time interval].

(c) Assume that basically all the dye will have flowed past the monitoring point during the 22 seconds. Explain why $D \approx (R/60)[c(t_1) + c(t_2) + \cdots + c(t_n)]\Delta t$, where the approximation improves as n gets large.

(d) Conclude that $D = \int_0^{22} (R/60)c(t)\, dt$, and solve for R.

Technology Exercises

35. In Fig. 16 a definite integral of the form $\int_a^b f(x)\, dx$ is approximated by the midpoint rule. Determine $f(x)$, a, b, and n.

Figure 16

36. In Fig. 17 a definite integral of the form $\int_a^b f(x)\, dx$ is approximated by the trapezoidal rule. Determine $f(x)$, a, b, and n.

Figure 17

In Exercises 37–40, approximate the integrals by the midpoint rule, the trapezoidal rule, and Simpson's rule with $n = 10$. Then find the exact value by integration and give the error for each approximation. Express your answers to the full accuracy given by the calculator or computer.

37. $\displaystyle\int_1^{11} \frac{1}{x}\, dx$

38. $\displaystyle\int_0^{\pi/2} \cos x\, dx$

39. $\displaystyle\int_0^{\pi/4} \sec^2 x\, dx$

40. $\displaystyle\int_0^1 2xe^{x^2}\, dx$

In Exercises 41 and 42, consider the definite integral $\displaystyle\int_0^1 \frac{4}{1+x^2}\, dx$, which has the value π.

41. Suppose the midpoint rule with $n = 20$ is used to estimate π. Graph the second derivative of the function in the window $[0, 1]$ by $[-10, 10]$ and then use the graph to obtain a bound on the error of the estimate.

42. Suppose the trapezoidal rule with $n = 15$ is used to estimate π. Graph the second derivative of the function in the window $[0, 1]$ by $[-10, 10]$ and then use the graph to obtain a bound on the error of the estimate.

Solutions to Practice Problems 9.4

1. $\Delta x = (3.4 - 1)/3 = 2.4/3 = .8$. Each subinterval will have length .8. A good way to proceed is first to draw two hatchmarks that subdivide the interval into three equal subintervals. [See Fig. 18(a).] Then label the hatchmarks by successively adding .8 to the left endpoint. [See Fig. 18(b).] The first midpoint can be obtained by adding one-half of .8 to the left endpoint. Then add .8 to get the next midpoint, and so on. [See Fig. 18(c).]

2. The midpoint rule uses only the midpoints of the subintervals:

$$\int_1^{3.4} (5x - 9)^2 \, dx \approx \big\{ (5[1.4] - 9)^2 + (5[2.2] - 9)^2$$
$$+ (5[3] - 9)^2 \big\} (.8)$$
$$= \big\{ (-2)^2 + 2^2 + 6^2 \big\} (.8)$$
$$= 35.2.$$

3. The trapezoidal rule uses only the endpoints of the subintervals:

$$\int_1^{3.4} (5x - 9)^2 \, dx \approx \big\{ (5[1] - 9)^2 + 2(5[1.8] - 9)^2$$
$$+ 2(5[2.6] - 9)^2$$
$$+ (5[3.4] - 9)^2 \big\} \left(\frac{.8}{2} \right)$$
$$= \big\{ (-4)^2 + 2(0)^2 + 2(4)^2 + 8^2 \big\} (.4)$$
$$= 44.8.$$

4. Using the formula $S = \dfrac{2M + T}{3}$, we obtain

$$\int_1^{3.4} (5x - 9)^2 \, dx \approx \frac{2(35.2) + 44.8}{3} = \frac{115.2}{3} = 38.4.$$

This approximation may also be obtained directly with formula (4), but the arithmetic requires about the same effort as calculating the midpoint and trapezoidal approximations separately and then combining them, as we do here.

5. $\displaystyle \int_1^{3.4} (5x - 9)^2 \, dx = \frac{1}{15} (5x - 9)^3 \Big|_1^{3.4}$

$$= \frac{1}{15} \big[8^3 - (-4)^3 \big]$$
$$= 38.4.$$

(Notice that here Simpson's rule gives the exact answer. This is so since the function to be integrated is a quadratic polynomial. Actually, Simpson's rule gives the exact value of the definite integral of any polynomial of degree 3 or less. The reason for this can be discovered from the **error of approximation theorem**.)

(a)   (b)   (c)

Figure 18

9.5 Some Applications of the Integral

Recall that the integral

$$\int_a^b f(t) \, dt$$

can be approximated by a Riemann sum as follows: We divide the t-axis from a to b into n subintervals by adding intermediate points $t_0 = a, t_1, \ldots, t_{n-1}, t_n = b$.

We assume that the points are equally spaced so that each subinterval has length $\Delta t = (b - a)/n$. For large n, the integral is very closely approximated by the

Riemann sum

$$\int_a^b f(t)\, dt \approx f(t_1)\Delta t + f(t_2)\Delta t + \cdots + f(t_n)\Delta t. \tag{1}$$

The approximation in (1) works both ways. If we encounter a Riemann sum like that in (1), we can approximate it by the corresponding integral. The approximation becomes better as the number of subintervals increases, that is, as n gets large. Thus, as n gets large, the sum approaches the value of the integral. This will be our approach in the examples that follow.

Our first two examples involve the concept of the present value of money. Suppose that we make an investment that promises to repay A dollars at time t (measuring the present as time 0). How much should we be willing to pay for such an investment? Clearly, we would not want to pay as much as A dollars. For if we had A dollars now, we could invest it at the current rate of interest and, at time t, we would get back our original A dollars plus the accrued interest. Instead, we should only be willing to pay an amount P that, if invested for t years, would yield A dollars. We call P the *present value of A dollars in t years*. We shall assume continuous compounding of interest. If the current (annual) rate of interest is r, then P dollars invested for t years will yield Pe^{rt} dollars (see Section 5.2). That is,

$$Pe^{rt} = A.$$

Thus the formula for the present value of A dollars in t years at interest rate r is

$$P = Ae^{-rt}.$$

EXAMPLE 1

Present value of an income stream Consider a small printing company that does most of its work on one printing press. The firm's profits are directly influenced by the amount of material that the press can produce (assuming that other factors, such as wages, are held constant). We may say that the press is producing a continuous stream of income for the company. Of course, the efficiency of the press may decline as it gets older. At time t, let $K(t)$ be the annual rate of income from the press. [This means that the press is producing $K(t) \cdot \frac{1}{365}$ dollars per day at time t.] Find a model for the present value of the income generated by the printing press over the next T years, assuming an interest rate r (with interest compounded continuously).

Solution We will divide the T-year period into n small subintervals of time, each of duration Δt years. (If each subinterval were 1 day, for example, Δt would equal $\frac{1}{365}$.)

We now consider the income produced by the printing press during a small time interval from t_{j-1} to t_j. Since Δt is small, the rate $K(t)$ of income production changes by only a negligible amount in that interval and can be considered approximately equal to $K(t_j)$. Since $K(t_j)$ gives an annual rate of income, the actual income produced during the period of Δt years is $K(t_j)\Delta t$. This income will be produced at approximately time t_j (t_j years from $t = 0$), so its present value is

$$[K(t_j)\Delta t]e^{-rt_j}.$$

The present value of the total income produced over the T-year period is the sum of the present values of the amounts produced during each time subinterval; that is,

$$K(t_1)e^{-rt_1}\Delta t + K(t_2)e^{-rt_2}\Delta t + \cdots + K(t_n)e^{-rt_n}\Delta t. \tag{2}$$

As the number of subintervals gets large, the length Δt of each subinterval becomes small, and the sum in (2) approaches the integral

$$\int_0^T K(t)e^{-rt}\,dt. \tag{3}$$

We call this quantity the *present value of the stream of income* produced by the printing press over the period from $t = 0$ to $t = T$ years. (The interest rate r used to compute present value is often called the company's *internal rate of return*.) ■

The concept of the present value of a continuous stream of income is an important tool in management decision processes involving the selection or replacement of equipment. It is also useful when analyzing various investment opportunities. Even when $K(t)$ is a simple function, the evaluation of the integral in (3) usually requires special techniques, such as integration by parts, as we see in the next example.

EXAMPLE 2

A company estimates that the rate of revenue produced by a machine at time t will be $5000 - 100t$ dollars per year. Find the present value of this continuous stream of income over the next 4 years at a 16% interest rate.

Solution We use formula (3) with $K(t) = 5000 - 100t$, $T = 4$, and $r = .16$. The present value of this income stream is

$$\int_0^4 (5000 - 100t)e^{-.16t}\,dt.$$

Using integration by parts, with $f(t) = 5000 - 100t$ and $g(t) = e^{-.16t}$, we find that the preceding integral equals

$$(5000 - 100t)\,\frac{1}{-.16}e^{-.16t}\,\Big|_0^4 - \int_0^4 (-100)\frac{1}{-.16}e^{-.16t}\,dt$$

$$\approx 16{,}090 - \frac{100}{.16}\cdot\frac{1}{-.16}e^{-.16t}\,\Big|_0^4$$

$$\approx 16{,}090 - 1847 = \$14{,}243. \qquad\blacksquare$$

A slight modification of the discussion in Example 1 gives the following general result.

> **Present Value of a Continuous Stream of Income**
>
> $$[\text{present value}] = \int_{T_1}^{T_2} K(t)e^{-rt}\,dt,$$
>
> where
> 1. $K(t)$ dollars per year is the annual rate of income at time t,
> 2. r is the annual interest rate of invested money,
> 3. T_1 to T_2 (years) is the time period of the income stream.

EXAMPLE 3

A demographic model It has been determined* that in 1940 the population density t miles from the center of New York City was approximately $120e^{-.2t}$ thousand

*C. Clark, "Urban Population Densities," *Journal of the Royal Statistical Society*, Series A, 114 (1951), 490–496. See also M. J. White, "On Cumulative Urban Growth and Urban Density Functions," *Journal of Urban Economics*, 4 (1977), 104–112.

people per square mile. Estimate the number of people who lived within 2 miles of the center of New York in 1940.

Solution Choose a fixed line emanating from the center of the city along which to measure distance. Subdivide this line from $t = 0$ to $t = 2$ into a large number of subintervals, each of length Δt. Each subinterval determines a ring. (See Fig. 1.)

Figure 1. A ring around the city center.

We will estimate the population of each ring and then add these populations together to obtain the total population. Suppose that j is an index ranging from 1 to n. If the outer circle of the jth ring is at a distance t_j from the center, the inner circle of this ring is at a distance $t_{j-1} = t_j - \Delta t$ from the center. The area of the jth ring is

$$\pi t_j^2 - \pi t_{j-1}^2 = \pi t_j^2 - \pi(t_j - \Delta t)^2$$
$$= \pi t_j^2 - \pi[t_j^2 - 2t_j \Delta t + (\Delta t)^2]$$
$$= 2\pi t_j \Delta t - \pi(\Delta t)^2.$$

Assume that Δt is very small. Then $\pi(\Delta t)^2$ is much smaller than $2\pi t_j \Delta t$. Hence the area of this ring is very close to $2\pi t_j \Delta t$.

Within the jth ring, the density of people is about $120e^{-.2t_j}$ thousand people per square mile. So the number of people in this ring is approximately

$$[\text{population density}] \cdot [\text{area of ring}] \approx 120e^{-.2t_j} \cdot 2\pi t_j \Delta t$$
$$= 240\pi t_j e^{-.2t_j} \Delta t.$$

Adding up the populations of all the rings, we obtain a total of

$$240\pi t_1 e^{-.2t_1} \Delta t + 240\pi t_2 e^{-.2t_2} \Delta t + \cdots + 240\pi t_n e^{-.2t_n} \Delta t,$$

which is a Riemann sum for the function $f(t) = 240t\pi e^{-.2t}$ over the interval from $t = 0$ to $t = 2$. This approximation to the population improves as the number n increases. Thus, the number of people (in thousands) who lived within 2 miles of the center of the city was

$$\int_0^2 240\pi t e^{-.2t}\, dt = 240\pi \int_0^2 t e^{-.2t}\, dt.$$

The last integral can be computed using integration by parts to obtain

$$240\pi \int_0^2 te^{-.2t}\, dt = 240\pi \frac{te^{-.2t}}{-.2}\Big|_0^2 - 240\pi \int_0^2 \frac{e^{-.2t}}{-.2}\, dt$$

$$= -2400\pi e^{-.4} + 1200\pi \left(\frac{e^{-.2t}}{-.2}\right)\Big|_0^2$$

$$= -2400\pi e^{-.4} + (-6000\pi e^{-.4} + 6000\pi)$$

$$\approx 1160.$$

Thus, in 1940, approximately 1,160,000 people lived within 2 miles of the center of the city. ∎

An argument analogous to that in Example 3 leads to the following result.

Total Population in a Ring around the City Center

$$[\text{population}] = \int_a^b 2\pi t D(t)\, dt,$$

where

1. $D(t)$ is the density of population (in persons per square mile) at distance t miles from city center,

2. Ring includes all persons who live between a and b miles from the city center.

EXERCISES 9.5

1. Find the present value of a continuous stream of income over 5 years when the rate of income is constant at \$35,000 per year and the interest rate is 7%.

2. A continuous stream of income is being produced at the constant rate of \$60,000 per year. Find the present value of the income generated during the time from $t = 2$ to $t = 6$ years, with a 6% interest rate.

3. Find the present value of a continuous stream of income over the time from $t = 1$ to $t = 5$ years when the interest rate is 10% and the income is produced at the rate of \$12,000 per year.

4. Find the present value of a continuous stream of income over 4 years if the rate of income is $25e^{-.02t}$ thousand dollars per year at time t and the interest rate is 8%.

5. Find the present value of a continuous stream of income over 3 years if the rate of income is $80e^{-.08t}$ thousand dollars per year at time t and the interest rate is 11%.

6. A continuous stream of income is produced at the rate of $20e^{1-.09t}$ thousand dollars per year at time t, and invested money earns 6% interest.

   (a) Write a definite integral that gives the present value of this stream of income over the time from $t = 2$ to $t = 5$ years.

   (b) Compute the present value described in part (a).

7. A growth company is one whose net earnings tend to increase each year. Suppose that the net earnings of a company at time t are being generated at the rate of $30 + 5t$ million dollars per year.

   (a) Write a definite integral that gives the present value of the company's earnings over the next 2 years using a 10% interest rate.

   (b) Compute the present value described in part (a).

8. Find the present value of a stream of earnings generated over the next 2 years at the rate of $50 + 7t$ thousand dollars per year at time t assuming a 10% interest rate.

9. **Demographic Model** In 1900 the population density of Philadelphia t miles from the city center was $120e^{-.65t}$ thousand people per square mile.

   (a) Write a definite integral whose value equals the number of people (in thousands) who lived within 5 miles of the city center.

   (b) Calculate the definite integral in part (a).

10. Use the population density from Exercise 9 to calculate the number of people who lived between 3 and 5 miles of the city center.

11. The population density of Philadelphia in 1940 was given by the function $60e^{-.4t}$. Calculate the number of people who lived within 5 miles of the city center. Sketch the graphs of the population densities for 1900 and 1940 (see Exercise 9) on a common graph. What trend do the graphs exhibit?

12. **Ecological Model** A volcano erupts and spreads lava in all directions. The density of the deposits at a distance t kilometers from the center is $D(t)$ thousand tons per square kilometer, where $D(t) = 11(t^2 + 10)^{-2}$. Find the tonnage of lava deposited between the distances of 1 and 10 kilometers from the center.

13. Suppose that the population density function for a city is $40e^{-.5t}$ thousand people per square mile. Let $P(t)$ be the total population that lives within t miles of the city center, and let Δt be a small positive number.

   (a) Consider the ring about the city whose inner circle is at t miles and outer circle is at $t + \Delta t$ miles. The text shows that the area of this ring is approximately $2\pi t \Delta t$ square miles. Approximately how many people live within this ring? (Your answer will involve t and Δt.)

   (b) What does
   $$\frac{P(t + \Delta t) - P(t)}{\Delta t}$$

   approach as Δt tends to zero?

   (c) What does the quantity $P(5 + \Delta t) - P(5)$ represent?

   (d) Use parts (a) and (c) to find a formula for
   $$\frac{P(t + \Delta t) - P(t)}{\Delta t}$$

   and from that obtain an approximate formula for the derivative $P'(t)$. This formula gives the rate of change of total population with respect to the distance t from the city center.

   (e) Given two positive numbers a and b, find a formula involving a definite integral, for the number of people who live in the city between a miles and b miles of the city center. [*Hint:* Use part (d) and the fundamental theorem of calculus to compute $P(b) - P(a)$.]

9.6 Improper Integrals

In applications of calculus, especially to statistics, it is often necessary to consider the area of a region that extends infinitely far to the right or left along the x-axis. We have drawn several such regions in Fig. 1. Note that the area under a curve that extends infinitely far along the x-axis is not necessarily infinite itself. The areas of such "infinite" regions may be computed using *improper integrals*.

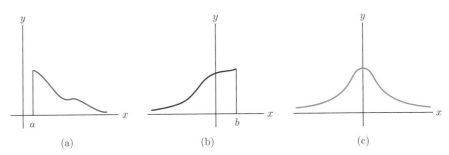

(a)   (b)   (c)

Figure 1

To motivate the idea of an improper integral, let us attempt to calculate the area under the curve $y = 3/x^2$ to the right of $x = 1$. (See Fig. 2.)

First, we compute the area under the graph of this function from $x = 1$ to $x = b$, where b is some number greater than 1. [See Fig. 3(a).] Then we examine how the area increases as we let b get larger. [See Figs. 3(b) and 3(c).] The area from 1 to b is given by

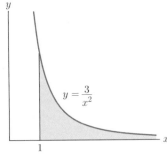

$$y = \frac{3}{x^2}$$

Figure 2

$$\int_1^b \frac{3}{x^2}\,dx = -\frac{3}{x}\Big|_1^b = \left(-\frac{3}{b}\right) - \left(-\frac{3}{1}\right) = 3 - \frac{3}{b}.$$

When b is large, $3/b$ is small and the integral nearly equals 3. That is, the area under the curve from 1 to b nearly equals 3. (See Table 1.) In fact, the area gets arbitrarily close to 3 as b gets larger. Thus it is reasonable to say that the region under the curve $y = 3/x^2$ for $x \geq 1$ has area 3.

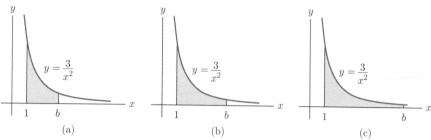

Figure 3

TABLE 1	Value of an "Infinitely Long" Area as a Limit
b	$Area = \displaystyle\int_1^b \frac{3}{x^2}\,dx = 3 - \frac{3}{b}$
10	2.7000
100	2.9700
1,000	2.9970
10,000	2.9997

Recall from Chapter 1 that we write $b \to \infty$ as shorthand for "b gets arbitrarily large, without bound." Then, to express the fact that the value of

$$\int_1^b \frac{3}{x^2}\,dx$$

approaches 3 as $b \to \infty$, we write

$$\int_1^\infty \frac{3}{x^2}\,dx = \lim_{b \to \infty} \int_1^b \frac{3}{x^2}\,dx = 3.$$

We call $\displaystyle\int_1^\infty \frac{3}{x^2}\,dx$ an *improper* integral because the upper limit of the integral is ∞ (infinity), rather than a finite number.

DEFINITION   Let a be fixed and suppose that $f(x)$ is a nonnegative function for $x \geq a$. If $\lim_{b \to \infty} \int_a^b f(x)\,dx = L$, we define

$$\int_a^\infty f(x)\,dx = \lim_{b \to \infty} \int_a^b f(x)\,dx = L.$$

We say that the improper integral $\int_a^\infty f(x)\,dx$ is *convergent* and that the region under the curve $y = f(x)$ for $x \geq a$ has area L. (See Fig. 4.)

Figure 4.   Area defined by an improper integral.

It is possible to consider improper integrals in which $f(x)$ is both positive and negative. However, we shall consider only nonnegative functions, since this is the case occurring in most applications.

EXAMPLE 1   Find the area under the curve $y = e^{-x}$ for $x \geq 0$. (See Fig. 5.)

Solution We must calculate the improper integral

$$\int_0^\infty e^{-x}\, dx.$$

We take $b > 0$ and compute

$$\int_0^b e^{-x}\, dx = -e^{-x}\Big|_0^b = (-e^{-b}) - (-e^0) = 1 - e^{-b} = 1 - \frac{1}{e^b}.$$

We now consider the limit as $b \to \infty$ and note that $1/e^b$ approaches zero. Thus

$$\int_0^\infty e^{-x}\, dx = \lim_{b \to \infty} \int_0^b e^{-x}\, dx = \lim_{b \to \infty}\left(1 - \frac{1}{e^b}\right) = 1.$$

Figure 5

Therefore, the region in Fig. 5 has area 1.   ■

EXAMPLE 2   Evaluate the improper integral

$$\int_7^\infty \frac{1}{(x-5)^2}\, dx.$$

Solution $\int_7^b \frac{1}{(x-5)^2}\, dx = -\frac{1}{x-5}\Big|_7^b = -\frac{1}{b-5} - \left(-\frac{1}{7-5}\right) = \frac{1}{2} - \frac{1}{b-5}.$

As $b \to \infty$, the fraction $1/(b-5)$ approaches zero, so

$$\int_7^\infty \frac{1}{(x-5)^2}\, dx = \lim_{b \to \infty} \int_7^b \frac{1}{(x-5)^2}\, dx = \lim_{b \to \infty}\left(\frac{1}{2} - \frac{1}{b-5}\right) = \frac{1}{2}.$$ ■

Not every improper integral is convergent. If the value $\int_a^b f(x)\, dx$ does not have a limit as $b \to \infty$, we cannot assign any numerical value to $\int_a^\infty f(x)\, dx$, and we say that the improper integral $\int_a^\infty f(x)\, dx$ is *divergent*.

EXAMPLE 3   Show that

$$\int_1^\infty \frac{1}{\sqrt{x}}\, dx$$

is divergent.

Solution For $b > 1$ we have

$$\int_1^b \frac{1}{\sqrt{x}}\, dx = 2\sqrt{x}\Big|_1^b = 2\sqrt{b} - 2.$$

As $b \to \infty$, the quantity $2\sqrt{b} - 2$ increases without bound. That is, $2\sqrt{b} - 2$ can be made larger than any specific number. Therefore, $\int_1^b \frac{1}{\sqrt{x}}\, dx$ has no limit as $b \to \infty$, so $\int_1^\infty \frac{1}{\sqrt{x}}\, dx$ is divergent.   ■

In some cases it is necessary to consider improper integrals of the form

$$\int_{-\infty}^{b} f(x)\,dx.$$

Let b be fixed and examine the value of $\int_{a}^{b} f(x)\,dx$ as $a \to -\infty$, that is, as a moves arbitrarily far to the left on the number line. If $\lim_{a \to -\infty} \int_{a}^{b} f(x)\,dx = L$, we say that the improper integral $\int_{-\infty}^{b} f(x)\,dx$ is *convergent*, and we write

$$\int_{-\infty}^{b} f(x)\,dx = L.$$

Otherwise, the improper integral is divergent. An integral of the form $\int_{-\infty}^{b} f(x)\,dx$ may be used to compute the area of a region such as that shown in Fig. 1(b).

EXAMPLE 4   Determine if $\int_{-\infty}^{0} e^{5x}\,dx$ is convergent. If convergent, find its value.

Solution $\displaystyle\int_{-\infty}^{0} e^{5x}\,dx = \lim_{a \to -\infty} \int_{a}^{0} e^{5x}\,dx = \lim_{a \to -\infty} \frac{1}{5}e^{5x}\Big|_{a}^{0} = \lim_{a \to -\infty}\left(\frac{1}{5} - \frac{1}{5}e^{5a}\right).$

As $a \to -\infty$, e^{5a} approaches 0, so $\frac{1}{5} - \frac{1}{5}e^{5a}$ approaches $\frac{1}{5}$. Thus the improper integral converges and has value $\frac{1}{5}$.   ∎

Areas of regions that extend infinitely far to the left *and* right, such as the region in Fig. 1(c), are calculated using improper integrals of the form

$$\int_{-\infty}^{\infty} f(x)\,dx.$$

We define such an integral to have the value

$$\int_{-\infty}^{0} f(x)\,dx + \int_{0}^{\infty} f(x)\,dx,$$

provided that both of the latter improper integrals are convergent.

An important area that arises in probability theory is the area under the so-called normal curve, whose equation is

$$y = \frac{1}{\sqrt{2\pi}}e^{-x^2/2}.$$

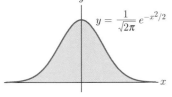

$y = \frac{1}{\sqrt{2\pi}}e^{-x^2/2}$

Figure 6. The standard normal curve.

(See Fig. 6.) It is of fundamental importance for probability theory that this area be 1. In terms of an improper integral, this fact may be written as

$$\int_{-\infty}^{\infty} \frac{1}{\sqrt{2\pi}}e^{-x^2/2}\,dx = 1.$$

The proof of this result is beyond the scope of this book.

INCORPORATING TECHNOLOGY

Improper Integrals   Although graphing calculators cannot tell you whether an improper integral converges, you can use the calculator to obtain a reliable indication of the behavior of the integral. Just look at values of $\int_{a}^{b} f(x)\,dx$ as b increases. Figures 7(a) and 7(b), which were created by setting $\mathbf{Y_1} = \mathbf{fnInt}(\mathbf{e}\wedge(-\mathbf{X}), \mathbf{X}, \mathbf{0}, \mathbf{X})$, give convincing evidence that the value of the improper integral in Example 1 is 1. The final $\mathbf{X}$ in $\mathbf{fnInt}(\mathbf{e}\wedge(-\mathbf{X}), \mathbf{X}, \mathbf{0}, \mathbf{X})$ is the upper limit of integration, that is, b.   ∎

Figure 7   (a)   (b)

Practice Problems 9.6

1. Does $1 - 2(1 - 3b)^{-4}$ approach a limit as $b \to \infty$?

2. Evaluate $\displaystyle\int_1^\infty \frac{x^2}{x^3 + 8} \, dx$.

3. Evaluate $\displaystyle\int_{-\infty}^{-2} \frac{1}{x^4} \, dx$.

EXERCISES 9.6

In Exercises 1–12, determine if the given expression approaches a limit as $b \to \infty$, and find that number when it does.

1. $\dfrac{5}{b}$ **2.** b^2 **3.** $-3e^{2b}$

4. $\dfrac{1}{b} + \dfrac{1}{3}$ **5.** $\dfrac{1}{4} - \dfrac{1}{b^2}$ **6.** $\dfrac{1}{2}\sqrt{b}$

7. $2 - (b+1)^{-1/2}$ **8.** $5 - (b-1)^{-1}$ **9.** $5(b^2 + 3)^{-1}$

10. $4(1 - b^{-3/4})$ **11.** $e^{-b/2} + 5$ **12.** $2 - e^{-3b}$

13. Find the area under the graph of $y = 1/x^2$ for $x \geq 2$.

14. Find the area under the graph of $y = (x+1)^{-2}$ for $x \geq 0$.

15. Find the area under the graph of $y = e^{-x/2}$ for $x \geq 0$.

16. Find the area under the graph of $y = 4e^{-4x}$ for $x \geq 0$.

17. Find the area under the graph of $y = (x+1)^{-3/2}$ for $x \geq 3$.

18. Find the area under the graph of $y = (2x+6)^{-4/3}$ for $x \geq 1$. (See Fig. 8.)

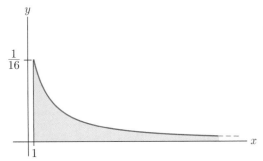

Figure 8

19. Show that the region under the graph of $y = (14x + 18)^{-4/5}$ for $x \geq 1$ cannot be assigned any finite number as its area. (See Fig. 9.)

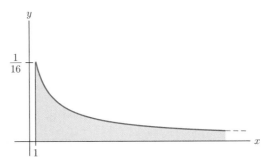

Figure 9

20. Show that the region under the graph of $y = (x-1)^{-1/3}$ for $x \geq 2$ cannot be assigned any finite number as its area.

Evaluate the following improper integrals whenever they are convergent.

21. $\displaystyle\int_1^\infty \frac{1}{x^3} \, dx$ **22.** $\displaystyle\int_1^\infty \frac{2}{x^{3/2}} \, dx$

23. $\displaystyle\int_0^\infty \frac{1}{(2x+3)^2} \, dx$ **24.** $\displaystyle\int_0^\infty e^{-3x} \, dx$

25. $\displaystyle\int_0^\infty e^{2x} \, dx$ **26.** $\displaystyle\int_0^\infty (x^2 + 1) \, dx$

27. $\displaystyle\int_2^\infty \frac{1}{(x-1)^{5/2}} \, dx$ **28.** $\displaystyle\int_2^\infty e^{2-x} \, dx$

29. $\displaystyle\int_0^\infty .01e^{-.01x} \, dx$ **30.** $\displaystyle\int_0^\infty \frac{4}{(2x+1)^3} \, dx$

31. $\displaystyle\int_0^\infty 6e^{1-3x} \, dx$ **32.** $\displaystyle\int_1^\infty e^{-.2x} \, dx$

33. $\displaystyle\int_3^\infty \frac{x^2}{\sqrt{x^3 - 1}} \, dx$ **34.** $\displaystyle\int_2^\infty \frac{1}{x \ln x} \, dx$

35. $\displaystyle\int_0^\infty xe^{-x^2} \, dx$ **36.** $\displaystyle\int_0^\infty \frac{x}{x^2 + 1} \, dx$

37. $\displaystyle\int_0^\infty 2x(x^2+1)^{-3/2}\,dx$ **38.** $\displaystyle\int_1^\infty (5x+1)^{-4}\,dx$

39. $\displaystyle\int_{-\infty}^0 e^{4x}\,dx$ **40.** $\displaystyle\int_{-\infty}^0 \frac{8}{(x-5)^2}\,dx$

41. $\displaystyle\int_{-\infty}^0 \frac{6}{(1-3x)^2}\,dx$ **42.** $\displaystyle\int_{-\infty}^0 \frac{1}{\sqrt{4-x}}\,dx$

43. $\displaystyle\int_0^\infty \frac{e^{-x}}{(e^{-x}+2)^2}\,dx$ **44.** $\displaystyle\int_{-\infty}^\infty \frac{e^{-x}}{(e^{-x}+2)^2}\,dx$

45. If $k>0$, show that $\displaystyle\int_0^\infty ke^{-kx}\,dx = 1$.

46. If $k>0$, show that $\displaystyle\int_1^\infty \frac{k}{x^{k+1}}\,dx = 1$.

47. If $k>0$, show that $\displaystyle\int_e^\infty \frac{k}{x(\ln x)^{k+1}}\,dx = 1$.

The capital value of an asset such as a machine is sometimes defined as the present value of all future net earnings. (See Section 9.5.) The actual lifetime of the asset may not be known, and since some assets may last indefinitely, the

capital value of the asset may be written in the form

$$[\text{capital value}] = \int_0^\infty K(t)e^{-rt}\,dt,$$

where r is the annual rate of interest compounded continuously.

48. Find the capital value of an asset that generates income at the rate of $5000 per year, assuming an interest rate of 10%.

49. Construct a formula for the capital value of a rental property that will generate a fixed income at the rate of K dollars per year indefinitely, assuming an annual interest rate of r.

50. Suppose that a large farm with a known reservoir of gas beneath the ground sells the gas rights to a company for a guaranteed payment at the rate of $10{,}000e^{.04t}$ dollars per year. Find the present value of this perpetual stream of income, assuming an interest rate of 12%, compounded continuously.

Solutions to Practice Problems 9.6

1. The expression $1-2(1-3b)^{-4}$ may also be written in the form

$$1 - \frac{2}{(1-3b)^4}.$$

When b is large, $(1-3b)^4$ is very large, so $2/(1-3b)^4$ is very small. Thus $1-2(1-3b)^{-4}$ approaches 1 as $b\to\infty$.

2. The first step is to find an antiderivative of $x^2/(x^3+8)$. Using the substitution $u=x^3+8$, $du=3x^2\,dx$, we obtain

$$\int \frac{x^2}{x^3+8}\,dx = \frac{1}{3}\int \frac{1}{u}\,du$$

$$= \frac{1}{3}\ln|u| + C$$

$$= \frac{1}{3}\ln|x^3+8| + C.$$

Now

$$\int_1^b \frac{x^2}{x^3+8}\,dx = \frac{1}{3}\ln|x^3+8|\Big|_1^b$$

$$= \frac{1}{3}\ln(b^3+8) - \frac{1}{3}\ln 9.$$

Finally, we examine what happens as $b\to\infty$. Certainly, b^3+8 gets arbitrarily large, so $\ln(b^3+8)$ must

also get arbitrarily large. Hence

$$\int_1^b \frac{x^2}{x^3+8}\,dx$$

has no limit as $b\to\infty$; so the improper integral

$$\int_1^\infty \frac{x^2}{x^3+8}\,dx$$

is divergent.

3. $\displaystyle\int_a^{-2} \frac{1}{x^4}\,dx = \int_a^{-2} x^{-4}\,dx$

$$= \frac{x^{-3}}{-3}\Big|_a^{-2} = \frac{1}{-3x^3}\Big|_a^{-2}$$

$$= \frac{1}{-3(-2)^3} - \left(\frac{1}{-3\cdot a^3}\right)$$

$$= \frac{1}{24} + \frac{1}{3a^3}$$

$$\int_{-\infty}^{-2} \frac{1}{x^4}\,dx = \lim_{a\to-\infty}\int_a^{-2}\frac{1}{x^4}\,dx$$

$$= \lim_{a\to-\infty}\left(\frac{1}{24}+\frac{1}{3a^3}\right) = \frac{1}{24}$$

REVIEW OF FUNDAMENTAL CONCEPTS

1. Describe integration by substitution in your own words.
2. Describe integration by parts in your own words.
3. Describe the change of limits rule for the integration by substitution of a definite integral.
4. State the formula for the integration by parts of a definite integral.
5. State the midpoint rule. (Include the meaning of all symbols used.)
6. State the trapezoidal rule. (Include the meaning of all symbols used.)

7. Explain the formula $S = \dfrac{2M + T}{3}$.
8. State the error of approximation theorem for each of the three approximation rules.
9. State the formula for each of the following quantities:
   (a) Present value of a continuous stream of income
   (b) Total population in a ring around the center of a city
10. How do you determine whether an improper integral is convergent?

SUPPLEMENTARY EXERCISES

Determine the following indefinite integrals:

1. $\displaystyle\int x \sin 3x^2 \, dx$

2. $\displaystyle\int \sqrt{2x + 1} \, dx$

3. $\displaystyle\int x(1 - 3x^2)^5 \, dx$

4. $\displaystyle\int \frac{(\ln x)^5}{x} \, dx$

5. $\displaystyle\int \frac{(\ln x)^2}{x} \, dx$

6. $\displaystyle\int \frac{1}{\sqrt{4x + 3}} \, dx$

7. $\displaystyle\int x\sqrt{4 - x^2} \, dx$

8. $\displaystyle\int x \sin 3x \, dx$

9. $\displaystyle\int x^2 e^{-x^3} \, dx$

10. $\displaystyle\int \frac{x \ln(x^2 + 1)}{x^2 + 1} \, dx$

11. $\displaystyle\int x^2 \cos 3x \, dx$

12. $\displaystyle\int \frac{\ln(\ln x)}{x \ln x} \, dx$

13. $\displaystyle\int \ln x^2 \, dx$

14. $\displaystyle\int x\sqrt{x + 1} \, dx$

15. $\displaystyle\int \frac{x}{\sqrt{3x - 1}} \, dx$

16. $\displaystyle\int x^2 \ln x^2 \, dx$

17. $\displaystyle\int \frac{x}{(1 - x)^5} \, dx$

18. $\displaystyle\int x(\ln x)^2 \, dx$

In Exercises 19–36, decide whether integration by parts or a substitution should be used to compute the indefinite integral. If substitution, indicate the substitution to be made. If by parts, indicate the functions $f(x)$ and $g(x)$ to be used in formula (1) of Section 9.2.

19. $\displaystyle\int x e^{2x} \, dx$

20. $\displaystyle\int (x - 3)e^{-x} \, dx$

21. $\displaystyle\int (x + 1)^{-1/2} e^{\sqrt{x+1}} \, dx$

22. $\displaystyle\int x^2 \sin(x^3 - 1) \, dx$

23. $\displaystyle\int \frac{x - 2x^3}{x^4 - x^2 + 4} \, dx$

24. $\displaystyle\int \ln\sqrt{5 - x} \, dx$

25. $\displaystyle\int e^{-x}(3x - 1)^2 \, dx$

26. $\displaystyle\int x e^{3 - x^2} \, dx$

27. $\displaystyle\int (500 - 4x)e^{-x/2} \, dx$

28. $\displaystyle\int x^{5/2} \ln x \, dx$

29. $\displaystyle\int \sqrt{x + 2} \ln(x + 2) \, dx$

30. $\displaystyle\int (x + 1)^2 e^{3x} \, dx$

31. $\displaystyle\int (x + 3)e^{x^2 + 6x} \, dx$

32. $\displaystyle\int \sin^2 x \cos x \, dx$

33. $\displaystyle\int x \cos(x^2 - 9) \, dx$

34. $\displaystyle\int (3 - x) \sin 3x \, dx$

35. $\displaystyle\int \frac{2 - x^2}{x^3 - 6x} \, dx$

36. $\displaystyle\int \frac{1}{x(\ln x)^{3/2}} \, dx$

Evaluate the following definite integrals:

37. $\displaystyle\int_0^1 \frac{2x}{(x^2 + 1)^3} \, dx$

38. $\displaystyle\int_0^{\pi/2} x \sin 8x \, dx$

39. $\displaystyle\int_0^2 x e^{-(1/2)x^2} \, dx$

40. $\displaystyle\int_{1/2}^1 \frac{\ln(2x + 3)}{2x + 3} \, dx$

41. $\displaystyle\int_1^2 x e^{-2x} \, dx$

42. $\displaystyle\int_1^2 x^{-3/2} \ln x \, dx$

Approximate the following definite integrals by the midpoint rule, the trapezoidal rule, and Simpson's rule.

43. $\displaystyle\int_1^9 \frac{1}{\sqrt{x}} \, dx; \; n = 4$

44. $\displaystyle\int_0^{10} e^{\sqrt{x}} \, dx; \; n = 5$

45. $\displaystyle\int_1^4 \frac{e^x}{x + 1} \, dx; \; n = 5$

46. $\displaystyle\int_{-1}^1 \frac{1}{1 + x^2} \, dx; \; n = 5$

Evaluate the following improper integrals whenever they are convergent.

47. $\displaystyle\int_0^\infty e^{6 - 3x} \, dx$

48. $\displaystyle\int_1^\infty x^{-2/3} \, dx$

49. $\int_{1}^{\infty} \dfrac{x+2}{x^2+4x-2}\,dx$

50. $\int_{0}^{\infty} x^2 e^{-x^3}\,dx$

51. $\int_{-1}^{\infty} (x+3)^{-5/4}\,dx$

52. $\int_{-\infty}^{0} \dfrac{8}{(5-2x)^3}\,dx$

53. It can be shown that $\lim_{b\to\infty} be^{-b} = 0$. Use this fact to compute $\int_{1}^{\infty} xe^{-x}\,dx$.

54. Let k be a positive number. It can be shown that $\lim_{b\to\infty} be^{-kb} = 0$. Use this fact to compute $\int_{0}^{\infty} xe^{-kx}\,dx$.

55. Find the present value of a continuous stream of income over the next 4 years, where the rate of income is $50e^{-.08t}$ thousand dollars per year at time t, and the interest rate is 12%.

56. Suppose that t miles from the center of a certain city the property tax revenue is approximately $R(t)$ thousand dollars per square mile, where $R(t) = 50e^{-t/20}$. Use this model to predict the total property tax revenue that will be generated by property within 10 miles of the city center.

57. Suppose that a machine requires daily maintenance, and let $M(t)$ be the *annual* rate of maintenance expense at time t. Suppose that the interval $0 \le t \le 2$ is divided into n subintervals, with endpoints $t_0 = 0, t_1, \ldots, t_n = 2$.

(a) Give a Riemann sum that approximates the total maintenance expense over the next 2 years. Then write the integral that the Riemann sum approximates for large values of n.

(b) Give a Riemann sum that approximates the present value of the total maintenance expense over the next 2 years using a 10% annual interest rate compounded continuously. Then write the integral that the Riemann sum approximates for large values of n.

58. The *capitalized cost* of an asset is the total of the original cost and the present value of all future "renewals" or replacements. This concept is useful, for example, when selecting equipment that is manufactured by several different companies. Suppose that a corporation computes the present value of future expenditures using an annual interest rate r, with continuous compounding of interest. Assume that the original cost of an asset is $80,000 and the annual renewal expense will be $50,000, spread more or less evenly throughout each year, for a large but indefinite number of years. Find a formula involving an integral that gives the capitalized cost of the asset.

APPENDIX

Areas under the Standard Normal Curve

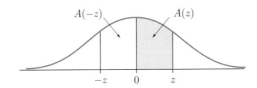

TABLE 1 Areas under the Standard Normal Curve

z	.00	.01	.02	.03	.04	.05	.06	.07	.08	.09
0.0	.0000	.0040	.0080	.0120	.0160	.0199	.0239	.0279	.0319	.0359
0.1	.0398	.0438	.0478	.0517	.0557	.0596	.0636	.0675	.0714	.0754
0.2	.0793	.0832	.0871	.0910	.0948	.0987	.1026	.1064	.1103	.1141
0.3	.1179	.1217	.1255	.1293	.1331	.1368	.1406	.1443	.1480	.1517
0.4	.1554	.1591	.1628	.1664	.1700	.1736	.1772	.1808	.1844	.1879
0.5	.1915	.1950	.1985	.2019	.2054	.2088	.2123	.2157	.2190	.2224
0.6	.2258	.2291	.2324	.2357	.2389	.2422	.2454	.2486	.2518	.2549
0.7	.2580	.2612	.2642	.2673	.2704	.2734	.2764	.2794	.2823	.2852
0.8	.2881	.2910	.2939	.2967	.2996	.3023	.3051	.3078	.3106	.3133
0.9	.3159	.3186	.3212	.3238	.3264	.3289	.3315	.3340	.3365	.3389
1.0	.3413	.3438	.3461	.3485	.3508	.3531	.3554	.3577	.3599	.3621
1.1	.3643	.3665	.3686	.3708	.3729	.3749	.3770	.3790	.3810	.3820
1.2	.3849	.3869	.3888	.3907	.3925	.3944	.3962	.3980	.3997	.4015
1.3	.4032	.4049	.4066	.4082	.4099	.4115	.4131	.4147	.4162	.4177
1.4	.4192	.4207	.4222	.4236	.4251	.4265	.4279	.4292	.4306	.4319
1.5	.4332	.4345	.4357	.4370	.4382	.4394	.4406	.4418	.4429	.4441
1.6	.4452	.4463	.4474	.4484	.4495	.4505	.4515	.4525	.4535	.4545
1.7	.4554	.4564	.4573	.4582	.4591	.4599	.4608	.4616	.4625	.4633
1.8	.4641	.4649	.4656	.4664	.4671	.4678	.4686	.4693	.4699	.4706
1.9	.4713	.4719	.4726	.4732	.4738	.4744	.4750	.4756	.4761	.4767
2.0	.4772	.4778	.4783	.4788	.4793	.4798	.4803	.4808	.4812	.4817
2.1	.4821	.4826	.4830	.4834	.4838	.4842	.4846	.4850	.4854	.4857
2.2	.4861	.4864	.4868	.4871	.4875	.4878	.4881	.4884	.4887	.4890
2.3	.4893	.4896	.4898	.4901	.4904	.4906	.4909	.4911	.4913	.4916
2.4	.4918	.4920	.4922	.4925	.4927	.4929	.4931	.4932	.4934	.4936
2.5	.4938	.4940	.4941	.4943	.4945	.4946	.4948	.4949	.4951	.4952
2.6	.4953	.4955	.4956	.4957	.4959	.4960	.4961	.4962	.4963	.4964
2.7	.4965	.4966	.4967	.4968	.4969	.4970	.4971	.4972	.4973	.4974
2.8	.4974	.4975	.4976	.4977	.4977	.4978	.4979	.4979	.4980	.4981
2.9	.4981	.4982	.4982	.4983	.4984	.4984	.4985	.4985	.4986	.4986
3.0	.4987	.4987	.4987	.4988	.4988	.4989	.4989	.4989	.4990	.4990
3.1	.4990	.4991	.4991	.4991	.4992	.4992	.4992	.4992	.4993	.4993
3.2	.4993	.4993	.4994	.4994	.4994	.4994	.4994	.4995	.4995	.4995
3.3	.4995	.4995	.4995	.4996	.4996	.4996	.4996	.4996	.4996	.4997
3.4	.4997	.4997	.4997	.4997	.4997	.4997	.4997	.4997	.4997	.4998
3.5	.4998	.4998	.4998	.4998	.4998	.4998	.4998	.4998	.4998	.4998

ANSWERS

CHAPTER 0

Exercises 0.1, page 12

1.   **3.**   **5.**   **7.** $[2,3)$ **9.** $[-1,0)$

11. $(-\infty, 3)$ **13.** $0, 10, 0, 70$ **15.** $0, 0, -\frac{9}{8}, a^3 + a^2 - a - 1$ **17.** $\frac{1}{3}, 3, \dfrac{a+1}{a+2}$ **19.** $a^2 - 1, a^2 + 2a$

21. (a) 1990 sales (b) 60 **23.** $x \neq 1, 2$ **25.** $x < 3$ **27.** Function **29.** Not a function **31.** Not a function **33.** 1 **35.** 3 **37.** Positive **39.** Positive **41.** $-1, 5, 9$ **43.** $.03$ **45.** $.04$ **47.** No

49. Yes **51.** $(a+1)^3$ **53.** $1, 3, 4$ **55.** $\pi, 3, 12$ **57.** $f(x) = \begin{cases} .06x & \text{for } 50 \leq x \leq 300 \\ .02x + 12 & \text{for } 300 < x \leq 600 \\ .015x + 15 & \text{for } 600 < x \end{cases}$

59.

61. You need parentheses $Y_1 = X^{\wedge}(3/4)$

63.
$[-2, 4]$ by $[-8, 5]$

65.
$[-4, 4]$ by $[-.5, 1.5]$

Exercises 0.2, page 20

1.   **3.**   **5.**   **7.** $\left(-\frac{1}{3}, 0\right), (0, 3)$ **9.** y-intercept $(0, 5)$

11. $(12, 0), (0, 3)$ **13.** (a) $K = \frac{1}{250}, V = \frac{1}{50}$ (b) $\left(-\dfrac{1}{K}, 0\right), \left(0, \dfrac{1}{V}\right)$ **15.** (a) \$58

(b) $f(x) = .20x + 18$ **17.** $300x + 1500$, $x = $ number of days **19.** The cost for another 5% is \$25 million. The cost for the final 5% is 21 times as much. **21.** $a = 3, b = -4, c = 0$ **23.** $a = -2, b = 3, c = 1$
25. $a = -1, b = 0, c = 1$ **27.**   **29.**   **31.**

33. 1 **35.** 10^{-2} **37.** 2.5 **39.** $-3985, 3008$ **41.** $-4.60569, 231.499$

A3

Exercises 0.3, page 25

1. $x^2 + 9x + 1$ **3.** $9x^3 + 9x$ **5.** $\dfrac{t^2 + 1}{9t}$ **7.** $\dfrac{3x + 1}{x^2 - x - 6}$ **9.** $\dfrac{4x}{x^2 - 12x + 32}$ **11.** $\dfrac{2x^2 + 5x + 50}{x^2 - 100}$

13. $\dfrac{2x^2 - 2x + 10}{x^2 + 3x - 10}$ **15.** $\dfrac{-x^2 + 5x}{x^2 + 3x - 10}$ **17.** $\dfrac{x^2 + 5x}{-x^2 + 7x - 10}$ **19.** $\dfrac{-x^2 + 3x + 4}{x^2 + 5x - 6}$ **21.** $\dfrac{-x^2 - 3x}{x^2 + 15x + 50}$

23. $\dfrac{5u - 1}{5u + 1}, u \neq 0$ **25.** $\left(\dfrac{x}{1 - x}\right)^6$ **27.** $\left(\dfrac{x}{1 - x}\right)^3 - 5\left(\dfrac{x}{1 - x}\right)^2 + 1$ **29.** $\dfrac{t^3 - 5t^2 + 1}{-t^3 + 5t^2}$ **31.** $2xh + h^2$

33. $4 - 2t - h$ **35. (a)** $C(A(t)) = 3000 + 1600t - 40t^2$ **(b)** \$6040   **37.** $h(x) = x + \dfrac{1}{8}$; $h(x)$ converts from British sizes to U.S. sizes.   **39.** The graph of $f(x) + c$ is the graph of $f(x)$ shifted up (if $c > 0$) or down (if $c < 0$) by $|c|$ units.   **41.**

43. $f(f(x)) = x, x \neq 1$

Exercises 0.4, page 32

1. $2, \frac{3}{2}$ **3.** $\frac{3}{2}$ **5.** No zeros   **7.** $1, -\frac{1}{5}$ **9.** $5, 4$ **11.** $2 + \dfrac{\sqrt{6}}{3}, 2 - \dfrac{\sqrt{6}}{3}$ **13.** $(x + 5)(x + 3)$

15. $(x - 4)(x + 4)$ **17.** $3(x + 2)^2$ **19.** $-2(x - 3)(x + 5)$ **21.** $x(3 - x)$ **23.** $-2x(x - \sqrt{3})(x + \sqrt{3})$

25. $(-1, 1), (5, 19)$ **27.** $(-1, 9), (4, 4)$ **29.** $(0, 0), (2, -2)$ **31.** $(0, 5), (2 - \sqrt{3}, 25 - 23\sqrt{3}/2),$
$(2 + \sqrt{3}, 25 + 23\sqrt{3}/2)$ **33.** $-7, 3$ **35.** $-2, 3$ **37.** -7 **39.** 16,667 and 78,571 subscribers   **41.** $-1, 2$

43. ≈ 4.56 **45.** $\approx (-.41, -1.83), (2.41, 3.83)$ **47.** $\approx (2.14, -25.73), (4.10, -21.80)$

49. $[-5, 22]$ by $[-1400, 100]$ **51.** $[-20, 4]$ by $[-500, 2500]$

Exercises 0.5, page 39

1. 27   **3.** 1   **5.** .0001   **7.** -16 **9.** 4   **11.** .01   **13.** $\frac{1}{6}$ **15.** 100   **17.** 16   **19.** 125   **21.** 1

23. 4   **25.** $\frac{1}{2}$ **27.** 1000   **29.** 10   **31.** 6   **33.** 16   **35.** 18   **37.** $\frac{4}{9}$ **39.** 7   **41.** $x^6 y^6$ **43.** $x^3 y^3$

45. $\dfrac{1}{x^{1/2}}$ **47.** $\dfrac{x^{12}}{y^6}$ **49.** $x^{12} y^{20}$ **51.** $x^2 y^6$ **53.** $16x^4$ **55.** x^2 **57.** $\dfrac{1}{x^7}$ **59.** x **61.** $\dfrac{27x^6}{8y^3}$

63. $2\sqrt{x}$ **65.** $\dfrac{1}{8x^6}$ **67.** $\dfrac{1}{32x^2}$ **69.** $9x^3$ **71.** $x - 1$ **73.** $1 + 6x^{1/2}$ **75.** $a^{1/2} \cdot b^{1/2} = (ab)^{1/2}$ (Law 5)

77. 16   **79.** $\frac{1}{4}$ **81.** 8   **83.** $\frac{1}{32}$ **85.** \$709.26   **87.** \$127,857.61   **89.** \$164.70   **91.** \$1592.75

93. \$3268.00   **95.** $\frac{500}{256}(256 + 256r + 96r^2 + 16r^3 + r^4) = A$ **97.** $\frac{1}{20}(2x)^2 = \frac{1}{20}(4x^2) = 4(\frac{1}{20}x^2)$

99. .0008103   **101.** .00000823

Exercises 0.6, page 47

1.   **3.**   **5.**   **7.** $P = 8x$; $3x^2 = 25$
9. $A = \pi r^2$; $2\pi r = 15$
11. $V = x^2 h$; $x^2 + 4xh = 65$

13. $\pi r^2 h = 100$; $C = 11\pi r^2 + 14\pi rh$ **15.** $2x + 3h = 5000$; $A = xh$ **17.** $C = 36x + 20h$

19. 75 cm^2 **21. (a)** 38   **(b)** \$40   **23. (a)** 200   **(b)** 275   **(c)** 25   **25. (a)** $P(x) = 12x - 800$
(b) \$640   **(c)** \$3150   **27.** 270 cents   **29.** A 100-in.^{3} cylinder of radius 3 in. costs \$1.62 to construct.
31. \$1.08   **33.** $R(30) = 1800$; $C(30) = 1200$ **35.** 40   **37.** $C(1000) = 4000$ **39.** Find the y-coordinate of

the point on the graph whose x-coordinate is 400.　**41.** The greatest profit, \$52,500, occurs when 2500 units of goods are produced.　**43.** Find the x-coordinate of the point on the graph whose y-coordinate is 30,000. **45.** Find $h(3)$. Find the y-coordinate of the point on the graph whose t-coordinate is 3.　**47.** Find the maximum value of $h(t)$. Find the y-coordinate of the highest point of the graph.　**49.** Solve $h(t) = 100$. Find the t-coordinates of the points whose y-coordinate is 100.　**51. (a)**

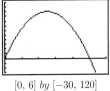

$[0, 6]$ *by* $[-30, 120]$

(b) 96 feet
(c) 1 and 4 seconds
(d) 5 seconds
(e) 2.5 seconds; 100 feet

53. (a)

$[200, 500]$ *by* $[42000, 75000]$

(b) 350 bicycles per year　**(c)** \$68,000　**(d)** \$5000　**(e)** No, revenue would only increase by \$4000.

Chapter 0: Supplementary Exercises, page 51

1. $2, 27\frac{1}{3}, -2, -2\frac{1}{8}, \frac{5\sqrt{2}}{2}$　**3.** $a^2 - 4a + 2$　**5.** $x \neq 0, -3$　**7.** All x　**9.** Yes　**11.** $5x(x-1)(x+4)$

13. $(-1)(x-6)(x+3)$　**15.** $-\frac{2}{5}, 1$　**17.** $\left(\dfrac{5+3\sqrt{5}}{10}, \dfrac{3\sqrt{5}}{5}\right), \left(\dfrac{5-3\sqrt{5}}{10}, -\dfrac{3\sqrt{5}}{5}\right)$　**19.** $x^2 + x - 1$

21. $x^{5/2} - 2x^{3/2}$　**23.** $x^{3/2} - 2x^{1/2}$　**25.** $\dfrac{x^2 - x + 1}{x^2 - 1}$　**27.** $-\dfrac{3x^2 + 1}{3x^2 + 4x + 1}$　**29.** $\dfrac{-3x^2 + 9x - 10}{3x^2 - 5x - 8}$

31. $\dfrac{1}{x^4} - \dfrac{2}{x^2} + 4$　**33.** $(\sqrt{x} - 1)^2$　**35.** $\dfrac{1}{(\sqrt{x}-1)^2} - \dfrac{2}{\sqrt{x}-1} + 4$　**37.** 27, 32, 4　**39.** $301 + 10t + .04t^2$

41. $x^2 + 2x + 1$　**43.** x

CHAPTER 1

Exercises 1.1, page 63

1. $m = -7, b = 3$　**3.** $m = \frac{1}{2}, b = \frac{3}{2}$　**5.** $m = \frac{1}{7}, b = -5$　**7.** $y - 1 = -(x - 7)$　**9.** $y - 1 = \frac{1}{2}(x - 2)$
11. $y - 5 = \frac{63}{10}(x - \frac{5}{7})$　**13.** $y = 0$　**15.** $y = 9$　**17.** $-\dfrac{x}{\pi} + y = 1$ or $y = \dfrac{x}{\pi} + 1$　**19.** $y = -2x - 4$
21. $y = x - 2$　**23.** $y = 3x - 6$　**25.** $y = x - 2$　**27.** Start at $(1, 0)$. To get back to the line, move one unit to the right and then one unit up.　**29.** Start at $(1, -1)$. To get back to the line, move

three units to the right and then one unit down.

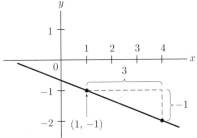

31. (a) C　**(b)** B

(c) D　**(d)** A　**33.** 1　**35.** $-\frac{3}{4}$　**37.** $(2, 5); (3, 7); (0, 1)$　**39.** $f(3) = 2$　**41.** l_1

43.

$\left(-\frac{1}{2}, 0\right)$

$(0, -1)$ $y = -2x - 1$

dollars per item.   **(c)** $C(11) - C(10) = $ marginal cost $= \$12$ **49.** $P(x) = .06x + 4.89$ dollars, where $P(x)$ is the price of one gallon x months since January 1st. Price of one gallon on April 1st is $P(3) = \$5.07$. Price of 15 gallons is $\$76.05$. Price of one gallon on September 1st is $P(8) = \$5.37$. Price of 15 gallons is $\$80.55$.
51. $C(x) = .03x + 5$ **53.** $G(x) = -\frac{5000}{3}x + \frac{25,000}{3}$, $G(4.35) \approx 1083.3$ gallons   **55. (a)** $C(x) = 7x + 1500$.
(b) Marginal cost is \$7 per rod.   **(c)** \$7   **57.** If the monopolist wants to sell one more unit of goods, then the price per unit must be lowered by 2 ¢ . No one is willing to pay \$7 or more for a unit of goods.   **59.** Let $A(x)$ be the amount in ml of the drug in the blood after x minutes, then $A(x) = 6x + 1.5$.
65. (a) $y = .38x + 39.5$ **(b)**

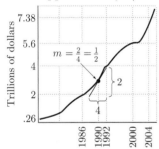

$[0, 30]$ *by* $[30, 60]$

(c) Every year, .38% more of the world becomes urban.

(d) 43.3%   **(e)** 2007   **(f)** 1.9%

Exercises 1.2, page 70

1. $-\frac{4}{3}$ **3.** 1   **5.** Small positive slope   **7.** Zero slope   **9.** Zero slope   **11.** Approximately \$22. The price was rising on both days.   **13.** \$104.50; $m \approx \dfrac{107 - 104.50}{2} = 1.25$ dollars/day   **15.** The slope of the tangent line is approximately 1/2. Thus the federal debt rose at the rate of .5 trillion dollars per year in 1990.

17. (a) The debt per capita was approximately \$1000 in 1950, \$15,000 in 1990, \$21,000 in 2000, and \$24,000 in 2004.

(b) The slope of the tangent line is 0. Thus, in 2000, the debt per capita held steady (at about \$21,000)   **19.** -1; $y - .16 = -1(x + .5)$ **21.** $\frac{2}{3}$; $y - \frac{1}{9} = \frac{2}{3}\left(x - \frac{1}{3}\right)$

23. $y - 6.25 = 5(x - 2.5)$ **25.** $\left(\frac{7}{4}, \frac{49}{16}\right)$ **27.** $\left(-\frac{1}{3}, \frac{1}{9}\right)$ **29.** 12   **31.** $\frac{3}{4}$ **33.** $a = 1$, $f(1) = 1$, slope is 2
35. $\left(\dfrac{1}{\sqrt{2}}, \dfrac{1}{2\sqrt{2}}\right)$, $\left(-\dfrac{1}{\sqrt{2}}, -\dfrac{1}{2\sqrt{2}}\right)$ **37. (a)** 3, 9   **(b)** increase   **39.** -3 **41.** Approximately .35.

Exercises 1.3, page 80

1. 3   **3.** $\frac{3}{4}$ **5.** $7x^6$ **7.** $\frac{2}{3}x^{-1/3} = \dfrac{2}{3\sqrt[3]{x}}$ **9.** $f(x) = x^{-5/2}$; $f'(x) = -\frac{5}{2}x^{-7/2}$ **11.** $\frac{1}{3}x^{-2/3}$

13. $f(x) = x^2$; $f'(x) = 2x$ **15.** 0   **17.** $\frac{3}{4}$ **19.** $-\frac{9}{4}$ **21.** 1   **23.** 2   **25.** 32   **27.** $f(-5) = -125$,
$f'(-5) = 75$ **29.** $f(8) = 2$, $f'(8) = \frac{1}{12}$ **31.** $f(-2) = -\frac{1}{32}$, $f'(-2) = -\frac{5}{64}$ **33.** $y - 8 = 12(x + 2)$
35. $y = 3x + 1$ **37.** $y - \frac{1}{3} = \frac{3}{2}(x - \frac{1}{9})$ **39.** $y - 1 = -\frac{1}{2}(x - 1)$ **41.** $f(x) = x^4$, $f'(x) = 4x^3$, $f(1) = 1$,
$f'(1) = 4$, then (6) (with $a = 1$) becomes $y - 1 = 4(x - 1)$, which is the given equation of the tangent line.
43. $P = (\frac{1}{16}, \frac{1}{4})$, $b = \frac{1}{8}$.   **45. (a)** $(16, 4)$ **(b)**   **47.** No, such a point would have to

have an x-coordinate satisfying $3x^2 = -1$, which has no real solutions.   **49.** $8x^7$ **51.** $\frac{3}{4}x^{-1/4}$ **53.** 0
55. $\frac{1}{5}x^{-4/5}$ **57.** 4, $\frac{1}{3}$ **59.** $a = 4$, $b = 1$ **61.** 1, 1.5; 2   **63.** $y - 5 = \frac{1}{2}(x - 4)$ **65.** $4x + 2h$
67. $-2x + 2 - h$ **69.** $3x^2 + 3xh + h^2$ **71.** $f'(x) = -2x$ **73.** $f'(x) = 14x + 1$ **75.** $f'(x) = 3x^2$
77. (a) and **(b) (c)** Parallel lines have equal slopes: Slope of the graph of $y = f(x)$ at the point $(x, f(x))$ is
equal to the slope of the graph of $y = f(x) + 3$ at the point $(x, f(x) + 3)$, which implies the given equation
79. .69315   **81.** .70711   **83.** .11111   **85.**

$[0, 4]$ by $[-5, 40]$

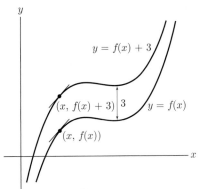

87. $y = \dfrac{x}{6} + \dfrac{3}{2}$ **89.** $y = -16x + 12$ **91.** 2   **93.** 4

Exercises 1.4, page 91

1. No limit   **3.** 1   **5.** No limit   **7.** -5 **9.** 5   **11.** No limit   **13.** 288   **15.** 0   **17.** 3   **19.** -4
21. -8 **23.** $\frac{6}{7}$ **25.** No limit   **27. (a)** 0   **(b)** $-\frac{3}{2}$ **(c)** $-\frac{1}{4}$ **(d)** -1 **29.** 6   **31.** 3

33. Step 1: $\dfrac{f(x + h) - f(x)}{h} = \dfrac{(x + h)^2 + 1 - (x^2 + 1)}{h}$ Step 2: $\dfrac{f(x + h) - f(x)}{h} = 2x + h$ Step 3: $f'(x) = 2x$.

35. Step 1: $\dfrac{f(x + h) - f(x)}{h} = \dfrac{(x + h)^3 - 1 - (x^3 - 1)}{h}$ Step 2: $\dfrac{f(x + h) - f(x)}{h} = 3x^2 + 3xh + h^2$

Step 3: $f'(x) = 3x^2$.   **37.** Steps 1, 2: $\dfrac{f(3 + h) - f(3)}{h} = 3$. Step 3: $f'(x) = \lim\limits_{h \to 0} 3 = 3$.

39. Steps 1, 2: $\dfrac{f(x + h) - f(x)}{h} = 1 + \dfrac{-1}{x(x + h)}$. Step 3: $f'(x) = \lim\limits_{h \to 0} 1 + \dfrac{-1}{x(x + h)} = 1 - \dfrac{1}{x^2}$.

41. Steps 1, 2: $\dfrac{f(x + h) - f(x)}{h} = \dfrac{1}{(x + 1)(x + h + 1)}$.

Step 3: $f'(x) = \lim\limits_{h \to 0} \dfrac{1}{(x + 1)(x + h + 1)} = \dfrac{1}{(x + 1)^2}$.

43. Steps 1, 2: $\dfrac{f(x + h) - f(x)}{h} = \dfrac{-2x - h}{((x + h)^2 + 1)(x^2 + 1)}$.

Step 3: $f'(x) = \lim\limits_{h \to 0} \dfrac{-2x - h}{((x+h)^2 + 1)(x^2 + 1)} = \dfrac{-2x}{(x^2 + 1)^2}.$

45. Steps 1, 2: $\dfrac{f(x+h) - f(x)}{h} = \dfrac{1}{\sqrt{x+h+2} + \sqrt{x+2}}.$

Step 3: $f'(x) = \lim\limits_{h \to 0} \dfrac{1}{\sqrt{x+h+2} + \sqrt{x+2}} = \dfrac{1}{2\sqrt{x+2}}.$

47. Steps 1, 2: $\dfrac{f(x+h) - f(x)}{h} = \dfrac{-1}{\sqrt{x}\sqrt{x+h}(\sqrt{x} + \sqrt{x+h})}.$

Step 3: $f'(x) = \lim\limits_{h \to 0} \dfrac{-1}{\sqrt{x}\sqrt{x+h}(\sqrt{x} + \sqrt{x+h})} = \dfrac{-1}{2x\sqrt{x}} = \dfrac{-1}{2x^{3/2}}.$ **49.** $f(x) = x^2; a = 1$

51. $f(x) = \dfrac{1}{x}; \; a = 10$ **53.** $f(x) = \sqrt{x}; \; a = 9$ **55.** Take $f(x) = x^2$, then the given limit is $f'(2) = 4.$

57. Take $f(x) = \sqrt{x}$, then the given limit is $f'(2) = \frac{1}{2\sqrt{2}}.$ **59.** Take $f(x) = x^{1/3}$, then the given limit is $f'(8) = \frac{1}{12}.$ **61.** 0   **63.** 0   **65.** 0   **67.** $\frac{3}{4}$ **69.** 0   **71.** 0   **73.** 0   **75.** .5

Exercises 1.5, page 98

1. No   **3.** Yes   **5.** No   **7.** No   **9.** Yes   **11.** No   **13.** Continuous, differentiable   **15.** Continuous, not differentiable   **17.** Continuous, not differentiable   **19.** Not continuous, not differentiable   **21.** $f(5) = 3$

23. Not possible   **25.** $f(0) = 12$ **27.** (a) $T(x) = \begin{cases} .15x & \text{for } 0 < x \le 27{,}050 \\ .275x - 3381.25 & \text{for } 27{,}050 < x \le 65{,}550 \\ .305x - 5347.75 & \text{for } 65{,}550 < x \le 136{,}750 \end{cases}$

(b)   (c) $T(65{,}550) - T(27{,}050) = 10{,}587.5$ dollars

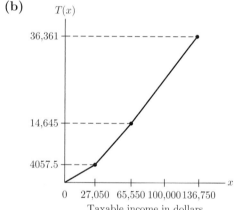

29. (a) $R(x) = \begin{cases} .07x + 2.5 & \text{for } 0 \le x \le 100 \\ .04x + 5.5 & \text{for } 100 < x \end{cases}$ (b) $P(x) = \begin{cases} .04x + 2.5 & \text{for } 0 \le x \le 100 \\ .01x + 5.5 & \text{for } 100 < x \end{cases}$ **31.** (a) 3

thousand dollars per hour   (b) 3 thousand dollars per hour, between 8 A.M and 10 A.M   **33.** $a = 1$

Exercises 1.6, page 106

1. $3x^2 + 2x$ **3.** $2(\frac{1}{2})x^{-1/2}$ or $\dfrac{1}{\sqrt{x}}$ **5.** $\dfrac{1}{2} - 2(-1)x^{-2}$ or $\dfrac{1}{2} + \dfrac{2}{x^2}$ **7.** $4x^3 + 3x^2 + 1$ **9.** $3(2x + 4)^2(2)$ or

$6(2x + 4)^2$ **11.** $7(x^3 + x^2 + 1)^6(3x^2 + 2x)$ **13.** $-\dfrac{8}{x^3}$ **15.** $3(\frac{1}{3})(2x^2 + 1)^{-\frac{2}{3}}(4x)$ or $4x(2x^2 + 1)^{-\frac{2}{3}}$

17. $2 + 3(x + 2)^2$ **19.** $\frac{1}{5}(-5)x^{-6}$ or $-\dfrac{1}{x^6}$ **21.** $(-1)(x^3 + 1)^{-2}(3x^2)$ or $-\dfrac{3x^2}{(x^3 + 1)^2}$ **23.** $1 - (x + 1)^{-2}$

25. $\dfrac{45x^2 + 5}{2\sqrt{3x^3 + x}}$ **27.** 3   **29.** $\frac{1}{2}(1 + x + x^2)^{-\frac{1}{2}}(1 + 2x)$ **31.** $10(1 - 5x)^{-2}$

33. $-45(1 + x + \sqrt{x})^{-2}(1 + \frac{1}{2}x^{-\frac{1}{2}})$ **35.** $1 + \frac{1}{2}(x + 1)^{-\frac{1}{2}}$ **37.** $\dfrac{3}{2}\left(\dfrac{\sqrt{x}}{2} + 1\right)^{1/2}\left(\dfrac{1}{4}x^{-1/2}\right)$ or

$\dfrac{3}{8\sqrt{x}}\left(\dfrac{\sqrt{x}}{2} + 1\right)^{1/2}$ **39.** 4   **41.** 15   **43.** $f'(4) = 48, \; y = 48x - 191$

45. (a) $y' = 2(3x^2 + x - 2) \cdot (6x + 1) = 36x^3 + 18x^2 - 22x - 4$ **(b)** $y = 9x^4 + 6x^3 - 11x^2 - 4x + 4$,
$y' = 36x^3 + 18x^2 - 22x - 4$ **47.** 4.8; 1.8   **49.** 14; 11   **51.** 10; $\frac{15}{4}$ **53.** $(5, \frac{161}{3})$; $(3, 49)$ **55.** $f(4) = 5$,
$f'(4) = \frac{1}{2}$ **57. (a)** $S(1) = 120.560$, $S'(1) = 1.5$.   **(b)** $S(3) = 80$, $S'(3) = -6$.
59. (a) $S(10) = \frac{372}{121} \approx 3.074$ thousand dollars, $S'(10) = -\frac{18}{113} \approx -.014$ thousand dollars per day.
(b) $S(11) \approx S(10) + S'(10) \approx 3.061$ thousand dollars. $S(11) = \frac{49}{16} = 3.0625$ thousand dollars.
61. (a) $A(8) = 12$, $A'(8) = .5$.   **(b)** $A(9) \approx A(8) + A'(8) = 12.5$. If the company spends $9000, it should
expect to sell 12,500 computers.   **63. (a)** $D(4) \approx 5.583$ trillion dollars, $D'(4) = .082$ or 82 billion dollars per
year.

Exercises 1.7, page 114

1. $10t(t^2 + 1)^4$ **3.** $8t + \frac{11}{2}t^{-\frac{1}{2}}$ **5.** $5T^4 - 16T^3 + 6T - 1$ **7.** $6P - \frac{1}{2}$ **9.** $2a^2t + b^2$ **11.** $y' = 1$, $y'' = 0$
13. $y' = \frac{1}{2}x^{-1/2}$, $y'' = -\frac{1}{4}x^{-3/2}$ **15.** $y' = \frac{1}{2}(x + 1)^{-\frac{1}{2}}$, $y'' = -\frac{1}{4}(x + 1)^{-\frac{3}{2}}$ **17.** $f'(r) = 2\pi r$, $f''(r) = 2\pi$
19. $f'(P) = 15(3P + 1)^4$, $f''(P) = 180(3P + 1)^3$ **21.** 36   **23.** 0   **25.** 34   **27.** $f'(1) = -\frac{1}{9}$, $f''(1) = \frac{2}{27}$
29. 0   **31.** 20   **33. (a)** $f'''(x) = 60x^2 - 24x$ **(b)** $f'''(x) = \dfrac{15}{2\sqrt{x}}$ **35. (a)** $2Tx + 3P$ **(b)** $3x$
(c) $x^2 + 2T$ **37.** When 50 bicycles are manufactured, the cost is $5000. For every additional bicycle
manufactured, there is an additional cost of $45.   **39. (a)** 2.60 dollars per unit   **(b)** 100 or 200 more units
41. (a) $R(12) = 22$, $R'(12) = .075$ **(b)** $P(x) = R(x) - C(x)$ so $P'(x) = R'(x) - C'(x)$. When 1200 chips are
produced, the marginal profit is $.75 - 1.5 = -.75$ dollars per chip.   **43.**

$[-4, 4]$ by $[-2, 2]$

Exercises 1.8, page 122

1. (a) 12; 10; 8.4   **(b)** 8   **3. (a)** 14   **(b)** 13   **5. (a)** $-\frac{7}{4}$ **(b)** $\frac{5}{6}$ **(c)** January 1, 1980   **7. (a)** 28
km/hour   **(b)** 96 km   **(c)** $\frac{1}{2}$hr   **9.** 63 units/hour   **11. (a)** $v(3) = 24$ ft/sec   **(b)** When $t > 5$ or when
$0 \le t < 2$.   **(c)** 131 ft.   **13. (a)** 160 ft/sec   **(b)** 96 ft/sec   **(c)** -32 ft/sec^2 **(d)** 10 sec
(e) -160 ft/sec   **15.** A–b; B–d; C–f; D–e; E–a; F–c; G–g   **17. (a)** 15 ft/sec   **(b)** No; positive velocity
indicates the object is moving away from the reference point   **(c)** 5 ft/sec   **19. (a)** 5010   **(b)** 5005
(c) 4990   **(d)** 4980   **(e)** 4997.5   **21.** Four minutes after it has been poured, the coffee is 120°. At that
time, its temperature is decreasing by 5°/ min; 119.5°   **23.** When the price of a car is $10,000, 200,000 cars are
sold. At that price, the number of cars sold decreases by 3 for each dollar increase in price.   **25.** When the
price is set at $1200, 60,000 computers are sold. For every $100 increase in price, the sales decrease by 2000
computers. 59,000 computers.   **27.** The profit from manufacturing and selling 100 luxury cars is $90,000.
Each additional car made and sold creates an additional profit of $1200; $88,800   **29. (a)** $C'(5) = 74$
thousand dollars per unit   **(b)** $C(5.25) \approx C(5) + C'(5)(.25) = 256.5$ thousand dollars   **(c)** $x = 4$
(d) $C'(4) = 62$, $R'(4) = 29$. If the production is increased by one unit, cost will rise by $62,000 and revenue will
increase by $29,000. The company should not increase production beyond the break-even point.
31. (a) $500 billion   **(b)** $50 billion/yr   **(c)** 1994   **(d)** 1994

33. (a)

$y = f(t)$

$y = f'(t)$

$[.5, 6]$ by $[-3, 3]$

(b) .85 sec   **(c)** 5 days   **(d)** $-.05$ sec/day   **(e)** 3 days

Chapter 1: Supplementary Exercises, page 127

1.

3.

5.

7.

9.

11.

13.

15. $7x^6 + 3x^2$ **17.** $\dfrac{3}{\sqrt{x}}$

19. $-\dfrac{3}{x^2}$ **21.** $48x(3x^2 - 1)^7$ **23.** $-\dfrac{5}{(5x-1)^2}$ **25.** $\dfrac{x}{\sqrt{x^2+1}}$ **27.** $-\dfrac{1}{4x^{5/4}}$ **29.** 0

31. $10[x^5 - (x-1)^5]^9[5x^4 - 5(x-1)^4]$ **33.** $\dfrac{3}{2}t^{-1/2} + \dfrac{3}{2}t^{-3/2}$ **35.** $\dfrac{2(9t^2-1)}{(t-3t^3)^2}$ **37.** $\dfrac{9}{4}x^{1/2} - 4x^{-1/3}$

39. 28   **41.** 14; 3   **43.** $\frac{15}{2}$ **45.** 33   **47.** $4x^3 - 4x$ **49.** $-\frac{3}{2}(1-3P)^{-1/2}$ **51.** 29   **53.** $300(5x+1)^2$

55. -2 **57.** $3x^{-1/2}$ **59.** Slope -4; tangent $y = -4x + 6$ **61.**

    63. $y = 2$

65. $f(2) = 3, f'(2) = -1$ **67.** 96 ft/s   **69.** 11 ft   **71.** $\frac{5}{3}$ ft/s   **73.** (a) \$16.10   (b) \$16   **75.** $\frac{3}{4}$in.
77. 4   **79.** Does not exist   **81.** $-\frac{1}{50}$ **83.** The slope of a secant line at (3, 9)

CHAPTER 2

Exercises 2.1, page 138

1. (a), (e), (f)   **3.** (b), (c), (d)   **5.** Increasing for $x < \frac{1}{2}$, relative maximum point at $x = \frac{1}{2}$, maximum value
$= 1$, decreasing for $x > \frac{1}{2}$, concave down, y-intercept $(0,0)$, x- intercepts $(0, 0)$ and $(1, 0)$.   **7.** Decreasing for
$x < 0$, relative minimum point at $x = 0$, relative minimum value $= 2$, increasing for $0 < x < 2$, relative maximum
point at $x = 2$, relative maximum value $= 4$, decreasing for $x > 2$, concave up for $x < 1$, concave down for $x > 1$,
inflection point at $(1,3)$, y-intercept $(0,2)$, x-intercept $(3.6,0)$ **9.** Decreasing for $x < 2$, relative minimum at
$x = 2$, minimum value $=3$, increasing for $x > 2$, concave up for all x, no inflection point, defined for $x > 0$, the
line $y = x$ is an asymptote, the y-axis is an asymptote   **11.** Decreasing for $1 \le x < 3.2$, relative minimum point
at $x = 3.2$, increasing for $x > 3.2$, maximum value $= 6$ (at $x = 1$), minimum value $= .9$ (at $x = 3.2$), inflection
point at $x = 4$, concave up for $1 \le x < 4$, concave down for $x > 4$, the line $y = 4$ is an asymptote.   **13.** Slope
decreases for all x.   **15.** Slope decreases for $x < 1$, increases for $x > 1$. Minimum slope occurs at $x = 1$.

17. (a) C, F (b) A, B, F (c) C **19.**

21.

23.

25.

27.

29. Oxygen content decreases until time a, at which time it reaches a minimum. After a, oxygen content steadily increases. The rate of increase increases until b, and then decreases. Time b is the time when oxygen content is increasing fastest.   **31.** 1960   **33.** The parachutist's speed levels off to 15 ft/sec

35.

37.

$y = \frac{3}{4}x + 5$

39. (a) Yes   (b) Yes   **41.** Relatively low

43. $x = 2$ **45.** $\frac{1}{6}$ unit apart

Exercises 2.2, page 147

1. (e)   **3.** (a), (b), (d), (e)   **5.** (d)   **7.**

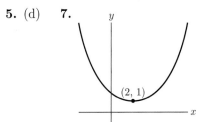
$(2, 1)$

9.

$(3, 5)$

11.

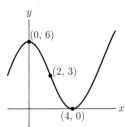
$(0, 6)$
$(2, 3)$
$(4, 0)$

13.

$(3, 4)$

15.

$(3, 1)$

17.

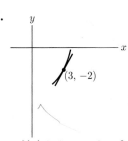
$(3, -2)$

19.

	f	f'	f''
A	POS	POS	NEG
B	0	NEG	0
C	NEG	0	POS

21. $t = 1$ **23.** (a) decreasing   (b) The function $f(x)$ is increasing for $1 \le x < 2$ because the values of $f'(x)$ are positive. The function $f(x)$ is decreasing for $2 < x \le 3$ because the values of $f'(x)$ are negative. Therefore, $f(x)$ has a relative maximum at $x = 2$. Coordinates: $(2, 9)$ **(c)** The function $f(x)$ is decreasing for $9 \leq x < 10$ because the values of $f'(x)$ are negative. The function $f(x)$ is increasing for $10 < x \leq 11$ because the values of $f'(x)$ are positive. Therefore, $f(x)$ has a relative minimum at $x = 10$. **(d)** concave down **(e)** at $x = 6$; coordinates: $(6, 5)$ **(f)** $x = 15$ **25.** The slope is positive because $f'(6) = 2$, a positive number. **27.** The slope is 0 because $f'(3) = 0$. Also, $f'(x)$ is positive for x slightly less than 3, and $f'(x)$ is negative for x slightly greater than 3. Hence $f(x)$ changes from increasing to decreasing at $x = 3$. **29.** $f'(x)$ is increasing at $x = 0$, so the graph of $f(x)$ is concave up. **31.** At $x = 1$, $f'(x)$ changes from increasing to decreasing, so the concavity of the graph of $f(x)$ changes from concave up to concave down. **33.** $y - 3 = 2(x - 6)$ **35.** 3.25 **37. (a)** $\frac{1}{6}$ in. **(b)** (ii), because the water level is falling. **39.** II **41.** I **43. (a)** 2 million **(b)** 30,000 farms per year **(c)** 1940 **(d)** 1945 and 1978 **(e)** 1960 **45.** rel. max: $x \approx -2.34$; rel. min: $x \approx 2.34$; inflection point: $x = 0$, $x \approx \pm 1.41$

Exercises 2.3, page 159

1. $f'(x) = 3(x + 3)(x - 3)$; relative maximum point $(-3, 54)$; relative minimum point $(3, -54)$.

Critical Values		-3		3	
	$x < -3$		$-3 < x < 3$		$3 < x$
$3(x + 3)$	$-$	0	$+$		$+$
$x - 3$	$-$		$-$	0	$+$
$f'(x)$	$+$	0	$-$	0	$+$
$f(x)$	Increasing on $(-\infty, -3)$	54	Decreasing on $(-3, 3)$	-54	Increasing on $(3, \infty)$

Local maximum $(-3, 54)$ Local minimum $(3, -54)$

3. $f'(x) = -3(x - 1)(x - 3)$; relative maximum point $(3, 1)$; relative minimum point $(1, -3)$.

Critical Values		1		3	
	$x < 1$		$1 < x < 3$		$3 < x$
$-3(x - 1)$	$+$	0	$-$		$-$
$x - 3$	$-$		$-$	0	$+$
$f'(x)$	$-$	0	$+$	0	$-$
$f(x)$	Decreasing on $(-\infty, 1)$	-3	Increasing on $(1, 3)$	1	Decreasing on $(3, \infty)$

Local minimum $(1, -3)$ Local maximum $(3, 1)$

5. $f'(x) = x(x - 2)$; relative maximum point $(0, 1)$; relative minimum point $(2, -1/3)$.

Critical Values		0		2	
	$x < 0$		$0 < x < 2$		$2 < x$
x	$-$	0	$+$		$+$
$x - 2$	$-$		$-$	0	$+$
$f'(x)$	$+$	0	$-$	0	$+$
$f(x)$	Increasing on $(-\infty, 0)$	1	Decreasing on $(0, 2)$	$-1/3$	Increasing on $(2, \infty)$

Local maximum $(0, 1)$ Local minimum $(2, -1/3)$

7. $f'(x) = -3x(x+8)$; relative maximum point $(0, -2)$; relative minimum point $(-8, -258)$.

Critical Values

Critical Values		-8		0	
		$x < -8$	$-8 < x < 0$		$0 < x$
$x + 8$		$-$	0 $+$		$+$
$-3x$		$+$	$+$	0	$-$
$f'(x)$		$-$	0 $+$	0	$-$
$f(x)$		Decreasing on $(-\infty, -8)$	-258 Increasing on $(-8, 0)$	-2	Decreasing on $(0, \infty)$

Local minimum $(-8, -258)$ Local maximum $(0, -2)$

9.

$(0, -8)$

11.

$\left(-1, -\frac{9}{2}\right)$

13.

$(3, 10)$

15.

$(-4, 6)$

17.

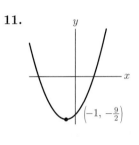

$(-3, 0)$

$(-1, -4)$

19.

$(-2, 16)$

$(2, -16)$

21.

$(9, 81)$

$(-3, -15)$

23.

$\left(4, -\frac{4}{3}\right)$

$(0, -12)$

25.

$(-1, 4)$

$(0, 2)$

$(1, 0)$

27.

$(2, 5)$

$(0, 1)$ $(1, 3)$

29.

$\left(-1, \frac{20}{3}\right)$

$\left(1, \frac{4}{3}\right)$

$(3, -4)$

31.

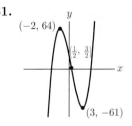

$(-2, 64)$

$\left(\frac{1}{2}, \frac{3}{2}\right)$

$(3, -61)$

33. No, $f''(x) = 2a \neq 0$ **35.** $(4, 3)$ min   **37.** $(1, 5)$ max   **39.** $(-.1, -3.05)$ min   **41.** $f(x) = g'(x)$
43. **(a)** f has a relative minimum   **(b)** f has an inflection point   **45.** **(a)** $A(x) = -893.103x + 460.759$ (billion dollars).   **(b)** Revenue is $x\%$ of assets or $R(x) = \frac{xA(x)}{100} = \frac{x}{100}(-893.103x + 460.759)$. $R(.3) \approx .578,484$ billion dollars or 578.484 million dollars; $R(.1) \approx .371, 449$ billion dollars or 371.449 million dollars.   **(c)** Maximum revenue when $R'(x) = 0$ or $x \approx .258$. Maximum revenue $R(.258) \approx .594,273$ or 594.273 million dollars.
47. $f'(x)$ is always nonnegative.   **49.** They both have a minimum point. The parabola does not have a vertical asymptote.

Exercises 2.4, page 165

1. $\left(\dfrac{3 \pm \sqrt{5}}{2}, 0\right)$ **3.** $(-2, 0)$, $\left(-\frac{1}{2}, 0\right)$ **5.** $\left(\frac{1}{2}, 0\right)$ **7.** The derivative $x^2 - 4x + 5$ has no zeros. No relative extreme points.

9.

11.

13.

15.

17.

19.

21.

23.

25.

27.

29.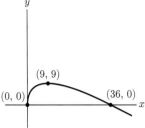

31. $g(x) = f'(x)$

33. $f(2) = 0$ implies $4a + 2b + c = 0$. Local maximum at $(0, 1)$ implies $f'(0) = 0$ and $f(0) = 1$. $a = -1/4$, $b = 0$, $c = 1$, $f(x) = -1/4x^2 + 1$.   **35.** If $f'(a) = 0$ and f' is increasing at $x = a$, then $f'(x) < 0$ for $x < a$ and $f'(x) > 0$ for $x > a$. By the first derivative test (case (b)), f has a local minimum at $x = a$.

37. (a)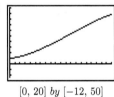

$[0, 20] \ by \ [-12, 50]$

(b) 15.0 g   **(c)** after 12.0 days   **(d)** 1.6 g/day
(e) after 6.0 days and after 17.6 days   **(f)** after 11.8 days

Exercises 2.5, page 172

1. 20   **3.** $t = 4$, $f(4) = 8$ **5.** $x = 1, y = 1$, maximum $= 1$ **7.** $x = 3, y = 3$, minimum $= 18$
9. $x = 6, y = 6$, minimum $= 12$ **11. (a)** Objective: $A = xy$, constraint: $8x + 4y = 320$
(b) $A = -2x^2 + 80x$ **(c)** $x = 20$ ft, $y = 40$ ft   **13. (a)**   **(b)** $h + 4x$ **(c)** Objective:

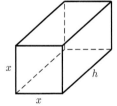

$V = x^2h$; constraint: $h + 4x = 84$ **(d)** $V = -4x^3 + 84x^2$ **(e)** $x = 14$ in., $h = 28$ in   **15.** Let x be the length of the fence and y the other dimension. Objective: $C = 15x + 20y$;
constraint: $xy = 75$; $x = 10$ ft, $y = 7.5$ ft   **17.** Let x be the length of each edge of the base and h the height.
Objective: $A = 2x^2 + 4xh$;
constraint: $x^2h = 8000$; 20 cm by 20 cm by 20 cm   **19.** Let x be the length of the fence parallel to the river
and y the length of each section perpendicular to the river. Objective: $A = xy$; constraint: $6x + 15y = 1500$;
$x = 125$ ft, $y = 50$ ft   **21.** Objective: $P = xy$; constraint: $x + y = 100$; $x = 50, y = 50$
23. Objective: $A = \dfrac{\pi x^2}{2} + 2xh$; constraint: $(2 + \pi)x + 2h = 14$; $x = \dfrac{14}{4 + \pi}$ ft

25. $w = 20$ ft, $x = 10$ ft

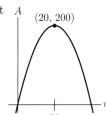
A
(20, 200)
20
w

27. (a) $C(x) = 6x + 10\sqrt{(20-x)^2 + 24^2};$

$$C'(x) = 6 - \frac{10(20-x)}{\sqrt{(20-x)^2 + 24^2}}; \ C'(x) = 0 \ (0 \le x \le 20) \text{ implies } x = 2. \text{ Use the first derivative test to conclude}$$

that the minimum cost is $C(2) = 312$ dollars.   **29.** $\left(\frac{3}{2}, \sqrt{\frac{3}{2}}\right)$ **31.** $x = 2$, $y = 1$

Exercises 2.6, page 181

1. (a) 90   **(b)** 180   **(c)** 6   **(d)** 1080 pounds   **3. (a)** $C = 16r + 2x$ **(b)** Constraint $rx = 800$
(c) $x = 80$, $r = 10$, minimum inventory cost $= \$320$ **5.** Let x be the number of cases per order and r the
number of orders per year. Objective: $C = 80r + 5x$; constraint: $rx = 10{,}000$ **(a)** $\$4100$ **(b)** 400 cases
7. Let r be the number of production runs and x the number of microscopes manufactured per run. Objective:
$C = 2500r + 25x$; constraint: $rx = 1600$; 4 runs   **11.** Objective: $A = (100 + x)w$; constraint: $2x + 2w = 300$;
$x = 25$ ft, $w = 125$ ft   **13.** Objective: $F = 2x + 3w$; constraint: $xw = 54$; $x = 9$ m, $w = 6$ m
15. (a) $A(x) = 100x + 1000$.   **(b)** $R(x) = A(x) \cdot (\text{Price}) = (100x + 1000)(18 - x)$ $(0 \le x \le 18)$. The graph of
$R(x)$ is a parabola looking downward, with a maximum at $x = 4$.   **(c)** $A(x)$ does not change,
$R(x) = (100x + 1000)(9 - x)$ $(0 \le x \le 9)$. Maximum value when $x = 0$.   **17.** Let x be the length of each edge
of the base and h the height. Objective: $C = 6x^2 + 10xh$; constraint: $x^2h = 150$; 5 ft by 5 ft by 6 ft   **19.** Let x
be the length of each edge of the end and h the length. Objective: $V = x^2h$; constraint: $2x + h = 120$; 40 cm by
40 cm by 40 cm   **21.** Objective: $V = w^2x$; constraint: $2x + w = 16$; $\frac{8}{3}$ in.   **23.** After 20 days
25. $2\sqrt{3}$ by 6   **27.** 10 in. by 10 in. by 4 in.   **29.** ≈ 3.77 cm

Exercises 2.7, page 190

1. $\$1$ **3.** 32   **5.** 5   **7.** $x = 20$ units, $p = \$133.33$ **9.** 2 million tons, $\$156$ per ton   **11. (a)** $\$2.00$
(b) $\$2.30$ **13.** Let x be the number of prints and p the price per print. Demand equation: $p = 650 - 5x$;
revenue: $R(x) = (650 - 5x)x$; 65 prints   **15.** Let x be the number of tables and p the profit per table.
$p = 16 - .5x$; profit from the cafe: $R = (16 - .5x)x$; 16 tables   **17. (a)** $x = 15 \cdot 10^5$, $p = \$45$.   **(b)** No. Profit
is maximized when price is increased to $\$50$.   **19.** 5%   **21. (a)** $\$75{,}000$ **(b)** $\$3200$ per unit   **(c)** 15 units
(d) 32.5 units   **(e)** 35 units

Chapter 2: Supplementary Exercises, page 194

1. (a) increasing: $-3 < x < 1, x > 5$; decreasing: $x < -3, 1 < x < 5$ **(b)** concave up: $x < -1, x > 3$; concave
down: $-1 < x < 3$ **3.**   **5.**   **7.** d, e   **9.** c, d   **11.** e   **13.** Graph

y
$y = f(x)$
x

y
$y = f(x)$
x

goes through $(1, 2)$, increasing at $x = 1$.   **15.** Increasing and concave up at $x = 3$.   **17.** $(10, 2)$ is a relative
minimum point.   **19.** Graph goes through $(5, -1)$, decreasing at $x = 5$.   **21. (a)** after 2 hours   **(b)** .8

(c) after 3 hours   **(d)** −.02 units per hour   **23.**

25.

27.

29.

31.

33.

35.

37.

39.

41.

43.

44.

45. $f'(x) = 3x(x^2 + 2)^{1/2}, f'(0) = 0$ **47.** $f''(x) = -2x(1 + x^2)^{-2}, f''(x)$ is positive for $x < 0$ and negative for $x > 0$.   **49.** A–c, B–e, C–f, D–b, E–a, F–d   **51. (a)** The number of people living between $10 + h$ and 10 mi from the center of the city.   **(b)** If so, $f(x)$ would be decreasing at $x = 10$ **53.** The endpoint maximum value of 2 occurs at $x = 0$.   **55.** Let x be the width and h the height. Objective: $A = 4x + 2xh + 8h$; constraint: $4xh = 200$; 4 ft by 10 ft by 5 ft   **57.** $\frac{15}{2}$ in.   **59.** Let r be the number of production runs and x the number of books manufactured per run. Objective: $C = 1000r + (.25)x$; constraint: $rx = 400{,}000$; $x = 40{,}000$ **61.** $x = 3500$ **63.** Let x be the number of people and c the cost. Objective: $R = xc$; constraint: $c = 1040 - 20x$; 25 people.

CHAPTER 3

Exercises 3.1, page 205

1. $(x + 1) \cdot (3x^2 + 5) + (x^3 + 5x + 2) \cdot 1$, or $4x^3 + 3x^2 + 10x + 7$

3. $(2x^4 - x + 1) \cdot (-5x^4) + (-x^5 + 1) \cdot (8x^3 - 1)$, or $-18x^8 + 6x^5 - 5x^4 + 8x^3 - 1$

5. $x(4)(x^2 + 1)^3(2x) + (x^2 + 1)^4(1)$ or $(x^2 + 1)^3(9x^2 + 1)$ **7.** $(x^2 + 3) \cdot 10(x^2 - 3)^9(2x) + (x^2 - 3)^{10} \cdot 2x$, or $2x(x^2 - 3)^9(11x^2 + 27)$ **9.** $(5x + 1) \cdot 2x + (x^2 - 1) \cdot 5 + \frac{2}{3}$, or $15x^2 + 2x - \frac{13}{3}$ **11.** $\dfrac{(x + 1) \cdot 1 - (x - 1) \cdot 1}{(x + 1)^2}$, or $\dfrac{2}{(x + 1)^2}$ **13.** $\dfrac{(x^2 + 1) \cdot 2x - (x^2 - 1) \cdot 2x}{(x^2 + 1)^2}$ or $\dfrac{4x}{(x^2 + 1)^2}$ **15.** $\dfrac{(2x + 1)^2 - 2(2x + 1)(2)(x + 3)}{(2x + 1)^4}$ or $\dfrac{-2x - 11}{(2x + 1)^3}$ **17.** $\dfrac{-4x}{(x^2 + 1)^2}$ **19.** $\dfrac{(3 - x^2) \cdot (6x + 5) - (3x^2 + 5x + 1) \cdot (-2x)}{(3 - x^2)^2}$, or $\dfrac{5(x + 1)(x + 3)}{(3 - x^2)^2}$ **21.** $2[(3x^2 + 2x + 2)(x - 2)][(3x^2 + 2x + 2) \cdot 1 + (x - 2) \cdot (6x + 2)]$, or $2(3x^2 + 2x + 2)(x - 2)(9x^2 - 8x - 2)$

23. $\dfrac{-1}{2\sqrt{x}(\sqrt{x}+1)^2}$ **25.** $\dfrac{3x^4 - 4x^2 - 3}{x^2}$ **27.** $\sqrt{x+2}\,2\,(2x+1)(2) + (2x+1)^2\dfrac{1}{2\sqrt{x+2}}$ or $\dfrac{(2x+1)(10x+17)}{2\sqrt{x+2}}$

29. $y - 16 = 88(x-3)$ **31.** $2, 7$ **33.** $0, \pm 2, \pm\frac{5}{4}$ **35.** $(\frac{1}{2}, \frac{3}{2}), (-\frac{1}{2}, \frac{9}{2})$ **37.** $\dfrac{dy}{dx} = 8x(x^2+1)^3$,

$\dfrac{d^2y}{dx^2} = 8(x^2+1)^2(7x^2+1)$ **39.** $\dfrac{dy}{dx} = \dfrac{x}{2\sqrt{x+1}} + \sqrt{x+1}$ or $\dfrac{dy}{dx} = \dfrac{3x+2}{2\sqrt{x+1}}, \dfrac{d^2y}{dx^2} = \dfrac{3x+4}{4(x+1)^{3/2}}$

41. $x \cdot f'(x) + f(x)$ **43.** $\dfrac{(x^2+1) \cdot f'(x) - 2x \cdot f(x)}{(x^2+1)^2}$ **45.** $2 \times 3 \times 1$ **47.** $150; AC(150) = 35 = C'(150)$

49. AR is maximized where $0 = \dfrac{d}{dx}(AR) = \dfrac{x \cdot R'(x) - R(x) \cdot 1}{x^2}$. This happens when the production level x

satisfies $xR'(x) - R(x) = 0$, and hence $R'(x) = R(x)/x = AR$.   **51.** 38 in.^{2}/s   **53.** 150,853,600 gal/yr

55. $(2, 10)$ **59.** $\dfrac{1 - 2x \cdot f(x)}{(1+x^2)^2}$ **61.** $\frac{1}{8}$ **63.** (b) $f'(x) \cdot g'(x) = -\dfrac{1}{x^2}(3x^2) = -3; [f(x)g(x)]' = 2x$

65. $f(x)g(x)h'(x) + f(x)g'(x)h(x) + f'(x)g(x)h(x)$ **67.** $\dfrac{h(t)w'(t) - 2w(t)h'(t)}{[h(t)]^3}$ **69.** (a)

$[0, 6]$ by $[-5, 20]$

(b) 10.8 mm^2 (c) 2.61 units of light
(d) $-.55$ mm^2/unit of light

Exercises 3.2, page 212

1. $\dfrac{x^3}{x^3+1}$ **3.** $\sqrt{x}(x+1)$ **5.** $f(x) = x^5, g(x) = x^3 + 8x - 2$ **7.** $f(x) = \sqrt{x}, g(x) = 4 - x^2$ **9.** $f(x) = \dfrac{1}{x}$,

$g(x) = x^3 - 5x^2 + 1$ **11.** $30x(x^2+5)^{14}$ **13.** $6x^2 \cdot 3(x-1)^2(1) + (x-1)^3 \cdot 12x$, or $6x(x-1)^2(5x-2)$
15. $2(x^3-1) \cdot 4(3x^2+1)^3(6x) + (3x^2+1)^4 \cdot 2(3x^2)$, or $6x(3x^2+1)^3(11x^3+x-8)$

17. $\dfrac{d}{dx}[4^3(1-x)^{-3}] = 192(1-x)^{-4}$ **19.** $3\left(\dfrac{4x-1}{3x+1}\right)^2 \cdot \dfrac{(3x+1) \cdot 4 - (4x-1) \cdot 3}{(3x+1)^2}$, or $\dfrac{21(4x-1)^2}{(3x+1)^4}$

21. $2xf'(x^2)$ **23.** $f'(-x)$ **25.** $\dfrac{2x^2 f'(x^2) - f(x^2)}{x^2}$ **27.**   **29.** $30(6x-1)^4$

31. $-(1-x^2)^{-2} \cdot (-2x)$, or $2x(1-x^2)^{-2}$ **33.** $[4(x^2-4)^3 - 2(x^2-4)] \cdot (2x)$, or $8x(x^2-4)^3 - 4x(x^2-4)$
35. $2[(x^2+5)^3+1]3(x^2+5)^2 \cdot (2x)$, or $12x[(x^2+5)^3+1](x^2+5)^2$ **37.** $6(4x+1)^{1/2}$

39. $\dfrac{dy}{dx} = \left(\dfrac{1}{2} - \dfrac{2}{u^2}\right)(1-2x)$ or $\dfrac{dy}{dx} = \left(\dfrac{1}{2} - \dfrac{2}{(x-x^2)^2}\right) \cdot (1-2x)$ **41.** $\dfrac{dy}{dt} = (2x-3)(2t)$ when $t = 0, x = 3$,

$\dfrac{dy}{dt} = 0$ **43.** $\dfrac{dy}{dt} = \dfrac{-2}{(x-1)^2} \cdot \dfrac{t}{2}$ when $t = 3, x = \dfrac{9}{4}, \dfrac{dy}{dt} = -\dfrac{48}{25}$ **45.** $y = 62x - 300$ **47.** $1; 2; 3$

49. (a) $\dfrac{dV}{dt} = \dfrac{dV}{dx} \cdot \dfrac{dx}{dt}$ (b) 2   **51.** (a) $\dfrac{dy}{dt}, \dfrac{dP}{dy}, \dfrac{dP}{dt}$ (b) $\dfrac{dP}{dt} = \dfrac{dP}{dy} \cdot \dfrac{dy}{dt}$ **53.** (a) $\dfrac{200(100-x^2)}{(100+x^2)^2}$

(b) $\dfrac{400[100 - (4+2t)^2]}{[100 + (4+2t)^2]^2}$ (c) Falling at the rate of \$480 per week   **55.** (a) $.4 + .0002x$ (b) increasing at

the rate of 25 thousand persons per year   (c) rising at the rate of 14 ppm per year   **57.** $x^3 + 1$ **59.** 24
61. (a) $t = 1.5, x = 40, W \approx 77.209$ million dollars; $t = 3.5, x = 30, W \approx 76.364$ million dollars

(b) $\dfrac{dx}{dt}\bigg|_{t=1.5} = 20$; the price of one share is \$40 and is rising at the rate of \$20 per month. $\dfrac{dx}{dt}\bigg|_{t=3.5} = 0$; the price of one share is steady at $30 a share. **63. (a)** $t = 2.5$, $x = 40$, $\left.\dfrac{dx}{dt}\right|_{t=2.5} = -20$; the price of one share is

$40 and is falling at the rate of $20 per month. $t = 4$, $x = 30$, $\left.\dfrac{dx}{dt}\right|_{t=4} = 0$; the price of one share is steady at $30

a share. **(b)** $\left.\dfrac{dW}{dt}\right|_{t=2.5} = \left.\dfrac{dW}{dx}\right|_{x=40} \left.\dfrac{dx}{dt}\right|_{t=2.5} = \dfrac{120}{1849} \cdot (-20) \approx -1.3$; the total value of the company is falling at

the rate of 1.3 million dollars per month. $\dfrac{dW}{dt} = \dfrac{dW}{dx}\dfrac{dx}{dt} = 0$; when $x = 30$, the total value of the company was

steady at $W \approx 76.364$ million dollars. **65.** The derivative of the composite function $f(g(x))$ is the derivative

of the outer function evaluated at the inner function and then multiplied by the derivative of the inner function.

Exercises 3.3, page 222

1. $\dfrac{x}{y}$ **3.** $\dfrac{1 + 6x}{5y^4}$ **5.** $\dfrac{2x^3 - x}{2y^3 - y}$ **7.** $\dfrac{1 - 6x^2}{1 - 6y^2}$ **9.** $-\dfrac{y}{x}$ **11.** $-\dfrac{y+2}{5x}$ **13.** $\dfrac{8 - 3xy^2}{2x^2 y}$ **15.** $\dfrac{x^2(y^3 - 1)}{y^2(1 - x^3)}$

17. $-\dfrac{y^2 + 2xy}{x^2 + 2xy}$ **19.** $\frac{1}{2}$ **21.** $-\frac{8}{3}$ **23.** $-\frac{2}{15}$ **25.** $y - \frac{1}{2} = -\frac{1}{16}(x - 4), y + \frac{1}{2} = \frac{1}{16}(x - 4)$

27. (a) $\dfrac{2x - x^3 - xy^2}{2y + y^3 + x^2 y}$ **(b)** 0 **29. (a)** $\dfrac{-y}{2x}$ **(b)** $\dfrac{-27}{16}$ **31.** $-\dfrac{x^3}{y^3}\dfrac{dx}{dt}$ **33.** $\dfrac{2x - y}{x}\dfrac{dx}{dt}$

35. $\dfrac{2x + 2y}{3y^2 - 2x}\dfrac{dx}{dt}$ **37.** $-\frac{15}{8}$ units per second **39.** Rising at 3 thousand units per week **41.** Increasing at

$20 thousand per month **43.** Decreasing at $\frac{1}{14}$ L per second **45. (a)** $x^2 + y^2 = 100$ **(b)** $\dfrac{dy}{dt} = -4$, so the

top of the ladder is falling at the rate of 4 ft/sec. **47.** Decreasing at $\dfrac{22}{\sqrt{5}}$ ft/sec (or 9.84 ft/sec)

Chapter 3: Supplementary Exercises, page 225

1. $(4x - 1) \cdot 4(3x + 1)^3(3) + (3x + 1)^4 \cdot 4$, or $4(3x + 1)^3(15x - 2)$ **3.** $x \cdot 3(x^5 - 1)^2 \cdot 5x^4 + (x^5 - 1)^3 \cdot 1$, or
$(x^5 - 1)^2(16x^5 - 1)$ **5.** $5(x^{1/2} - 1)^4 \cdot 2(x^{1/2} - 2)(\frac{1}{2}x^{-1/2}) + (x^{1/2} - 2)^2 \cdot 20(x^{1/2} - 1)^3(\frac{1}{2}x^{-1/2})$, or
$5x^{-1/2}(x^{1/2} - 1)^3(x^{1/2} - 2)(3x^{1/2} - 5)$ **7.** $3(x^2 - 1)^3 \cdot 5(x^2 + 1)^4(2x) + (x^2 + 1)^5 \cdot 9(x^2 - 1)^2(2x)$, or
$12x(x^2 - 1)^2(x^2 + 1)^4(4x^2 - 1)$ **9.** $\dfrac{(x - 2) \cdot (2x - 6) - (x^2 - 6x) \cdot 1}{(x - 2)^2}$, or $\dfrac{x^2 - 4x + 12}{(x - 2)^2}$

11. $2\left(\dfrac{3 - x^2}{x^3}\right) \cdot \dfrac{x^3 \cdot (-2x) - (3 - x^2) \cdot 3x^2}{x^6}$, or $\dfrac{2(3 - x^2)(x^2 - 9)}{x^7}$ **13.** $-\frac{1}{3}, 3, \frac{31}{27}$ **15.** $y + 32 = 176(x + 1)$

17. $x = 44$ m, $y = 22$ m **19.** $\dfrac{dC}{dt} = \dfrac{dC}{dx} \cdot \dfrac{dx}{dt} = 40 \cdot 3 = 120$. Costs are rising $120 per day. **21.** $0; -\frac{7}{2}$

23. $\frac{3}{2}; -\frac{7}{8}$ **25.** $1; -\frac{3}{2}$ **27.** $\dfrac{3x^2}{x^6 + 1}$ **29.** $\dfrac{2x}{(x^2 + 1)^2 + 1}$ **31.** $\frac{1}{2}\sqrt{1 - x}$ **33.** $\dfrac{3x^2}{2(x^3 + 1)}$

35. $-\dfrac{25}{x(25 + x^2)}$ **37.** $\dfrac{x^{1/2}}{(1 + x^2)^{1/2}} \cdot \frac{1}{2}(x^{-1/2})$, or $\dfrac{1}{2\sqrt{1 + x^2}}$ **39. (a)** $\dfrac{dR}{dA}, \dfrac{dA}{dt}, \dfrac{dR}{dx}$, and $\dfrac{dx}{dA}$

(b) $\dfrac{dR}{dt} = \dfrac{dR}{dx}\dfrac{dx}{dA}\dfrac{dA}{dt}$ **41. (a)** $-y^{1/3}/x^{1/3}$ **(b)** 1 **43.** -3 **45.** $\frac{3}{5}$ **47. (a)** $\dfrac{dy}{dx} = \dfrac{15x^2}{2y}$ **(b)** $\frac{20}{3}$

thousand dollars per thousand unit increase in production **(c)** $\dfrac{dy}{dt} = \dfrac{15x^2}{2y}\dfrac{dx}{dt}$ **(d)** 2 thousand dollars per

week **49.** Increasing at the rate of 2.5 units per unit time **51.** 1.89 m²/year

CHAPTER 4

Exercises 4.1, page 233

1. $2^{2x}, 3^{(1/2)x}, 3^{-2x}$ **3.** $2^{2x}, 3^{3x}, 2^{-3x}$ **5.** $2^{-4x}, 2^{9x}, 3^{-2x}$ **7.** $2^x, 3^x, 3^x$ **9.** $3^{2x}, 2^{6x}, 3^{-x}$ **11.** $2^{(1/2)x}$, **13.** $2^{-2x}, 3^x$ **15.** $\frac{1}{9}$ **17.** 1 **19.** 2 **21.** -1 **23.** $\frac{1}{5}$ **25.** $\frac{5}{2}$ **27.** -1 **29.** 4 **31.** 1

33. 1 or 2   **35.** 1 or 2   **37.** 2^h **39.** $2^h - 1$ **41.** $3^x + 1$ **43.** 0.6931   **45.** 2.7

$[-1, 2]$ by $[-1, 4]$

Exercises 4.2, page 237

1. 1.16, 1.11, 1.10   **3.** (a) $2m$ (b) $m/4$ where $m \approx .693$ **5.** (a) e (b) $1/e$ **7.** 1.005, 1.001, 1.000
9. $10e^{10x}$ **11.** (a) $4e^{4x}$ (b) ke^{kx} **13.** e^{2x}, e^{-x} **15.** e^{-6x}, e^{2x} **17.** e^{6x}, e^{2x} **19.** $x = 4$
21. $x = 4, -2$ **23.** 1 or -1 **25.** $3e^x - 7$ **27.** $xe^x + e^x = (x+1)e^x$
29. $8e^x(2)(1 + 2e^x)(2e^x) + 8e^x(1 + 2e^x)^2$ or $8e^x(6e^x + 1)(1 + 2e^x)$ **31.** $\dfrac{e^x(x+1) - e^x}{(x+1)^2}$ or $\dfrac{xe^x}{(x+1)^2}$
33. $\dfrac{e^x(e^x + 1) - e^x(e^x - 1)}{(e^x + 1)^2}$ or $\dfrac{2e^x}{(e^x + 1)^2}$ **35.** $\dfrac{e^x}{\sqrt{2e^x + 1}}$ **37.** $y' = e^x(1 + x)^2$, $y' = 0$ when $x = -1$. The
point is $(-1, 2/e)$.   **39.** 1   **41.** $y - \frac{1}{3} = \frac{1}{9}x$ **43.** $f'(x) = e^x(x^2 + 4x + 3)$, $f''(x) = e^x(x^2 + 6x + 7)$
45. (a) 800 g/cm^2 (b) 14 km   (c) -50 g/cm^2 per km   (d) 2 km   **47.** $y = x + 1$
49.   **51.** $\left.\dfrac{d}{dx}(10^x)\right|_{x=0} \approx 2.3026$; $\dfrac{d}{dx}(10^x) = m10^x$ where $m = \left.\dfrac{d}{dx}(10^x)\right|_{x=0}$

$[-1, 3]$ by $[-3, 20]$

Exercises 4.3, page 242

1. $8e^{2x}$ **3.** $4 - 2e^{-2x}$ **5.** $e^{t^2}(2t^2 + 2t + 1)$ **7.** $3(e^x + e^{-x})^2(e^x - e^{-x})$ **9.** $\frac{1}{8}e^{\frac{t}{2}+1}$ **11.** $-\frac{3}{t^2}e^{\frac{3}{t}}$
13. $e^{x^2-5x+4}(2x-5)$ **15.** $\dfrac{1 - x + e^{2x}}{e^x}$ **17.** $2e^{2t} - 4e^{-4t}$ **19.** $2(t+1)e^{2t} + 2(t+1)^2e^{2t} = 2(t+1)(t+2)e^{2t}$
21. $e^x e^{e^x}$ **23.** $\dfrac{(15 + 4x)e^{4x}}{(4 + x)^2}$ **25.** $(-x^{-2} + 2x^{-1} + 6)e^{2x}$ **27.** Max at $x = -2/3$ **29.** Min at $x = 5/4$
31. Max at $x = 9/10$ **33.** \$54,366 per year   **35.** (a) 45 m/sec   (b) 10 m/sec^2 (c) 4 sec   (d) 4 sec
37. 2 in./week   **39.** $.02e^{-2e^{-.01x}}e^{-.01x}$ **41.** $y = Ce^{-4x}$ **43.** $y = e^{-.5x}$ **47.** 1   **49.** (a) Volume seems
to stabilize near 6 ml   (b) 3.2 ml   (c) 7.7 wk   (d) .97 ml/wk   (e) 3.7 wk   (f) 1.13 ml/wk

Exercises 4.4, page 247

1. $\frac{1}{2}$ **3.** $\ln 5$ **5.** $\frac{1}{e}$ **7.** -3 **9.** e **11.** 0   **13.** x^2 **15.** $\frac{1}{49}$ **17.** $2x$ **19.** $\frac{1}{2}\ln 5$ **21.** $4 - e^{1/2}$
23. $\pm e^{9/2}$ **25.** $-\dfrac{\ln .5}{.00012}$ **27.** $\frac{5}{3}$ **29.** $\dfrac{e}{3}$ **31.** $3\ln\frac{9}{2}$ **33.** $\frac{1}{2}e^{8/5}$ **35.** $\frac{1}{2}\ln 4$ **37.** $-\ln\frac{3}{2}$
39. $x = \ln 5$, $y = 5(1 - \ln 5) \approx -3.047$ **41.** (a) $y' = -e^{-x}$, $y' = -2$ when $x = -\ln 2 = \ln(\frac{1}{2})$. The point is
$(\ln(\frac{1}{2}), 2)$.   (b)   **43.** $(-\ln 3, 3 - 3\ln 3)$, minimum   **45.** Max at $t = 2\ln 51$

$y = e^{-x}$

Slope $= -2$

$-\ln 2 \approx -.69$

47. $\ln 2$ **49.** The graph of $y = e^{\ln x}$ is the same as the graph of $y = x$ for $x > 0$.

Answers

51. $\frac{1}{5}e^2 \approx 1.4778$

$[-1, 3]$ by $[-3, 3]$

Exercises 4.5, page 251

1. $\frac{1}{x}$ **3.** $\frac{1}{x+3}$ **5.** $\frac{e^x - e^{-x}}{e^x + e^{-x}}$ **7.** $\left(\frac{1}{x}+1\right)e^{\ln x + x}$ **9.** $\frac{1}{x}(\ln x + \ln 2x)$ **11.** $\frac{2\ln x}{x} + \frac{2}{x}$ or $\frac{2}{x}(\ln x + 1)$

13. $\frac{1}{x}$ **15.** $2x(1 + \ln(x^2 + 1))$ **17.** $\frac{2}{x^2 - 1}$ **19.** $\frac{5e^{5x}}{e^{5x} + 1}$ **21.** $3 + 2\ln t$ **23. (a)** $t > 1$.   **(b)** $t > e$.

25. $y = 1$ **27.** $\left(e^2, \frac{2}{e}\right)$ **29.** $y' = x(1 + 2\ln x)$, $y'' = 3 + 2\ln x$, $y' = 0$ when $x = e^{-1/2}$, $y''(e^{-1/2}) > 0$,

relative minimum at $x = e^{-1/2}$, $y = -e^{-1}/2$.   **31.** $1 + 3\ln 10$ **33.** $R'(x) = \frac{45(\ln x - 1)}{(\ln x)^2}$; $R'(20) \approx 10$

35. $\frac{1}{7}$ **37.**

$[-5, 5]$ by $[-2, 2]$

Exercises 4.6, page 255

1. $\ln 5x$ **3.** $\ln 3$ **5.** $\ln 2$ **7.** x^2 **9.** $\ln \frac{x^5 z^3}{y^{1/2}}$ **11.** $3\ln x$ **13.** $3\ln 3$ **15. (a)** $2\ln 2 = 2(.69) = 1.38$

(b) $\ln 2 + \ln 3 = 1.79$ **(c)** $3\ln 3 + \ln 2 = 3.99$ **17. (a)** $-\ln 2 - \ln 3 = -1.79$ **(b)** $\ln 2 - 2\ln 3 = -1.51$

(c) $-\frac{1}{2}\ln 2 = -.345$ **19.** d   **21.** d   **23.** 3   **25.** $\sqrt{e}$ **27.** e or e^{-1} **29.** 1 or e^4 **31.** $\frac{1 + 2e}{e - 1}$

33. $\frac{1}{x+5} + \frac{2}{2x-1} - \frac{1}{4-x}$ **35.** $\frac{2}{1+x} + \frac{3}{2+x} + \frac{4}{3+x}$ **37.** $\frac{1}{2x} + x$ **39.** $\frac{4}{x+1} - 1$

41. $\ln(3x+1)\frac{5}{5x+1} + \ln(5x+1)\frac{3}{3x+1}$ **43.** $(x+1)^4(4x-1)^2\left(\frac{4}{x+1} + \frac{8}{4x-1}\right)$

45. $\frac{(x+1)(2x+1)(3x+1)}{\sqrt{4x+1}}\left(\frac{1}{x+1} + \frac{2}{2x+1} + \frac{3}{3x+1} - \frac{2}{4x+1}\right)$ **47.** $2^x \ln 2$ **49.** $x^x[1 + \ln x]$

51. $y = cx^k$ **53.** $h = 3$, $k = \ln 2$

Chapter 4: Supplementary Exercises, page 257

1. 81   **3.** $\frac{1}{25}$ **5.** 4   **7.** 9   **9.** e^{3x^2} **11.** e^{2x} **13.** $e^{11x} + 7e^x$ **15.** $x = 4$ **17.** $x = -5$ **19.** $70e^{7x}$

21. $e^{x^2} + 2x^2 e^{x^2}$ **23.** $e^x \cdot e^{e^x} = e^{x+e^x}$ **25.** $\frac{(e^{3x}+3)(2x-1) - (x^2 - x + 5)3e^{3x}}{(e^{3x}+3)^2}$ **27.** $y = Ce^{-x}$

29. $y = 2e^{1.5x}$ **31.**   **33.**   **35.**

$\frac{1}{4}x + \frac{1}{2}$ **39.** $\sqrt{5}$ **41.** $\frac{2}{3}$ **43.** 1   **45.** $e, \frac{1}{e}$ **47.** $\frac{1}{2}\ln 5$ **49.** $e^{5/2}$ **51.** $\frac{6x^5 + 12x^3}{x^6 + 3x^4 + 1}$

53. $\dfrac{5}{5x-7}$ **55.** $\dfrac{2\ln x}{x}$ **57.** $\dfrac{1}{x}+1-\dfrac{1}{2(1+x)}$ **59.** $\ln x$ **61.** $\dfrac{1}{x\ln x}$ **63.** $\dfrac{e^x}{x}+e^x\ln x$

65. $\dfrac{x}{x^2+1}-\dfrac{1}{2x+3}$ **67.** $2x-\dfrac{1}{x}$ **69.** $\ln 2$ **71.** $\dfrac{1}{x-1}$ **73.** $-\tfrac{1}{2}x^{-\frac{1}{2}}$ or $\dfrac{-1}{2\sqrt{x}}$

75. $\sqrt[5]{\dfrac{x^5+1}{x^5+5x+1}}\left[\dfrac{x^4}{x^5+1}-\dfrac{x^4+1}{x^5+5x+1}\right]$ **77.** $x^{\sqrt{x}-\frac{1}{2}}[1+\tfrac{1}{2}\ln x]$

79. $(x^2+5)^6(x^3+7)^8(x^4+9)^{10}\left[\dfrac{12x}{x^2+5}+\dfrac{24x^2}{x^3+7}+\dfrac{40x^3}{x^4+9}\right]$ **81.** $10^x\ln 10$ **83.** $\dfrac{1}{2}\sqrt{\dfrac{xe^x}{x^3+3}}\left[\dfrac{1}{x}+1-\dfrac{3x^2}{x^3+3}\right]$

85. $e^{x+1}(x^2+1)x\left[1+\dfrac{2x}{x^2+1}+\dfrac{1}{x}\right]$ or $e^{x+1}(x^3+3x^2+x+1)$ **87.**

$(0, \ln 2)$

89.

$(0, \ln 2)$

91.

$(e, 1)$
$(1, 0)$

93.

$(1, 1)$

95. $\dfrac{e^x}{3+e^x}$ **97.** 1

CHAPTER 5

Exercises 5.1, page 266

1. (a) $P(t)=3e^{.02t}$ **(b)** 3 million   **(c)** .02   **(d)** 3.52 million   **(e)** 80,000 people per year   **(f)** 3.5 million   **3. (a)** 5000   **(b)** $P'(t)=.2P(t)$ **(c)** 3.5 h   **(d)** 6.9 h   **5.** .017   **7.** 22 yr   **9.** 27 million cells **11.** 34.0 million   **13. (a)** $P(t)=8e^{-.021t}$ **(b)** 8 g   **(c)** .021   **(d)** 6.5 g   **(e)** .021 g/yr   **(f)** 5 g **(g)** 4 g; 2 g; 1 g   **15. (a)** $f'(t)=-.6f(t)$ **(b)** 14.9 mg   **(c)** 1.2 h   **17.** 30.1 yr   **19.** 176 days **21.** $f(t)=8e^{-.014t}$ **23. (a)** 8 g   **(b)** 3.5 h   **(c)** .6 g/h   **(d)** 8 h   **25.** 13,412 years   **27.** 58.3% **29.** 10,900 years ago   **31.** a–D, b–G, c–E, d–B, e–H, f–F, g–A, h–C   **33. (a)** $y-10=-5t$ or $y=-5t+10$ **(b)** $P(t)=10e^{-.5t}$ **(c)** $T=2$.

Exercises 5.2, page 273

1. (a) $5000   **(b)** 4%   **(c)** $7459.12   **(d)** $A'(t)=.04A(t)$ **(e)** $298.36 per year   **(f)** $7000 **3. (a)** $A(t)=4000e^{.035t}$ **(b)** $A'(t)=.035A(t)$ **(c)** $4290.03   **(d)** 6.4 yr   **(e)** $175 per year   **5.** $378 per year   **7.** 15.3 yr   **9.** 29.3%   **11.** 17.3 years   **13.** 7.3%   **15.** 2002   **17.** 2006   **19.** $786.63 **21.** $7985.16   **23.** 15.7 yr   **25.** a–B, b–D, c–G, d–A, e–F, f–E, g–H, h–C   **27. (a)** $200   **(b)** $8 per year **(c)** 4%   **(d)** 30 yr   **(e)** 30 yr   **(f)** $A'(t)$ is a constant multiple of $A(t)$ since $A'(t)=rA(t)$. **29.**

$[-.5, .5]$ by $[-.5, 4]$

31. .06

Exercises 5.3, page 282

1. 20%, 4% **3.** 30%, 30% **5.** 60%, 300% **7.** $-25\%, -10\%$ **9.** 12.5% **11.** 5.8 yr **13.** $p/(140 - p)$, elastic **15.** $2p^2/(116 - p^2)$, inelastic **17.** $p - 2$, elastic **19. (a)** inelastic **(b)** raised **21. (a)** elastic **(b)** increase **23. (a)** 2 **(b)** yes **29. (a)** $p < 2$ **(b)** $p < 2$

Exercises 5.4, page 291

1. (a) $f'(x) = 10e^{-2x} > 0$, $f(x)$ increasing; $f''(x) = -20e^{-2x} < 0$, $f(x)$ concave down **(b)** As x becomes large, $e^{-2x} = \dfrac{1}{e^{2x}}$ approaches 0. **(c)** y **3.** $y' = 2e^{-x} = 2 - (2 - 2e^{-x}) = 2 - y$

5. $y' = 30e^{-10x} = 30 - (30 - 30e^{-10x}) = 30 - 10y = 10(3 - y)$, $f(0) = 3(1 - 1) = 0$ **7.** 4.8 h **11. (a)** 2500 **(b)** 500 people/day **(c)** day 12 **(d)** day 6 and day 13.5 **(e)** at time 9.78 days **(f)** $f'(t) = .00004f(t)(10,000 - f(t))$ **(g)** 1000 people per day **13. (a)**

[0, 12] by [−20, 75]

(b) 30 units **(c)** 25 units per hour **(d)** 9 hr **(e)** 65.3 units after 2 hr **(f)** 4 hr

Chapter 5: Supplementary Exercises, page 293

1. $29.92e^{-.2x}$ **3.** $5488.12 **5.** .058 **7. (a)** $17e^{.018t}$ **(b)** 20.4 million **(c)** 2011 **9. (a)** $36,693 **(b)** The alternative investment is superior by $3859. **11.** .02; 60,000 people per year; 5 million people **13.** a–F, b–D, c–A, d–G, e–H, f–C, g–B, h–E **15.** 6% **17.** 400% **19.** 3%, decrease **21.** increase **23.** $100(1 - e^{-.083t})$ **25. (a)** $400°$ **(b)** decreasing at a rate of $100°/\text{sec}$ **(c)** 17 sec **(d)** 2 sec

CHAPTER 6

Exercises 6.1, page 304

1. $\frac{1}{2}x^2 + C$ **3.** $\frac{1}{3}e^{3x} + C$ **5.** $3x + C$ **7.** $x^4 + C$ **9.** $7x + C$ **11.** $\dfrac{x^2}{2c} + C$ **13.** $2\ln|x| + \dfrac{x^2}{4} + C$ **15.** $\frac{2}{5}x^{5/2} + C$ **17.** $\frac{1}{2}x^2 - \frac{2}{3}x^3 + \frac{1}{3}\ln|x| + C$ **19.** $-\frac{3}{2}e^{-2x} + C$ **21.** $ex + C$ **23.** $-e^{2x} - 2x + C$ **25.** $-\frac{5}{2}$ **27.** $\frac{1}{2}$ **29.** $-\frac{1}{5}$ **31.** -1 **33.** $\frac{1}{15}$ **35.** 3 **37.** $\frac{2}{5}t^{5/2} + C$ **39.** C **41.** $-\frac{5}{2}e^{-.2x} + \frac{5}{2}$ **43.** $\dfrac{x^2}{2} + 3$ **45.** $\frac{2}{3}x^{3/2} + x - \frac{28}{3}$ **47.** $2\ln|x| + 2$ **49.** Testing all three functions reveals that (b) is the only one that works. **51.** **53.** $\frac{1}{4}$ **55. (a)** $-16t^2 + 96t + 256$ **(b)** 8 sec **(c)** 400 ft

$= 60t + t^2 - \frac{1}{12}t^3$ **59.** $20 - 25e^{-.4t}\ °C$ **61.** $-95 + 1.3x + .03x^2 - .0006x^3$ **63.** $5875(e^{.016t} - 1)$

65. $C(x) = 25x^2 + 1000x + 10,000$ **67.** $F(x) = \frac{1}{2}e^{2x} - e^{-x} + \frac{1}{6}x^3$

$[-2.4, 1.7]$ *by* $[-10, 10]$

Exercises 6.2, page 313

1. .5; .25, .75, 1.25, 1.75   **3.** .6; 1.3, 1.9, 2.5, 3.1, 3.7   **5.** 8.625   **7.** 15.12   **9.** .077278   **11.** 40   **13.** 15
15. 5.625; 4.5   **17.** 1.61321; error = .04241   **19.** 1.08 L   **21.** 2800 ft   **23.** Increase in the population (in millions) from 1910 to 1950; rate of cigarette consumption t years after 1985; 20 to 50   **25.** Tons of soil eroded during a 5-day period   **29.** 9.5965   **31.** 1.7641   **33.** .8427

Exercises 6.3, page 323

1. $\int_{1/2}^{2} \frac{1}{x} dx$ **3.** $\int_{1}^{3} (1-x)(x-3)dx$ **5.**   **7.** 0   **9.** 5   **11.** 1   **13.** $\frac{4}{3}(1-e^{-3})$

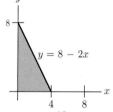

$y = 8 - 2x$

15. $e - e^{-1}$ **17.** $\frac{4}{9}$ **19.** $\ln 2$ **21.** $4\ln 5$ **23.** $\frac{13}{4} - 2\sqrt{e}$ **25.** $3\frac{3}{4}$ **27.** $\frac{1}{5}$ **29.** 10   **31.** $2(e^{1/2} - 1)$
33. $6\frac{3}{5}$ **35.** Positive   **37.** 95.4 trillion   **39. (a)** 30 ft   **(b)**   **41. (a)** $1185.75

$y = v(t)$

(b) The area under the marginal cost curve from $x = 2$ to $x = 8$.   **43.** The increase in profits resulting from increasing the production level from 44 to 48 units   **45. (a)** $368/15 \approx 24.5$ **(b)** The amount the temperature falls during the first 2 h   **47.** 2088 million m^3 **49.** 10   **51.** $\ln \frac{3}{2}$ **53.** $\frac{80}{3}$

Exercises 6.4, page 331

1. $\int_{1}^{2} f(x)\,dx + \int_{3}^{4} -f(x)\,dx$ **3.**   **5.** $\frac{64}{3}$ **7.** $\frac{52}{3}$ **9.** $e^2 - e - \frac{1}{2}$ **11.** $\frac{1}{6}$ **13.** $\frac{32}{3}$

$y = h(x)$
$y = f(x)$
$y = g(x)$

15. $\frac{1}{2}$ **17.** $\frac{1}{24}$ **19. (a)** $\frac{9}{2}$ **(b)** $\frac{19}{3}$ **(c)** $\frac{79}{6}$ **21.** $\frac{3}{2}$ **23.** $2 + 12\ln\frac{3}{2}$
25. $\int_{0}^{20} (76.2e^{.03t} - 50 + 6.03e^{.09t})dt$ **27.** No; 20; the additional profit from using the original plan
29. (a) The distance between the two cars after 1 h   **(b)** after 2 h   **31.** 1.4032   **33.** 1.4293

Exercises 6.5, page 340

1. 3 **3.** $50(1-e^{-2})$ **5.** $\frac{3}{4}\ln 3$ **7.** $55°$ **9.** ≈ 82 g **11.** $20 **13.** $404.72 **15.** $200 **17.** $25
19. Intersection $(100, 10)$, consumers' surplus $= 100, producers' surplus $= 250 **21.** $3236.68 **23.** $75,426
25. 13.35 y **27. (b)** $1000e^{-.04t_1}\Delta t + 1000e^{-.04t_2}\Delta t + \cdots + 1000e^{-.04t_n}\Delta t$ **(c)** $f(t) = 1000e^{-.04t}; 0 \le t \le 5$

(d) $\int_0^5 1000e^{-.04t}dt$ **(e)** $4531.73 **29.** $\frac{32}{3}\pi$ **31.** $\frac{31\pi}{5}$ **33.** 8π **35.** $\frac{\pi}{2}(1-e^{-2})$ **37.** $n = 4, b = 10,$

$f(x) = x^3$ **39.** $n = 3, b = 7, f(x) = x + e^x$ **41.** The sum is approximated by $\int_0^3 (3-x)^2 dx = 9$.

43. (a) $\frac{1000}{3r}(e^{3r}-1)$ **(b)** 4.5% **45. (a)** $\frac{1000}{r}(e^{6r}-1)$ **(b)** 5%

Chapter 6: Supplementary Exercises, page 343

1. $9x + C$ **3.** $\frac{2}{3}(x+1)^{3/2} + C$ **5.** $2(\frac{1}{4}x^4 + x^3 - x) + C$ **7.** $-2e^{-x/2} + C$ **9.** $\frac{3}{5}x^5 - x^4 + C$
11. $-\frac{2}{3}(4-x)^{3/2} + C$ **13.** $\frac{8}{3}$ **15.** $\frac{14}{3}\sqrt{2}$ **17.** $\frac{15}{16}$ **19.** $\frac{3}{4}$ **21.** $\frac{1}{3}\ln 4$ **23.** $\frac{1}{2}$ **25.** $\frac{80}{81}$ **27.** $\frac{5}{32}$
29. $\frac{1}{3}$ **31.** $\frac{1}{2}$ **33.** $\frac{28}{3}$ **35.** $\frac{e}{2} - 1$ **37.** 8 **39.** $\frac{1}{3}(x-5)^3 - 7$ **41. (a)** $2t^2 + C$ **(b)** Ce^{4t}
(c) $\frac{1}{4}e^{4t} + C$ **43.** $.02x^2 + 150x + 500$ dollars **45.** The total quantity of drug (in cubic centimeters) injected
during the first 4 min **47.** 25 **49.** $.68571; .69315$ **51.** $433.33 **53.** 15 **55.** $.26; 1.96$
57. (a) $f(t) = Q - \frac{Q}{A}t$ **(b)** $\frac{Q}{2}$ **59. (a)** The area under the curve $y = \frac{1}{1+t^2}$ from $t = 0$ to $t = 3$

(b) $\frac{1}{1+x^2}$ **63.** $\frac{15}{4}$ **65.** True **67.** $65,000$ km³ **69.** $f(x) = x^3 - x^2 + x$

CHAPTER 7

Exercises 7.1, page 352

1. $f(5, 0) = 25, f(5, -2) = 51, f(a, b) = a^2 - 3ab - b^2$ **3.** $g(2, 3, 4) = -2, g(7, 46, 44) = \frac{7}{2}$
7. $C(x, y, z) = 6xy + 10xz + 10yz$ **9.** $f(8, 1) = 40, f(1, 27) = 180, f(8, 27) = 360$ **11.** $\approx $50. $50 invested
at 5% continuously compounded interest will yield $100 in 13.8 years **13. (a)** $1875 **(b)** $2250; yes
15. **17.** **19.** $f(x, y) = y - 3x$ **21.** They correspond to the points

having the same altitude above sea level. **23.** d **25.** c

Exercises 7.2, page 361

1. $5y, 5x$ **3.** $4xe^y, 2x^2e^y$ **5.** $\frac{1}{y} - \frac{y}{x^2}; \frac{-x}{y^2} + \frac{1}{x}$ **7.** $4(2x - y + 5), -2(2x - y + 5)$ **9.** $(2xe^{3x} + 3x^2e^{3x})\ln y,$

x^2e^{3x}/y **11.** $\frac{2y}{(x+y)^2}, -\frac{2x}{(x+y)^2}$ **13.** $\frac{3}{2}\sqrt{\frac{K}{L}}$ **15.** $\frac{2xy}{z}, \frac{x^2}{z}, -\frac{1+x^2y}{z^2}$ **17.** $ze^{yz}, xz^2e^{yz}, x(yz+1)e^{yz}$

19. 1, 3 **21.** -12 **23.** $\frac{\partial f}{\partial x} = 3x^2y + 2y^2, \frac{\partial^2 f}{\partial x^2} = 6xy, \frac{\partial f}{\partial y} = x^3 + 4xy, \frac{\partial^2 f}{\partial y^2} = 4x, \frac{\partial^2 f}{\partial y \partial x} = \frac{\partial^2 f}{\partial x \partial y} = 3x^2 + 4y$
25. (a) Marginal productivity of labor $= 480$; of capital $= 40$ **(b)** $480h$ **(c)** Production decreases by 240
units. **27.** If the price of a bus ride increases and the price of a train ticket remains constant, fewer people will
ride the bus. An increase in train-ticket prices coupled with constant bus fare should cause more people to ride
the bus. **29.** If the average price of DVDs increases and the average price of a DVD player remains constant,
people will purchase fewer DVDs. An increase in average DVD player prices coupled with constant DVD prices
should cause a decline in the number of DVD players purchased. **31.** $\frac{\partial V}{\partial P}(20, 300) = -.06, \frac{\partial V}{\partial T}(20, 300) = .004$

33. $\frac{\partial f}{\partial r} > 0, \frac{\partial f}{\partial m} > 0, \frac{\partial f}{\partial p} < 0$ **35.** $\frac{\partial^2 f}{\partial x^2} = -\frac{45}{4}x^{-5/4}y^{1/4}$. Marginal productivity of labor is decreasing.

Exercises 7.3, page 369

1. $(-2, 1)$ **3.** $(26, 11)$ **5.** $(1, -3), (-1, -3)$ **7.** $(\sqrt{5}, 1), (\sqrt{5}, -1), (-\sqrt{5}, 1), (-\sqrt{5}, -1)$ **9.** $\left(\frac{1}{3}, \frac{4}{3}\right)$
11. Relative minimum; neither relative maximum nor relative minimum **13.** Relative maximum; neither relative maximum nor relative minimum; relative maximum **15.** Neither relative maximum nor relative minimum **17.** $(0, 0)$ min **19.** $(-1, -4)$ max **21.** $(0, -1)$ min **23.** $(-1, 2)$ max; $(1, 2)$ neither max nor min **25.** $\left(\frac{1}{4}, 2\right)$ min; $\left(\frac{1}{4}, -2\right)$ neither max nor min **27.** $\left(\frac{1}{2}, \frac{1}{6}, \frac{1}{2}\right)$ **29.** 14 in. $\times$ 14 in. $\times$ 28 in.
31. $x = 120$, $y = 80$

Exercises 7.4, page 377

1. 58 at $x = 6$, $y = 2$, $\lambda = 12$ **3.** 13 at $x = 8$, $y = -3$, $\lambda = 13$ **5.** $x = \frac{1}{2}$, $y = 2$ **7.** 5, 5 **9.** Base 10 in., height 5 in. **11.** $F(x, y, \lambda) = 4xy + \lambda(1 - x^2 - y^2)$; $\dfrac{\sqrt{2}}{2} \times \dfrac{\sqrt{2}}{2}$
13. $F(x, y, \lambda) = 3x + 4y + \lambda(18,000 - 9x^2 - 4y^2)$; $x = 20$, $y = 60$
15. (a) $F(x, y, \lambda) = 96x + 162y + \lambda(3456 - 64x^{3/4}y^{1/4})$; $x = 81$, $y = 16$ **(b)** $\lambda = 3$ **17.** $x = 12$, $y = 2$, $z = 4$
19. $x = 2$, $y = 3$, $z = 1$ **21.** $F(x, y, z, \lambda) = 3xy + 2xz + 2yz + \lambda(12 - xyz)$; $x = 2$, $y = 2$, $z = 3$
23. $F(x, y, z, \lambda) = xy + 2xz + 2yz + \lambda(32 - xyz)$; $x = y = 4$, $z = 2$

Exercises 7.5, page 384

1. $E = 6.7$ **3.** $E = (2A + B - 6)^2 + (5A + B - 10)^2 + (9A + B - 15)^2$ **5.** $y = 4.5x - 3$ **7.** $y = -2x + 11.5$
9. $y = -1.4x + 8.5$ **11. (a)** $y = .2073x + 2.7$ **(b)** \$4773 **(c)** 2006 **13. (a)** $y = .497x + 11.2$
(b) 22.6 percent **(c)** 2002 **15. (a)** $y = -4.24x + 22.01$ **(b)** $y = 8.442$ degrees Celsius

Exercises 7.6, page 390

1. $e^2 - 2e + 1$ **3.** $2 - e^{-2} - e^2$ **5.** $309\frac{3}{8}$ **7.** $\frac{5}{3}$ **9.** $\frac{38}{3}$ **11.** $e^{-5} + e^{-2} - e^{-3} - e^{-4}$ **13.** $9\frac{1}{3}$

Chapter 7: Supplementary Exercises, page 391

1. $2, \frac{5}{6}, 0$ **3.** ≈ 19.94. Ten dollars increases to 20 dollars in 11.5 y. **5.** $6x + y, x + 10y$ **7.** $\dfrac{1}{y}e^{x/y}$, $-\dfrac{x}{y^2}e^{x/y}$ **9.** $3x^2, -z^2, -2yz$ **11.** 6, 1 **13.** $20x^3 - 12xy, 6y^2, -6x^2, -6x^2$ **15.** $-201, 5.5$. At the level $p = 25$, $t = 10,000$, an increase in price of \$1 will result in a loss in sales of approximately 201 calculators, and an increase in advertising of \$1 will result in the sale of approximately 5.5 additional calculators. **17.** $(3, 2)$
19. $(0, 1), (-2, 1)$ **21.** Min at $(2, 3)$ **23.** Min at $(1, 4)$; neither max nor min at $(-1, 4)$ **25.** 20; $x = 3$, $y = -1$ **27.** $x = \frac{1}{2}$, $y = \frac{3}{2}$, $z = 2$ **29.** $F(x, y, \lambda) = xy + \lambda(40 - 2x - y)$; $x = 10$, $y = 20$ **31.** $y = \frac{5}{2}x - \frac{5}{3}$
33. $y = -2x + 1$ **35.** 5160 **37.** 40

CHAPTER 8

Exercises 8.1, page 396

1. $\dfrac{\pi}{6}, \dfrac{2\pi}{3}, \dfrac{7\pi}{4}$ **3.** $\dfrac{5\pi}{2}, -\dfrac{7\pi}{6}, -\dfrac{\pi}{2}$ **5.** 4π **7.** $\dfrac{7\pi}{2}$ **9.** -3π **11.** $\dfrac{2\pi}{3}$
13.

15.

Answers

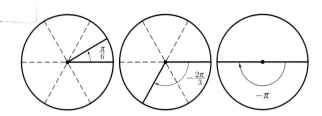

Exercises 8.2, page 401

1. $\sin t = \dfrac{1}{2}$, $\cos t = \dfrac{\sqrt{3}}{2}$ **3.** $\sin t = \dfrac{2}{\sqrt{13}}$, $\cos t = \dfrac{3}{\sqrt{13}}$ **5.** $\sin t = \dfrac{12}{13}$, $\cos t = \dfrac{5}{13}$ **7.** $\sin t = \dfrac{1}{\sqrt{5}}$,

$\cos t = -\dfrac{2}{\sqrt{5}}$ **9.** $\sin t = \dfrac{\sqrt{2}}{2}$, $\cos t = -\dfrac{\sqrt{2}}{2}$ **11.** $\sin t = -.8$, $\cos t = -.6$ **13.** .4   **15.** 3.6   **17.** 10.9

19. $b = 1.3$, $c = 2.7$ **21.** $\dfrac{\pi}{6}$ **23.** $\dfrac{3\pi}{4}$ **25.** $\dfrac{5\pi}{8}$ **27.** $\dfrac{\pi}{4}$ **29.** $\dfrac{\pi}{3}$ **31.** $-\dfrac{\pi}{6}$ **33.** $\dfrac{\pi}{4}$ **35.** Here $\cos t$
decreases from 1 to -1.   **37.** 0, 0, 1, -1 **39.** .2, .98, .98, $-.2$ **41. (a)**   **(b)** 46°

$[0, 365]$ by $[-10, 75]$

(c) 45° coldest, 73° warmest   **(d)** January 26   **(e)** July 27   **(f)** October 27 and April 27

$[0, 365]$ by $[-10, 75]$

Exercises 8.3, page 410

1. $4\cos 4t$ **3.** $4\cos t$ **5.** $-6\sin 3t$ **7.** $1 - \pi\sin\pi t$ **9.** $-\cos(\pi - t)$ **11.** $-3\cos^2 t\sin t$
13. $\dfrac{\cos\sqrt{x-1}}{2\sqrt{x-1}}$ **15.** $\dfrac{\cos(x-1)}{2\sqrt{\sin(x-1)}}$ **17.** $-8\sin t(1+\cos t)^7$ **19.** $-6x^2\cos x^3\sin x^3$ **21.** $e^x(\sin x + \cos x)$

23. $2\cos(2x)\cos(3x) - 3\sin(2x)\sin(3x)$ **25.** $\cos^{-2} t$ **27.** $-\dfrac{\sin t}{\cos t}$ **29.** $\dfrac{\cos(\ln t)}{t}$ **31.** -3 **33.** $y = 2$

35. $\frac{1}{2}\sin 2x + C$ **37.** $-\dfrac{7}{2}\sin\dfrac{x}{7} + C$ **39.** $\sin x + \cos x + C$ **41.** $\cos x + \sin 3x + C$

43. $-\frac{1}{4}\cos(4x+1) + C$ **45.** $-\frac{7}{3}\cos(3x-2) + C$ **47.** **(a))** $\max = 120$ at 0, $\dfrac{\pi}{3}$; $\min = 80$ at $\dfrac{\pi}{6}$, $\dfrac{\pi}{2}$
(b) 57   **49.** 0   **51. (a)** 69°   **(b)** increasing 1.6°/wk   **(c)** weeks 6 and 44   **(d)** weeks 28 and 48
(e) week 25, week 51   **(f)** week 12, week 38

Exercises 8.4, page 414

1. $\sec t = \dfrac{\text{hypotenuse}}{\text{adjacent}}$ **3.** $\tan t = \dfrac{5}{12}$, $\sec t = \dfrac{13}{12}$ **5.** $\tan t = -\dfrac{1}{2}$, $\sec t = -\dfrac{\sqrt{5}}{2}$ **7.** $\tan t = -1$, $\sec t = -\sqrt{2}$
9. $\tan t = \dfrac{4}{3}$, $\sec t = -\dfrac{5}{3}$ **11.** $75\tan(.7) \approx 63$ ft   **13.** $\tan t\sec t$ **15.** $-\csc^2 t$ **17.** $4\sec^2(4t)$
19. $-3\sec^2(\pi - x)$ **21.** $4(2x+1)\sec^2(x^2 + x + 3)$ **23.** $\dfrac{\sec^2\sqrt{x}}{2\sqrt{x}}$ **25.** $\tan x + x\sec^2 x$ **27.** $2\tan x\sec^2 x$

29. $6[1 + \tan(2t)]^2 \sec^2(2t)$ **31.** $\sec t$ **33.** (a) $y - 1 = 2(x - \frac{\pi}{4})$ (b)

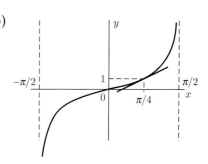

35. $\frac{1}{3}\tan 3x + C$ **37.** 2   **39.** $\tan x + C$

Chapter 8: Supplementary Exercises, page 416

1. $\dfrac{3\pi}{2}$ **3.** $-\dfrac{3\pi}{4}$ **5.**

7. $\frac{4}{5}, \frac{3}{5}, \frac{4}{3}$ **9.** $-.8, -.6, \frac{4}{3}$ **11.** $\pm\dfrac{2\sqrt{6}}{5}$

13. $\dfrac{\pi}{4}, \dfrac{5\pi}{4}, -\dfrac{3\pi}{4}, -\dfrac{7\pi}{4}$ **15.** Negative   **17.** 16.3 ft   **19.** $3\cos t$ **21.** $(\cos\sqrt{t}) \cdot \frac{1}{2}t^{-1/2}$

23. $x^3 \cos x + 3x^2 \sin x$ **25.** $-\dfrac{2\sin(3x)\sin(2x) + 3\cos(2x)\cos(3x)}{\sin^2(3x)}$ **27.** $-12\cos^2(4x)\sin(4x)$

29. $[\sec^2(x^4 + x^2)](4x^3 + 2x)$ **31.** $\cos(\tan x)\sec^2 x$ **33.** $\sin x \sec^2 x + \sin x$ **35.** $\cot x$

37. $4e^{3x}\sin^3 x \cos x + 3e^{3x}\sin^4 x$ **39.** $\dfrac{\tan(3t)\cos t - 3\sin t \sec^2(3t)}{\tan^2(3t)}$ **41.** $e^{\tan t}\sec^2 t$ **43.** $2(\cos^2 t - \sin^2 t)$

45. $\dfrac{\partial f}{\partial s} = \cos s \cos(2t)$, $\dfrac{\partial f}{\partial t} = -2\sin s \sin(2t)$ **47.** $\dfrac{\partial f}{\partial s} = t^2 \cos(st)$, $\dfrac{\partial f}{\partial t} = \sin(st) + st\cos(st)$

49. $y - 1 = 2\left(t - \dfrac{\pi}{4}\right)$ **51.**

53. 4   **55.** $\dfrac{\pi^2}{2} - 2$ **57.** (a) $V'(t) = 8\pi\cos\left(160\pi t - \dfrac{\pi}{2}\right)$

(b) 8π L/min   (c) 16 L/min   **59.** $\cos(\pi - x) + C$ **61.** 0   **63.** $\dfrac{\pi^2}{2}$ **65.** $2\tan\dfrac{x}{2} + C$ **67.** $\sqrt{2} - 1$

69. $\sqrt{2}$ **71.** 1   **73.** $1000 + \dfrac{400}{3\pi}$ **75.** $\tan x - x + C$ **77.** $\tan x + C$ **79.** $1 - \dfrac{\pi}{4}$

CHAPTER 9

Exercises 9.1, page 425

1. $\frac{1}{6}(x^2 + 4)^6 + C$ **3.** $2\sqrt{x^2 + x + 3} + C$ **5.** $e^{(x^3 - 1)} + C$ **7.** $-\frac{1}{3}(4 - x^2)^{3/2} + C$ **9.** $(2x + 1)^{1/2} + C$

11. $\frac{1}{2}e^{x^2} + C$ **13.** $\frac{1}{2}(\ln 2x)^2 + C$ **15.** $\frac{1}{5}\ln|x^5 + 1| + C$ **17.** $-(2 - 12x + 2x^2)^{-1} + C$

19. $\frac{1}{4}(\ln x)^2 + C = (\ln\sqrt{x})^2 + C$ **21.** $\frac{1}{3}\ln|x^3 - 3x^2 + 1| + C$ **23.** $-4e^{-x^2} + C$ **25.** $\frac{1}{2}\ln|\ln x^2| + C$

27. $-\frac{1}{10}(x^2 - 6x)^5 + C$ **29.** $\frac{1}{6}(1 + e^x)^6 + C$ **31.** $\frac{1}{2}\ln(1 + 2e^x) + C$ **33.** $\ln|1 - e^{-x}| + C$

35. $-\ln(e^{-x} + 1) + C$ **37.** $f(x) = (x^2 + 9)^{1/2} + 3$ **39.** $2e^{\sqrt{x+5}} + C$ **41.** $\frac{1}{2}\tan x^2 + C$ **43.** $\frac{1}{2}(\sin x)^2 + C$

45. $2\sin\sqrt{x} + C$ **47.** $-\frac{1}{4}\cos^4 x + C$ **49.** $-\frac{2}{3}\sqrt{2 - \sin 3x} + C$ **51.** $\ln|\sin x - \cos x| + C$

53. $\frac{1}{2}(x^2 + 5)^2 + C = \frac{1}{2}x^4 + 5x^2 + \frac{25}{2} + C$; $\frac{1}{2}x^4 + 5x^2 + C_1$

1. $\frac{1}{5}xe^{5x} - \frac{1}{25}e^{5x} + C$ **3.** $\frac{x}{5}(x+7)^5 - \frac{1}{30}(x+7)^6 + C$ **5.** $-xe^{-x} - e^{-x} + C$

7. $2x(x+1)^{1/2} - \frac{4}{3}(x+1)^{3/2} + C$ **9.** $(1-3x)(\frac{1}{2}e^{2x}) + \frac{3}{4}e^{2x} + C$ **11.** $-2xe^{-3x} - \frac{2}{3}e^{-3x} + C$

13. $\frac{2}{3}x(x+1)^{3/2} - \frac{4}{15}(x+1)^{5/2} + C$ **15.** $\frac{1}{3}x^{3/2}\ln x - \frac{2}{9}x^{3/2} + C$ **17.** $x\sin x + \cos x + C$

19. $\frac{x^2}{2}\ln 5x - \frac{1}{4}x^2 + C$ **21.** $4x\ln x - 4x + C$ **23.** $-e^{-x}(x^2 + 2x + 2) + C$

25. $\frac{1}{5}x(x+5)^5 - \frac{1}{30}(x+5)^6 + C$ **27.** $\frac{1}{10}(x^2+5)^5 + C$ **29.** $3(3x+1)e^{x/3} - 27e^{x/3} + C$

31. $\frac{1}{2}\tan(x^2+1) + C$ **33.** $\frac{1}{2}xe^{2x} - \frac{1}{4}e^{2x} + \frac{1}{3}x^3 + C$ **35.** $\frac{1}{2}e^{x^2} - x^2 + C$ **37.** $2x\sqrt{x+9} - \frac{4}{3}(x+9)^{3/2} + 38$

Exercises 9.3, page 434

1. $\frac{1}{15}$ **3.** 312   **5.** $\frac{8}{3}$ **7.** $\frac{64}{3}$ **9.** 0   **11.** $\frac{1}{2}\ln(\frac{4}{3})$ **13.** $\frac{1}{3}(e^{27} - e)$ **15.** $\frac{1}{2}$ **17.** 0   **19.** $\frac{1}{\pi}$ **21.** $\frac{\pi}{2}$

23. $\frac{9\pi}{2}$ **25.** $\frac{16}{3}$

Exercises 9.4, page 442

1. $\Delta x = .4$; 3, 3.4, 3.8, 4.2, 4.6, 5   **3.** $\Delta x = .5$; $-.75, -.25, .25, .75$ **5.**

7. $(n=2)$ 40, $(n=4)$ 41, exact: $41\frac{1}{3}$ **9.** .63107, exact: .63212   **11.** .09375, exact: .08333   **13.** 1.03740, exact: .8   **15.** $M = 72$, $T = 90$, $S = 78$, exact: 78   **17.** $M = 44.96248$, $T = 72.19005$, $S = 54.03834$, exact: 53.59815   **19.** $M = 573.41797$, $T = 612.10806$, $S = 586.31466$, exact: 586.26358   **21.** 3.24124   **23.** 1.61347   **25.** 25,750 ft^2 **27.** 2150 ft   **29.** **(a)** $f''(x) = x^2 + 6$ **(b)** $A = 10$ **(c)** .0333

(d) $-.0244$, satisfies the bound in (c)   **(e)** quartered   **35.** $f(x) = \frac{1}{25 - x^2}$, $a = 3$, $b = 5$, $n = 10$

37. Midpoint rule: 2.361749156; trapezoidal rule: 2.474422799; Simpson's rule: 2.399307037; Exact value: 2.397895273; Error using midpoint rule: $-.036146117$; Error using trapezoidal rule: .076527526; Error using Simpson's rule: .001411764   **39.** Midpoint rule: .9989755866; trapezoidal rule: 1.00205197; Simpson's rule: 1.000001048; Exact value: 1; Error using midpoint rule: $-.0010244134$; Error using trapezoidal rule: .00205197; Error using Simpson's rule: .000001048   **41.** bound on the error is .0008333

Exercises 9.5, page 449

1. $147,656   **3.** $35,797   **5.** $182,937   **7.** (a) $\int_0^2 (30+5t)e^{-.10t}\,dt$ (b) $63.1 million

9. (a) $240\pi \int_0^5 te^{-.65t}\,dt$ (b) $1,490,623   **11.** $\approx 1,400,000$;

People moved away from
the center of the city

13. (a) $80\pi te^{-.5t}\Delta t$ thousand people   (b) $P'(t)$ (c) the number of people who live between 5 mi and
$5 + \Delta t$ mi from the city center   (d) $P'(t) = 80\pi te^{-.5t}$ (e) $P(b) - P(a) = \int_a^b P'(t)\,dt = \int_a^b 80\pi te^{-.5t}\,dt$

Exercises 9.6, page 454

1. 0   **3.** No limit   **5.** $\frac{1}{4}$ **7.** 2   **9.** 0   **11.** 5   **13.** $\frac{1}{2}$ **15.** 2   **17.** 1   **19.** Area under the graph
from 1 to b is $\frac{5}{14}(14b+18)^{1/5} - \frac{5}{7}$. This has no limit as $b \to \infty$.   **21.** $\frac{1}{2}$ **23.** $\frac{1}{6}$ **25.** Divergent   **27.** $\frac{2}{3}$
29. 1   **31.** $2e$ **33.** Divergent   **35.** $\frac{1}{2}$ **37.** 2   **39.** $\frac{1}{4}$ **41.** 2   **43.** $\frac{1}{6}$ **45.** $\frac{K}{r}$ **49.** $\frac{K}{r}$

Chapter 9: Supplementary Exercises, page 456

1. $-\frac{1}{6}\cos(3x^2) + C$ **3.** $-\frac{1}{36}(1-3x^2)^6 + C$ **5.** $\frac{1}{3}[\ln x]^3 + C$ **7.** $-\frac{1}{3}(4-x^2)^{3/2} + C$ **9.** $-\frac{1}{3}e^{-x^3} + C$
11. $\frac{1}{3}x^2\sin(3x) - \frac{2}{27}\sin(3x) + \frac{2}{9}x\cos(3x) + C$ **13.** $2(x\ln x - x) + C$ **15.** $\frac{2}{3}x(3x-1)^{1/2} - \frac{4}{27}(3x-1)^{3/2} + C$
17. $\dfrac{1}{4}\left[\dfrac{x}{(1-x)^4}\right] - \dfrac{1}{12}\left[\dfrac{1}{(1-x)^3}\right] + C$ **19.** $f(x) = x,\ g(x) = e^{2x}$ **21.** $u = \sqrt{x+1}$ **23.** $u = x^4 - x^2 + 4$
25. $f(x) = (3x-1)^2,\ g(x) = e^{-x}$; then integrate by parts again.   **27.** $f(x) = 500 - 4x,\ g(x) = e^{-x/2}$
29. $f(x) = \ln(x+2),\ g(x) = \sqrt{x+2}$ **31.** $u = x^2 + 6x$ **33.** $u = x^2 - 9$ **35.** $u = x^3 - 6x$ **37.** $\frac{3}{8}$
39. $1 - e^{-2}$ **41.** $\frac{3}{4}e^{-2} - \frac{5}{4}e^{-4}$ **43.** $M = 3.93782,\ T = 4.13839,\ S = 4.00468$
45. $M = 12.84089,\ T = 13.20137,\ S = 12.96105$ **47.** $\frac{1}{3}e^6$ **49.** Divergent   **51.** $2^{7/4}$ **53.** $2e^{-1}$
55. $137,668   **57.** (a) $M(t_1)\Delta t + \cdots + M(t_n)\Delta t \approx \int_0^2 M(t)\,dt$

(b) $M(t_1)e^{-.1t_1}\Delta t + \cdots + M(t_n)e^{-.1t_n}\Delta t \approx \int_0^2 M(t)e^{-.1t}\,dt$

INDEX

Index of Applications